Third Edition

Practical
Toxicology

Evaluation, Prediction, and Risk

Third Edition

Practical Toxicology

Evaluation, Prediction, and Risk

David Woolley
Adam Woolley

CRC Press
Taylor & Francis Group
Boca Raton   London   New York

CRC Press is an imprint of the
Taylor & Francis Group, an **informa** business

CRC Press
Taylor & Francis Group
6000 Broken Sound Parkway NW, Suite 300
Boca Raton, FL 33487-2742

First issued in hardback 2019

© 2017 by Taylor & Francis Group, LLC
CRC Press is an imprint of Taylor & Francis Group, an Informa business

No claim to original U.S. Government works

ISBN-13: 978-1-4987-0928-6 (hbk)

Library of Congress Cataloging-in-Publication Data

Names: Woolley, David (Toxicologist), author. | Woolley, Adam, author. |
Preceded by (work): Woolley, Adam. Guide to practical toxicology.
Title: Practical toxicology : evaluation, prediction, and risk / David
Woolley and Adam Woolley.
Description: Third edition. | Boca Raton : Taylor & Francis, 2017. | Preceded
by A guide to practical toxicology : evaluation, prediction, and risk /
Adam Woolley. 2nd ed. c2008. | Includes bibliographical references and
index.
Identifiers: LCCN 2016034981| ISBN 9781498709286 (hardback : alk. paper) |
ISBN 9781498709316 (e-book) | ISBN 9781498709323 (e-book) | ISBN
9781498709309 (e-book)
Subjects: | MESH: Toxicity Tests | Toxicology--methods | Models, Chemical |
Data Interpretation, Statistical | Risk Assessment
Classification: LCC RA1199 | NLM QV 602 | DDC 615.9/07--dc23
LC record available at https://lccn.loc.gov/2016034981

Visit the Taylor & Francis Web site at
http://www.taylorandfrancis.com

and the CRC Press Web site at
http://www.crcpress.com

All things are poison and nothing without poison;
only the dose determines that a thing is not poison.

Philippus Theophrastus Aureolus Bombastus von Hohenheim
(Paracelsus, 1493–1541), alchemist, physician, and astrologer

If a man will begin with certainties, he shall end in doubts;
but if he will be content to begin with doubts,
he shall end in certainties.

Francis Bacon
Philosopher, statesman, scientist, jurist, orator, and author,
from *The Advancement Of Learning* (1605);
Bacon has been considered to be the father
of the empirical approach and of critical scientific thought.

Contents

Preface

Toxicology is a dynamic subject with unique relevance to the public and, as a result, places heavy responsibility on the people who practice it, those who register or accredit toxicologists, and those who regulate or manage it. As a consequence, toxicologists and toxicology are inherently conservative, and the practice of the discipline reflects this. In this case, practice includes both the conduct of the experiments and their interpretation by regulatory authorities or the toxicological community. Whatever its purpose, good-quality toxicological evaluation is essential. Although there is a wide-ranging bibliography in toxicology, the textbooks do not necessarily tell the reader how to conduct a toxicological evaluation and then to extend that process to risk assessment. In addition, the procedures with which toxicity studies are designed and conducted are not easily accessible to the reader who wishes to learn more about the subject and, perhaps, to understand the processes of evaluation and risk assessment, and their application in the real world.

The first and second editions of this book arose out of our perception that there was a need for a practical, user-friendly introductory text for those coming to toxicology from related fields or professions, and who need some insight into how toxicity studies and investigations are carried out. We thought that the book should be informative but readable and should also act as a gateway to the subject, indicating where further information can be found, including the use of websites for literature searches and other areas, such as regulations and guidelines.

The book is set out as a guide on how to evaluate toxicity and then how to handle and use the data that are generated. After an introduction to the concepts of toxicology, the book takes the reader through the processes of toxicity testing and interpretation before looking at the concepts of hazard prediction and risk assessment and management. Two final chapters look at the evaluation of different chemical classes and at the future of toxicity testing and risk assessment.

The audience for this book includes new graduates starting careers in toxicology, those coming to the subject from different fields, and specialists in particular areas of toxicology who need some background on the other areas.

Toxicology has been evolving especially quickly since the second edition was published, and this new edition includes several new chapters or sections that address *in silico* toxicology, nanotoxicology, immunotoxicity of biological products, and risk assessment of extractables, leachables, and other impurities in drug substances and products. All the existing chapters have been thoroughly reviewed and revised to bring them up to date with current practice, including comment on add-ons to conventional studies (e.g., safety pharmacology and bone marrow micronucleus tests) and new test designs (e.g., OECD 422: Combined Repeated Dose Toxicity Study with the Reproduction/Developmental Toxicity Screening Test).

Toxicological evaluation of chemicals is conducted within a number of overlapping regulatory frameworks, covering chemical or product classes such as agrochemicals or plant protection products, industrial chemicals, medical devices, cosmetics and consumer products, and pharmaceuticals. However, the basic principles of toxicology are common to all areas, whether you are evaluating pharmaceuticals or industrial chemicals. The differing frameworks result in slightly different study designs and durations, while the objectives are subtly oriented toward the product class. Broadly, there are two types of chemical exposure: unintentional (e.g., agrochemicals) and intentional (primarily pharmaceuticals). These two categories drive the objectives and regulatory frameworks for evaluation. We are largely pharmaceutical toxicologists, so there is some bias toward this area of toxicology, but the principles outlined here are applicable to all areas of toxicological investigation.

Much toxicology is conducted for safety evaluation; however, it should be remembered that safety is a negative and, as such, cannot be proven. As stated in the front pages, Paracelsus realized the basic principle of toxicology that the elusive concept of safety is solely dependent on dose; water is toxic, if taken to excess.

Throughout the book, words like "may," "could," and "however" appear frequently; this is tacit recognition that there are few certainties in life beyond the single gold standard that wherever a statement is made, there will be someone to disagree with it. As with any walk of life, if a situation is seen as black and white, it simply means that the intervening shades of gray have not been discovered or are not understood. This is particularly true for any aspect of judgment or interpretation; differing opinions between toxicologists, especially toxicological pathologists, can be extremely frustrating for anyone needing a definitive answer to a question of safety. Getting a decision wrong in toxicology can be associated with far-reaching adverse effects and with consequent litigation or (politically far worse) loss of votes. For this reason, toxicologists (especially those in regulatory agencies) tend to be conservative in their opinions; this is not necessarily a bad thing. However, conservatism is made more likely by poor, incomplete, or poorly understood data or results, which may lead to imposition of inappropriately restrictive exposure limits.

The intention of this book is to provide a basis of knowledge—a series of pointers—which can be expanded through use of the references given. There are many different ways of achieving an objective in toxicological study and evaluation, and this book cannot pretend to address them all or to be absolutely definitive in any one area.

The future of toxicology is assured; the means of its future investigation is changing, and it behooves us to think about what we are doing or what we are asking toxicologists to do (or, more importantly, not to do). As toxicologists, we should do nothing without thought, without considering the impact of our actions on the animals we use, the public, or a host of other stakeholders. In many ways, if we cause the reader to think more about toxicology and its importance and impact on this world, we will have achieved one of our unwritten objectives.

Acknowledgments

Such a wide-ranging text could not be prepared in isolation. Many thanks go to those who assisted with the first and second editions, whose contributions are still valid and provided valuable guidance and comment. In addition, Annette Dalrymple provided assistance with the overview of the Pig-a assay and other tobacco-related texts. Having said that, all the opinions expressed are our own, as is responsibility for any errors that may still lurk in the text.

We also would like to thank Sandy Woolley, for obtaining the necessary permissions. Finally, I would like to thank Barbara Norwitz, who initiated this third edition, and Steve Zollo for taking up the baton when Barbara retired.

About the Authors

Adam Woolley, MSc DABT FRCPath CBiol MRSB ERT, fellow ATS, is a director and consultant toxicologist at ForthTox Ltd, Linlithgow, UK, with 40 years' experience in both contract research and the pharmaceutical industry. Adam is certified as a toxicologist in both Europe (European Registered Toxicologist) and the USA (DABT), and is a Fellow of the Academy of Toxicological Sciences and a Fellow of the Royal College of Pathologists. He has been involved in a number of training initiatives and has published a number of papers and given presentations at meetings around the world. He has been a director on the board of the American Board of Toxicology (currently European liaison officer to ABT) and chair of the Registration Panel to the UK Register of Toxicologists. He is chief examiner in toxicology for the Royal College of Pathologists.

David Woolley, PhD BScBioMedSci (Pharmacology) CBiol MRSB ERT, has been a consultant and director of *In Silico* Toxicology at ForthTox since 2008 following the completion of his PhD in neuropharmacology. He has extensive experience with the use of Derek Nexus, Meteor, Leadscope, ToxTree, Vega, and TEST and has prepared reports for submission to regulatory authorities, particularly on the potential mutagenicity and genotoxicity of pharmaceutical impurities, leachates, and degradants. He has also prepared nonclinical overviews on behalf of major pharmaceutical companies and prepared numerous toxicological overview reports and biological evaluations for medical devices. David is a European Registered Toxicologist and a Chartered Biologist. He is a member of the British Toxicology Society, the British Pharmacology Society, the Royal Society of Biology, the Society of Toxicology, and the American College of Toxicology.

1 Introduction to Toxicology
The Necessity of Measurement

INTRODUCTION

Exposure to chemicals is unavoidable—we live in a chemical world. We are composed of, and, in the course of our daily lives, exposed to, a wide range of chemicals, the vast majority of which are naturally occurring. However, an increasing number of those found in our bodies are persistent man-made chemicals such as siloxanes, polychlorinated biphenyls, or bisphenol A. The substances of concern themselves vary over time and are dependent on popular anxiety; in previous times, organochlorine pesticides such as DDT and its metabolite DDE were the substances of concern, but as their use as fallen, so has the general fear of exposure. It is a comforting and often-held belief that all man-made chemicals are poisonous and all natural chemicals are safe. Sadly, this is not the case; for example, botulin toxin, the active principle in Botox injections, is one of the most poisonous chemicals known but is found naturally. The ancient Greeks and Romans killed each other with natural poisons such as hemlock. Lead and cadmium are natural elements but with no levels that can be described as safe.

It is not simply these obvious toxicants, however, that should be considered to be poisonous. Caffeine is widely available and naturally occurring in coffee and tea, but is also increasingly an ingredient in energy drinks. Consumers of these drinks may not be aware of this and may overdose on caffeine without intending to. While caffeine may be tolerable at relatively high doses in humans, pets such as dogs may not tolerate such doses. In a similar vein, some dogs are more sensitive than others to the effects of chocolate and grapes, both of which can be very toxic to them. While it is easy to understand that something like coffee can have adverse effects if consumed to excess, extrapolating the same concept to water may appear more difficult. However, it has been realized that drinking water to avoid dehydration during marathons can result in overcompensation by the runners, which is then followed by toxicity due to overconsumption of water.

The idea that an exposure to a chemical or substance may lead to toxicity is not a new concept. The furore over the presence of lead in the paint for children's toys was presciently foreseen in the *Treatise on Adulterations* by Frederick Accum, published in 1820. In a concluding paragraph, he observed that children are apt to mouth any toys or objects and that the practice of painting toys with poisonous coloring substances should be abolished. This is so obviously self-evident that it is a wonder that action has been taken only recently and, even then, is effective only by post-manufacture testing.

Toxicology is a very broad discipline, requiring broad expertise in a number of areas, including chemistry, pharmacology, physiology, biochemistry, anatomy, and numerous others. Definitions of toxicology tend to emphasize the role of exogenous substances or xenobiotics (literally foreign chemicals), while implicitly ignoring the contribution to toxic effect that can be seen with endogenous substances. The overproduction of endorphins in athletes and resulting runners' high is an example of this; the storage of various proteins in Alzheimer's disease is another. While many drug development candidates are clearly nonendogenous (xenobiotic), it may be prudent to consider unnaturally high concentrations of an endogenous substance in an unusual location to be xenobiotic too. It should also be considered that toxicity of a substance varies from subject to subject and a toxic reaction can occur from exposure to a normally nonnoxious substance, such as peanuts.

Absence of a chemical (see Focus Box 1.1 for terminology) can also have an effect; vitamin deficiency or decreased sensitivity to insulin (or its reduced production) may also be seen as an effect associated with chemicals. Liebler has written a broad-based review that considers the place of toxicology in the wider context of health and medicine, and also considers the role of endogenous chemicals (Liebler 2006). He points out that the implicit link between toxicology and exposure to xenobiotics ignores the role of endogenous chemicals and produces an unwarranted separation between toxicology, and health and medical practice (although the role of the occupational toxicologist comes closer to this than other branches of the science). Endogenous

FOCUS BOX 1.1   A NOTE ON TERMINOLOGY

The words *chemical* and *substance* are used interchangeably; other words used to suggest the same concept include *compound, test item*, and *test substance*. While the term *substance* may include mixtures, the words *compound* and *chemical* tend to be more specific, meaning a material composed (chiefly) of a single molecule. "Chiefly" is needed here because 100% purity is rare, especially with low-grade industrial chemicals; in this respect, much of toxicology is about mixtures. There is no specific rule for this usage, but the context of the remarks should make it clear whether a single compound or chemical is being described.

Other terms that are important are *biological*—usually used to describe a drug or pesticide—and *nanoparticle* and *nanotech*. *Biological* usually means a drug or pesticide composed of protein-related substances or nucleic acids, although it may also refer to pesticides that are bacteria or fungi. *Nanoparticle* is a particle that is up to 100 nm in size and that behaves as a unit; *nanotechnology* is the discipline that studies and uses these particles. Examples of use include in sunscreens, delivery of drugs, and so forth.

substances are important in disease but can also be generated in response to xenobiotic exposure. While we can control exogenous exposures to a certain extent, some more than others, there is no easy escape from disease such as cancer, diabetes, and others that can be related to diet, lifestyle, and so forth.

Toxicology is a science that has a direct impact on, and responsibility to, the public in a way that other subjects, for instance, astronomy or particle physics, do not. This responsibility arises from the role that toxicology has in assessing the safety of chemicals that have been or will be in daily use or to which the public are exposed. If the assessment is wrong, there is a distinct probability that adverse effects will be seen in exposed populations or that the benefits of a new chemical will be denied to people who would be advantaged by its use.

The public perception of toxicity is very important to people who conduct or interpret toxicological investigations. A change may be incorrectly perceived as adverse through incomplete access to all information, and this will provoke questions: When is a cluster of disease patients significant? How do you investigate? Whom should we believe? Why? What is the true, unprejudiced significance of this finding for the exposed population? Major concerns of the public are cancer, loss of special senses (especially sight), general debilitation, reproductive effects, disease, and shortened life span. Frame of reference is everything; much emphasis is placed on exposure to pesticides in food, without concern about the natural chemicals that occur in the same plants (e.g., green potatoes or broccoli), or on exposure to low-level radiation but not on sunbathing and consequent increased risk of skin cancer, including melanoma.

How a problem is expressed is also very important. Often, communication of toxicological information is one sided, emphasizing the apparently beneficial elements, while ignoring others. For instance, if the incidence of a particular fatal disease is 20 people in 1000, and this could be reduced by a novel (possibly hazardous) treatment by 20%, the new incidence would be 16 in 1000. However, the initial situation is that 98% of people are free of the disease; the new treatment would increase that to 98.4%. An increase of 0.4% is much less attractive than a decrease of 20%.

Accidents and emergencies, whether involving human, animal, or plant life, often provide salutary lessons. After the discharge of dioxins at Seveso, Italy, in 1976, prolonged investigations told us that dioxin is very toxic to animals in various ways; it is clear that humans suffer chloracne but other effects in humans are unproven or unknown. In another example, the discharge of inorganic mercury waste at Minimata Bay in Japan shows that nature does not always make things safer—in fact, quite the opposite: it can increase hazard, in this case, by methylation and increasing lipophilicity of the mercury, such that the human food chain was affected. It is automatically assumed by many that synthetic chemicals are harmful, but this assumption may ignore significant benefits.

More recently, pet food and infant formula were deliberately adulterated with melamine, which resulted in kidney toxicity and a number of deaths in both pets and human consumers, including children.

In the developed world, pharmaceutical standards and purity are assumed and are regulated; however, these same standards are not applied to the quality of designer drugs or the diluent of street cocaine. The expanding market for herbal extracts and remedies provides real cause for concern; for example, are the sources correctly identified, processed, stored, and labeled?

Treatment of some foods, such as peanuts, with mold-preventing chemicals carries some small risk from the chemical exposure, but importantly, the absence of mold markedly reduces the risk of liver cancer due to aflatoxin, which is produced by *Aspergillus flavus* growing on damp-stored peanuts. Aflatoxin is a particularly potent hepatotoxin and carcinogen, which may induce cancer at levels as low as 1 ppb; it has been found in trace amounts in peanut butter prepared from untreated peanuts. This could be sold as "organic" peanut butter; does this support the campaign for organic production?

THE BEGINNINGS OF TOXICOLOGICAL MEASUREMENT

Although not always known as toxicology, this fascinating amalgam of different disciplines has had a long history, stemming from the Ebers Papyrus of the ancient Egyptians and progressing steadily through ancient Greece and Rome. In Greece and Rome, the knowledge of poisons was crucial in eliminating unwanted politicians, rivals, or relatives. This use was particularly noted in some Roman wives who used contract poisoners to do away with rich husbands so that they could inherit the wealth and move on to the next hapless, but temptingly rich, victim. This cheerful habit was revived in Renaissance Italy, where dwarves were created by feeding known growth inhibitors to children, a practice noted by Shakespeare in *A Midsummer Night's Dream*, where he writes, "Get you gone, you dwarf … of hindering knotgrass made."

It was clear to the practitioners of the day that the dose, that is, the level of exposure, was the critical factor determining success or failure. However, it was Paracelsus, born in 1493, who linked dose with effect by stating that everything is a poison; only the dose differentiates between a poison and a remedy. Although this has been quoted extensively, the full context of the quotation is instructive (Ball 2006). The religious context is clear, as Paracelsus asks what God has created that was not blessed with some gift beneficial to man. Possibly without appreciating its latter-day significance, he says that to despise a poison is to be ignorant of what it contains, which might be interpreted as indicating ignorance of potential toxicity or therapeutic benefit. Having said that, it has to be admitted that, while his treatments did not have the precision that he may have wanted, he was at the forefront of the movement to formulate new medicines.

To put dose relationship in a modern context, a daily glass of red wine may be considered to be therapeutically beneficial (depending on which epidemiological study you wish to believe); increase that to a bottle or more a day, and cirrhosis of the liver beckons.

SAFETY AS A CONCEPT

It is usually fairly easy to say what dose of a chemical is toxic or harmful but much more difficult to predict safety. In fact, as it is not possible to prove a negative, the question "Is it safe?" is effectively impossible to answer affirmatively as biological responses differ between individuals and some may respond to low levels of a chemical when the majority are unaffected. While the concept of *poisonous* was understood, for instance, in the seventeenth century, as the effect of poisons or of an excess of something, the concept of safety was of little concern. The work of people like Percival Potts, who linked scrotal cancer in former chimney sweeps with prior exposure to soot, led to gradual recognition of safety as a concept. However, it took many years to do anything about it, in line with the speed of change in regulation in the twenty-first century.

With the enormous increase in the use of chemicals that has taken place during the late nineteenth century and in the twentieth century, it became apparent that there should be an increasing emphasis on demonstration of safety. This concern for safety is applied in many areas, ranging from novel or genetically modified (GM) foods to industrial chemicals and from pharmaceuticals to leachables from medicinal packaging. In some cases, where a traditional or long-used chemical is known to be unsafe, efforts are made to find a substitute. When the search is successful, it is sometimes the case that the substitute removes the old problems while introducing new ones. However, it is generally accepted that to predict safety, given that there is no such thing as a safe chemical or a risk-free existence, it is necessary first to demonstrate what dose of the study chemical is toxic and how that toxicity develops as dose increases.

In the modern context, there is public recognition that there are chemicals to which people are exposed voluntarily (for example, cigarette smoke, medicines, and alcohol) and those to which exposure is involuntary (pesticides in vegetables, other people's cigarette smoke, pollution, food preservatives, antibiotics in food animals, and so on). There is a lively public debate on many of these substances, which often takes extreme views due to lack of knowledge or willful misinformation or misinterpretation by interested parties. Such fine lines are drawn by politicians, but it is the responsibility of the toxicological community to define safe doses or inclusion limits for these various chemicals. Above all, this must be done in a credible manner, within the existing framework of regulation and ethical behavior.

In addition, there is a growing body of scientific work investigating the effects of chemicals that occur naturally in our food. For instance, it has been shown in several papers that some constituents of mushrooms can cause cancer in mice when given at high dosages. It should be borne in mind that, in the correct circumstances, administration of water might be capable of inducing cancer in mice (although it is more often a cause of drowning). If a study was conducted that demonstrated that water was a carcinogen, would this mean that we should give up drinking water or, maybe, convert it to beer? The relationship between dosage and harmful effects is crucial in the assessment of chemicals, including those that occur in a natural diet. Given that much of the exposure of people to individual chemicals is at low levels, the fact that many may cause cancer at high levels is probably not significant for everyday life. Furthermore, it is important to remember that the majority of testing is performed on single substances, whereas the majority of exposure is to many substances simultaneously, for example, in a normal diet. Life is about mixtures.

STRONG TOXICANTS AND WEAK TOXICANTS

A reasonably clear ranking of potency among chemicals can be established when appropriate compounds are selected for comparison (Table 1.1). Thus, tetrachlorodibenzo-*p*-dioxin (TCDD) is one of the most potent chemicals known, and can be lethal to guinea pigs at 1 µg/kg of body weight, while the lethal dose of an everyday substance such as paracetamol (acetaminophen) is very much higher. However, this type of ranking is, in some ways, distorting, as the potency of any chemical can change markedly depending on the species under consideration, TCDD being 20 to 50 times less toxic in rats (Table 1.2) (Russell and Burch 1959). Organophosphate insecticides are much more toxic, by design, in insects than in

TABLE 1.1
Comparative Toxicity—Approximate Lethal Doses (mg/kg) for Chemical Class, Route, and Species

| Compound (Class) | Species | Route of Administration | | |
		Oral	Parenteral[a]	Dermal
Botulin toxin	Mice		0.000002 (IP)	
Ethanol	Man (est.)	7000		
	Mice	10,000		
Digitoxin (cardiac glycoside)	Cat	0.18		
	Guinea pig	60		
DDT (OC insecticide)	Rats	113		
Methoxychlor (OC insecticide)	Rats	6000		
Nicotine	Rat	50		
	Rabbit			50
Paracetamol (analgesic)	Man (est.)	250		
	Mice	340	500 (IP)	
Pentobarbital (barbiturate)	Mice	280	80 IV; 130 (IM, IP)	
Phenytoin (anticonvulsant)	Mice	490	92 (IV)	110
Malathion (OP insecticide)	Rat	1000		>4000
Parathion (OP insecticide)	Male rat	4		
	Female rat	13		
Soman/VX (OP nerve gas)	Man (est.)		0.007 (IV)	0.142
	Rat		0.012 (SC)	
	Guinea pig		0.008 (SC)	

Source: Adapted from Woolley A, *A Guide to Practical Toxicology: Evaluation, Prediction and Risk*, New York and London: Informa Healthcare, 2008.

Note: est., estimated; OC, organochlorine; OP, organophosphorus.

[a] Parenteral routes: IM, intramuscular; IP, intraperitoneal; IV, intravenous; SC, subcutaneous.

TABLE 1.2

Acute Toxicity of TCDD in Different Species

Species	LD_{50} (µg/kg)
Guinea pig	1
Male rat	22
Female rat	45
Mouse	114
Rabbit	115
Hamster	5000
Monkey	70

Source: Adapted from Poland A, Knutson JC, 2,3,7,8-Tetrachlorodi-benzo-*p*-dioxin and related halogenated aromatic hydrocarbons: Examination of the mechanism of toxicity, *Ann Rev Pharmacol Toxicol*, 22: 517–554, 1982.

mammals. Table 1.1 also shows the differing toxicities according to species and route of administration.

The toxicity of a substance is determined by the following factors:

- Dose (usually expressed in mg/kg of body weight, sometimes as mg/m^2 body surface)
    - Frequency of dosing: single or repeated
    - Duration of exposure or administration
    - Route of exposure
    - Physical form or formulation
        - Gas or aerosol
        - Liquid: viscous or free flowing; volatile or inert; aqueous or organic
        - Solution: concentrated or not
        - Solid: dust, inert mass, crystalline, or amorphous
    - Absorption, metabolism, distribution, excretion
- Species
    - Phenotypic variations; the individual exposed
    - Presence of receptors
    - Protein binding and disturbance due to competitive binding or deficiency of sites
- Presence of other chemicals: synergistic, additive, or inhibitory effects

When a substance is a potent toxicant, it is usually readily apparent from its effects on humans or other animals. The majority of debate comes at the lower end of the potency spectrum, particularly with synthetic chemicals such as pesticides, to which people are exposed at extremely low levels in everyday life. The bottom line is that we are exposed daily to thousands of chemicals, the majority of which occur naturally in our food and environment and about which very little is known in terms of toxicity. As far as food is concerned, as a result of culture and tradition, foods that are harmful are avoided or are treated specially before consumption. Thus, red kidney beans and cassava root are harmful, if they are not properly prepared before eating, due to the presence of toxins in the raw food. Fugu fish, a delicacy in Japan, requires careful removal of the skin, liver, and ovaries, which contain a potent nerve poison, tetrodotoxin, to which there is no antidote. Equally, there are ancient remedies, such as some herbal teas, which were given to people who were ill; the modern tendency to use these teas daily as health supplements can result in unwanted side effects. Thus, a traditional remedy may be safe when used as tradition indicates but becomes harmful, if used incorrectly or in combination. For instance, ginseng and gingko is a popular combination that was not used in history and can be associated with a range of side effects, including interaction with prescribed medicines.

THE NECESSITY FOR TOXICOLOGICAL ASSESSMENT

It is a moral and legal requirement that new drugs, pesticides, or food additives should be as safe as is reasonably possible, when they are made available to the doctors, workers, farmers, or consumers. The degree to which a product must be safe is determined by its intended use. Pesticides or food additives should be entirely safe at the levels at which consumers are exposed. Pesticides, by their nature, are toxic to the target species but should be safe for nontarget species. Food additives, whether added for processing reasons or flavor or as preservatives, have a lower margin of tolerance for safety than pesticides, for which an interval between treatment and harvest can be set together with an acceptable daily intake. With drugs, the acceptable margin of safety, the difference between therapeutic and toxic doses, is dependent to a large extent on the indication for which they are intended. Toxicity, seen as side effects, is more tolerable in an anticancer drug than in an analgesic sold over the counter.

To ensure that the required margin of safety is demonstrated, for instance, for a new drug, it is essential to conduct a program of experiments to assess the toxicity of the new molecule, specifically to describe a dose–response curve. From these data, it should be possible, with appropriate interpretation and experimental support, to extrapolate the effects seen to humans. For chemicals that are already marketed, toxicological assessment becomes necessary when effects are seen in consumers that have not been seen or noticed before. In this case, the intent is to establish a mechanism for the toxicity observed and to recommend appropriate changes, in the way the chemical is sold or in the way in which it is used. It is a fact of life that it is not considered ethical (in almost all circumstances) to administer new drugs to humans without some assessment of their effects in other test systems. With chemicals that are not intended primarily to be administered to people (e.g., pesticides), the restraints on giving them to humans are even greater. As a consequence of this, it is routine and, in many cases, mandatory to use animals in toxicological safety evaluation programs or in experiments that investigate mechanisms of toxicity seen in humans. At the current development of the science, a reduction in animal usage is always favorable but not always possible.

With this in mind, in using animals, it is increasingly understood that their use should be regulated to high ethical standards. A plethora of evidence indicates that animals can be good models for the behavior of chemicals in humans and that they are the only ethical and valid test system. However, evidence also suggests that they are not good models and that their use should be discontinued completely. Both positions are extremes, and inevitably the truth probably lies between them. In other words, an ideal situation is where the use of animals is reduced to an extent where there is sufficient scientific "comfort" to make a sensible and secure assessment of the risks. For some chemicals, one species will be a better model for humans than another, and for this reason, it has become normal practice to study toxicity of new drugs or pesticides in two mammalian species. Considerable savings (in terms of both animals and money) could arise from validating one or another species, or at least not requiring work in "invalid species." Exceptions to this two-species rule occur when there is only one valid target species (other humans). This generally occurs in special cases, such as biologicals, and must have a valid scientific rationale behind it.

TOXICOLOGY AND TOXICITY DEFINED

Toxicology is many things to many people. Toxicology has been defined by the US Society of Toxicology as "the study of the adverse effects of chemical, physical, or biological agents on living organisms and the ecosystem, including the prevention and amelioration of such adverse effects" (http://www.toxicology.org). In *Casarett and Doull's Toxicology: The Basic Science of Poisons* (Klaassen 2001), it is defined as "the study of the adverse effects of xenobiotics."

The *Dictionary of Toxicology* (2nd edition, Macmillan Reference Ltd, 1998) defines toxicity as "the ability of a chemical to cause a deleterious effect when the organism is exposed to the chemical." The *Oxford English Dictionary* indicates that toxicity is a "toxic or poisonous quality, especially in relation to its degree or strength." These definitions contain a number of important concepts, such as deleterious effect, exposure of an organism, and degree or strength. They also suggest that toxicity is only seen following exposure to externally applied chemicals. In many ways, they throw up more questions than they answer. What is a deleterious effect or a toxic or poisonous quality? A simpler definition might be that it is an adverse change from normality, which may be irreversible; but this requires definition of adverse change and, crucially, of normality.

An adverse change is one that affects the well-being of the organism, either temporarily or permanently, while normality is probably best considered in statistical terms of the normal distribution with a mean ±2 standard deviations. Toward the upper and lower limits of such a population, the decision as to whether a value is normal or abnormal may become more complex and open to debate. In the absence of quantitative or semiquantitative data, the decision as to what constitutes normality becomes subjective and dependent on the judgment or prejudices of the decision maker. In crude terms, it is easy to define some changes as adverse; for instance, cirrhosis of the liver is an irreversible change, which is often associated with early death. In cases where this is brought on by drinking excessive amounts of alcohol, it is easy to conclude that alcohol is toxic. There has been extensive debate as to whether low doses of alcohol are beneficial, with current opinion leaning toward the duller side of the fence. Some substances, such as vitamin A, are clearly essential but are associated with toxicity, including reproductive effects, at high doses; in addition, arctic explorers learnt that polar bear liver is toxic due to excessive vitamin A storage.

How, then, to separate toxicity from the norm or from effects that are potentially beneficial? This question is valid for all substances to which we are exposed, whether these are natural constituents of our diet or synthetic chemicals such as pharmaceuticals or pesticides. At what point does vitamin A cease to be beneficial and start to have an adverse effect? With medicines, the question becomes much more complex because beneficial effects such as treatment to kill a tumor may be associated with unpleasant side effects that, in a person without cancer, would be clearly adverse and unacceptable. Nausea and vomiting would be unacceptable as routine side effects of an analgesic for headaches but are accepted in cancer treatment, where the cost of side effects is offset by the benefit of a potential cure.

Much of the effort that is put into the determination of toxicity has the ultimate motive of assessing or predicting safety in terms of daily exposure levels that can be expected to have no long-term adverse effect. Some toxicity investigations are undertaken to elucidate the mechanism by which a substance is toxic, when effects have been shown in toxicity studies or through epidemiological investigation or clinical experience. The discovery of the mechanism by which paracetamol (acetaminophen) is toxic has greatly contributed to the successful treatment of overdose.

For synthetic chemicals, whether they are intended to be pharmaceuticals, pesticides, industrial chemicals, or intermediates used in the synthetic pathways for these substances, there is a clear need to define toxicity so that any adverse effects can be understood and their effects in humans can be predicted. This need is relevant as much to the people producing the chemical as to the eventual consumers. The definition of toxicity is important for natural chemicals as well, although the usual reaction to the toxicity of such substances, for instance, vitamin A, is one of surprise and disbelief. It is assumed that if a small amount is good for you (essential even), then getting a large dose must be particularly beneficial.

The necessity for the study of toxicity becomes less clear when the chemical in question is a natural constituent of a normal diet. Going back to the mushrooms example, much effort has been invested in the various chemical constituents of mushrooms. At high doses, it has been shown that it is possible to induce cancer in Swiss mice but only when they are fed unrealistically high concentrations of the individual chemicals found in mushrooms or of whole or processed mushrooms. Once again, it is necessary to invoke Paracelsus and point out that response is dependent on dose. We should question the conclusion of research that implies that we should be careful about or give up eating a vegetable because a constituent can cause cancer at high levels of ingestion in rodents. There is also the paradox that fruit and vegetables are known to be good for you, but there is also the realization that they contain many chemicals that may be toxic if enormous doses are taken. This is especially true when they are taken or administered in isolation from their natural source or context.

In assessing safety, one of the prime concerns is whether the test chemical is capable of causing or promoting cancer. However, one of the basic problems here is that cancer is expected to develop in between 25% and 40% of the population, depending on source of the estimate, and that it is very diverse in form and causation. Cancer's origins are multifactorial and often cannot readily be ascribed individually to a particular cause. Even with an apparently cut-and-dry association, such as lung cancer and smoking, it is not often possible to say that smoking has caused the cancer, because of the influence of other factors such as alcohol consumption, occupational exposures, and so forth. Thus, if someone regularly consumed 100 g of mushrooms each day, as part of an otherwise balanced diet, it would probably not be possible to ascribe a stomach cancer diagnosed in old age to the ingestion of unusually large amounts of mushrooms. In addition, many cancers have latency periods that may last many years, and a tumor such as mesothelioma, due to asbestos, may occur 30 to 40 years after the causative exposure. Many dietary constituents are potentially carcinogenic at high doses, but that does not mean that we should give up eating the normal foods in which they occur. Clearly, it is important that we should understand such effects and attempt to reduce the risk as far as possible.

IT IS NATURAL, SO IT MUST BE SAFE— EVERYDAY TOXICOLOGICAL CONUNDRUMS

It is not sensible to assume that natural chemicals are safe, just as it is not sensible to assume that a synthetic chemical is inevitably toxic. The website of the American Council on Science and Health (ACSH: http://www.acsh.org) has published a number of items on the presence (in our normal diet) of various chemicals and carcinogens that are found naturally in an everyday diet. The listing included allyl isothiocyanate in broccoli spears, hydrazines in mushroom soup, aniline and caffeic acid in carrots, psoralens in celery, and finally, a long (and incomplete) list of chemicals found in coffee. No one is suggesting that consumption of normal quantities of an everyday diet is going to be associated with unacceptable toxicity, but the list gives some perspective on the relevance of chemical intake and the fact that many toxins cannot be avoided. The ACSH make the point that more than 99% of the chemicals that people ingest occur naturally in a normal diet. However, the chemicals listed on the ACSH website, although safe when eaten in a normal diet, are variously mutagenic and carcinogenic in rodents, associated with contact hypersensitivity and phototoxicity, or are simply toxic when given in their pure form at high concentrations to rodents. If the Delaney clause—a notorious piece of US legislation banning synthetic chemicals from foods if they were shown to cause cancer in animals—were applied to chemicals naturally present in food, our diet would be immediately impoverished and probably unhealthy.

For the majority of chemicals, it is possible to plot increasing toxic effect against increasing dose to produce a dose–response curve that is sigmoid in shape. For some chemicals, the response curve is U-shaped. For chemicals that are essential for the well-being of an organism, such as vitamin A, there is an optimum range of dose over which normality is found. Doses lower than this show increasing evidence of deficiency; higher doses show evidence of toxicity that increases with dose. This type of finding is common to many vitamins or other essential naturally occurring chemicals. Another type of response curve is shown by aspirin, which inhibits platelet aggregation at low dose (reducing the incidence of heart attacks), is active at normal doses for inflammation or pain, but becomes toxic at high doses. The difference with aspirin is that it is not essential and absence from a normal diet will not be associated with adverse effect.

One of the most important questions to consider is, at what point should findings in toxicological experiments alert us to hazards arising from routine exposure to individual chemicals? To answer this requires that the toxicological hazard is actually due to the chemical under study coupled with confirmation that the mechanism of toxicity is relevant to humans. For some rodent carcinogens, this question is easy to answer. Where there is a direct effect on DNA that can lead to cancer, often at low doses, as for aflatoxin, there is a clear human-relevant hazard. Where the mechanism of carcinogenicity is not related to direct DNA damage but to a nongenotoxic effect, the answer is less clear. However, this cozy distinction between genotoxic and nongenotoxic has been thrown into doubt by the realization that epigenetic change (such as changes in DNA methylation or histone changes) can lead to heritable genomic change.

Many rodent carcinogens achieve their effects through nongenotoxic mechanisms that are not relevant in humans. D-limonene is carcinogenic in the kidney of male rats through formation of slowly degraded complexes with α2u-globulin, a protein found at high concentrations in the urine only of male rats. This protein is normally degraded in lysosomes in the kidney, but when it is complexed with d-limonene, this degradation is slowed, resulting in overload of the lysosomes and necrosis in the proximal tubule cells and regenerative cell division. The resulting hyperplasia can lead to the formation of cancers. But because α2u-globulin is specific to male rats, this effect is of no relevance to human health.

A further example of nongenotoxic carcinogenicity in animals that is not relevant to human health is peroxisome proliferation and the subsequent induction of liver tumors in rats and mice. Other species have been shown not to respond to these agents in this way, notably in a 7-year study with a peroxisome-proliferating hypolipidemic compound in marmosets (see the ciprofibrate case study in Chapter 12). This same study indicated that stomach tumors were seen only in rats. The risks of peroxisome proliferation and the relevance of this to humans are looked at in greater detail in the chapters on risk assessment.

This should not be taken to imply, however, that nongenotoxic carcinogenicity is irrelevant to humans, as a large number of human cancers, such as colon or breast cancer, are attributable to such mechanisms. However, in general, if a chemical is carcinogenic by a nongenotoxic mechanism in one species of rodent (perhaps in one sex), at doses that are very much higher than those found in routine human exposure, it is probable that this effect is not relevant to humans.

It is relatively easy to identify chemicals that damage DNA and are mutagenic in tests *in vitro*. Detection of human-relevant nongenotoxic effect is more complicated as there are many more endpoints, but this would seem to be the route of the future for investigations of carcinogenic potential.

The fear of cancer is very real as it is a widespread condition that almost always has unpleasant side effects and is frequently fatal. However, this fear is usually reserved for new untried factors or for occupational exposures that become associated over a period of years with cancer. Furthermore, cancer is not the only hazard of which people should be aware; there are many toxic properties contained in apparently innocuous preparations and foods, which are used routinely and without concern.

NATURAL MEDICINES AND POISONS

A wide range of herbs has been used in traditional medicines, often in teas and infusions that were taken as indicated by a physician or herbalist. There has been an unfortunate tendency to drink these teas regularly as a tonic, and this overfrequent use can result in serious unwanted effects. Herbs often contain pharmacologically active compounds of great potency and, apart from toxicity arising from excessive pharmacological action, can have carcinogenic and teratogenic properties (Focus Box 1.2).

FOCUS BOX 1.2   TOXICITIES ASSOCIATED WITH NATURAL REMEDIES

Traditional remedies, often taken as herbal teas, sometimes have highly pharmacologically active constituents, and innocent overuse can have significant adverse effects.

- Ginseng is used in Chinese medicine for impotence, fatigue, ulcers, and stress. It contains active compounds that produce central nervous system (CNS) stimulation and increase gastrointestinal motility. Chronic or excessive use can be associated with diarrhea, nervousness, cardiac effects and nervous system disturbances, and imbalance of fluids and electrolytes.
- Gingko (*Gingko biloba*) is used in combination with ginseng. Although relatively safe, it has effects such as decreasing platelet aggregation, meaning that it can increase spontaneous bleeding and can exacerbate the effects of drugs such as aspirin taken to "thin the blood."
- Comfrey (*Symphytum* sp.), which has been used as a wound healer, anti-irritative, antirheumatic, and anti-inflammatory, contains pyrrolizidine alkaloids, which are highly hepatotoxic and potentially carcinogenic through damage to DNA. Daily consumption of comfrey over several years in salads or teas can lead to liver toxicity. Teas derived from the roots are particularly hazardous; in addition, the preparations may be fetotoxic.
- Fresh garlic has wide antimicrobial activity and fibrinolytic activity, reduces blood cholesterol and lipid concentrations, and reduces formation of atherosclerotic plaque. However, taken to excess, it can induce nausea, vomiting, diarrhea, and bronchospasm; it is also associated with contact dermatitis attributed to the presence of antibacterial sulfides.

Source: Adapted from Dart R et al., Medical Toxicology,
3rd ed., Lippincott Williams and Wilkins, 2004

There are many instances of interactions between herbal remedies and prescribed drugs, either through increased or decreased effect. Even simple dietary components can have unexpected effects; grapefruit juice consumption is known to be associated with inhibition of cytochrome P450 (CYP3A4), which is responsible for the metabolism of a wide range of drugs. This has been associated with increased plasma concentrations of cisapride, a drug given for irritable bowel syndrome. Inhibition of cisapride metabolism, which probably takes place in the small intestine, can increase the likelihood of life-threatening cardiac arrhythmias in some patients. A similar effect has been reported with carbamazepine, a drug given in epilepsy. Equally, administration of metabolism inhibitors can have useful effects, for instance, in reducing the doses of some drugs needed to achieve therapeutic effect.

With the commonplace example of grapefruit juice, it becomes clear that there are unsuspected risk factors in everyday existence; equally, given the multitude of such risks, attempts to account for all of them and lead a risk-free existence are probably doomed to failure. Another unsuspected source of risk is honey. There would normally be no reason to suspect honey as potentially harmful, but when it is produced from rhododendron flowers, it is toxic, two teaspoons being enough for adverse effect in some subjects.

Many natural substances or mixtures have been associated with abuse and resultant toxicity, prime examples being tobacco, cannabis, and opium.

NATURAL VERSUS SYNTHETIC

Digitalis from the foxglove (*Digitalis* spp.) has been known for hundreds of years and has been commonly used in cases of edema (dropsy), essentially as a diuretic with cardiac side effects that gradually came to be appreciated as a primary action of the drug. The foxglove contains a number of pharmacologically potent cardiac glycosides, which at low doses have a complex range of actions on the cardiovascular system. They are used in congestive heart failure and are sometimes used to decrease ventricular rate in atrial fibrillation. The initial plant mixture of numerous active constituents has been refined to the extent that single compounds are now used, for example, digoxin. However, digoxin oral absorption tends to be variable, and this, together with a steep dose–response curve (i.e., narrow margin of safety), makes therapy more hazardous than is desirable. With the advent of modern pharmaceutical research, new cardioactive agents were discovered that are safer than digitalis, especially the calcium channel blockers, such as verapamil, diltiazem, or nifedipine. These have dose–response curves that are less steep than digitalis-like drugs and so are easier to use because the toxic dose is appreciably higher than the therapeutic dose. In this instance, the synthetic drug is safer to use than the naturally derived agent, having a more targeted action and a much smaller range of adverse side effects. The original problem with digitalis extract, that of administering an imprecisely defined mixture of highly potent alkaloids with wide-ranging effects, has been gradually circumvented by purification and finally by synthesis of carefully targeted molecules.

The effects of endogenous chemicals are, in some cases, mimicked by those of xenobiotics. Compounds such as opium and morphine have well-known addictive properties and share some properties with neuropeptides that are present naturally in mammals. Opium is a mixture of alkaloids that includes morphine, of which codeine is a methyl derivative; as with digitalis, use of opium, which was known to the ancient Greeks, gave way to the use of the individual compounds. The complex range of actions of opioids is explained by the presence of several receptor types, for which endogenous peptides have been discovered. These peptides—the endogenous opiates—of which endorphins are one example are produced in reaction to stress, such as exercise, and there is increasing evidence that, like their natural plant-derived counterparts, they have addictive properties. Exercise-induced euphoria (runner's high) is a relatively frequent term in the literature, and endorphin release is associated with alterations in pain perception, feelings of well-being, and lowered appetite. It has been suggested that the euphoria leads to altered perception of risk and may be associated with some accidents where joggers are hit by cars. Another possible side effect of excessive exercise is that addiction to the endogenous opioids may be associated with eating disorders, including anorexia.

The natural-versus-synthetic debate should not be left without consideration of the issue of transgenic materials or GM organisms. Transgenic indicates the transfer of genetic material from one species to another, often from another taxonomic phylum. One example is transgenic maize produced by inclusion of a gene from *Bacillus thuringensis* that expresses an insecticidal protein, which kills maize borers, a significant source of damage to crops. Other insertions delay deterioration of fruit and vegetables or seek to improve flavor. There is concern that the novel foods thus produced may not be safe either in terms of human use or their environmental safety. One of the cited environmental advantages of the modified maize was the reduced use of pesticides, including herbicides, which has not been confirmed in practice. Where a novel gene is inserted with the intention of expressing a protein or peptide, this type of inclusion is highly unlikely to be of any danger to consumers of the products due to the normal process of digestion and consequent low absorption of intact peptides from the gastrointestinal tract. Although the individual nucleotides of the inserted genes and the resulting amino acids from their protein products will be absorbed, they will be biochemically equivalent to the natural nucleotides and amino acids. As such, they would be indistinguishable from them and should be safe. However, where insertion is intended to express a small molecule that can be absorbed intact or have local effects in the gastrointestinal tract, there may be greater risks to consumers.

The issue of GM crops is so polarized that irrational belief may drive interpretation of marginal data. This *a priori* belief overcomes dispassionate assessment of difference and any sensible judgment on how that difference has been achieved; difference may be due to poor study design; small group size; and wide, natural, biological variation amongst a host of other factors. Unfortunately, there is pressure to conduct research, even if the available funds inevitably mean that the study design is exiguous and interpretation biased by preinterest in a particular outcome. Similar bias has, in the views of some, tainted research into the effects of tobacco smoke; there is no doubt that early studies were poorly designed and that interpretation of the differences seen was biased toward the prevailing expectation of toxicity. Review of these early papers shows how difficult it was to elicit toxicity in the test animals;

this was partly a function of dose of nicotine relative to the other, less acutely toxic, components of tobacco smoke. (It should be said that tobacco smoke is clearly toxic and has an assortment of adverse effects. However, its many components clearly work with a range of other factors to exert its effects on health; there are few heavy smokers who eat good diets, do not drink, and/or have no family history of cancer.)

A case study on GM corn and potatoes is presented in Chapter 20. These show that toxicological research in this area has to be conducted to very high standards. The risks of getting it wrong—either in overstating the risk with attendant effects on future food production levels or in understating risk that might result in unacceptable toxicity in consumers—are clearly high. Either way, the toxicologists concerned will be at the forefront of the debate, and their independence and scientific standing must be unimpeachable.

Environmental risks are another matter; genetic transfer between herbicide-resistant crops and weeds has been shown to require the use of more toxic chemicals, which the original insertions were supposed to make redundant. This is the type of situation where indirect toxicity could result. If pesticide use increases, there could well be environmental detriments that affect the public indirectly, for instance, through increased concentrations in drinking water or through less easily measured changes in the ecosystem. Another type of effect is shown by the widespread use of warfarin as a rodenticide and the subsequent emergence of rats that are resistant to its effects, through evolution in its metabolism.

GENERAL OBJECTIVES OF TOXICOLOGICAL STUDY

Given that toxicology is a wide-ranging subject with many applications, there are many reasons for starting a toxicological investigation. Among these are the following:

- To assure safety of new chemicals for use as pesticides, drugs, or food additives before they are registered for general use in industry or doctors' clinics. This type of toxicity study is regulated by government and international guidelines that describe minimum study designs and the types of study that must be conducted and the test systems that may be used. Toxicologists in regulatory authorities tend to be conservative in their approach, as they have a responsibility to the public to ensure that the safety of new chemicals is thoroughly investigated before significant human exposure is allowed.
- To define a dose–response curve—the quantitative relationship between dose and response. It is important to define the steepness of the response; for some drugs such as phenytoin, digitalis, or warfarin, a small increase in dose can produce very large increases in adverse side effects that can be life threatening. The safety margin for these compounds is very small; for other classes of drug, it can be much greater, and a doubling of dose will have little additional effect.
- To establish the mode of action or mechanism for a toxic effect that may have been seen in other studies.

- To elicit epidemiological studies to explain observations in the population, for instance, the long investigation into the association of smoking with lung cancer and other diseases. This type of study may also be used to seek explanations for toxicities seen in patients or workers in particular industries.
- To investigate or validate new methods of testing or investigation, particularly those conducted *in vitro* rather than in animals.

The last point is particularly important, as the extensive use of animals in toxicological experiments is increasingly questioned. There is a considerable dilemma here; animals offer a whole multiorgan system in which to conduct experiments, and the interrelationships of the various organs can be investigated in a way that cannot be done (at present) in a single cell or tissue system in a static vessel. However, the emergence of new techniques such as organ or body on a chip may begin to mitigate these weaknesses. Substances do not limit themselves to the target section of the environment or body that they are introduced to. For instance, an oral diuretic does not transit directly to the kidney where it is expected to have its effect. In order to get to the site of action in the kidney, the drug has to pass through the gastrointestinal tract and then the liver and the bloodstream. In the blood, on the way to the kidney, the drug can pass through every other organ in the body where there are opportunities for unwanted effects. For example, certain classes of antibiotic (e.g., gentamycin) and diuretic (e.g., furosemide) have been associated with effects on the hair cells in the inner ear, leading to hearing impairment. Similar off-target effects can be seen in the case of the pesticide DDT, the extensive use of which was associated with accumulation up the food chain and a decline in predatory birds due to eggshell thinning. Another example is provided by the use of diclofenac in cattle in India and a subsequent decline in the population of vultures. Any program of work that seeks to investigate the effects of a novel compound intended for extensive, regulated use in man must, at some point, include experiments in animals that will examine these interrelationships.

The downside of this is that animal research is expensive and may not give a wholly reliable and relevant result when related to humans. The best test system for humans would appear to be other humans, but even this is complicated by differences between ethnic groups and individuals and by ethical questions as to whether human volunteers should be asked to take potentially toxic pesticides and/or new drugs for which the toxicity is completely unknown. Knowledge of the genetic differences in human populations also indicates, because of genetic diversity, that one group of humans may not be a good model for another. For instance, indigenous Canadians tend to be able to metabolize the antituberculosis drug isoniazid faster than Egyptians and would therefore give erroneous results were they to be used as a test system to investigate safety of use in Egyptians.

Finally, there is no doubt, in the absence of exceptional circumstances (such as biological drugs for which the only relevant species is humans), that giving even small doses of a chemical with unknown toxicological potential to humans is ethically unacceptable. For these reasons, experiments that attempt to find new methods of investigation and to validate them for use in safety evaluation comprise one of the more important avenues of toxicological exploration. A consequence of validation and acceptance (two different concepts) of a new method should be that there is also evolution in the risk assessment process. Such evolution continues as understanding of the new method increases, highlighting its weaknesses as well as its strengths; this translates into new protocols and practices, a point illustrated by the evolution of the local lymph node assay for prediction of sensitization (see Chapter 9).

BIOLOGY AND OUTCOME OF TOXIC REACTIONS

Chemicals interact with cells in numerous ways and can be broadly categorized as having effects at the level of individual cellular components, cell organelles, the cell, the tissue, organs (or part of an organ—different lobes of the liver may show more or less effect compared to others), or the whole organism. This can be further expanded in environmental toxicology to cover effects on populations. Interactions at the cellular level are often associated with a precise molecular target, such as a pharmacologically important receptor, DNA, an enzyme, or another molecular component, such as membrane lipids. Such interactions may result simply in changes in the biochemistry of the cell, the effects of which may or may not be visible under the microscope, or in effects so severe that they result in cell death. Cell organelles, such as the endoplasmic reticulum, which carries many enzymes responsible for metabolism of xenobiotic chemicals, can be disrupted by lipid peroxidation brought about by free radicals generated by metabolism of the chemical. This autodestructive process can be so extensive that the whole cell is affected and dies through necrosis. Equally, the endoplasmic reticulum may become more extensive as a result of increased amounts of enzymes produced to metabolize a particular chemical. This process of enzyme induction is classically associated with hepatocytic hypertrophy around the central vein in the liver, which produces a characteristic appearance in histological sections and is often associated with an increase in liver size. In many cases, this is considered to be evidence of adaptive change rather than toxicity, as it is usually readily reversible.

When many cells in a tissue are affected, the whole organ may be changed in functional terms or have a different appearance under the microscope when compared with controls or expectation (or normality—see Chapter 2). The kidney is a good example of an organ in which particular parts may be affected while leaving the rest of the tissue apparently untouched. Damage to the glomerulus or the proximal convoluted tubule may not be reflected by visible change in other parts of the organ. However, the function of the kidney as a whole may be affected by influences such as blood pressure or hormonal diuresis, which may result in unwanted side effects in the rest of the organism.

All toxic reactions have a biochemical basis, which may be more or less precise. Cyanide specifically inhibits cytochrome oxidase in the mitochondria, preventing oxidative phosphorylation through inhibition of mitochondrial electron transport. Fluorocitrate, a metabolite of the rodenticide fluoroacetate, is bound by aconitase, an enzyme in the tricarboxylic acid cycle. Thus, the central cycle of carbohydrate metabolism is inhibited by blocking the conversion of citrate to isocitrate, leading to the death of the recipient.

The toxicity of many chemicals is due to their metabolites rather than to the parent molecule—inhibition of which may mitigate or remove the toxic effects. This toxicity-inducing metabolism may be a normal fate for the molecule; fluoroacetate is always metabolized to fluorocitrate. Paracetamol, however, is normally eliminated from the body by conjugation with sulfate or glucuronide, while a small proportion (approximately 4%) is metabolized via cytochrome P450 to a metabolite (N-acetyl-p-benzoquinone imine) that is normally conjugated with glutathione (GSH). During conditions of overdose, the major routes of metabolism become saturated, and thus, the proportion of paracetamol catalyzed to the reactive metabolite increases. This leads to GSH depletion, and although GSH is present in the liver at high concentrations relative to other peptides, such depletion frees the toxic metabolite to bind covalently to liver proteins, leading to liver necrosis. Paracetamol overdose is often fatal, but in survivors, the effects may be transient; liver biopsies taken a few months after overdose may reveal no evidence of previous damage, due to the liver's enormous powers of self-repair.

The toxic reactions discussed above are examples of effects that are immediately obvious and that usually follow a high, or relatively high, dose. Other toxic reactions may be expressed slowly, through gradual reduction of functional reserve. The kidney of a young adult has an excess of functional capacity, which declines with age. If high doses of nonsteroidal, anti-inflammatory drugs (NSAIDs) are taken, through prescription or abuse, this normal age-related decline can be accelerated to a point where renal failure occurs. The same situation can exist in the nervous system, where normal age-related decline may be accelerated by constant exposure to doses of a chemical that are not individually toxic but have a disastrous additive effect. This has been seen following daily work-related exposure over long periods to n-hexane or methyl n-butyl ketone. The long-term result is a neuropathy and muscular weakness, which begins in the extremities and progresses toward the center with continued exposure.

Another type of reaction to toxins is one characterized by a biochemical and then a morphological response to a chemically induced imbalance in the organism, often hormonal. Drugs such as some hypolipidemic fibrates, which in rats inhibit secretion of gastric acid, have been associated with carcinoid tumors of the rodent stomach. This is due to an increase in the plasma concentration of gastrin, which stimulates the neuroendocrine cells, resulting in their hyperplasia. This hyperplastic response is translated, in a proportion of the animals, to malignant tumors. Similarly, hormonal imbalance can result in an increased incidence of breast cancer or prostate cancer.

Through effects in the reproductive tract or on the reproductive cycle, which may not be apparent at the time of exposure to the responsible chemical, toxicity can also be expressed in succeeding generations. Thalidomide is the classic example of this and is interesting from several standpoints. A sedative given to pregnant women to reduce nausea in early pregnancy, thalidomide was associated with a range of defects in the offspring mainly due to its inhibition of angiogenesis, the most apparent being shortening of the long bones of the limbs (phocomelia). The defects were closely dependent on the day the drug was taken, usually in the fourth or fifth week of pregnancy. Furthermore, it is a chiral molecule, and the $S(-)$ form is more embryotoxic (teratogenic) than the $R(+)$ form. Thalidomide illustrates the precision of effects on the developing fetus and the importance of time of exposure relative to the stage of embryonic development. A further example of unsuspected reproductive effect is diethyl stilbestrol, which was given to pregnant women as an antiabortion agent but was associated with the appearance of clear cell adenocarcinomas in the vagina and testicular defects in their offspring. These effects did not become evident until puberty of the child.

Toxic reactions, severe or trivial, may be reversible or irreversible according to which tissue is affected. As indicated above for paracetamol, the liver has a large capacity for repair of extensive lesions; other tissues associated with easy repair are those that can divide and replicate themselves quickly, such as the skin and gastrointestinal tract. The kidney falls into both the reparative and the next, nonreparative category. The epithelium of the proximal tubule is a common target for toxic attack, but as long as the basement membrane on which these cells rest is not breached, repair can be very rapid, if exposure is stopped. Other parts of the kidney, notably the glomerulus and the renal pelvis, do not repair so readily in parts due to differences in embryonic origin. As suggested by the example of the kidney, tissues that do not divide readily do not repair easily or, in some cases, at all. Of these, the usual example is the central nervous system, another tissue with a large functional reserve that can be overwhelmed by insidious toxic actions over a period of years.

Toxicity in the environment has a number of analogies with toxicity in individual organisms or species. Following introduction into the environment, a chemical is, ideally, absorbed, changed, and eliminated. However, it may also accumulate to the point where it has adverse effects on individuals (analogous to the cell in a tissue or organ), on species (organs), and on ecosystems (analogous to the individual). The result is imbalance in the ecosystem, which can have long-term effects, including an adverse impact on humans living in it.

CELLULAR BASIS AND CONSEQUENCES OF TOXIC CHANGE

With the exception of a few substances that are corrosive, direct effects of toxic substances are expressed in individual cells. According to the extent of exposure, the number of cells affected increases to the point where the whole organ or tissue is changed, biochemically or morphologically. It should be borne in mind that dose is not necessarily the same as exposure as not all of a given dose may be absorbed to result in exposure at the target site. There are relatively few substances that have direct effects in cells without first being metabolized; these include reactive substances such as alkylating agents and metals such as lead and cadmium. Those that are active without prior metabolism are often intrinsically reactive or have activity at specific receptors on the cell membrane or in the cell itself.

Corrosive substances tend to act nonselectively from outside the cell and have widespread effects that result in the deaths of many cells but in a different manner to toxicants that work selectively on and within the cell. In other words, in corrosion, there is no initial molecular event that could be said to be the initiator of toxicity. With something like an acid, the corrosive effect is equal on all molecules in the cell and the effects devastating for the exposed tissue, which is often the skin. Although the corrosive effects may be local initially, depending on the substance involved, the toxicity expressed may become systemic as the protective barrier of the skin is broken down and absorption takes place; for example, dermal exposure to phenol may lead to systemic effects.

In individual cells, toxicity may be classified in broad terms as either reversible or irreversible, and there may be a change in functional competence or morphological or biochemical lesions that impair the well-being of the cell. Irreversibility may not be associated with immediate expression of effect, as with the development of adenocarcinoma in young women following maternal exposure to diethyl stilbestrol or with mesothelioma years after exposure to asbestos. In the case of minor functional changes, which might be associated with the activity of cellular pumps or signal transmission capabilities, the changes may be repairable and so prove to be reversible. Where repair is not possible due to the extent of the lesions, the cell may die through two possible routes—necrosis or apoptosis. Necrosis is a process over which the organism's biochemistry has no control and consequently is accompanied by characteristic morphological changes indicating an almost violent death of the cell. It is often associated with the presence of inflammation, typified by the presence of leukocytes that have migrated into the tissue to the site of damage. Apoptosis (programmed cell death), on the other hand, involves a series of defined biochemical events that result in the removal of the cell contents and membranes in a manner that, in comparison with necrosis, leaves little morphological evidence. Apoptosis is a physiologically normal and essential process that occurs in the absence of noxious stimuli. For instance, correct regulation of apoptosis is essential in normal embryonic development.

Where a cell is damaged in a way that does not result in immediate necrosis or later apoptosis, the effects may persist for years, without causing any further damage. In cases where there is an unrepaired change in the DNA, this may lie dormant until the cell is stimulated to divide, and if this division is repeated and the process not controlled, the result may be a tumor—benign or malignant. In this way, an apparently benign or invisible change may have devastating effects, years after the relevant exposure is forgotten.

Receptors, as distinct from the active sites of enzymes, play an important role in many toxicities. Binding of a foreign chemical to a receptor, instead of or in competition with the natural ligand, can be expected to result in adverse effects, if the receptor is inappropriately activated, inactivated, blocked, or modulated. As with other mechanisms of toxicity, the effects of receptor binding may be acute, as with tetrodotoxin to sodium channels in the neuronal axons, or delayed, as in tumor promotion by phorbol esters. In the latter case, the phorbol ester binds to protein kinase C, which triggers a cascade response, ultimately resulting in cell division in which existing DNA damage can be fixed, leading to tumor growth from previously initiated cells. Where effect is dependent on a substance binding to a receptor, the extent of the effect may be influenced by changes in the expression of that receptor. Simplistically, effect may be proportional to changes in receptor expression or receptor signaling (the efficiency with which the receptor passes on the results of binding).

For toxicities that are expressed through reversible, noncovalent binding to pharmacological receptors, it is probable, for short-duration exposure at least, that when the stimulus or ligand for the receptor is removed, the undesirable effect will cease.

Receptor effects can be induced indirectly by toxicants reacting at the active sites of enzymes. The reversibility of effect, driven by strength of binding to the site of action, is illustrated by the difference in toxicity between organophosphate and carbamate insecticides. Both these classes of insecticide bind to the active site of acetylcholinesterase, the enzyme that hydrolyzes the neurotransmitter acetylcholine. Whereas carbamate binding to the active site is relatively transient, the binding of organophosphates lasts much longer, to the extent that it is, in some cases, considered irreversible. Where the target cholinesterase is the neuropathy target esterase, "aging" of the enzyme takes place, resulting in a permanent change. This is the basic mechanism behind organophosphate-induced delayed neuropathy, a persistent effect of exposure to some organophosphates. The role of organophosphate-based sheep dips in the occupational health of farm workers is discussed in Chapter 14 (see Focus Box 14.1).

Where cellular homeostasis is affected, there can be severe consequences for the cell, but where there is a widespread effect, consequences are for the whole organism. A classic example of this is the effects of ricin, from the castor bean plant, which is a mixture of enzymatic proteins that fragment ribosomes, thus inhibiting protein synthesis. Although ricin is known as a parenteral toxin, ingestion of castor beans can be associated with severe toxicity, particularly in the gastrointestinal tract, with large ingestions producing hemorrhagic gastritis, diarrhea, and dehydration. The effects are exacerbated by the presence of ricinoleic acid in the oil of the seed, which increases the peristalsis in the intestine; beyond the intestine, target organs are the kidney and liver. Thus, action at a vital cellular target produces adverse effects in the whole organism.

The interaction of one cell with another is another aspect of the cellular basis of toxicity. In normal tissue, adjacent cells have channels between them through which small molecules can pass—these intercellular channels are termed gap junctions. Cells in which these gap junctions are still patent are less prone to proliferation than when they are closed. Several tumor promoters, for example, phorbol esters and phenobarbital, reduce intercellular communication through gap junctions, and this is thought to lead to transformation of the cells and so to neoplasia.

The axons of neurons can be extremely long and are dependent on the transport of nutritional components from the neuron cell body. When this transport is disrupted, the axon dies back in the type of reaction that is seen in response to chronic exposure to n-hexane. In this case, metabolism of the hexane to 2,5-hexanedione is associated with cross-linking of neurofilaments in the axon and subsequent blockage of transport at the nodes of Ranvier. Here, effects on one type of cell are associated with adverse change (progressive peripheral paralysis) in the rest of the organism.

Toxic attack on specific cell types is characteristic of many chemicals, for instance, the effects of paraquat and 4-ipomeanol in the lung. Paraquat is a widely available herbicide, which has two nitrogen atoms that are the same distance apart as in two endogenous polyamines (putrescine and spermine). This similarity allows it to be taken into the type I and II pneumocytes in the lung via an active transport process, where it accumulates, and with the high local concentration of oxygen, undergoes redox cycling (Figure 1.1). This process involves reduction of paraquat by an electron donor (e.g., NADPH) and its reoxidation by transfer of an electron to oxygen. This results in the

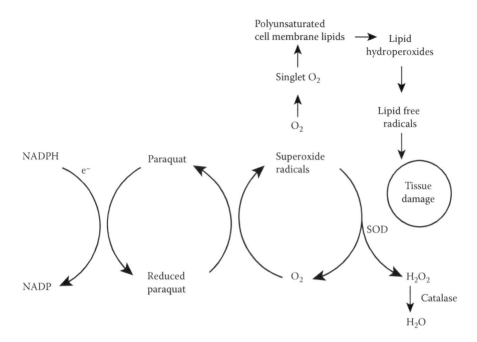

FIGURE 1.1 Mechanism of paraquat toxicity. NADP, nicotinamide adenine dinucleotide phosphate; NADPH, reduced form of NADP; SOD, superoxide dismutase. (From Timbrell JA, *Introduction to Toxicology*, London: Taylor & Francis, 1999.)

generation of superoxide radicals, which go on to form hydrogen peroxide through the action of the protective enzyme superoxide dismutase. Where this enzyme activity is too low to remove all the available superoxide, reactive hydoxyl radicals are produced, with subsequent attack on the cellular lipid membranes through lipid peroxidation. The ability of these cells to concentrate paraquat and the locally high concentration of oxygen work together to cause severe local toxicity, affecting the whole animal. The fact that the molecule is not metabolized to a less toxic form but is regenerated, leaving it free to repeat the cycle many times, leads to depletion of NADPH, with significant effects on cellular homeostasis. Paraquat also has effects in the kidney and other organs, but it is the lesion in the lung that is responsible for the death of the patient. The effect of the kidney change is to reduce renal function, slowing excretion and thus exacerbating the toxicity.

The mold *Fusarium solani*, found on sweet potatoes, produces 4-ipomeanol (Figure 1.2), which attacks the Clara cells of the lung specifically, through production by cytochrome P450 of an epoxide on its furan ring. This enzyme is also present in the liver, although it is present there in larger amounts than in the Clara cells but is less active. In addition, the liver has large concentrations of GSH, a tripeptide that is crucial in protection against oxidative attack. Consequently, the Clara cells are at a disadvantage in comparison with the hepatocytes and show the effects caused by binding of the reactive intermediates to cellular macromolecules, leading to necrosis and pulmonary edema.

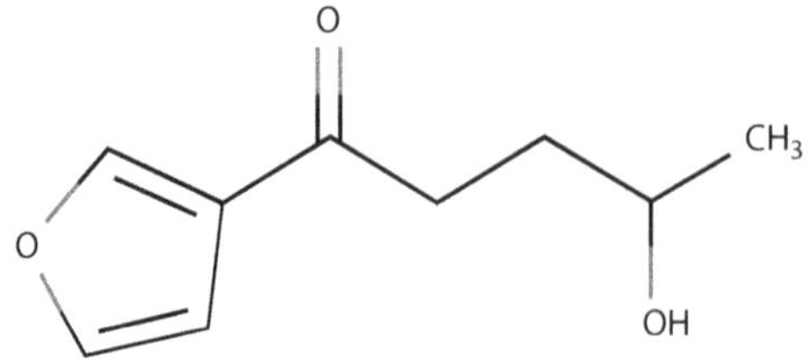

FIGURE 1.2 Structure of 4-ipomeanol.

The most significant target in the cell is the DNA. This is subject to a wide range of direct attacks, such as covalent binding and formation of adducts, intercalation of planar molecules, and radiation damage. Indirect effects result from the upregulation or downregulation of gene expression, which can be detected through the burgeoning sciences of genomics and proteomics. The ability to relate changes in gene expression or protein levels to specific toxicities will become a powerful tool in the earlier detection of toxicity.

EXPRESSION OF TOXICITY

The following is a brief review of the ways in which toxicity can be expressed, or its expression influenced, in individuals exposed to unusual concentrations of any chemical, natural or synthetic; it is not intended to be exhaustive. The expression of toxicity is influenced by factors inherent in the exposed subject, in addition to the factors listed in the section "Strong Toxicants and Weak Toxicants," including age, disease, pregnancy, genetics, nutrition, lifestyle, sex, and occupation.

AGE

The ability to metabolize and eliminate chemicals at the two extremes of age is notably different from that seen during the majority of a life span. Neonates and geriatric people show different sensitivities to drugs due to differences in liver and kidney function; this can be extrapolated to other chemicals with which they may come in contact. Benoxaprofen was introduced as a new drug for use in arthritis, but due to the normal age-related decline in kidney function and metabolizing capability, it was associated with serious toxicity in some geriatric patients. Neonates also show lower drug-metabolizing capabilities; in some cases, this deficiency is protective, while in others, it is not.

GENETICS

Genetics may affect the response to a chemical through differences in the way in which it is handled through the processes of absorption, distribution, metabolism, and elimination (ADME). The most common factor is variation or deficiency in metabolic enzyme activity. Early discoveries of this were made with isoniazid and debrisoquine, both of which were associated with genetic polymorphisms—slower acetylation of isoniazid and deficient hydroxylation of debrisoquine. There is great variation in paracetamol (acetaminophen) metabolism among individuals by a factor of up to 10, the highest rate being comparable to that in the most sensitive animal species (hamsters) and the lowest to that in the least sensitive (rats). These differences have been exploited in the development of animal models with deficiencies in particular enzyme systems. For instance, the Gunn rat has a deficiency in glucuronosyl transferase, which is responsible for conjugation (phase 2 metabolism) of initial metabolites of chemicals (phase 1 metabolism).

Ethnic differences are important; for instance, native Canadians have a lower capacity for ethanol metabolism than Caucasians. Sex is another factor, especially in hormonal terms; sensitivity to chloroform's renal toxicity is much greater in male mice than in females. This difference is removed by castrating the males and restored by administration of male hormones. There are significant differences between male and female rats in drug metabolism and physiology, which can result in different toxicological responses.

PREGNANCY

Pregnancy is associated with an increase in plasma volume and consequent borderline anemia and with changes in protein binding. The extracellular space is increased with an associated increase in the amount of fluid available for dissolution of drugs (increased volume of distribution). There are increased cardiac output and changes in respiratory parameters, seen as increased tidal volume, increased distribution, and faster gaseous equilibrium. Retention of contents in the upper gastrointestinal tract is prolonged.

DIET

Diet and nutrition are also significant. The importance of diet is illustrated by the tumor profile for Japanese people in Japan, which is different from that found in west coast Americans. However, Japanese people living in California show a profile of tumors similar to that of their American neighbors. Food restriction in rats and use of a low-protein diet produces an increase in life span and a reduction in the incidence of several tumors. Lipid content of diet may be important in affecting absorption of lipophilic chemicals, and fiber content of diet affects the bioavailability of toxicants by binding, and thus reducing absorption.

Nutrition, as distinct from diet, is also important; anorexia or a diet low in protein can result in lower synthesis of enzymes responsible for metabolism and elimination of chemicals. Related to diet are lifestyle factors such as smoking and alcohol consumption, which can also influence expression of toxicity.

OCCUPATION

Occupation can determine the likelihood of toxic expression, either directly or indirectly. Exposure to vinyl chloride was associated with a rare liver cancer, hemangiosarcoma; its rarity and presence in a clearly defined segment of the population lead to epidemiological identification of the cause. In uranium miners, there was a greatly increased probability of lung cancer, in those who smoked. Occupational toxicity can be found in more mundane forms. People harvesting parsnips or celery may show phytophotodermatitis, which results from the transfer of psoralens from the plants to the skin and exposure to sunlight, a relationship that may not be readily identified clinically.

DISEASE

Disease is also a factor to be considered. This may be preexisting or induced by the exposure to the toxicant. The two major organs of concern are the kidney and the liver, in an analogous way to the effects of age. Where there is preexisting cirrhosis of the liver, hepatic function will differ significantly from normal. Damage in the liver inflicted by paracetamol has the effect of prolonging exposure to the drug. Where liver disease is associated with bilirubinemia, drugs bound to plasma proteins may be displaced by bilirubin, markedly increasing the plasma concentration of free drug. For drugs such as phenytoin or warfarin, which are highly protein bound and have steep dose–response curves, a small decrease in protein binding can more than double the amount of free drug available for pharmacological effect. Protein binding can also be affected in kidney disease, where reduced filtration rates lead to slower elimination and vital blood flows can be affected by other factors such as cardiac disease. It should also be remembered that exposure to some chemicals may increase susceptibility to infection, as with polychlorobiphenyls, which are immune suppressants.

LIFESTYLE

Although lifestyle clearly includes some of the factors above such as diet, it also includes aspects such as tobacco and alcohol consumption, exercise, overeating, and so on. For example, smoking in miners has been associated with greater incidence of lung cancers, and liver disease associated with alcohol may result in differences in drug handling through the liver.

From the above, it should be evident that toxicity is manifested in many different ways and can be seen as changed organ function, reproductive effects (sterility, impotence, teratogenicity, loss of libido, transplacental cancer), changes in normal biochemistry, excess pharmacological action, phototoxicity, or cancer. There are indirect effects, for instance, due to stress, or nonspecific changes, for which no direct cause can be identified, such as lowered appetite and associated loss of body weight. The end result, however, is that toxicity is usually expressed through specific organs, known as target organs. The problem with this approach is that many of the more potent, often less biochemically specific, chemicals affect a wide range of organs or tissues.

TARGET ORGAN EXPRESSION OF TOXICITY

The concept of *target organ* has a broad range of definitions or possibilities: intended targets in pharmacology or therapy; organs in which unintended effects are shown [such as with the liver in paracetamol (acetaminophen) overdose]; major target organs (those with significant change) versus minor target tissues. The problem with this is that the focus of investigation tends to be the target, while the rest of the organism is ignored (or the various targets are discussed separately without being considered together). Target organs are too often considered in isolation from other organs or from the rest of the organism. It is convenient to teach toxicology in this way, but it means that it can be difficult to cross-link information so that interrelationships are evident. In considering the toxicity of a chemical, it is important to keep the general view in mind; equally, in looking at the effects in one organ or tissue, it is important to remember the rest of the organism. Paracetamol is one of the classic hepatotoxins, but it also affects the kidney. Phenytoin, used in the control of epilepsy, can result in convulsions in overdose (it has a low therapeutic index or safety margin); chronic use is associated with gingival hyperplasia; it is a teratogen and can cause hypersensitivity with extensive dermal reactions. Lead has effects on learning ability, in the nervous system, in the blood, and in the kidney and is associated with reproductive changes and may be carcinogenic. The susceptibility of organs to the effects of chemicals is influenced by a number of factors, some of which are discussed below.

There are a number of factors that influence the extent to which the effects of a chemical are expressed in particular tissues, and these are summarized in Table 1.3 and expanded in the following text.

TABLE 1.3
Factors in Target Organ Toxicity

Factor	Examples
Blood supply	Liver, kidney, and lung have greater blood supplies than adipose or muscle tissue.
Oxidative exposure	Lung and paraquat toxicity.
Cell turnover	Gastrointestinal mucosa, bone marrow, and toxicity of cytotoxic chemotherapies.
Repair and reversibility	Hepatic change may be easily repaired while change in the CNS is not.
Physiology	Concentration effects in the distal renal tubule.
Morphology	Length and diameter of axons in the peripheral nervous system.
Processing ability or metabolic activity	Liver and xenobiotic metabolism. Renal proximal tubule versus the loop of Henle. Oxygen concentrations.
Hormonal control	Reproductive tract and endocrine organs. Induction of hepatic metabolism and increased clearance of thyroid hormones.
Accumulation	Lung and paraquat; adipose tissue and TCDD; cadmium in kidney; lead in bone. Environmental accumulation of pesticides such as DDT.
Protection mechanisms	High concentrations of antioxidant GSH in the liver. DNA repair differences or deficiencies between organs/tissues.

Source: Adapted from Woolley A, *A Guide to Practical Toxicology: Evaluation, Prediction and Risk*, New York and London: Informa Healthcare, 2008.

Blood Supply

The blood is the main vehicle for distributing chemicals of all kinds around the body, and it is logical that the blood supply is important in defining the degree of exposure of individual tissues to the chemicals in the blood, endogenous as well as foreign. The liver receives all the blood supply from the gastrointestinal tract via the inferior vena cava, from where it goes to the heart and thence to the lungs. Thus, chemicals absorbed in the gut go to the liver, the main site of xenobiotic metabolism, and are distributed with any persistent metabolites to the heart and then on to the lungs and then the kidneys, which receive 25% of the cardiac output via the abdominal aorta and then the renal arteries.

Oxidative Exposure

Much toxicity is due to oxidative attack on macromolecules, and this is affected by blood supply and, of course, in the lung, where the locally high concentration of oxygen is partly responsible for the high toxicities seen with compounds like paraquat.

Cell Turnover

Tissues that have an intrinsically high turnover of cells are at risk from chemicals that inhibit cell division. These include the mucosa of the gastrointestinal tract, the skin, the bone marrow, and the testes. Inhibition of the division in the bone marrow can affect the whole organism through induction of anemia and/or reductions in the numbers of circulating leukocytes, in turn leading to reduced immune competence. Where there is a high level of apoptosis, for instance, in developing embryos, disturbances in cell turnover have the potential to result in malformations in the fetus. Where cell turnover is increased, for instance, through necrosis with replacement through increased cell division, there are inherent risks of DNA replication errors, which can lead in the long term to tumor formation.

Repair Ability and Reversibility

An important aspect of assessing the significance of toxic effect is whether it is reversible, either on removal of the stimulus or through repair of tissue damage. The extent to which tissues can repair themselves differs markedly according to tissue type and, to an extent, embryonic origin. Some tissues are able to repair themselves readily, especially the liver. In rats given toxic doses of carbon tetrachloride, early evidence of liver damage seen in the plasma a few days after administration is frequently not reflected in histopathological evidence of damage after 14 days. This considerable capacity for self-repair means that it is possible to miss toxicologically significant hepatotoxicity in standard acute toxicity tests, which require single administration followed by 14 days' observation before autopsy. This repair capability is seen in humans following overdose with paracetamol (acetaminophen), where there is often severe liver toxicity; in survivors, biopsy of the liver 3 months after the overdose sometimes shows no evidence of persisting liver damage.

Equally, some tissues do not readily repair themselves, especially the nervous system. In these tissues, regeneration does not take place or is very slow. Whereas a necrotic hepatocyte can be quickly replaced, a necrotic neuron is lost completely and the function of that part of the nervous system reduced proportionately. In some organs, particularly the kidney, different parts have different capabilities for repair. Thus, damage to the glomerulus and the renal pelvis is not readily repaired, but the proximal tubule epithelium shows considerable repair capability, provided the basement membrane (on which the cells lie) is not breached.

PHYSIOLOGY

Cells or tissues with specific characteristics are susceptible to toxicants, which disrupt or take advantage of those characteristics. Paraquat is an example of this, through its accumulation in the lung, via the uptake mechanism for the endogenous polyamines. In the kidney, the passage of the urine through the distal tubule can lead to toxicity, as the toxins increase in concentration as water is reabsorbed.

MORPHOLOGY

The length and small diameter of axons in the peripheral nervous system contribute to the axonopathy induced by *n*-hexane due to cross-linking of the microfilaments and subsequent poor nutrition of the distal parts of the cell. This is an instance where physiology is also important, as the axon depends on transport of nutrients from the neuronal body and appears to be unable to acquire them from elsewhere. With the passage of nutrients blocked, the axon dies distally from the blockage. Gross morphology is also a factor to be considered, if only rarely. When a rodent is fed, its stomach may press on particular lobes of the liver, restricting circulation in that lobe; this has been known to affect the distribution of liver tumors among the lobes, seen in response to carcinogens fed in the diet.

PROCESSING ABILITY

Tissues that have high processing or metabolic activity are also frequent targets of toxicity. The liver has high activities of enzymes responsible for chemical metabolism, and therefore, if toxic metabolites are produced, they are likely to be produced in higher concentrations than in other tissues, increasing the risk of local effect. The difference between the liver and the lung in terms of enzymic activity is one of the determining factors in the toxicity of 4-ipomeanol. The proximal tubule of the kidney is another site of high metabolic activity and is a frequent target. High metabolic activity may also mean greater potential for oxidative attack through oxygen radicals, which can be produced as a result of normal metabolic processes. The kidney is also at risk through its normal physiological function of producing concentrated urine; this can increase the exposure of cells in the lower reaches of the nephron to a point at which toxicity is elicited.

HORMONAL CONTROL

Tissues that are subject to hormonal control will be affected when the concentrations of the relevant hormones are increased or decreased. When hepatic enzymes are induced in rats, there is often an increase in follicular hypertrophy or hyperplasia in the thyroid due to increased removal of thyroid hormones from the plasma as a result of the increased hepatic metabolism. The plasma levels of thyroid-stimulating hormone are controlled by circulating thyroid hormone concentrations by negative feedback; where this feedback is reduced, the pituitary is stimulated to produce more thyroid-stimulating hormone, which acts on the thyroid. The endocrine system is extremely complex, and effects in one part can have a number of knock-on changes in other tissues.

ACCUMULATION

Tissues that are able to accumulate specific toxins are also frequent targets for toxicity. As discussed, paraquat accumulates in the lung. Cadmium is widespread in the environment and accumulates in shellfish and plants. In mammals, cadmium is complexed with a metal-binding protein, metallothionein, which accumulates in the kidney. When a critical level of cadmium content in the kidney is reached—generally quoted as being approximately 200 µg/g of kidney tissue in humans—nephrotoxicity becomes evident, and renal failure follows. Constant low intake at slightly raised levels can produce gradual accumulation over many years, which ultimately results in renal failure.

Accumulation in bone is a feature of toxicity of lead and strontium, which is a cause for concern if the strontium is the radioactive isotope. Bisphosphonates, used in the treatment of osteoporosis, also bind tightly to bone, and this is a source of some of their toxicity. Environmental accumulation is also a factor to consider because it can have dire consequences, as illustrated by concentrations of fat-soluble compounds such as DDT, which increase in concentration up the food chain, as in bird-of-prey populations. DDT, and similar compounds such as TCDD, tends to accumulate in lipid tissue, from which they are released very slowly. This is particularly a problem in species at the top of the food chain and has recently been acknowledged to be a factor in marine mammal toxicology. At one time, Americans were, by their own regulatory standards, inedible due to the amounts of DDT they had accumulated in their adipose tissue. For such compounds, toxicity can be expressed if there is a sudden loss of weight, reducing adipose tissue and releasing large amounts of toxin into the plasma where it can pass to the target organs, as in migration or pregnancy in malnourished people. In the development of brown-field sites, care must be taken that the residues of any industrial waste are considered in licensing use of the land, especially for growing food crops. This was a major source of toxicity at Love Canal in New York State, where a housing estate was built on a toxic waste dump.

PROTECTION MECHANISMS

Some tissues, particularly the liver, have high concentrations of endogenous compounds that have protective functions, normally against reactive oxygen species.

GSH, which may be present in the liver at a concentration of up to 5 mM, is a good example of this. Enzymes such as superoxide dismutase, which catalyzes the conversion of superoxide to hydrogen peroxide, are also important in protection of the cell. There are also differences between tissues in the activity of DNA repair mechanisms. For example, the brain is less able than the liver to excise the DNA base guanine methylated at the O^6 position, making it more susceptible to tumor formation following administration of dimethylnitrosamine. Defective DNA repair is also seen in patients with xeroderma pigmentosum, which gives a high incidence of skin cancer in response to exposure to ultraviolet (UV) light.

The problem in drawing such distinctions is the same as with describing toxicity in terms of target organs. A single chemical may be associated with effects in several organs and have a different mechanism of toxicity in each due to the differences in tissue susceptibility. This breadth of possible effects makes testing for toxicity extremely complex, until a mechanism is suspected. The enormous range of potencies of the chemicals to which we are exposed is an additional complication.

ETHICS OF TOXICOLOGICAL ASSESSMENT

The use of animals instead of humans in toxicity testing has been considered briefly in the earlier sections of this chapter, and it could be said that early studies in humans with new drugs are simply toxicity studies in a more awkward species. However, while it is normal to give animals the highest feasible dose to elicit toxicity, early studies in humans are conducted at lower doses. These early studies are to assess the pharmacokinetics and any adverse effects in healthy volunteers (phase I clinical studies) and then to study safety together with some aspects of efficacy in greater detail in patients (phase II studies). However, no administration to volunteers is allowed until an initial battery of toxicity tests has been conducted to assess the new molecule and to provide data from which to calculate a safe starting dose in humans. However, as indicated earlier, there are circumstances in which this may not apply. The costs of such clinical trials bear out the contention that using humans as test models for humans is hugely expensive (as patients can be scattered across centers around the globe), inconvenient, and slow. There is the additional concern, as with animal models, that experiments in some groups of humans may not be relevant to other groups, as indicated above for isoniazid in Native Canadians and Egyptians. In fact, the inherent variability of humans is being exploited in pharmacogenetics, which studies the genetic basis for variations in drug metabolism and toxicity, and should eventually give data that allow design of specific treatment regimens for patients with specific phenotypes. If a valid cross section of the human global population were used, the variability present would be so great as to obscure subtle but important changes from normality.

The traditional view is that there is no escape from the premise that for effective toxicity testing, the variability in the test system must be controlled to allow a satisfactory definition of normality. Change from a carefully defined and understood baseline can be more readily detected than in a diverse population of unmatched individuals. The test system must also be inexpensive and easy to look after, as the volume of toxicity testing is so great that the expense would become prohibitive otherwise. The volume of toxicological research and experiment is partly due to the numbers of new chemicals under development and partly due to regulatory guidelines imposed by governments. Mostly it is a compromise, balancing cost against failure; some might say that science has lost out, but science and politics are uneasy bedfellows.

The cost or risk benefit of the chemical has to be assessed in deciding the necessity or ethical acceptability of undertaking toxicity studies in animals. In the case of cosmetics, in the European Union, it has been decided that the use of animals to test cosmetic ingredients or products is now unacceptable, meaning that cosmetic toxicity has to be assessed in other ways. It has to be said that this restriction has slowed the bringing to market of new cosmetic ingredients. However, although such methods of safety assessment may be acceptable in testing of voluntarily used cosmetics, they are not applicable to food additives, medical devices, pharmaceuticals, or pesticides to which human exposure may be involuntary or indicated by illness. Furthermore, nearly all cosmetics are for use on the skin or in the mouth, and the risks of their use are very different from those encountered with a pharmacologically active drug given by mouth or injection. Some of the most aggressive chemicals used in our personal lives are fragrances.

There are, at present, few ethical concerns about using long-lived cell cultures in toxicity testing, assuming consent to use has been granted in the case of human tissue. Some cells for use as test systems can be obtained by taking a blood sample from a healthy volunteer, for example, leukocytes, or from tissues obtained in operations from patients; these also pose no ethical problem, unless the tissues are obtained without patient consent.

Primary cell cultures and preparations of cell organelles such as microsomes, however, require fresh cells derived from a freshly killed animal. Is an experiment that uses an animal in this way more ethical than one that uses a complete and conscious animal? Indeed, some primary cell culture experiments require a greater number of animals than a comparable *in vivo* study. In assessing the relevance of the data for humans, the process of extrapolation from an isolated culture of rodent hepatocytes to humans is much more precarious than making the same leap from a complete animal in which all the organ interrelationships are intact.

It is fundamental that if some toxicity studies are ethically essential, there are some that should not be conducted. The ban on the use of animals to investigate cosmetics and their ingredients has been discussed. Other examples are more complex; is it necessary to investigate the acute toxicity of a household cleaner if it is thought that it may be ingested accidentally by children or purposefully by adults? In the case of natural constituents of a traditional food, for example, mushrooms, is it ethical to investigate the toxicity of individual components? Possibly, especially if there has been evidence gathered from humans that there is a problem for which it is considered that the food or its constituent is responsible. Is it ethical to undertake an experiment on such a constituent that exposes a single group of mice to an unrealistically high dose in the drinking water for 2 years? The inescapable conclusion from the published data for the mouse study by McManus and coworkers (Focus Box 1.3) is that the experimental design was deficient (at the very least), the study was not reported adequately, and it served no useful purpose; consequently, it may be said that the animals were needlessly used. In the absence of dose–response information,

FOCUS BOX 1.3   STUDIES WITH THE
CONSTITUENTS OF MUSHROOMS

The natural constituents of mushrooms have been widely investigated for their carcinogenic potential and the results published in a number of papers.

- One study (Toth et al. 1989) looked at the carcinogenicity of a constituent of *Agaricus xanthodermus*, an inedible species, albeit related to the common field mushroom; this experiment used subcutaneous administration and, as a result, is of dubious relevance to people who ingest their food. Production of tumors at the site of administration could have been expected.
- In another study (McManus et al. 1987), 4-hydrazinobenzoic acid, a constituent of the common mushroom, was given in drinking water at 0.125% to a single group of Swiss mice, equivalent to approximately 5 mg/day per mouse or about 125 mg/kg/day for a 40 g mouse. There was "substantial" early mortality due to rupture of the aorta. In survivors, a proportion of the mice developed unusual tumors in the aorta—the site of the initial toxicity—diagnosed as leiomyomas or leiomyosarcomas.
- No dose–response information could be gained from this experiment (there was only one group); the formulation of the chemical in the drinking water did not mimic the natural occurrence of the chemical.
- The only conclusion that could be drawn directly was that hydrazinobenzoic acid given in the drinking water caused tumors in Swiss mice in this study. In the absence of dose–response information, especially a no-effect level, no risk assessment for human use is possible from these data, and conclusions about human use cannot sensibly be drawn.
- Other studies have failed to show carcinogenic potential for the common mushroom or its constituents (Pilegaard et al. 1997; Toth et al. 1997). The study by Toth and coworkers in 1997 suggested that the negative finding was due to insufficient mushroom intake, acknowledging the fact that if you are sufficiently dedicated, you can induce cancer in mice eventually, so long as the strain, dose levels, and design are chosen appropriately.

no risk assessment for humans was possible, yet one of the conclusions was that humans should not eat "edible" mushrooms, despite the many years of human consumption without any serious suggestion from epidemiological studies of any hazard, when eaten at normal amounts. If this flawed reasoning were rigorously applied to the whole human diet, we would soon die of starvation or boredom. Hopefully, this type of ill-conceived experiment would not be considered ethical in the current, or indeed, in any, climate.

In assessing the ethical need for a toxicological experiment using animals, it is important to question the objective of the study and to assess the design in light of that objective. If the experiment is not being undertaken to answer a specific question of human concern or the design will not allow the generation of meaningful data, the need for that experiment should be carefully questioned.

THE THREE Rs: REDUCE, REFINE, AND REPLACE

One of the cornerstones of modern toxicological investigations is the concept of the three Rs—reduce, refine, and replace, put forward by Russell and Burch (1959). The intention is to reduce the numbers of animals used in toxicological experiment, to refine the methods by which they are used, and to replace the use of animals as appropriate alternative methods become available. A successful example of this approach is the replacement of rabbits in pyrogen testing. Pyrogens (sterilization-resistant components of bacteria especially in parenteral solutions) may be detected in batches of finished product by injection into the ear veins of rabbits and monitoring the temperature response. This test is costly in terms of labor and animals and is also subject to interference by the presence of pharmaceuticals in some preparations; in addition, it does not give a measure of the amount (only the potency) of pyrogenic substances present. Now, the rabbit has been replaced to a large extent by an *in vitro* system using a lysate of amebocytes from the horseshoe crab *Limulus;* this test is considerably more sensitive than the rabbit test and is one of the best examples of a fully validated alternative assay in use in toxicology. Even so, the output from the two tests is not identical. A high endotoxin level does not necessarily lead to pyrexia.

The use of rabbits or mice in bioassays for potency of various biological pharmaceutical preparations, such as insulin, has been greatly reduced by the use of more precisely targeted pharmacological or analytical tests. Replacement of animals in toxicological testing is inherently simpler if a specific endpoint is being investigated, whether pyrogen content or DNA damage. The more complex or uncertain the endpoint, the more difficult it is to devise a simple test that will answer the question. Refinement of testing methods and protocols is also a long-term goal; the use of guinea pigs in allergenicity testing has been largely superseded by the local lymph node assay in mice. Animal experimentation has also been refined by careful application of statistics in the design of experiment protocols.

The pursuit of the three Rs is an ongoing process, but it can be expected to be long and, in all probability, ultimately incomplete. The achievement of the first 10% to 20% has been relatively easy; the next 20% will be much more difficult, as the endpoints to be studied become less amenable to simplification. It follows that the next 20% beyond that will be harder and slower still, but this should not be taken as a reason for not investigating further. Much is made of the minimal study design, which results from use of the three Rs; this approach is fraught with risk as, like a net with a wider mesh, more errors can creep through. Much better would be an optimized study design.

In all the current enthusiasm for the three Rs, the purpose and objectives of toxicological investigation should not be forgotten. Mostly, these tests are undertaken to elucidate potential safety of the use of chemicals by humans. While the use of animals should be questioned, controlled, and reduced, it should be remembered that use of too few animals in an ill-conceived experiment is as morally questionable as the use of a study design that will achieve the objectives of the study or experimental program.

If it is accepted that the development of new chemicals is necessary or inevitable (or the use of naturally occurring chemicals in an unnatural context or quantity is proposed), it should be accepted that their safety, or otherwise, should be investigated up to state of the art. Such investigation is essential in the development of new drugs, food additives, medical devices, pesticides, or veterinary medicines. While the first three categories are expected only to be in contact with humans, pesticides and veterinary products are not intended for direct human exposure. Pesticides are subject to a wide range of other tests to examine their effects on beneficial species, such as bees and fish, in an assessment of their likely environmental effects. Toxicity testing is also useful in investigating the effects of chemicals or dietary components that are believed to be associated with, or that exacerbate, human disease. The necessity of such testing is decided in part by regulatory guidelines (for the three major categories of registered chemicals) and partly by scientific need judged from the expected properties of the chemical. Where animals are used, there should be local ethical review committees that monitor numbers and procedures, and ensure that the highest possible standards of ethical research are maintained.

The basic tenets of the 3Rs are clearly laudable but have been hijacked by groups who believe that all animal testing should cease and have, in some ways, been pursued with the objective in mind but without clear regard to the objective of assessing safety in humans. Clearly, there is an element of opinion at the other end of the spectrum that insists that animal tests are essential. If it is agreed that safety evaluation or toxicity testing of some kind is essential, it would be better to think not in terms of the 3Rs but of devising and consciously evolving a test paradigm that would deliver the best results for the purpose of extrapolation to humans or whatever other target is relevant. If this approach is taken, achievement of the 3Rs will follow as a natural consequence.

Toxicity testing has to assess the probability of an enormously wide range of reactions—covalent, noncovalent, hormonal, and metabolic, the bases for which are at the cellular level, although this extends in many instances to the whole organism and is dependent on interrelationships between organs or tissues. The simplest way to do this for unknown or unpredicted mechanisms is to examine the response of a whole organism to the test chemical and to screen for as many endpoints as possible in a set of general tests. Where an unexpected reaction is seen, mechanistic studies can be undertaken to investigate precise endpoints, and it is in these precisely targeted experiments that *in vitro* systems become powerful and effective.

In the final analysis, toxicological data cannot be interpreted unless the significance and meaning of detected change is thoroughly understood. When the meaning of the test results is known, it should be possible to take the relationships shown and the targets identified and relate these to expected effects in humans. For chemicals that are coming to market and significant public exposure for the first time, the process of evaluation continues after sales begin, to monitor for unexpected effects, which can then be investigated in appropriate mechanistic or epidemiological studies. It should be borne in mind that correct interpretation of data is often lacking in the immediate aftermath of a crisis.

SUMMARY

This book looks at how toxicity tests are conducted, how the results are interpreted, and then at the process of how these conclusions are used in risk assessment. However, the first step in all this is to define normality, so that change from expectation can be detected in the first place. This is the subject of Chapter 2.

In some ways, the ethics of toxicological testing are driven by the suggestion that if you do not know what you will do with the answer, you should not ask the question in the first place. It is too easy to fall into the trap of routine regulatory requirement. All tests conducted should have some degree of scientific justification, although it must be acknowledged that the sometimes arcane requirements laid down by regulatory authorities make this difficult to achieve.

One of the objectives of this book is to make toxicologists and the users of toxicology think about the subject and use of toxicology and to question its conduct at every stage.

REFERENCES

Ball P. *The Devil's Doctor: Paracelsus and the World of Renaissance Magic and Science.* London: William Heinemann, 2006.

Dart RC, Caravati EM, McGuigan MA, Whyte IM, Dawson AH, Seifert SA, Schonwald S, Yip L, Keyes DC, Hurlbut KM, Erdman AR. *Medical Toxicology*, 3rd ed. Lippincott Williams and Wilkins, 2004.

Klaassen CD. *Casarett & Doull's Toxicology; The Basic Science of Poisons*, 6th ed. New York: McGraw-Hill, 2001.

Liebler DC. The poisons within: Application of toxicity mechanisms to fundamental disease processes. *Chem Res Toxicol* 2006; 19: 610.

McManus BM, Toth B, Patil KD. Aortic rupture and aortic smooth muscle tumours in mice: Induction by p-hydrazinobenzoic acid hydrochloride of the cultivated mushroom *Agaricus bisporus. Lab Invest* 1987; 57(1): 78–85.

Pilegaard K, Kristiansen E, Meyer OA, Gry J. Failure of the cultivated mushroom (*Agaricus bisporus*) to induce tumors in the A/J mouse lung tumour model. *Cancer Lett* 1997; 120(1): 79–85.

Poland A, Knutson JC. 2,3,7,8-Tetrachlorodibenzo-*p*-dioxin and related halogenated aromatic hydrocarbons: Examination of the mechanism of toxicity. *Ann Rev Pharmacol Toxicol* 1982; 22: 517–554.

Russell WMS, Burch RL. *The Principles of Humane Experimental Technique*. London: Methuen, 1959.

Timbrell JA. *Introduction to Toxicology*. London: Taylor & Francis, 1999.

Toth B, Patil K, Taylor J, Stessman C, Gannett P. Cancer induction in mice by 4-hydroxybenzenediazonium sulphate of the *Agaricus xanthodermus* mushroom. *In Vivo* 1989; 3(5): 301–305.

Toth B, Erickson J, Gannett P. Lack of carcinogenesis by the baked mushroom *Agaricus bisporus* in mice: Different feeding regimen. *In Vivo* 1997; 11(3): 227–231.

Woolley A. *A Guide to Practical Toxicology: Evaluation, Prediction and Risk*. New York and London: Informa Healthcare, 2008.

2 Normality
Definition and Maintenance

INTRODUCTION

The detection of differences from experimental normality that are attributable, with reasonable certainty, to the influence of the substance under investigation is the sole basis of toxicological investigation. This simplistic overview, however, then begs the question as to what is normal or, by association, natural. Dictionary definitions of *normal* use words such as usual, typical, or expected—the normal state or condition or being in conformity with a standard, for instance, as shown by body temperature. Natural is defined as existing in or derived from nature; not artificial; in accordance with nature; or normal. In terms of public perception, normality is seen so routinely that it may be more useful to think of abnormality, which can provoke a reaction that is not seen in response to the normal. Normal is not a single value or phenotype; there are ranges of normality. It should also be considered that genetic and thus phenotypic polymorphisms exist naturally, as a matter of course, and thus, normality can vary between population groups—what is normal in one population may be considered abnormal in another. Equally, there are degrees of abnormality or difference from the normal. In addition, there is the paradox that some abnormalities may actually be engineered into test systems, as in strains of transgenic mice, compared with the wild type or strains of *Salmonella* used in bacterial reversion (Ames) tests. In these situations, the abnormal becomes the experimental normality. Such simple judgment is based on perception, which may not be readily supportable in scientific terms. Using a simplistic example, someone permanently in a wheelchair, as a result of a severe spinal injury, may be seen as more different from the normal than a person in a wheelchair following a broken leg; the former will always be disabled, while the latter's injury may be reversible, and he/she can revert to the state he/she were before the insult. In toxicological terms, normality can usually be defined by numerical data, means, or incidence data or, less verifiably, experience. For characteristics that are defined by presence or absence or narrow ranges of values, definition of normality is relatively simple in comparison with those that are present on a graduated scale or have a wider range of values.

Deviation from normality may be determined through the circumstances of the observation and according to experience or expectation; left-hand-drive cars are not normal in the United Kingdom but are clearly normal in France. Similarly, a tumor may be expected routinely in old age (when it could be said to be normal) but unexpected at the age of 21. It was this characteristic that indicated diethylstilbestrol as a carcinogen—vaginal adenocarcinoma is unusual in young women but was seen in the young daughters of women who had taken diethylstilbestrol in pregnancy to prevent abortion.

Another example is that given by the antihistamine terfenadine, which was responsible for approximately 430 adverse cardiac reactions and 98 deaths due to severe cardiac arrhythmias between 1985 and 1986. To put this apparently alarming figure in perspective, this represented roughly 0.25 adverse reactions per million daily doses sold (Fermini and Fossa 2003). Such a low incidence is likely to be subsumed in the background for some time before coming to light as clinical experience or an epidemiological database grows and someone makes a connection (which may be due to some serendipitous insight). Cardiac arrhythmia that is not related to drug use is found in the general population—it occurs normally—and therein lies the problem. One of the greatest challenges in toxicology is the detection of minor differences from normality or background incidence and a meaningful assessment of their significance in the real world of clinical or consumer use.

WHAT IS NORMALITY?

Normality, or abnormality, may also be indicated by the presence or absence of an observation. The thalidomide tragedy was shown by the presence of severely shortened limbs, phocomelia, in babies born to mothers who had taken thalidomide in the first 40 days of pregnancy, although other abnormalities became apparent as investigations continued. This was a demonstration of an increase in a very rare observation in a specific population, associated with exposure that was definable in terms of dose and day or week of pregnancy. Abnormality may be defined in biochemical terms, usually by the absence of enzymes responsible for some aspect of basal metabolism. Phenylketonuria is associated with a recessive deficiency for phenylalanine hydroxylase. This leads to increased excretion of metabolites of phenylalanine that are responsible for neurological effects, including mental retardation and low IQ; many mutations have been reported for this gene.

Deviation from normality is also definable through changes in the incidences of observations, which may then be associated with exposure (in epidemiological studies) or with treatment (in toxicity studies). Phocomelia and hemangiosarcoma (due to occupational exposure to vinyl chloride) are both seen in a normal population but at incidences that are so low as to make them abnormal. In contrast to this, lung cancer may be seen in people who do not smoke and might be seen as part of the background and normal tumor incidence in a nonsmoking population. However, there is a clear association of lung cancer with cigarette smoking, marking this as a deviation from normality that is attributable to a toxic exposure. In a similar way, excessive consumption of alcohol during pregnancy is associated with fetal alcohol syndrome seen in the babies born to these mothers.

The above are examples where the incidences of presence or absence of data have changed in response to exposure or treatment. Where toxicological (as opposed to epidemiological) investigation is concerned, the most usual way of demonstrating deviation from normality is through statistical analysis of continuous data from the tests conducted. Continuous data, as opposed to discontinuous data (e.g., positive or negative, presence or absence, or grades of presence), can be readily exemplified by height in people. It is relatively easy to say if someone is abnormally short or tall, for example, 1.2 or 2.5 m. However, this judgment becomes much more difficult if the subject is between 1.5 and 1.8 m. The average male height in a population may be around 1.7 m, but that population would probably encompass the very short and very tall, grouped in a classic bell-shaped or normal distribution (Figure 2.1) (Heath 2000). This distribution is characterized by a mean that falls at the center of the distribution but that does not reflect its width or range of values in the population. The range of values present is defined by the standard deviation of the mean, which tends to be large when there is a wide range of values and smaller with a narrow range. The standard deviation will also tend to get smaller with increasing sample size (n). In a group of 500 men taken from different sports, the mean heights of two samples of 100 randomly chosen individuals are unlikely to be significantly different. Equally, the mean height of a sample of 50 football goalkeepers is likely to be fairly uniform and so have a small standard deviation. In fact, in comparison with the mean height of 50 jockeys, the goalkeepers might be seen as unusually taller, and the two samples would probably be significantly different when compared statistically. If the two samples of 50 were combined, the mean would fall somewhere in the middle, and the standard deviation would increase to reflect the wider range of values in the more diverse sample. In this instance, two samples of defined individuals have been extracted from a relatively undefined population. As a further complication, it should be remembered that normality can change with time; over a period of centuries, the average height of the population increased due to better nutrition and other factors. In addition, of course, the average life span also increased, so what is normal now might be seen as abnormal with respect to values five centuries previously. In the same way, toxicological normality may drift with time.

The characteristics of the normal distribution mean that 95% of values lie within 1.96 standard deviations of the mean; these are the 95% confidence limits. However,

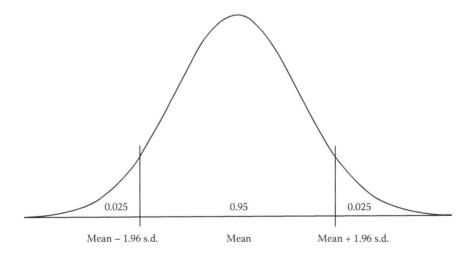

FIGURE 2.1 Characteristics of the normal distribution. (Adapted from Heath D, *An Introduction to Experimental Design and Statistics for Biologists*, London: Taylor & Francis, 2000, and Woolley A, *A Guide to Practical Toxicology*, 2nd ed., London: Informa Healthcare, 2008.)

this does not necessarily mean that the values falling outside these limits are abnormal; caution should be used when assessing data points that fall outside these limits to make sure they represent a true value or one that has been generated as a result of experimental or user error. The values for a parameter, such as alanine aminotransferase concentration in blood plasma, may be expressed as a mean with 95% confidence limits and also as a range of values, which will be wider. It is quite usual to look at a set of values and see that one is larger or smaller than the others and to question its relevance to the rest of the data set. This is often the case for enzyme activity data, which can be subject to very wide variation due to individual circumstances. There are several tests for so-called outliers, and it is legitimate to use these (although care should be taken if they are in a treated group and especially if they are in the highest dose group).

The normal distribution plays a central role in assessment of numerical data in toxicology because it forms a basis of making comparisons of treated groups against controls. Effect is demonstrated by showing the null hypothesis (that there is no difference between controls and treated groups) to be false (i.e., that there is a difference between the two populations, treated and control). However, demonstrating such a difference does not necessarily imply abnormality; two samples of 21-year-old males may have mean heights that are significantly different, but this does not mean that one group is abnormal.

Although it is useful to consider normality within populations, it is also applicable to individuals and is affected by factors that influence normal function, including disease. During pregnancy, there are important changes from individual normality that affect the composition of the blood and how chemicals are cleared from the body. Paradoxically, pregnancy is a normal condition with its own limits of normality. Age is also an important consideration, and normality for an individual, defined by an assortment of parameters, alters with increasing age. Factors that change with age include protein binding, clearance, metabolism, renal, and central nervous system (CNS) function. There is a general tendency for normal function to decline with age, for example, in the kidneys or CNS. Normal age-related decline in clearance may result in abnormal reaction to drugs, such as that seen in adverse reactions to benoxaprofen. These affected a number of elderly patients and eventually led to the drug's withdrawal from the market. Such decline is normal, but it may be accelerated in some cases by chronic exposure to chemicals; this acceleration clearly results in abnormality at an earlier stage than expected. Normality in the very young is also different from that in adults. Neonatal absorption and metabolism for different chemicals can be more or less extensive than that seen in adults, young or old; this clearly has an effect on the way babies handle drugs or other chemicals.

Abnormality may be singular, in that an individual may be normal in every respect except one, deformity of one arm for instance, or multiple. However, a single genetic defect can have multiple effects as in trisomy 21 (Down's syndrome) or cystic fibrosis, the abnormalities of which are interconnected.

At some point, a judgment becomes necessary as to what is normal and what is not; ideally, this is based on numerical data or a scientific appreciation of qualitative data. Frequently, however, a subjective judgment is made, and the perceived abnormal is treated with suspicion. This tendency has a huge influence on the public perception of the effects of chemicals of all kinds—food additives, pesticides, components of genetically modified crops, new drugs, etc. Normality may be judged against difference from expectation; this is simple for presence or absence phenomena but more complex for a sliding scale such as height. A distinction may be made between an objective, mathematical definition and the subjective perception of normality. At what point is a tall man abnormally tall, and is this perception the same for everyone? A very tall person might consider people of average height to be short, when in fact they are mathematically normal in height. The people of the Netherlands are generally taller than those of other European countries, but this does not indicate abnormality in either population. Clarity of difference makes these judgments much simpler; however, such clarity is a luxury that is not often available in toxicology. Some treatment-related effects may not render the treated test system abnormal but could significantly affect life expectancy if allowed to continue. It is this subtlety of effect that makes definition of normality in toxicology so important.

NORMALITY AS A RANGE

The example of the goalkeepers and jockeys, given above, is an example of the fact that biological normality is seen as a range of values that is defined by its low and high values. Typically, this takes the form of a population mean and standard deviation; a simplistic definition of normality might then be that any value that falls within the 95% confidence limits of the mean is normal; values outside that may be statistically defined as outliers and, by implication but not necessarily fact or judgment, to be abnormal.

The crucial point here is that the extent of variation within the normal population dictates how easy it is to detect a toxicologically or biologically significant change from normal. Some parameters of clinical pathology, such as plasma concentrations of sodium and potassium, are tightly defined with a low coefficient of variation; this is a reflection of their physiological importance. A normal range for blood sodium might be between 140 and 150 mmol/L; a deviation from these limits of only 10% may be associated with functional change and adverse effects. On the other hand, enzymes such as lactate dehydrogenase can be very variable between individual controls and between occasions in the same animal; it is quite possible that a twofold change in activity in the plasma would not be associated with toxicity; it also means that change due to treatment is much easier to see in electrolytes than in enzyme activities or other very variable parameters.

EXPERIMENTAL NORMALITY

In any toxicological experiment, it is vital that the test system should be as clearly defined as possible, always bearing in mind that it is a biological organism. For the most part, this means the exclusion of any factor that might confound the experimental objective. In animal experiments, it is important to ensure that the animals are free from the effects that may be induced by disease or parasitic infection or poor husbandry practices, among numerous sources of abnormality. It should also be realized, however, that normality in a test system may not equate to normality in the wild-type organism, particularly in the case of bacteria. The bacterial reverse mutation assay (previously known as the Ames test) is conducted with strains of *Salmonella typhimurium* (and some forms of *Escherichia coli*) that have had specific characteristics designed into them; deviation from these characteristics invalidates the test results. Equally, the laboratory rat may be said to be normal in laboratory terms but is clearly not so in comparison with the wild rat from which it is derived. In an analogous manner to bacterial strains, there are strains of rat bred with specific abnormalities for purposes of metabolism investigation, or transgenic mice for investigation of specific toxicological mechanisms or for accelerated assessment of carcinogenicity. Furthermore, some "standard" rat strains have predisposition to certain conditions and diseases.

In a review of the role of environmental stress on the physiological responses to toxicants, Gordon (2003) reiterated the ancient truth that toxic response is a function of the poison itself, the situation of exposure, and the subject exposed. It is the variability inherent in the third of these factors that is a large factor in influencing the variation in experimental normality that may be seen in a group of microorganisms, animals or, particularly, humans. This variation may be due to the individual test organism or test system or to factors that are external to the organism, principally, the environment in which it finds itself. Variation due to the biology of the test system is a function of the complexity and genetic diversity of the test system; an outbred mouse strain is inherently more variable genetically than an inbred strain. Broadly speaking, nonhuman primates are more variable than dogs, which are more variable than rodents. Factors that are external to the test system include husbandry (e.g., nutrition, housing, sanitation) and other environmental factors such as temperature and humidity.

Gordon points out that the majority of toxicological experiments are conducted under conditions that are optimized to the well-being of that particular species. Thus, animals are housed in standard cages, at a standard temperature and humidity range, with a 12-hour light/12-hour dark cycle and are offered standard diet and water. Sanitation and hygiene are carefully controlled to obviate infection. The animals are generally resting or get little exercise. From this, it is clear that a distinction should be drawn between wild-type normality and experimental normality.

There are many factors that can affect the test system and so have direct or indirect effects on the responses to treatment with chemicals. This can be reflected in a lack of reproducibility of data between laboratories or in data that mislead interpretation, particularly if an abnormality common to the whole experimental population conceals a response to treatment that is present in a "normal" test system.

Of these factors, one of the most difficult to assess is stress. Although this refers particularly to animals, it should be noted that incorrect storage or preparation of *in vitro* test systems would also compromise the experiments undertaken. In animals, a degree of stress is normal in everyday life, but this leaves the question as to what is a normal level of stress. Assessment of excessive stress is relatively easy, as the changes seen in comparison with an unstressed control are usually plain. Low-grade chronic stress and its effects are much more difficult to distinguish, and as a result, it may be impossible to separate subtle effects of treatment from the effects of stress. Apart from stress due to the effects of treatment, stress may be incorporated into toxicity experiments through a number of different ways, particularly husbandry and environment, although these are clearly intimately related.

HUSBANDRY

Aside from treatment itself, one of the most important sources of stress in animal studies is inappropriate husbandry. In general, the animals used in toxicological experiments are social and benefit from housing in groups, which reduces stress and so produces a more normal, and physiologically relevant, animal. Studies of animals implanted with telemetry devices that relay data on heart rate, blood pressure, or body temperature have shown that heart rate and blood pressure increase when cage mates are separated, when a new cage is provided, or when an unfamiliar person enters the room. Recently, the increasing practice of housing primates in groups, instead of in individual 1 m^3 steel cages, has shown the benefits of attention to good husbandry technique. Individually housed primates are likely to show atypical behavior, including self-mutilation and repetitive movements that can in no way be said to be normal. In gang-housed experiments, the animals are housed together in single-sex treatment groups and have the opportunity to socialize and interact with each other; they are visibly more relaxed and outwardly normal. The downside of this is that a particularly dominant animal may pick on one at the bottom of the social chain and cause undesirable levels of stress, leading to separation from the rest of the group. In a similar way, it is generally accepted that rats housed together are better models for toxicity experiments than singly housed animals.

In minipigs, feeding is an unexpected source of changes in blood pressure and heart rate, and these effects may persist for several hours. In fact, the changes brought about by feeding are more extensive than those due to administration of doses. It has been found appropriate in some laboratories, therefore, to feed the animals in the evening so that the effects of feeding do not mask the effects of treatment.

One argument for single housing is that food consumption data are more precise; however, this perceived precision may be illusory due to inherent inaccuracies in food records, where spillage is usually difficult to assess. Lack of food consumption data can be offset by regular measurement of body weight and, to some extent, by observation. Generally, the benefits of individual food consumption data of dubious accuracy or utility are outweighed by the advantages of group housing in providing a less stressed animal. For rodents, individual housing has the downside that the animals are subject to low-grade chronic stress, which has the potential to produce hypertrophy in the adrenal cortex with increased adrenal weight. In studies of up to 13 weeks, the adrenal changes are often accompanied by reductions in thymus weight, which may mask immunotoxic change, although stress itself can induce immunosuppression. In longer studies with individually housed rats, the life span of the animals is somewhat shorter than with group housing, and the animals tend to be more difficult to handle and have a different tumor profile. Group housing alleviates much of this stress. Contrary to this, some animals are better housed individually, usually due to fighting, as in male mice and hamsters. That being said, there is some anecdotal evidence that male mice can be housed in groups if they are kept together from a very early age; this means allocating them to cages at delivery and then randomizing cages to treatment groups rather than individual animals.

ENVIRONMENT

While husbandry is clearly the progenitor of the environment in which the test system is kept, it is worth remembering that purely physical factors can have profound effects on a test system's response to treatment. High temperature tends to exacerbate toxicity; for some chemicals, as temperature increases, so does toxicity [as measured by acute median toxicity (Lethal Dose 50; LD_{50}) values]. Other physical factors to consider, although they may have less impact, are relative humidity, the number of air changes (and whether it is filtered), and less obvious aspects such as noise. Do rats enjoy heavy metal music, or do they prefer Sinatra?

Another factor increasingly used in animal experiments is environmental enrichment, where toys are provided or activities such as foraging are encouraged. Although group or pair housing is good environmental enrichment in itself, provision of cage furniture for dogs and primates or cardboard cage inserts for rodents is also used increasingly. For primates, the use of small bits of food hidden in the bedding or a honey-and-seed mixture smeared on wooden perches is a very good way of promoting behavioral patterns that are closer to normality. In addition, some primate facilities allow, on occasion, their charges access to a separate "playroom" to further enrich their environment.

OTHER SOURCES OF STRESS

While husbandry is a potential source of continuous low-grade stress, other causes of stress may be transient. These include handling, treatment (or the prospect of treatment), irregular examinations such as recording of electrocardiograms, blood or urine collection, or simply removal from the home cage for examination. While stress is generally accepted as being an important factor to avoid in toxicological experiments, its effects are inherently difficult to quantify, as the investigation is itself often stressful and reaction to stresses, especially hormonal ones, can be very rapid. Experimental design and conduct should be optimized to reduce stress as far as possible. Stress reduction can start with the choice of study personnel. In studies with animals, the choice of people who will carry out the majority of the handling and procedures is very important because their attitude and approach will affect the behavior of the animals. A relaxed, caring technician will have animals that are themselves relaxed and easy to handle, while the opposite is true for those who are impatient or bad tempered.

In routine toxicity studies with dogs or primates, it is usual to record electrocardiograms from unanesthetized animals. Unless the animals are thoroughly acclimatized to the procedure, heart rate will be significantly increased above normal values, along with blood pressure; careful handling of the animals is vital to minimize stress. Procedure-related effects can conceal reductions in blood pressure and smaller increases in heart rate that can result in histopathological lesions in the heart, an effect to which the papillary muscle of the laboratory beagle dog is known to be sensitive. Although it is possible to perform these measurements on anesthetized animals, the choice of anesthetic becomes important as it too can affect blood pressure or heart rate. However, such experimentally induced abnormality can be avoided by the use of jacket-based telemetry, in which the animal wears a jacket with sensors and a transmitter. This allows real-time tracking of changes in this type of parameter. Alternatively, telemetric implants can be used for some studies, especially safety pharmacology studies of the cardiovascular system, though these implant types require surgical implantation and are expensive for routine safety studies. Either option gives a remote reading of data in an unrestrained animal and an indication of change from "true" normality.

It may also be useful to draw a distinction between psychological, biochemical, and physiological stress, the latter two being more likely to be treatment induced in a well-designed experiment. Poor protocol design can be a source of unwarranted stress, especially when an excessive blood-sampling regimen is pursued, producing a marginal anemia that may itself make the animal more susceptible to the toxicity of the test substance.

OTHER SOURCES OF EXPERIMENTAL VARIATION

The distribution of animals in the treatment room is also worth considering so as to avoid minor environmental differences that may become significant in longer studies. In this respect, rodents on the top row of cage racking can show ocular abnormalities due to greater levels of light in that position, relative to the lower rows, which are inevitably more shaded. Incautious distribution of controls or treated groups can produce an uneven distribution of ocular changes, which may be misinterpreted as treatment related.

The source from which animals are obtained is also important. In terms of continuity and comparability of historical control ranges, a consistent supplier for each species and strain should be used wherever possible. In the study, significant differences were apparent between wild-caught primates and those bred in captivity in special facilities. The wild-caught animals were of unknown age, except in general terms, and were of unknown medical history, especially with respect to parasitic infections. The results of this included behavioral abnormalities and preexisting histopathological lesions, which could confound interpretation of the study findings. These were exacerbated by the practice of housing them singly. Such problems have been largely circumvented by the use of captive breeding, which produces an animal that is notably more relaxed and which, with group housing, is much easier to work with. In addition, group housing has had significant effects on animal welfare and study quality. Husbandry practices at dog breeders have also improved, with people employed to familiarize the animals with handling and interaction with technicians. For both species, these practices produce a more relaxed animal that is easier to work with and, possibly, also safer.

PROTOCOL DESIGN AND PROCEDURE AS SOURCES OF ABNORMALITY

Many of the issues such as husbandry refer to abnormalities that may be present in the test system before the study starts or are present as a function of test facility management. However, it is also possible to introduce abnormality into the study by poor protocol design. Taking samples for toxicokinetic determinations from dogs and primates is often performed only on treated animals or, to a much lesser extent, in the controls. For this reason, it is vital that samples for clinical pathology are collected before the toxicokinetic samples. The extensive sampling normally associated with collection of samples for toxicokinetic analysis may induce a marginal anemia, which will be more apparent in treated animals than in the controls, which were sampled less extensively than treated animals. This can give a false impression of a treatment-related effect. The order and time of day at which investigations are undertaken must be consistent, in order to ensure that diurnal variations are discounted. It may be useful to consider large studies as a number of replicates, with equal numbers from each treatment group in each replicate; this helps to ensure that the investigations are evenly spaced across the groups and that the variation between replicates is accounted for.

CONTROL GROUPS AS NORMALITY

The contemporary control or untreated test system is always assumed to represent the experimental standard of normality from which treatment-related deviation may be assessed. However, the group sizes in toxicity experiments are usually quite small, and in a typical rat study, a control group of 10 males would usually be randomly allocated from a group of about 45 males or fewer, allowing some animals as replacements, the remaining animals going to the treatment groups. Within the four groups, it is quite possible to have differences in various parameters that are statistically significant before the effects of treatment are added; this level of difference tends to increase as the diversity or heterogeneity of the original population increases. To assess the normality of the control group, their data may be compared with those from similar groups from other studies conducted in the same facility. This gives a valuable check that the controls have behaved in a similar manner to those on previous (recent) studies (i.e., that the controls of the current study are producing data that are in line with expectation). This type of check is routine for *in vitro* studies, particularly for positive control groups, and in reproductive studies, where the incidences of rare variants or malformations is a critical factor in determining the relevance of any differences seen from the contemporary control. These checks involve the maintenance and use of historical control data, which are discussed in detail below. One disadvantage of using historical control data, and why the comparator studies used should be recent, is that "normal" parameters can change over time—what was within the range of normality from an animal in the 1950s may be significantly different from that of an animal today.

A protocol or experimental plan defines the endpoints or parameters to be examined and, therefore, defines the areas in which the controls must be normal. However, it should be appreciated that normality in scheduled parameters does not necessarily imply normality in others. In other words, the test system should be normal in all respects in order to be valid for that experiment.

For the majority of experiments, it is necessary to have a control group because of the relatively small differences that may be seen and also to facilitate the evaluation of dose–response curves and the presence of a no-observed-effect level. However, it may not be necessary to have controls where the change from normality is expected to be large or clearly evident, as in dose range-finding studies early in an experimental program. The control group becomes essential in longer studies where experimental normality can drift as the study progresses. *Longer* may be a relative term, as this can apply *in vitro*, especially with primary cell cultures, as well as in animals. Such change with time is seen typically with the plasma activity of alkaline phosphatase (ALP), which decreases with age. Increases in ALP activity in the plasma are seen with some compounds that affect the liver, and when these increases are small, they can be countered by the normal decrease in ALP activity, leaving the enzyme activity essentially unchanged from the previous examination. This lack of change, seen in treated animals, is an effect of treatment even though the ALP activity in those animals has not risen; the controls should show the expected decrease in activity.

In studies in which there is a statistically viable group size, it is not normally necessary to collect data before treatment, with the exception of noninvasive data such as body weight or food consumption. This is because the number of animals in each group allows a sensible statistical comparison to be made with the controls. Where the group size is smaller, usually seen in studies with dogs or nonhuman primates, or *in vitro* studies, the scope for statistical analysis is greatly reduced because statistical power is affected by the sample size. In studies *in vitro*, it is normal to compare the contemporary control against expectation as contained in the historical control ranges for the testing laboratory. In animal studies using dogs or nonhuman primates, it is usual to collect data before treatment begins to indicate the baseline data for individual animals; these also act as a health check as individuals with abnormal results can be excluded or treated as appropriate. These data provide a within-animal control that, when evaluating data collected during the treatment period, can be used with the contemporary control data in a dual comparison to indicated treatment-related changes. Equally, change in the control groups in some parameters from one examination to the next may be contrasted with the absence of change in treated groups, thereby allowing the difference from controls to be interpreted appropriately. Carcinogenicity bioassays frequently have two control groups, allowing separate comparisons with treated groups to be made. If the high-dose group shows a difference from one control group but not the other, the difference may be dismissed as unrelated to treatment.

Inevitably, there will be occasions when individual control cultures or animals show results that are clearly abnormal, either high or low; such results that lie grossly outside normal ranges are also occasionally seen in treated individuals at any treatment level. Examples of such deviation from expectation might include high colony counts *in vitro* in untreated controls or lack of response in positive control groups in bacterial mutagenicity tests or markedly high activities for single enzymes in the plasma of individual rodents. Some strains of rat are prone to early kidney failure, which is seen in controls and may be exacerbated or accelerated by treatment in test groups. In untreated controls, it is obvious that the result has arisen by chance; where the individual is treated, such dismissal is more difficult but can be achieved by reference to other data from that individual and to other members of the group, or by the absence of dose response.

In some animal experiments, it is normal to use the animal as its own control. This is true for experiments where individual data are collected before treatment, but is a much more significant factor in some short-term tests. In studies such as skin sensitization in guinea pigs, the mouse ear swelling test, or in irritation studies, an untreated area of skin or the untreated ear or eye is assessed. This use of within-animal control data is based on the assumption that the chemical administered will not enter the systemic circulation and that all effects are therefore confined to the site of administration.

ESTABLISHING AND MAINTAINING NORMALITY
IN EXPERIMENTAL GROUPS

An experiment has been planned, and the test system has been delivered, prepared, or taken from the freezer. As a first step, it may be assumed, initially at least, that the population of cells, bacteria, or animals that has been provided for the experiment is itself normal. However, some degree of heterogeneity in this population must be assumed, and some heterogeneity of response must be expected from the individuals. Before the experimental variable of treatment can be applied, it is necessary to distribute any inherent variability in the individuals of the test system evenly among the groups, in order to avoid experimental bias that may be seen as spurious differences between treatment groups. With cellular systems, the stock culture may be mixed and subsampled to provide the cultures for the different control and treated groups, a process made more valid by the huge numbers of cells that are involved.

With animals, it is slightly more difficult because of the smaller numbers and the greater extent of genetic diversity that is to be expected. There are several methods available for random allocation of animals to treatment groups, which will achieve this even distribution.

For small animals, such as rodents, it is normal to allocate them to cages as they are removed from their transport boxes. The allocation to cages may be decided randomly or in a set sequence so that cages from each group are filled in a rotation ensuring, for instance, that the controls are not all selected from the animals first out of the boxes. The study plan may stipulate a weight distribution across the groups or within groups so that no animal that is, for example, more than 20% outside the mean body weight is selected. It is common also to stipulate that the group mean weights at the start of treatment should be similar. For larger animals, although body weight is still a factor to consider, other factors such as social groups or housing may be taken into account. Group size in these studies is smaller than in studies with rodents, and the mean starting body weights are likely to be more variable than for more homogenous species such as rats or mice.

In some studies, especially those with large animals or in reproductive studies, it is important to ensure that litter mates are not grouped together, to avoid any genetic bias that may be introduced to a single group. Where a specific parameter is considered of importance, it is good practice to examine it before treatment starts and to randomly allocate the animals to the treatment groups using these data. Use of pretreatment cholesterol or triglyceride values as the basis for allocation to groups would be appropriate for a drug expected to affect plasma lipid concentrations, for example. The intention and overall result is to produce the correct number of treatment groups that are as similar as possible before treatment begins, in particular, that the control groups are comparable with the groups destined for treatment.

Having completed the process of random allocation to groups or treatment, it is now necessary to ensure that the various groups remain comparable to each other, at least until treatment is applied. The groups should also be comparable to previous control groups from similar experiments to allow comparison when necessary. For most test systems, but particularly for animals, an acclimatization period is necessary to acclimatize them to the laboratory environment. For cell cultures, this may mean that they are allowed to go through a few cycles of divisions to check for viability. For animals, a period of between 1 and 4 or more weeks is allowed—the former for rodents, the latter for larger animals such as dogs. This allows them to settle after transport to the testing laboratory and to become accustomed to the new procedures or cage mates. It is during this period that health checks are performed and the distribution among the groups is confirmed.

The importance of husbandry for test systems has already been mentioned, and it is vital that standards of husbandry should not change during experiments, especially short ones. In some cases, it is necessary to change housing, but this should not be undertaken lightly as the animals will be subject to stress as a result, and this may affect their responses to the test compound. It makes sense to avoid cage changes around the time of critical investigations on the study.

In animal studies, one of the most important factors to consider is the diet offered. The aim is to provide a diet that is well balanced and nutritious without compromising health. Correct nutrition has a critical role in toxicology; low-protein diets are known to affect metabolic capability and may increase the toxicity of directly toxic chemicals and decrease the effects of chemicals that require metabolism. Diets have been formulated traditionally, to maximize growth and reproductive performance. However, a high-protein diet is not suited to mature rats and gives a different tumor profile in comparison with low-protein diets. In fact, the growth of tumors may be enhanced by high levels of fat and protein. Conversely, the presence of antioxidants and trace elements can decrease tumor growth.

There has been gradual realization that high-protein diets can be responsible for excessive weight gain and adverse tumor profiles and poor survival, which has resulted in the use of low-protein diets or dietary restriction. The problem here is that a low-protein diet may not be appropriate for a growing animal. The apparently logical choice of changing from a high-protein to low-protein diet at the end of the growth phase is fraught with difficulty, as this might affect the absorption of the test chemical and lead to increased or decreased exposure levels. Equally, the use of dietary restriction to reduce weight gain, for instance, by offering food for 2 hours/day, means that the animals soon learn to eat only at that time of day. As a result, the food intake is not significantly reduced in comparison with studies where diet is freely available. Also, the practical considerations of giving and removing food each day adds to the study workload and increases costs. These factors are of particular concern in rodent studies, especially long ones, but are less of a problem with dogs, which are generally given a set amount each day. In some laboratories, food is offered to dogs at a set interval after treatment and is then removed after 2 or 3 hours; this may have implications in studies in which the test substance has been added to the diet. A potential problem here is that if treatment has a daily transient effect that reduces food consumption while food is present, there is likely to be a spurious reduction in food consumption. In any study, there is a general preference for using the same batch of diet throughout; in any case, the batches offered should be known. It should be understood, however, that changing the diet offered or the feeding regimen will have a knock-on effect on the historical control ranges, which will take some time to reestablish.

When studies are performed by mixing the test material in the diet, care must be taken that the inclusion levels do not affect the nutritional value of the diet. This is particularly difficult to achieve when testing novel foods; studies with genetically modified potatoes used dietary inclusion levels of potato that have been associated with pathological findings in the intestine. The presence of these changes was wrongly attributed to the genetic modification of the potato.

These various comments may also apply to consistency of media used for *in vitro* work, although, as these are largely synthetic, the overall significance may not be so great. Changing media routinely used in a particular type of *in vitro* study may be expected to change the response of the test system and will require the establishment of new historical control ranges.

One of the aims of study conduct should be to maintain consistency of design from one study to the next in order to obtain control data that may be referenced to the historical control ranges of the laboratory. This allows the performance of the controls to be measured against expectation at the laboratory and facilitates an answer to the question, "Are the controls normal?" As indicated above, the establishment and use of these historical control ranges or background data is a critical part of laboratory management.

BASELINE DATA AND HISTORICAL CONTROLS

In any scientific experiment, as many variables as possible are controlled in order to assess the truth or otherwise of the test hypothesis—in an ideal world, the only difference between control and treated groups is the presence or absence of treatment. The use of concurrent controls gives a baseline against which change in treated groups can be measured; they become the experimental definition of normality *for that experiment*. Within this narrow definition, it is possible for each experiment in a series to stand alone, providing that controls remain valid as comparators for their test group or groups. However, when it becomes necessary to compare results across a series of experiments, it is important that the control groups are comparable historically with each other; otherwise, deviation from expected control values in one experiment cannot be assessed with any confidence.

In toxicology, as in most scientific disciplines, the presence of a suitable control group, or groups, allows a judgment to be made as to the presence or absence of a treatment-related effect. It is possible, however, for control groups to provide data that are not consistent with normal expectation in the testing laboratory and to give results that are not consistent with previous experience. Even with the large, randomly allocated, double control groups typically used in carcinogenicity bio-assays, each of 50 animals of each sex, it is possible to see statistically significant differences between the two control groups, especially in body weight or food consumption. It is therefore important to have a database of historical control data that will allow an assessment of the normality of the controls, particularly when group size is small.

USES FOR HISTORICAL CONTROL DATA

Historical control data can be used to answer questions on the control data or on the deviation of treated group data from expectation, if the concurrent control is in some way inappropriate, for example, if the data are unexpectedly low. Equally, they can be used to show that the occasional outlying result in treated groups is within the expected range of values. They are particularly important in tests such as those for mutagenicity, when they are used to check that the controls have produced the expected number of colonies or that the positive controls have shown a significant increase over untreated control values. In reproductive studies, they are used routinely to compare the incidences of rare malformations with normal incidences in previous control groups.

If part of a study is finished earlier than the concurrent controls, it is possible (although tricky) to use historical control data to assess any treatment-related changes in the "uncontrolled" test group. This type of assessment is best when gross change from expectation is sought, as subtle changes cannot be confidently assessed in these circumstances. Changes in these cases can be assessed against the presence or absence of similar change in groups that complete the treatment or exposure period.

TYPES OF HISTORICAL CONTROL DATA

It is important to consider the data type before trying to set up historical control data bases, as some data are not appropriate for this treatment. Data that are suitable are objectively derived and are typically for continuously variable parameters or are incidence data for findings that are present or absent. The numeric type includes colony counts, mitotic index, clinical pathology parameters, body weight, and similar measurements. Incidence data are typically for reproductive malformations or variants and tumors. Subjective data—those that require some judgment to grade—are not generally suitable for historical control data. Nonneoplastic pathology findings are a good example of this because of their presence in a number of severity grades, which are dependent on the judgment of the individual pathologist. The inconsistency of nomenclature, between pathologists and with passing time, is just another aspect of this. In rats, a limited degree of basophilic tubules may be normal in the kidney, but because of the probability of different interpretation and grading between pathologists, it is not generally possible to assign a numeric value that would be consistent from one study to the next. With tumor data, it is much simpler because either the animal has a tumor or it does not (always allowing for pathological disputes about what constitutes a tumor, for instance, with hepatic foci and adenomas).

It is also impossible to have background data for clinical signs, especially those that are largely subjective such as hypoactivity or the extent of thinning of background lawn in the Ames test.

DEFINING A NORMAL RANGE

In defining normal values, the most important characteristic to be determined is the mean of the control values for the parameter in question. However, this single figure does not indicate by itself the extent of the variability of the data about that mean. While a range, minimum and maximum, will give some indication of that variability, it is usual to apply 95% confidence limits to the data and give the mean ± 1.96 standard deviations, which excludes any outliers at the extremes. This gives reasonable confidence that 95% of values will be inside the normal range and also indicates the inherent variability of the parameter. However, this basic approach does not take account of the variable sample size for the different parameters, which is typical of historical control databases. Thus, a commonly measured enzyme such as alanine aminotransferase activity may have 200 or more values, while a less common enzyme such as sorbitol dehydrogenase may only have 30. In this case, the use of the simple mean ± 2 standard deviations may give a false impression of security for the lower sample size. To take account of this, it is possible to use a formula that takes account of the sample size by applying degrees of freedom. The lower and upper limits of a normal range may be calculated from individual values using the following formula:

$$\text{Mean} - (t_{n-1} \times \text{SD}) \text{ to mean} + (t_{n-1} \times \text{SD}),$$

where t_{n-1} is the value of t for the number of samples (n) less one degree of freedom and SD is the standard deviation. This takes the number of data points into account and gives a statistically more secure value. An important assumption in this is that the data follow a normal distribution.

When such a range has been calculated, the data reported for that parameter should include the number of samples (n), the mean, and the calculated normal range and the actual range of values. All these values indicate the reliability of the figures generated; a normal range quoted from 30 values of an inherently variable parameter, for instance, an enzyme activity, will be less reliable than from a similar population of a consistent parameter, such as plasma sodium concentration.

A vital aspect of normal range definition is the origin of the data contained in relation to what they are being compared with. For valid comparisons to be made, the control groups have to be as similar as possible, and, if the experiment is in any significant way atypical, its control data should not be included in the historical control ranges for the laboratory.

DRIFT IN HISTORICAL CONTROL RANGES

With any biological organism, some drift in normality is to be expected. With humans, the normal height of the general population has increased significantly since medieval times, and the health profile of the population has also changed as diet, medical knowledge, patient care, and a host of other factors have improved (or deteriorated, according to viewpoint). In toxicological test systems, reasons for drift or evolution in historical control values include change in genetics, storage or husbandry, culture media or diet, environmental factors, and subtle variations in methodology. It is unusual to be able to pinpoint a precise cause for such drift, as various factors tend to interact to produce values that may be far from expectation, leading to historical control subsamples that may not be consistent with the rest of the data.

Suppliers have huge influence over historical control data through the pressure to select for particular characteristics in cell lines or animals that have large litters; these animals tend to be larger and to grow more quickly (which may be associated with earlier sexual maturity). In the late 1980s, it was noted that some strains of rat were eating more and gaining more weight than previously. This had the effect of reducing survival in long-term studies below 50% at 2 years—the magic cutoff figure produced by regulatory authorities to define an "acceptable" carcinogenicity bioassay. This was due in part to unexplained deaths but also to increases in renal disease and increased incidences of mammary and pituitary tumors. The weights of some organs had also increased over previous historical controls (Nohynek et al. 1993).

Drift, which may be due to supplier influences, has also been important in the historical control data seen with reproductive studies in fetuses of rats and rabbits. A good example of this is the reduction in the number of extra ribs seen in some rat strains from more than 20% of fetuses to fewer than 5% in a 5-year period at one laboratory. Litter size has tended to increase in line with breeder pressures, and this may be expected to influence the historical background incidences of study findings, quite apart from the expected reductions in fetal weights.

Another source of drift can be in the diagnostic criteria and nomenclature, used by pathologists in evaluating the study tissues, especially in carcinogenicity studies. To a large extent, this is an artifact, and comparison of tumor rates between laboratories is also made more complex by differences in these factors.

The consequence of this variability is that it is necessary to update historical control ranges at regular intervals to ensure that they stay relevant to contemporary studies. The best way of achieving this is to use a data capture system, which has the facility to take control values from suitable studies for recent historical control data and from which studies are excluded as needed. Selection of studies for inclusion needs some care to avoid data bias.

How Many Data Points?

The number of data points needed, to give a viable historical control database for any one parameter, is dependent on the data type and the inherent variability of the parameter in question. For a continuously variable parameter with low variability, a smaller number of points will be adequate than with a highly variable parameter. Although it may be feasible to define the number of values necessary for statistical security in light of the variability in a parameter, it may not actually be sensible to do so. The reason for this is that all too frequently, the smaller data sets relate to rarely examined parameters, particularly enzymes, and in these cases, any historical control data may be a bonus and cannot be ignored. The weakness of the comparison has to be taken into account in the interpretation, and it may become necessary to refer to similar data from dissimilar studies or from other laboratories.

For incidence data, Tables 2.1 and 2.2 (Glaister 1986) give some indication of the sample sizes needed for a viable comparison. In this case, which is typical for tumor incidence data in carcinogenicity bioassays, the more data available, the better the security of the conclusions drawn from them. The principal caveat is that they should be recent data, as some drift in tumor incidence may be expected with time.

The minimum incidence of a tumor that is possible in a treated group in a standard carcinogenicity bioassay is 2% or 1 animal in 50. However, although the control groups in modern carcinogenicity bioassays are typically of 100 animals (in two groups of 50) for each sex, this does not necessarily give enough statistical sensitivity to dismiss a single tumor as incidental. This is because a control group of this size is only likely to show a tumor with a minimum incidence of between 3% and 4%. The implication of the data in the tables, especially Table 2.2, is that to assign a tumor found in one treated animal to incidental causes, a historical database of a minimum of 200 animals would be desirable; in practice, it is much larger than this. With tumors, the statistical device of combining males and females, which would give a control group of 200 animals, is only viable when there is no sex-related bias in incidence. This is clearly not the case for prostate, testicular, mammary, or other sex-specific tumors and cannot even be applied to liver tumors, due to sex-related differences in metabolism. For analysis of rare tumors, it may be necessary to resort to outside sources of historical control incidences, which are likely to be larger, although more diverse in origin and derived from different study designs.

Reproductive developmental toxicity studies provide similar problems, and it is likely that a minimum of 50 litters would be needed to give a secure historical control database. The litter is the unit of evaluation in reproductive studies, and the number of individual pups is less significant.

TABLE 2.1
Probability of a Given Sample of Animals Containing at Least One Case for Different Background Incidence Levels

Sample Size	True Population Incidence Level (%)				
	2	5	10	20	40
10	0.183	0.402	0.652	0.893	0.994
20	0.333	0.642	0.879	0.989	1.0
30	0.455	0.786	0.958	0.999	1.0
50	0.636	0.923	0.995	1.0	1.0

Source: Adapted from Glaister JR, *Principles of Toxicological Pathology*, London: Taylor & Francis, 1986.

TABLE 2.2
Maximum Background Incidence That Would Yield a Zero Result in a Sample ($P < .05$)

Sample size (n)	5	10	15	20	50
Background incidence (%)	45	26	18	14	6

Source: Adapted from Glaister JR, *Principles of Toxicological Pathology*, London: Taylor & Francis, 1986.

For numerical data, such as clinical pathology or colony counts, 100 data points may be acceptable for an evenly distributed parameter with limited variability between individuals. For parameters where the variance about the mean is lower, for instance, electrolyte concentrations in plasma, a smaller number may be acceptable. Where the parameter is highly variable among individuals and, in some cases, among occasions of examination, as for ALP activity in the plasma, it is desirable to have a much larger data set for comparison.

TRANSFERABILITY OF NORMALITY—DATA FROM OTHER LABORATORIES

In the same way that the contemporary control data is the best estimate of normality for any particular experiment, control data generated in similar experiments at the same laboratory are the best source of historical control data. However, there are occasions when the in-house experience is insufficient for confident interpretation of the results. This is so for parameters that are not frequently measured, or when it becomes necessary to compare in-house data with those from outside, in order to verify correct performance of the test. It also becomes necessary when a laboratory changes location or a new animal facility is opened; this is especially true for reproductive data.

There are a number of factors that work against this, however, including differing environmental factors such as diet or noise, differences in analytical technique or instrumentation, and pathological nomenclature. Despite these challenges, data from other laboratories may be better than nothing at all, particularly if the same supplier is used. The criteria for deciding the relevance of data from other laboratories for purposes of historical control use are the same as for in-house data, as discussed below. The greater lack of comparability of such data with those produced in-house must not be forgotten in the process of interpretation. This is illustrated by data published for control groups from a number of studies using Charles River rats and mice (Giknis and Clifford 2000).

(Note: There are sources in the literature for other parameters such as those of clinical chemistry or hematology; however, these are probably less useful than tumor incidences and may well be subject to undefined methodological differences, rendering them essentially useless. Animal suppliers should have background data on their websites, especially for parameters such as growth and food consumption or specialist data such as glucose concentrations in diabetic models. Some animals, particularly transgenic animals, can only be sourced from a single supplier. Such animals should have a good amount of historical control data, which should be investigated before choosing the model.)

CRITERIA FOR COMPARISON OF HISTORICAL CONTROLS

The following list uses a fairly arbitrary order to indicate the importance of the individual factors that need to be considered in deciding how much one set of data is comparable with another. The comparability of the data sets will increase as the number of similarities from the listings below increases. Many, if not all, of these parameters should be controlled with the use of standard operating procedures (SOPs) to ensure conformity across a site. It is vital that SOPs are followed to ensure the production of valid data sets.

- *Species and strain or derivation of culture or test system:* Use of the same species is clearly essential, but different strains of mouse or rat, as with strains of *S. typhimurium* or *E. coli*, have their own characteristics, which significantly influence the values for their various parameters. This distinction extends to primary or secondary cell cultures derived from animals; for strict comparability, they should be from the same strain, similarly treated or derived from the same original culture source. Hepatocytes from rats pretreated with Aroclor 1254 (used rarely now) will probably not be the same as those from pentobarbitone-treated animals.
- *Experimental protocol:* The procedures used in producing the data must be as similar as possible, extending to culture methods, husbandry, treatment or exposure period, and data collection methods. Where the protocol delineates the criteria for data collection, it is critical that data from similar protocols are compared as small shifts in definition can produce large differences in incidence. This is particularly relevant in epidemiological toxicology and may be an Achilles' heel for future evaluation of records collected under outdated regulations. Data collected from restrained animals may well differ significantly from those collected by telemetry, for example, heart rate and blood pressure.
- *Age of the test systems must be comparable:* Many parameters change with age. In primary cultures of hepatocytes, metabolic capability declines over a period of hours so that data from a fresh culture may be radically different from those given by older cells. In animals, the most obvious change with age is seen in body weight, but other parameters (such as sexual maturity) move in concert with this. In the plasma, ALP activity reduces as the growth phase slows, and the bone-derived isoenzyme becomes less important. Renal function in older animals is significantly lower than in the young, and this may influence response to toxins through slower clearance and possible changes in the importance of metabolic pathways. In addition, greater variability among individuals is a characteristic of old age.

- *Dose route or method of exposure:* Although this is less important *in vitro*, there is still scope for significant difference, as with exposure in the vapor phase or in a solvent such as DMSO. In animals, Absorption, Distribution, Metabolism and Excretion (ADME) and toxicity following the various routes of administration can be very different. There may also be effects on parameters due to the procedure itself; muscle damage at the site of injection can be associated with increased release of enzyme markers of muscle damage into the plasma; e.g., creatine kinase, lactate dehydrogenase, and aspartate aminotransferase. Such damage may be due to the vehicle used in the formulation.
- *Vehicle or culture media:* With some compounds, especially poorly soluble pharmaceuticals, there is considerable pressure to use more complex vehicles in an attempt to produce a solution and enhance bioavailability. A significant number of excipients used in formulations, such as suspending agents and, particularly, solubilizing or wetting agents such as polysorbates (e.g., Tween 80), have their own toxicity, which complicates comparisons between them. Such toxicities can generate false-positive (i.e., not treatment related) results in *in vitro* systems. In addition, formulation differences in terms of percentage content of individual agents give an extra layer of uncertainty.
- *Same source or supplier of test system:* The importance of environmental factors and genetic differences between suppliers may have effects that are determined by the inherent stability of test systems.
- *Similar environmental and housing conditions:* These include factors such as stocking density, breeding, culture procedures, etc. Housing and stocking densities, which affect stress levels, can have wide-ranging effects, including effects on tumor profile. Changes in husbandry practices, for instance, to exclude or reduce disease, have far-reaching effects on animals; this has been seen with rabbits where improvements in housing conditions greatly reduced the mortality seen in older facilities.
- *Similarity of data recording:* Occasionally, new procedures or personnel can lead to a new observation that has been present before but simply missed due to lack of observation at the critical time or because the personnel or method of observation changed. The overall quality of data is also a factor but is one that cannot be readily established from tables in isolation.
- *Laboratory instrumentation and analytical procedures:* This is an important factor to consider (but not always available) when comparing analytical data from different laboratories. It may be a factor in published differences in the plasma activity of alanine aminotransferase in marmosets from different laboratories. Change from manual or visual data collection, for example, differential cell counts, to instrumental methods can produce large changes in incidences. Inception of new analytical kits or data treatments can likewise have significant effects that make it impossible to compare early data with new.

At some point, the decision must be made as to whether the historical control database offers sufficient comparability with the experimental controls to make a meaningful contribution to the interpretation of the results. If a significant number of the factors given above are not matched, it would probably be better to rely on the experimental controls rather than to use historical control data of dubious similarity.

SUMMARY

Definition of normality is fundamental to all toxicity testing, which relies on the ability to detect change or difference from controls (which represent normality). The ability to detect minor differences from normality or background incidence, induced by chemicals, is critical to meaningful assessment of their impact in the real world of clinical or consumer use.

- Contemporaneous controls define normality for that experiment.
- For numerical data, normality is best defined as a range.
- It is possible for controls and their group mean to fall anywhere within a normal range.
- If a parameter has a wide normal range, it will be more difficult to detect difference from that background.
- Normal ranges tend to drift with time, particularly in tumor incidence and the incidences of malformations in fetal developmental studies. In long experiments in animals, the control characteristics will change as they grow older and put on weight.
- Normality may be affected by experimental procedures, such that change related to treatment may be obscured by change related to the process of collecting the data.
- If the controls fall outside the normal range, they should be considered to be suspect.
- If there is doubt about the performance of the contemporaneous controls, historical background data may be useful.
- Historical control ranges, from other experiments in the laboratory or, less securely, from suppliers or other facilities, need to be carefully matched to the test system of the experiment to which they are compared. Factors to be considered include test system strain or cell identity and source, study plan or protocol, age, treatment route, solvents, and husbandry.

REFERENCES

Fermini B, Fossa AA. The impact of drug-induced QT interval prolongation on drug discovery and development. *Nat Rev Drug Discov* 2003; 2: 439–447.

Giknis MLA, Clifford CB. *Spontaneous neoplastic lesions in the Crl: CD-1® (ICR) BR mouse.* Charles River Laboratories, Tewksbury, MA, 2000.

Glaister JR. *Principles of Toxicological Pathology.* London: Taylor & Francis, 1986.

Gordon CJ. Role of environmental stress in the physiological response to chemical toxicants. *Environ Res* 2003; 92: 1–7.

Heath D. *An Introduction to Experimental Design and Statistics for Biologists.* London: Taylor & Francis, 2000.

Nohynek GJ, Longeart L, Geffray B, Provost JP, Lodola A. Fat frail and dying young: Survival, bodyweight, and pathology of the Charles River Sprague–Dawley-derived rat prior to and since the introduction of the VAF variant in 1988. *Human Exp Toxicol* 1993; 12: 87–98.
Woolley A. *A Guide to Practical Toxicology*, 2nd ed. London: Informa Healthcare 2008.

3 Determination of Toxicity
The Basic Principles

INTRODUCTION

This chapter sets out the basic concepts that need to be considered in the conduct of any toxicological experiment, including a review of regulatory influences, especially Good Laboratory Practices (GLPs).

CIRCUMSTANCES OF TOXICITY TESTING

Nothing in this world is without toxicity of some kind. Investigations to determine the extent or severity of toxicity are undertaken for a number of different reasons and with various objectives in mind, often with evaluation of risk for human use or extent of exposure as an ultimate goal. The endpoint of such investigations can be precise, as in attempting to clarify the relationship between an effect and exposure to the test substance or to elucidate mechanisms of toxicity. Very frequently, there is no such readily definable endpoint, merely the broad question, how toxic is this substance? The answer to this question may well be used to predict effects in humans and to assess safe dose levels (drugs) or amounts to which people may be exposed (acceptable or tolerable daily intakes, permitted daily exposure, benchmark, or reference dose) without adverse effect (pesticides, food additives, or chemicals at work).

In terms of studies needed before and after release of a substance intended for public use or leading to human exposure, the investigations of a chemical's safety fall into two phases and, broadly, two types. The first set of investigations addresses two objectives, the first of which is to identify intrinsic hazard in the test chemical and the second to estimate where and when that hazard may be manifested. The purpose of the second type of investigation is to confirm the predictions arising out of the first set in the real world. In other words, before release to the public, the first step is to predict and suggest ways of controlling or preventing effects or events that might require investigation. In the second step, epidemiological or marketing vigilance studies are undertaken after distribution of the chemical has started, to ensure that hazards to human health or the environment have been correctly predicted and assessed. The first set of studies is undertaken in a laboratory environment, and the second, conducted in the field in patients, consumers, or the environment.

Before a marketing authorization for sale to the public is granted, the testing of a chemical falls into the category of safety evaluation. The extent of such testing and the type of study conducted are closely regulated by government and international guidelines. Compliance with these guidelines is not optional; it is also a general proviso that the studies be conducted according to the scientific needs of the particular chemical under evaluation.

Once sales of a chemical—be it a drug, a pesticide, a cosmetic, or an industrial reagent—have started and significant public exposure has occurred, it is possible for untoward and unpredicted effects to be detected that may be associated with that chemical. It is popular in this increasingly litigious society to suggest that any abnormality in a population is attributable to chemical exposure, for instance, to local use of agrochemicals. These effects become the subject of epidemiological investigation that is intended to establish a link, or otherwise, between the chemical and the effect. With drugs, for which the dose is given and the patient history and comedication are known (theoretically) and for which the exposed population is easily defined, such investigations are relatively simple. For low-level exposures to pesticides where the dose is unknown or only roughly estimated, and for which there are numerous other exposures to contend with, certain establishment of a relationship between effect and the chemical is much more difficult. In both cases, once a relationship has been proven, specific mechanistic studies can be undertaken to show exactly how the substance exerts its toxicity and, in some cases, under which circumstances it has its undesired effect.

In order to bring a new chemical to the marketplace, it is necessary to design a program of investigation according to complex sets of regulatory guidelines specific to the area of use of the chemical. Although the studies prescribed for each class of chemical may be different in type or duration, they are all conducted to the same basic set of principles or protocol. In essence, these are to define an objective, choose a test system, and design and conduct a study according to a predetermined study plan or protocol, having a set of investigations and endpoints that are chosen to meet the objective. The evaluation of specific classes of chemicals is covered in Chapter 19.

EFFECTS SOUGHT IN TOXICITY STUDIES

It is an inherent handicap of toxicity measurement that programs of evaluation are necessarily designed to detect historically known effects, especially where a particular change is expected due to prior knowledge of similar chemicals. Unexpected and, therefore, generally unwanted effects may be found during the course of routine toxicity investigations and are then subject to mechanistic investigations to assess their relevance to humans. Occasionally, unpredicted effects are found in patients or consumers after marketing has started, e.g., cardiac arrhythmias seen with the gastrointestinal drug cisapride, which were due to prolongation of the QT interval in the electrocardiogram, in some patients. This has induced a general examination of new drug candidates for effects on QT interval prolongation in safety evaluation programs. Such triggers for additional investigations are relatively frequent, adding to the guidelines for the range of tests that must be conducted before marketing is authorized. The classic example of a new previously unconsidered effect was that of thalidomide, in which serious reproductive effects were found. The thalidomide disaster had an enormous influence on the practice of toxicity testing and safety evaluation, ultimately leading to a rigid framework of investigative studies but also encouraging a conservative response to toxicological innovation that is still with us.

One of the objectives of premarketing toxicity evaluation must be to search for mechanisms of possible epidemiological concern. It is relatively straightforward to identify severe effects, such as acute renal toxicity, but much more challenging to find or predict the minimal effect that may accumulate in human subjects over years or a lifetime and lead to insidiously progressive renal failure. The longest study routinely conducted in evaluation of medicines is the carcinogenicity bioassay in mice or rats, over the majority of the life span of the chosen species. These studies are primarily designed to evaluate the potential for tumor formation and are not necessarily, therefore, very good for revealing subtle changes due to true long-term toxicity.

BASIC PRINCIPLES OF TOXICOLOGICAL INVESTIGATION

OBJECTIVES

Investigations of toxicity may be carried out in isolation as single studies, or part of a program for the safety evaluation of a new chemical. In the case of novel drugs, the purpose of testing is to support initial entry to humans in volunteer studies and subsequent trials in patients. The amount of testing required for a pharmaceutical increases as the intended duration of treatment in humans increases. Other objectives include assessment of environmental impact for pesticides or the toxicity evaluation of intermediate chemicals used in the production of others, usually for occupational purposes. Broadly, the objectives and philosophy of a program of testing are driven by the intended use and genre of the test chemical. It is quite conceivable that a substance may fall into two categories and that studies are designed to cover both sets of guidelines.

Whether part of a program or as a stand-alone study, there should be a robust, preferably simple, study plan with a clearly stated objective. In animal-based tests, this objective is likely to be broadly stated without precise endpoints, usually "to determine the toxicity of..." This is in recognition of the complexity of animals and the large number of endpoints that are addressed in these studies. With simpler test systems, the objectives are more precise, for instance, to determine chromosome damage. This objective should be strictly adhered to, if the integrity and interpretability of the study are to be maintained. It is not sensible to incorporate an assessment of reproductive toxicity into a 13-week general toxicity study, if an animal becomes pregnant unintentionally. Reallocation of animals from one purpose to another may produce unwanted bias in the data, making interpretation more complex.

In regulatory toxicology, there is considerable emphasis on demonstrating toxicity rather than safety, as a negative outcome (safety) cannot be proved with confidence. Safety is a relative concept. Comparison of cyanide and water shows that both are toxic but, consistent with Paracelsus, that this is dependent on dose. Safety, therefore, has to be inferred, and demonstration of a toxic dose supports this inference. Consequently, for chemicals of low toxicity, there is only reluctant acceptance of limit tests or doses, which may be maximally practicable due to the physical characteristics of the chemical, for example, low solubility. However, there is little sense in giving excessively high doses, as there is frequently a dose beyond which absorption is saturated and no further systemic exposure can be achieved. As a consequence, toxicity expressed at high dose may be due simply to local effects in the gastrointestinal tract, which will probably not be relevant to humans, in whom significantly lower doses or exposures can be expected.

TEST SUBSTANCE CONSIDERATIONS

The quality and characteristics of the test substance are crucial to the successful conduct of a toxicity investigation or to a development program. Impurities can have far-reaching effects on the toxicity of substances, both for study programs and when marketed for human consumers or users. Impurities in tryptophan, used as a nutritional supplement in fitness programs, were responsible for muscle wasting in many consumers and for a number of deaths; this debacle occurred after production changes, which introduced new impurities that were not present in the early batches. Production processes or the pathways of synthesis often involve the use of solvents that can be difficult to remove. The use of dichloroethane is controversial because it is known to be carcinogenic in rodents at high doses. Although the levels present in the final material may be far below the level for carcinogenicity to be evident, it is better to be safe and to remove it from the synthesis. Genotoxicity studies for mutagenic potential are particularly at risk from low levels of contaminants that may give a false-positive result for a non-mutagenic substance.

It is normal to use the purest possible material in definitive toxicological investigations to avoid false reactions due to impurities. Impurities may express greater toxicity than the test molecule; for instance, TCDD was a significant contaminant in Agent Orange, used as defoliant in Vietnam by the US armed forces. When effects are receptor mediated, impurities may compete with the parent molecule at binding sites, thereby affecting toxicity or pharmacological response. Having said that, for pharmaceuticals, there is some sense in using slightly less pure material than the clinical intention because this may "qualify" the impurities in toxicological terms; the test animals will be exposed to the same impurities as patients. The problem here is that the scale-up of production may produce a different impurity profile in terms of both quantity and identity. When a toxicological program as a whole is considered, it is best to use the same batch of test substance or batches of similar or better purity that are consistent with the product to be sold.

The quality of the test substance should be defined by suitable certificates of analysis or by information from the supplier in the case of studies that are not part of a development program. By the time that early studies in patients begin, new pharmaceuticals should be accompanied by analytical certificates of Good Manufacturing Practice standards, and the supporting toxicity studies should be conducted in material of the same standard.

The physicochemical characteristics of the test substance should also be considered, as these can have significant impact on the choice of vehicle or carrier for toxicity studies and may determine whether such studies can be conducted. For example, volatile compounds may be tested by inhalation in animals but need a special testing apparatus for *in vitro*, genotoxicity experiments such as the Ames test.

The presence of enantiomers of a molecule can also affect toxicity or action, as they may be associated with different effects, either in degree or type; for example, thalidomide has two enantiomers, only one of which was associated with the characteristic developmental toxicity. For this reason, it is sometimes advisable to develop only one of the enantiomers; in fact, this may well be the preferred strategy of the regulatory authorities. Development of the racemic mixture may be acceptable, if there is rapid conversion between the two forms *in vivo*, making distinction between them impossible. One final point to consider is that the International Conference on Harmonisation (ICH) M7 guideline requires that all starting products, intermediates, by-products, and impurities in pharmaceutical products must undergo evaluation for bacterial mutagenicity and be controlled below a threshold. The mutagenic evaluation can be either via literature review or by *in silico* analysis using a rule-based and statistical structural activity relationship program—if both systems indicate a negative result, no further action is required. If a substance is predicted to be mutagenic, it should be assayed for bacterial mutagenicity or controlled below a threshold.

CARRYING SYSTEM AND ROUTE OF ADMINISTRATION

In order to bring the test substance together with the test system, it is usual to formulate it with a solvent or carrier that allows the concentration or dose of the substance to be varied in a controlled manner. The choice of carrying system or vehicle in which the test substance will be dissolved or suspended is closely interlinked with the choice of test system and with the characteristics of the test substance. In an ideal world, such carriers are simple and aqueous for use with easily soluble compounds; reality is often sadly different, and increasingly complex formulations may be devised to achieve adequate exposure of the test system.

The intended route of administration has a profound influence on the choice of vehicle and is chosen according to the purpose and objective of the intended study. To examine the potential toxicity of a drug or agrochemical for which the expected route of therapy or exposure is oral, the most appropriate route of administration is also oral. For studies to assess hazards of occupational exposure, dermal administration or inhalation are also likely to be relevant routes of exposure. For drugs, clinical intentions will drive the choice of route in the toxicity studies, although almost all drugs are also given intravenously at some point in order to generate comparative pharmacokinetic and bioavailability data. For dermal preparations that are not absorbed significantly, greater systemic exposure can be achieved by parenteral or oral administration, allowing worst-case systemic exposure to be investigated. This is an important consideration if the barrier functions of the patient's skin have been compromised by abrasion or some other breakdown, allowing greater absorption than through intact skin.

The best vehicles or carrier systems are the simplest, usually aqueous, or with suspending agents that are toxicologically inert. For test substances that are poorly soluble in aqueous media, corn oil or similar oil may be used. However, this is not suitable for *in vitro* experiments and cannot be used orally in some animals such as rabbits, where the physiology of the gut is not compatible with large amounts of lipid. Wetting agents such as polysorbates (e.g., Tween 80) can have their own effects in the gastrointestinal tract, leading to fecal abnormalities and possible effects on absorption of the test substance. Use of simple aqueous media is unlikely to be associated with long-term effects. Use of corn oil in rats or mice, at up to 5 mL/kg, may be associated with a compensatory decrease in food consumption and slight functional change in plasma lipid levels, which is unlikely to be of toxicological significance.

Increasingly complex vehicle systems, devised for insoluble compounds with a view to increasing test system exposure, can be associated with their own toxicity, which can mask the effects of the test substance. For substances with low solubility in water, it is more difficult to design a sensible *in vitro* assay, as these are generally water based. In these cases, water-miscible solvents such as ethanol or dimethyl sulfoxide can be used, provided the concentration is kept low; even then, there is the possibility that the test substance will precipitate when it hits the water of the experimental medium. In some *in vitro* tests, it is possible to claim exposure of the test system if the test substance is present as a precipitate. Broadly speaking, the more complex the vehicle, the more likely it is to be associated with its own toxicity, although the effects of simple solvents such as dimethyl sulfoxide or ethanol have their own distinctive toxicities.

Preparations for intravenous administration pose a particular problem and often mean that, with limited solubility, studies are conducted at low doses that are nearly meaningless unless continuous infusion techniques are used. It is important to avoid intravenous use of solvents or excipients that have their own toxicity; for example, histamine release may be triggered in dogs by injection of polysorbates.

For dermal administration, the vehicle is usually chosen to be similar to the circumstances of exposure in the human target population. For drugs, this means that the formulation tested in toxicology studies should be as close as possible, preferably identical, to that intended for use in patients. Vehicles are usually chosen to enhance absorption across the skin, and this applies equally to the clinical situation as to the toxicity investigation. For *in vitro* investigation of dermal effects using skin replacement systems, the vehicle should be chosen to be compatible with the test system chosen.

For agrochemicals and food additives, the most usual carrier is the diet, as this corresponds to the expected route of human exposure. Admixing exogenous chemicals with diet is usually straightforward, although it is crucial to establish that the final mixture is homogeneous. In difficult cases, the test substance may be dissolved in a solvent or food-grade oil before adding to the diet. With organic solvents, caution must be exercised, as residues can persist despite efforts to remove them, acetone being a case in point. With dietary administration, there exists the possibility that the diet may become unpalatable at high concentrations. In any case, care should be exercised that the additions to the diet do not affect the nutritional status of the animals. The maximum inclusion rate for nutritionally inert test substances is generally taken to be 5%, which is approximately equivalent to 2 g/kg/day in rats.

For food additives, the inclusion level may be raised to 10%, but care has to be taken with nutrition of the animals. This factor is likely to be a problem in testing high concentrations of individual genetically modified foods, as it is unlikely that effects will be expressed at normal dietary inclusion levels. However, the relevance of effects expressed at unrealistically high dietary inclusion of such foods should be questioned before the study begins, there being no point in generating data that are irrelevant. As an alternative to the diet, the drinking water may be used as a route of administration, but this has a number of disadvantages over the diet, especially in tracking the amounts of water, and therefore, chemical, ingested as distinct from that spilled.

With both water and dietary administration, there is the ever-present problem of recording the amounts consumed. With unpalatable diets, there is usually a significant degree of "food scatter" (especially in the first few days) where the diet is discarded by the animals without consumption. Estimation of this, especially for low amounts, has usually been only semiquantitative; housing animals on sawdust bedding makes estimation of scatter practically impossible. With drinking water, evaporation of water spilled or lost through playing with the outlet means that there is no hope of estimating amounts lost. Without such estimation, there can be no accurate calculation of dose levels in terms of mg/kg-bw/day, and hence no extrapolation to humans (or any species of concern).

Dietary administration of pharmaceuticals in long-term carcinogenicity studies may be used and is an attractive option in cases where the pharmacokinetics are such that admixture with the diet may produce a greater systemic exposure to the test substance than once-a-day dosing by the oral route. A more pragmatic reason is that this route of dosing is simple and less labor intensive than oral intubation (gavage). The calculations for dietary concentration required to achieve the desired dose levels are straightforward; the dietary concentration increases with continued body weight gain by the animals. To calculate dietary concentrations (ppm) to achieve constant dose, the following equation is used:

Dietary concentration

$$= \frac{\text{target dose (mg/kg/day)} \times \text{estimated midweek bodyweight (g)}}{\text{estimated food consumption (g/day)}}$$

To calculate the achieved dose level from food consumption and dietary concentration, the formula is as follows:

Dose (mg/kg/day)

$$= \frac{\text{dietary concentration (ppm)} \times \text{food consumed (g/day per animal)}}{\text{midweek bodyweight (g)}}$$

Inhalation toxicity is a specialized branch in its own right, due to the technical complexity of generating respirable atmospheres at precise concentrations and monitoring them. Because of the vast amount of equipment needed to generate, administer, and then dispose of waste air, it is an expensive method of administration. A related route is intranasal administration, which is increasingly popular for some drugs; in this case, the vehicle used can prove to be irritant to the nasal epithelia, thereby compromising absorption.

CHOICE OF TEST SYSTEM

The test system should be selected on scientific grounds, bearing in mind its suitability for achieving the experimental objective and whether it is ethical to use it. Only small amounts of test material may be available, especially in the early development of a chemical, and this may be a factor that influences the choice of test species. If the objective is to study potential effects in a nonhuman species, it is usual to use that species or to use an *in vitro* preparation from that species. Although it is politically correct to think in terms of moving away from animal experiments to studies in humans or to *in vitro* techniques, many *in vitro* methods require fresh tissue or cell preparations from animals. In most cases, animal use may be reduced (though in some assays, it can be increased); it is not eliminated. *In vitro* methods are particularly useful in screening a large number of chemicals during a process of lead candidate selection, usually for a specific activity or toxicity that is of interest.

In cases where the ultimate objective is to predict safety in humans, the only viable rationale is to choose a test system that is sufficiently close to humans in terms of pharmacological susceptibility, metabolism, pharmacokinetics, pharmacodynamics, and physiology. This is especially so for pharmaceuticals. However, it is highly unlikely that a perfect match will be available, and therefore, the choice of test system must be a pragmatic compromise. The factors to be considered are summarized in Focus Box 3.1. In addition, there are also other factors to remember, including housing, diet, genetic homogeneity, life span, reproductive cycle and litter size, size of the individual, and amount of test material needed.

A test system may be chosen according to similarity of metabolism in the available species in comparison with that expected in man by using *in vitro* preparations of hepatocytes, microsomes, or isolated enzymes. A typical experiment would use hepatocytes from rats, dogs, nonhuman primates, and humans; minipigs may be added or substituted, as appropriate. This type of experiment is often used in pharmaceutical testing where there is a regulatory requirement for two species—a rodent and a nonrodent. One of the caveats here is to look at metabolism in quantitative terms as well as qualitative. Although all the metabolites predicted to be present in man might be present in the test species selected, the quantities produced may be very different. Metabolism in man may be largely by one pathway and in laboratory species by a completely different route. The result is likely to be that a major metabolite in man has not been properly examined in the experimental animals. Some careful decisions will be needed, on what to test and in which species; this may result in stand-alone tests of the metabolite.

The question being asked in the proposed experiment will, in most cases, determine the type of test system. The shortest question—Is it toxic?—requires the most complex test system, as the range of possible interactions is so large that they cannot be encompassed in a simple *in vitro* preparation. On the other hand, a question addressing a single defined endpoint, such as DNA damage, can be answered *in vitro* in a simple test system.

The choice of *in vivo* test systems is usually straightforward, as the numbers of laboratory species are quite small, the mouse and rat, in various guises, being the favorites; Focus Box 3.1 explores this. Unless the more exotic transgenic strains are

FOCUS BOX 3.1   FACTORS IN TEST SYSTEM SELECTION

The following has been compiled with ultimate human use in mind; similar criteria can be used for other target species.

- Scientific justification: Using a test system without scientific reason is not sensible, although regulatory guidelines or advice may result in questionable choices on occasion.
- Similarity of absorption, distribution, metabolism, and excretion, and pharmacokinetics to humans (but beware of variation in humans): Choice may be influenced by comparative studies of metabolism *in vitro*, especially using hepatocytes from possible test species and humans.
- Genetic homogeneity: Inhomogeneity means that the system is likely to be more variable and that normality is more difficult to define. Increased variability reduces the ability of the study to detect minor adverse change especially if present at low incidence. This us made worse by small group size.
- Strain, outbred versus inbred, especially for rodents.
- Availability, feasibility, and cost: There is no point in commencing an extensive testing program if the most suitable species is not readily available at sensible cost.
- Regulatory acceptance.
- Purpose of test and applicability of species: For a new veterinary drug for use in dogs, it is essential to carry out the evaluation program in dogs.
- Validation: Investigations of genotoxicity normally require the use of validated cellular or bacterial systems or exposure of tissues such as bone marrow in rodents. Validation applies to newer *in vitro* systems in particular, but also to any new *in vivo* tests such as the local lymph node assay in mice.

selected, they are also inexpensive and easy to maintain. A further factor in favor of the use of rodents is that because of their ready availability, ease of husbandry, and price, it is relatively easy to design a statistically sound experiment without breaking the bank. Specific strains of rodent can be selected to answer questions of mechanism; for example, the Gunn rat is deficient in glucuronosyl transferase and thus can be used in mechanistic studies of toxicity due to phase 2 metabolism. For *in vitro* systems, the choice is more complex and can be more driven by the question asked; for example, the potential for nephrotoxicity can be investigated in isolated nephrons or kidney slices.

Another significant factor in test system choice is regulatory acceptance and expectation. In general terms, toxicologists are conservative and regulatory toxicologists especially so. New test systems are adopted slowly and with caution; an example of this is the gradual adoption and recognition of the minipig as an acceptable model for human skin and its increasing predominance in safety evaluation of topical pharmaceuticals.

As indicated previously, the simpler and more direct the objective of the experiment, the simpler the test system can be. For a broad investigation of toxicity for a new compound, it is normal to use an animal system because whole animals have all the complex interrelationships in place, between organs and tissues. Absorption from the gut and passage through the liver can result in metabolites that affect the kidney or other tissues. Such interrelationships cannot, at the present state of the science, be reproduced *in vitro*, although progress is being made in the construction of individual tissue systems. Liver bioreactors are under development for investigation of drug metabolism, and the possibility of adaptation for toxicity screening will no doubt follow on from this work.

A variation on the *in vivo/in vitro* theme is the use of *ex vivo* systems. In these animals are dosed once or twice with the substance of interest and are killed at an appropriate interval after administration. Tissues can then be removed for study *in vitro*. This type of system is used frequently in the unscheduled DNA synthesis test, in preference to the exposure, *in vitro* of hepatocyte cultures. The comet assay is another application of this strategy.

In comparison with whole animals, *in vitro* systems are usually inexpensive and quick to complete, and tend to give simple data sets. However, they can be technically challenging and, as a consequence, may not be reproducible in inexperienced hands or from one laboratory to another. Animal test systems, in contrast, tend to be expensive, both in terms of the facilities needed to maintain animals and in study conduct, and also tend to be slower in completion. They cannot be performed without permits or licenses from governmental authorities, which cover both the facilities and the experimenters. In addition, the data sets produced can be very large and complex to interpret; a 4-week rat study with 80 animals produces more than 10,000 individual data points. A research program based on animals will be more expensive and less flexible than one based on *in vitro* techniques, but because of regulatory reliance on animal tests, this choice is not available for every study type. Focus Box 3.2 examines the use of humans as the ultimate species of choice.

The test systems available for the various types and areas of investigation will be reviewed in greater detail in succeeding chapters, under the sections dealing with investigation of specific toxicities.

STUDY DESIGN BASICS AND CONFOUNDING FACTORS

The best toxicological study designs are simple but robust—this facilitates conduct and interpretation of the final data sets. Design, whether of generic guideline protocols or of tailor-made experiments, should take into account the three Rs—reduction, refinement, and replacement—as well as any regulatory requirements. The presence in the testing laboratory of relevant historical control data relating to the test system to be used has to be considered in deciding the size of control groups.

The classic design for a toxicological investigation has at least three treated groups and a control group, which receives the same treatment regimen as the treated groups

FOCUS BOX 3.2   HUMANS AS A TEST SPECIES

One of the objectives of toxicological study is to predict effects in humans. There has been considerable criticism of the accuracy of extrapolation of animal data to humans, although as understanding of the various animal models improves with our understanding of their relevance to humans, this is becoming less of a problem. However, the extrapolation of results to humans becomes more difficult the further one moves taxonomically from humans or from a whole animal model. It is one thing to extrapolate from rats to humans but quite another to use an *in vitro* system of liver slices from rats or, on an even less complex plane, a rodent hepatocyte culture, to achieve the same predictive power.

Although the best predictive model for humans would appear to be humans, this has difficulties, not least in ethics. By any of the above considerations, humans fail as a test system candidate. They are large, requiring huge amounts of test material, which, in the early stages of development of a chemical, will probably not be available. They have a long life span and slow growth patterns, making assessment of toxic effects on growth difficult. They require and consume a varied diet, in large amounts, that is not easy to standardize (diet affects the absorption of chemicals and their toxicity). Investigation of reproductive effects is long winded, due to the long period of gestation and low litter size. They are extremely heterogeneous genetically, meaning that normality is difficult to define. The data obtained from a randomly selected group could be so diverse as to be uninterpretable.

Above all, the moral arguments against the use of humans have been used in recent years to slow the acceptance of volunteer tests with pesticides, conducted to investigate human metabolism and pharmacokinetics. Possession of these data would make risk assessment of selected pesticides much more secure and would help to reduce the perception of pesticides as demon chemicals that do nothing but harm. Recent advances in microdosing using low doses of radiolabeled compound have great potential in the design of sensible and ethical experiments in humans with agrochemicals. While such studies may have limitations for pharmaceutical development due to their low dose relative to clinical doses, this would not be a problem for agrochemicals where the expected human dose should be expected to be very low. The exception to this would be where humans are the only relevant species (particularly in biologic pharmaceuticals), in which case an escalating dose paradigm may be appropriate—in such cases, it is vital to speak with regulators and have a good scientific rationale for your proposals.

but without the test substance. Positive controls treated with reference compounds may be included to check the sensitivity of the assay (particularly in experiments *in vitro*). It is also important to demonstrate exposure of the test system to the substance under investigation. This is not a problem *in vitro*, where the test system is exposed to the test substance in the culture medium usually as a solution. In animals, however, it is necessary to take blood samples for toxicokinetic (i.e., high-dose pharmacokinetics) analyses. While collection at several time points after administration is not a problem in dogs or nonhuman primates, such as cynomolgus monkeys, rats and other small animals are too small for this, and it may be necessary to add groups of satellite animals to the study solely for this purpose. This means that the animals allocated to the study itself are not stressed to the extent that the toxicity of the test substance is exacerbated or masked. However, with increasing sophistication and sensitivity of analytical methods, the use of satellite animals may not be necessary, especially in early screening studies using sparse sampling techniques or, simply, low sample volume (e.g., blood spots). Toxicokinetic data demonstrate the extent of exposure to the test substance and are vital in the full interpretation of the data that result from the study. A further consideration is the metabolism of the test substance and the toxicity or otherwise of the metabolites, as indicated by absorption, distribution, metabolism, and elimination (ADME) studies that are conducted as a separate program of experiments. The data from these studies should be considered in confirming the choice of test system.

If the test substance is carried in a particularly unusual vehicle, it may be necessary to have a further, negative control that is treated with water or a similarly benign substance, to take account of any toxicity due to the vehicle. Although there may be statistical arguments for having larger numbers of control animals than in the individual treatment groups, it is normal toxicological practice to use equal numbers. Having said that, there are cases where larger numbers of controls are used, particularly in carcinogenicity bioassays where two control groups of equal size to the test groups are often used. Use of greater numbers of controls can help to obviate the effects of exiguous historical control databases.

In routine toxicological investigations, the choice of controls is simple; a suitable number of cultures or animals is allocated to the control group or groups from the same stock set, as for the remaining groups in the study. In epidemiological investigations, where the subjects of investigation are humans exposed to the substance under investigation, the choice of controls becomes much more critical and complicated. For such studies, it is vital that confounding factors (such as smoking, excess alcohol consumption, or family history) are avoided and that the correct comparators are chosen.

Although confounding factors are well known in epidemiology, they should also be considered in routine toxicological investigation. One such factor is high heart rate in response to restraining procedures for recording electrocardiograms in unanesthetized animals. Unless an animal is accustomed to the procedure, the unfamiliarity with the situation will result in abnormally high heart rates, which can obscure a milder compound related effect, possibly associated with reduced blood pressure. Such effects can result in minor heart lesions that are not detected until histopathological examination; the use of external telemetry jackets or surgical implants can avoid such problems. Another frequent cause of confounding arises when blood samples for clinical pathology are taken immediately after extensive sampling for toxicokinetic investigations. As the controls may not be subject to the same sampling regimen for toxicokinetics as the treated animals, the samples taken for clinical pathology will probably show an apparent treatment-related anemia and associated effects. Experimental investigations should be timed so that they will not affect future examinations or be affected by previous procedures. Husbandry can also be a source of confounding factors, especially with regard to stress, which can have a variety of effects. As a result, care should be taken to avoid unnecessary levels of stress, which might complicate the interpretation of the data. Do not change cages immediately before critical examinations on a study.

Correct housing is one means of avoiding stress in laboratory animals, and there is increasing emphasis on cage size and environmental enrichment. Group housing is generally considered to be the best husbandry system but is not always appropriate if the benefits are negated by aggressive behavior as in male mice (see comment in Chapter 2). Lesions in group-housed male mice have been shown to lead to subcutaneous sarcoma when allowed to persist, an effect that can be exacerbated by treatments that increase aggression (Cordier et al. 1990). This may give a false impression of carcinogenic potential.

Although husbandry is clearly an important factor in the maintenance of animal test systems, its equivalent should not be neglected for *in vitro* systems, where inappropriate storage or preparation of cell cultures can lead to test systems drifting from expectation.

In all toxicological investigations, the quality of data that are generated is paramount. Poor data—incomplete, badly recorded, or ill-chosen timings—will probably result in poor interpretation of the experiment. GLP has made a huge contribution to the quality of data recorded in regulatory studies. Although it has been suggested that GLP and basic research do not mix well, the principles of GLP have a lot to offer to this area as well, as it should be a duty of all toxicologists to record data as accurately as possible. Secure interpretation can only be made from a complete set of appropriate data; in this, it is probably better to overrecord slightly, as recorded data can be dismissed as irrelevant after the event, but unrecorded data cannot be interpreted. Having said that, it is easy to get carried away on a wave of enthusiasm and record everything; it can be difficult to draw a sensible line between underrecording and overrecording, or to overburden the protocol with too much detail that endangers the interpretability of the final product.

CHOICE OF DOSE LEVELS OR TEST CONCENTRATIONS

The correct choice of dose level is crucial to the successful completion of every toxicological experiment and is normally achieved by use of a small preliminary study or by looking at the results of other studies. The normal procedure is to use a sighting or dose range–finding study or experiment, to assess the toxicity of the test substance in the same test system and under the same conditions as for the main study. Typically, a small quantity of the test system is exposed to increasing doses or concentrations of the test substance until a reaction is seen that indicates that a maximum tolerated dose (MTD) level or concentration has been reached. For studies in animals, the MTD either is used as the high dose in the main study or is used to select a slightly lower level that is expected to result in some toxicity but also to result in the survival of the animals until the treatment period has been completed. In *in vitro* systems, the maximum concentration is usually chosen according to degrees of toxicity seen; for example, the highest test concentration in cytotoxicity studies for registration is usually selected as the one that causes 50% to 80% toxicity in the test system. In the Ames test for bacterial mutagenicity, the thinning of background lawn (the minimal growth of bacteria resulting from a small amount of histidine in the culture medium) is taken as a measure of toxicity.

Sighting studies tend to use a lower number of animals or test replicates than the main study, and because of this they are sometimes unreliable in their prediction of high dose, either more or less. A typical design for an *in vitro* study is to use a wide range (e.g., 1000-fold) of concentrations up to a maximum of 5 mg/plate or 5 mg/mL, using two plates or cultures at each concentration. In animals, it is usual to have an initial phase in a small group that is treated daily, with dose levels that are increased at twice-weekly intervals, until effects are seen that indicate that an MTD has been reached. A second group of previously untreated animals is then dosed for 7 or 14 days at the chosen high dose to ensure that reactions are consistent; this second phase is important in ensuring that tolerance has not built up gradually in the animals receiving the rising dose levels. Having completed a sighting study, it is still possible to obtain unexpected results, either excessive toxicity or none at all when the main or definitive study is started. Although the three-dose-level design is calculated to mitigate the effects of the loss of the high-dose or high-concentration groups, it is better to avoid this as far as possible. Some of the explanations for unexpected results are explored in Focus Box 3.3.

Such short sighting studies are usual only in the early stage of development, or when there is a need to investigate a new formulation or form of the test substance. As a general rule of thumb, it is good practice not to administer more than half of a dose that has caused death in a previous group or study. Having said that, this approach is likely to be conservative in some cases, and it is probably best to use it as a starting point in choosing the new dose level. Factors to take into account are any existing dose response and the types of signs and toxicity seen at the lethal dose.

A program of toxicological investigation in animals normally progresses from sighting studies to 2- or 4-week studies and then on to studies of 13 weeks or longer. At each stage, the results of the previous study are used to select dose levels for the next study. Successful dose level selection, in this case, depends on continuity of the factors discussed above. In other words, do not make radical changes to a formulation during a development program without further sighting studies; use the same strain of test system throughout and so on. It is also wise to go through a toxicological program systematically; data from a 2-week study may well prove unsuitable for selection of dose levels for a 26-week study. In general, animals can tolerate high dose levels of chemicals only for a short period but tolerate lower doses for much longer. Hence, a high dose of 300 mg/kg/day selected for a 4-week study might have decreased to 100 mg/kg/day, when the dose levels are chosen for the chronic toxicity or carcinogenicity studies.

DURATION OF TREATMENT

The process of harmonization of testing requirements for pharmaceutical products—enshrined in the ICH—has arrived at agreed maximum lengths of toxicity study in the conventional species, assuming that they are proven relevant to the target species (usually human) (ICH M3(R2), 2009). In rats, this is 6 months for each of the main

FOCUS BOX 3.3   TROUBLESHOOTING
DOSE RANGE–FINDING RESULTS

Unexpected results may arise from a wide range of factors, including the following:

- *Different age of animals used in the pilot and main studies:* The metabolic capability of the liver in rats increases at about 6 and 8 weeks of age; it is possible that a sighting study in 7-week-old animals will give results different from those in animals of 10 weeks, for instance, if the test substance induces its own metabolism to a toxic metabolite that is not present in younger animals.
- *Different strain or supplier:* Responses vary from one strain to another and can also vary between suppliers.
- *Dose range–finding data from another laboratory:* There is no guarantee that the conditions in the original laboratory will be the same as in the facility chosen for the main study. In particular, husbandry can be significantly different, leading to different absorption profiles.
- *Differences in formulation or form:* Significant differences may arise with changes in formulation; if solubility and absorption of the compound are poor, it is tempting to change the formulation to increase absorption. Changing to an isotonic parenteral formulation can result in very much quicker bioavailability and greater toxicity. Micronizing is often used to enhance bioavailability and can lead to greater toxicity. If a new form or formulation is suggested, a short study should be performed to check consistency with the earlier work.
- *Differences between naive animals and those dosed over several days where an animal is dosed at a low initial level that is gradually increased:* Tolerance to the test compound can develop that is not evident in naive animals. Dosing naive animals from the outset at the chosen MTD can result in unexpectedly high toxicity because they have not had the chance to acclimatize to gradually increasing dose levels.
- *Differences in time of day of dosing:* This is increasingly recognized as a source of differential toxicity or effect and is being investigated both to avoid toxicity and to increase therapeutic effect.

jurisdictions covered: United States, Japan, and European Union. For nonrodents (usually dogs) the guideline gives the maximum duration as 9 months; however, in the European Union, 6-month studies are accepted. Six-month studies are accepted in the United States and Japan in certain circumstances such as when exposure to the drug is intermittent and short term, rather than chronic, or for drugs for the chronic prevention of cancer or when life expectancy is short. Other reasons for shortening studies include development of immunotoxicity or tolerance to the drug.

Demonstration of Test System Exposure—Toxicokinetics

It is a basic tenet of toxicology that effect is dependent on exposure; if there is no exposure, there can be no effect, either directly or indirectly, immediate or delayed. Equally, if it is not possible to demonstrate direct or indirect exposure to a chemical, it is very difficult to assert that an effect is due to that chemical.

For *in vitro* systems, it is usually appropriate to assume exposure of the test system, at least externally, in the case of cell cultures. The presence of precipitate in some genotoxicity assays may still mean that the system has been exposed, although it may not be to the nominal concentration. A flattening of the response curve may indicate that absorption is limited at higher concentrations. In view of the apparent simplicity of *in vitro* exposure assessment, the following discussion is largely limited to the demonstration of exposure in animals.

Exposure may be demonstrated by direct measurement of the chemical itself in an appropriate matrix, usually plasma or serum, sometimes in the urine. For substances that are rapidly and completely metabolized, it may be acceptable to measure a metabolite, particularly if that metabolite is the active moiety. Occasionally, it may be necessary to measure an indirect marker of exposure, for example, plasma glucose levels in response to insulin treatment (although this is a bad example because there are plenty of assay methods for insulin itself and it may be possible to distinguish between insulin given as opposed to endogenous insulin).

The systemic exposure, the internal dose, is dependent on the amount of test substance that is absorbed from the external dose, i.e., the amount of compound administered. Note that the term *internal dose* is also used in radiation to differentiate the radiation dose derived from an internally placed source of radiation as opposed to external sources such as x-rays, gamma radiation, etc. For a substance that is administered orally, the internal dose is a function of the bioavailability of the material from the formulation delivered. As the lumen of the gastrointestinal tract is theoretically exterior to the systemic circulation, the dose is partitioned by absorption into absorbed material (the internal dose) and the nonabsorbed material (the residue of the external dose). Following absorption, the internal dose is further partitioned by the effects of metabolism in the gastric mucosa or (principally) the liver, so that in some cases, only a fraction of the given dose of parent compound will reach the systemic circulation to be transported to other tissues, which may include the target site of action or toxicity.

If the substance is given by injection, then it is usually reasonable to assume that all the material has entered the body. This confident assumption may break down in some cases, however. If an inappropriate formulation is injected intramuscularly, it is possible for some of the material to be lodged at the injection site and not to be released into the systemic circulation, except very slowly. (This may actually be a desired effect if you are studying the kinetics of release from a depot formulation.) Equally, intravenous injection of a test substance as a bolus is usually associated with complete availability of the whole dose, given from the moment the injection stops. If, however, the injection technique is poor and some material is injected extravasally, it is possible for the immediate systemic exposure to be less than the given dose. Local precipitation of compound, as the formulation enters the blood, can have a similar delaying effect. Formulation is critical to the bioavailability even of a parenteral dose; variations in formulation, including physical form of the chemical, can make the difference between severe toxicity and nothing.

This discussion is straying into the realms of toxicokinetic interpretation, which is dealt with in Chapter 7. However, it serves to underline the often-fickle nature of exposure to chemicals by whatever route of exposure. The practical result is that internal exposure is a function of a number of factors, which include route (and sometimes rate) of exposure, formulation, rate, and extent of absorption and metabolism, as well as the dose administered. Additionally, the permeability of the barrier to absorption can change (e.g., via injury or disease) and thus alter uptake. From consideration of these factors, it will become apparent that a single blood sample may show that the animals have been exposed, but it will give no other information than the plasma concentration at the time of sampling and, if several doses are tested, of any proportionality of that concentration to dose. Kinetics of a compound's distribution and elimination from the body are dependent on many factors, of which dose is just one. The critical element in this is time, and so it is necessary to take a number of samples from animals in toxicity studies to show the time course and extent of exposure.

For most toxicity studies, the chemical is given as a bolus dose, and there is a clear point at which the complete dose is inside the animal and from when the collection of samples can be timed and concentrations assessed. This is not necessarily the case when the chemical is given over a prolonged period. In studies using intravenous infusion or inhalation, the chemical is administered at a constant concentration over a number of minutes or hours; it is normal in these cases to start collection of blood samples from the time that administration ceases until the start of the next dosing session or up to 24 hours later. In studies where the substance is given in the diet, particularly in rodents, sampling is complicated by the feeding habits of the animals. As rodents eat at night, it is normal to collect samples at intervals during the dark period; however, switching on the lights may bring about a burst of feeding activity, and this may distort blood levels. In addition, disturbance during the day may precipitate some feeding as a displacement activity. For some studies, it may be useful to invert the light–dark period to allow samples to be collected during the rodent night/toxicologist day.

TOXICOKINETIC DESIGN

Except in certain study types where the animals are killed a few hours after dosing, it is normal to take between six and eight samples from before and close to the time of dosing to 24 hours afterward. A typical sample progression might be predose and 30 minutes and 1, 2, 4, 8, 12, and 24 hours postdose; following a period of repeated dosing, the 24-hour sample may be taken to be equivalent to a predose sample. For intravenous injection, it is normal to take a sample within 5 minutes of completion of dosing; the same principle can be applied to inhalation studies. In early studies, it is often possible to use a sparse sampling regimen, meaning that fewer samples are taken from fewer animals at carefully chosen time points; the danger of such an approach with an unknown chemical is that the samples may be timed before or after peak plasma concentrations have been achieved and the value of the samples is reduced.

It is normal to take samples on the first day of dosing and at the end of the study. The first set gives an indication of kinetics in naive animals, and the second set should show if there has been any increase or decrease in exposure over the treatment period. In chronic studies, it may be useful to take additional samples midway through the treatment period. If the dose levels are changed during the study, additional samples should be taken on the first day of the new dose administration. For chemicals with long half-lives, it may be expedient to take additional samples at 48 hours postdose or later. When a depot formulation is used, samples should be taken at appropriate intervals over the expected lifetime of the implant.

It is also normal to take samples from control animals; this is now a requirement in some jurisdictions and is a check for cross-contamination, which may be seen even if the controls have not been given the test substance directly.

The volume of blood that can be taken from an animal is usually limited by ethical constraints that may be set by local legislation. As an approximate rule of thumb, the laboratory animals routinely used in toxicity testing have blood volumes of approximately 70 mL/kg body weight (Derelanko 2000). Suggested percentages vary according to source; however, a working group in the United Kingdom (BVA/FRAME/RSPCA/UFAW Joint Working Group 1993) suggested that it should be acceptable to take up to 10% of an animal's circulating blood on one occasion. This assumes that the animal is healthy and normal and not too old. Sampling regimens should be designed to take this type of guideline into account and to allow adequate time for recovery, if sampling is to be repeated.

For larger experimental species, including nonhuman primates (but not marmosets), it is normal to take the toxicokinetic samples from the animals in the main study, as they should be large enough to cope with the volumes of blood withdrawn. It is sensible to ensure that samples for clinical pathology are taken before the toxicokinetic samples to avoid artifactual anemias due to excessive blood collection. For smaller animals, principally rodents but also marmosets, it has been normal to include additional animals in the study design, which are used solely for collection of blood for toxicokinetic analyses. In definitive studies, it is normal to use three animals per time point and to sample them at no more than two or three points. As analytical methodology becomes more sensitive, it is possible that low sample volumes (0.3 mL) can be collected from main study animals; this reduces the numbers of animals required for the study and is a useful application of the principles of the three Rs. Care should be taken when using microsampling techniques to ensure that the results of the chosen method are valid for the test compound and are comparable to the traditional methods.

REGULATORY FRAMEWORK AND INFLUENCES

In the second half of the twentieth century, the regulation of toxicity testing increased tremendously in terms of practice and the types and duration of study conducted. Frequently, slightly different study designs were required in the United States, in

Japan, or in Europe; this is true across all forms of regulatory toxicology. This resulted in duplication of tests and greater use of animals, with associated increases in development costs, which were not associated with significantly better quality of safety evaluation. In pharmaceutical regulation, the progress of the ICH process has contributed greatly to the streamlining of testing programs. This has been driven by industry and the regulatory authorities from the three main pharmaceutical markets. The process is likely to continue for the foreseeable future, as our knowledge base develops and as new testing paradigms are produced and debated.

As new or unexpected effects have been detected, especially in the patient or consumer population, new studies or investigative programs have been added to the guidelines. The detection of unexpected effects in the human population, particularly for new drugs, inevitably has a huge influence on the practice of toxicological safety evaluation. In particular, the thalidomide tragedy had far-reaching effects on testing regimens and methods, and the marked growth in regulatory toxicity and greater regulation can be charted from that point onward.

LEGISLATION, GUIDELINES, AND ANIMAL USE

Animal welfare legislation is also an important factor to consider in toxicity testing, in terms of attainment of ideal housing standards and in the prevention or curtailment of suffering. In order to minimize the number of animals needed, it is important that testing is undertaken in healthy, unstressed animals, factors that have been reviewed in Chapter 2. The restraints on excessive use of animals have become stronger over the last 30 years. In the United Kingdom, it is necessary to seek special Home Office approval for the use of nonhuman primates, and the use of great apes is forbidden; furthermore, since 1986, all laboratories, projects, and personnel directly involved in animal experimentation are required to have a Home Office–issued license. Testing is discouraged in cases where the work is a simple repeat of a study, although the testing of drugs that have come "off patent" is a case where tests may be repeated because the original data files are not available to the generic producer. However, in many cases, the new studies are conducted to the higher standards of the most recent legislation. Throughout this, the concept of the three Rs is central, the overall aim being to reduce the use of animals through refinement of investigative programs and, where possible, by introduction of methods that replace them altogether.

The move away from animal experiments for routine toxicological safety evaluation has been slow and will continue to be so while the science is still growing and, in particular, while regulatory acceptance of the alternative methodology is minimal. However, progress is being made. Replacement may be exemplified by the use of *in vitro* screening tests for dermal or ocular irritancy and the replacement of the rabbit in pyrogen testing by use of the *Limulus* amebocyte lysate test for endotoxins. Reduction has been achieved through regulatory acceptance of the local lymph node assay, instead of the guinea pig maximization tests in sensitization. The use of sparse sampling regimens in rodent studies reduces the need for large numbers of satellite animals for toxicokinetic evaluation—an effect that can also be achieved by the use of smaller sample volumes and more sensitive analytical techniques. Refinement of technique includes offering environmental enrichment (particularly important in primates), careful selection of dose levels, and the reduction of pain and stress. In particular, the evaluation of acute toxicity testing toward the evaluation of severe toxicity rather than death, as required by the classic LD_{50} test, has markedly reduced animal use and suffering. The LD_{50} is still a useful concept, however, and may be estimated from the results of early studies and from the single dose toxicity studies that are still required.

LEGISLATIVE INFLUENCES ON THE CONDUCT OF TESTING

The way in which tests are conducted is influenced by a raft of legislation, aiming to protect workers from occupational exposure to hazardous chemicals and, in animal facilities, to allergens and diseases originating from the test system itself. Regulation for the workplace is provided by the requirements of health, safety, and environmental control, which are present in many jurisdictions around the world. These regulations are aimed at protecting the workers involved in the testing or production of new chemicals, and are the object of occupational toxicology. With some new medicines having therapeutic activity in microgram amounts, the protection of the workforce becomes of paramount importance, even if some of the purpose is slightly cynical and aimed at the avoidance of litigation.

REGULATORY CONSERVATISM—THE WAY FORWARD

The inherent conservatism of regulatory authorities with regard to test models or methods is also a factor to consider. The general requirement is to use test systems or models that have been thoroughly validated, so that the data will provide a secure and fully understood basis for interpretation of the data and their significance for humans. This process of validation is complex and is often tied to extensive ring experiments—a series of similar protocols or replicates of the same protocol performed in different laboratories around the world. There is an understandable tendency to use methods that are understood and have been shown to be reliable in scientific literature. In this way, the local lymph node assay has gradually become more accepted in prediction of sensitivity reactions as a replacement for guinea pigs. The use of *in vitro* models in regulatory toxicology is likewise made difficult; at this point, they are clearly acceptable in genotoxicity testing and may become so in some safety pharmacology tests, which can be considered to be a branch of toxicity testing. The Registration, Evaluation, Authorisation and restriction of Chemicals (REACH) process in Europe has added a welcome boost to the efforts to validate and expand *in vitro* methods, as there is explicit guidance on avoiding unnecessary testing in animals (although there is debate on how successful this latter aim has been).

PUBLIC PERCEPTIONS

A further, and important, influence is provided by public and hence political pressure. This is rarely influenced by complete appreciation of the scientific data relating to a set of circumstances and is often based on a part set of the data. Partial understanding of a complete data set or complete understanding of a partial data set is unlikely to lead to a satisfactory understanding of the mechanisms, leading to a particular set of effects or variation from normal, however normality is defined. Politics and science are often poor bedfellows, and the use of science in political judgments, or vice versa, has to be carefully assessed according to the circumstances of the situation. The results may be overreaction and production of a set of regulations that can have worse consequences than the status quo.

REGULATION OF STUDY CONDUCT— GOOD LABORATORY PRACTICE

GLP grew from the experience of the Food and Drug Administration (FDA) in the United States in the early 1970s, when it was discovered that a number of irregularities had been perpetrated in undertaking studies for new medicines. Inspection of submissions for pharmaceutical registration became more detailed following the thalidomide disaster of the early 1960s; this may be seen as the origin of many of the regulatory influences that affect the development and testing of chemicals of all types and classes. The FDA found many problems in submissions, most of which were unintentional deficiencies, inaccuracies, or simply ill-informed bad practice; however, a proportion were found to be due to intentional fraud.

In an investigation of study quality, it was revealed that experiments were poorly conceived, badly performed, and inaccurately interpreted or reported. The importance of the protocol was not understood, nor the need to keep to it. Study personnel were not aware that administration of test substances and subsequent observations should be performed and recorded accurately. Managerial deficiencies were found; the study designs were often poor, which hindered evaluation of the data. Training and experience of the people involved in the conduct of studies were not assured. The surgical removal of tumors from animals and putting them back on study is the classic example of the type of practice that was loudly vilified.

Problems were found in two companies, Industrial Bio-Test and Biometric Testing Inc., which were so severe that 36% of studies from the former laboratory were invalid. Looking at the same laboratory, the US Environmental Protection Agency found that 75% of the studies were invalid. Consequently, a draft GLP regulation was published in 1976 and finalized in 1979. This has spread across the globe, becoming internationally acknowledged with the publication of the Principles of Good Laboratory Practice by the Organization for Economic Co-operation and Development (OECD), which was revised in 1997.

In summary, the GLP regulations were set up to increase accountability and prevent fraud in the safety evaluation of drugs; they have since been adopted globally and are applied to all toxicological testing performed for regulatory submission. They define the requirements that a testing facility must fulfill in order to produce studies that are acceptable to the various authorities, and the term *GLP* has become pervasive, as an adjective attached to facilities or studies.

The following is a brief review of the main points of GLP; for more detailed information, the reader is referred to the various GLPs published by organizations such as the OECD (1998), FDA (2007), and UK GLP Monitoring Authority (Department of Health 2000). This is based on the UK Department of Health Guide to UK GLP Regulations

1999, which are very similar to European and OECD guidelines. The account here is necessarily brief and so does not reproduce the stirring text of the original.

THE BASIC ELEMENTS OF GLP

The following are the basic requirements of GLP:

1. Facility management: responsible for setting up the Quality Assurance Unit (QAU) and ensuring that the basic requirements for GLP are in place.
2. A QAU that is independent of study conduct.
3. A complete set of standard operating procedures (SOPs) that describe, in simple language, how each task or procedure is performed.
4. Facilities, equipment (including computers and their software), and reagents that are appropriate.
5. A designated study director and personnel who have been trained appropriately for their various roles; training records should be maintained and updated regularly.
6. A study plan or protocol for each study.
7. Characterization of the test item and test system.
8. Raw data generated in the course of GLP studies.
9. A study report.
10. An archive.

The phrase "fit for purpose" is used frequently in GLP and covers every aspect of GLP inception, management, and study conduct. Facilities should be suitable for the purpose for which they are intended. In other words, laboratories or animal facilities have to be appropriately designed and maintained; storage facilities for test items must be effective and working to specifications. This applies equally to the personnel performing the study; if they are not appropriately trained, they are not fit for that purpose. However, this overarching concept does not mean that facilities should be overengineered or procedures carried out by overqualified personnel.

These basic elements come together, so that the quality of any GLP-compliant study is reasonably assured. While occasional problems emerge, the likelihood is much reduced compared to the era before GLP; in particular, the system of inspection and penalties for infringement of the regulations ensure effective enforcement of the regulations.

FACILITY MANAGEMENT

Management is pivotal in the implementation and management of GLP. Their responsibilities may be summarized as follows:

- Test facility management must be identified.
- They must ensure that there are sufficient numbers of qualified personnel to conduct studies and associated tasks and procedures.

- They must provide appropriate facilities with properly maintained equipment; they must ensure that computer systems and their software are validated.
- They ensure that records of the qualifications, training, experience, and job description are maintained for all staff involved in the conduct of GLP studies and that these people understand their duties.
- Management is also responsible for setting up a QAU staffed by personnel who report directly to them and are independent of study management. An archivist must be appointed and an archive designated.
- Management must appoint an appropriately qualified study director for each study, who signs the protocol and makes it available to the QAU.
- A principal investigator should be appointed by management for parts of studies that are conducted at different sites, for example, analysis of blood samples for toxicokinetics.
- They should also maintain a master schedule of all the studies that have been conducted, are in progress, or are planned.

QUALITY ASSURANCE

The QAU reports directly to the facility management and is completely independent of study management; it has no practical role in studies. Its roles are as follows:

- To keep copies of all protocols and amendments
- To review the study protocol
- To ensure compliance with GLP by carrying out a program of inspections of study procedures and facilities
- To audit the final report and sign a Quality Assurance (QA) statement for it, confirming that the report accurately reflects the protocol, the methods used, and the data collected
- To report all QA activities directly to management and the study director

A critical aspect of the QAU role is the inspection of critical procedures on GLP studies. These are activities such as colony counting in Ames tests, cell culture in genotoxicity studies, or recording of body weights in a long-term toxicity study. For long studies, a program of inspections is planned, which may include any aspect of study conduct; for short studies, it is normal to inspect procedures from a sample of studies rather than to look at each one individually.

STANDARD OPERATING PROCEDURES

SOPs serve as guidelines for the conduct of good science. The language should be clear and concise while containing enough detail to complete the procedure. They form the basis for conducting a study or maintaining a facility but for procedures common to most studies or facilities. SOPs should be reviewed and updated regularly; furthermore, it is essential that facilities' staff are aware of them and follow them. Where the protocol goes into more detail than the SOP or differs in some respect, it is the protocol that takes precedence. SOPs should be written to cover all of the following areas:

- Test and reference items
- Test system
- Apparatus, materials, and reagents
- Computer systems and validation of software
- Study procedures
- Documents, data, and records
- QA

FACILITIES AND EQUIPMENT

Facilities and equipment must be fit for purpose. Facilities should be of suitable design and build, suitably located, and large enough for the purpose. For equipment, there should be SOPs and records to cover all aspects of maintenance and use, including operator training. Equipment should be appropriate for the purpose; a rat should not be weighed on a balance intended for weighing dogs as it is unlikely to be sensitive enough, quite apart from the issue of cross-contamination between species. Equipment may also be seen to include reagents, which should have similar standards of preparation and labeling, and, importantly, a valid expiry date. Computer software should be validated, especially if used for collection or manipulation of data.

THE STUDY DIRECTOR AND STUDY PERSONNEL

The study director is the single point of control for the study and the final report. He or she must

- Approve, sign, and distribute the protocol and any amendments to study personnel, ensuring that the QAU also receives copies.
- Define the roles and identity of any principal investigator at other sites where subsidiary investigations such as bioanalysis are conducted.
- Ensure that the study plan is followed and assess, document, and take appropriate action on any deviations either from the protocol or from SOPs.
- Ensure that all data are fully documented for the validation of any computer software.
- Sign and date the final report and arrange for the data and the final report to be archived. The study director's signature indicates the validity of the report and compliance with GLP.

Study personnel involved should be adequately trained in GLP and the procedures they are expected to perform. They must have the protocol, amendments, and relevant SOPs, and tell the study director promptly of any deviation from any of them. They must record all data promptly and in the proper place. All study personnel should have a training record; people who have not been trained in a particular procedure should not be allowed to perform it unless they are being supervised and trained.

THE STUDY PLAN OR PROTOCOL

These two terms tend to be used interchangeably, although the former tends to be used in the GLP guidance and the latter to be used in industry. The basic GLP practice is that there should be one protocol and one study number per study. The protocol should include the following information:

- Study title and number
- Identity of the test and any reference item
- Name and address of the sponsor and test facility
- The study director's name and any principal investigators
- Dates for approval, experimental start, and finish
- Test methods
- The test system and justification for its use
- Dose levels or concentrations and duration(s) of treatment
- Timings of study activities
- A list of records to be retained, where, and for how long

Any intended change to the protocol should be covered by a protocol amendment; this details any change made after the study director has signed the protocol, which is taken as the initiation date of the study. An amendment cannot be retrospective; where there is a deviation from the protocol that is unplanned, this should be recorded separately in a study or file note. The temptation is to produce a study note for every small deviation from the protocol; however, common sense says that some deviations, such as a lower temperature than target or higher humidity, are better reported in the final report.

TEST ITEM AND TEST SYSTEM

The characteristics of the test item—batch number, purity and concentration, stability, and homogeneity in formulations—must be given in the protocol. The test item has to be labeled and stored appropriately, and all occasions of its use must be recorded so that the quantities used can be reconciled with the amount of material remaining at the end of the study or series of studies. The same applies to any reference item that may be used in the study.

In a similar manner, it is important that the test system be correctly characterized and identified. A beagle dog is clearly a beagle; it is more difficult to distinguish between two strains of rat. There should be confirmation that transgenic mice have the appropriate genotype and that strains of bacteria or mammalian cell lines used in *in vitro* studies have the appropriate characteristics that are necessary for the intended studies. Details of the test system should include

- Source, species, strain, substrain, number, batch, and, if given, age and weight
- Details of storage or husbandry
- Details of order and receipt

Raw Data

This is the basis of GLP, which the FDA has defined as follows:

> ...any laboratory worksheets, records, memoranda, notes, or exact copies thereof, that are the result of original observations and activities of a nonclinical laboratory study and are necessary for the reconstruction and evaluation of the report of that study. In the event that exact transcripts of raw data have been prepared (e.g., tapes which have been transcribed verbatim, dated, and verified accurate by signature), the exact copy or exact transcript may be substituted for the original source as raw data. Raw data may include photographs, microfilm or microfiche copies, computer printouts, magnetic media, including dictated observations, and recorded data from automated instruments. (FDA 2007)

Measurements, observations, or activities recorded in laboratory notebooks, used in research settings, including academia, should conform to the same standards as those for regulatory studies.

- Raw data should be recorded promptly and permanently and have the unique study identification code.
- Records must be signed and dated by the person making the record; any changes should not obscure the original and should be signed, dated, explained, and auditable.
- Copies should be verified by date and signature.
- The study director is responsible for the raw data, which should be transferred as soon as possible to the study files to avoid losses.

The Study Report

There should be one report for each study, although it is acceptable to produce interim reports for longer studies. The final report should reflect the raw data and the methods used. The overall intention is to facilitate the reconstruction of the study, if necessary. As with most aspects of study conduct, the final responsibility rests with the study director.

The study report should

- Identify the study, test, and reference items
- Identify the sponsor and test facility
- Give dates of start, experimental phase, and completion
- Contain a GLP statement from the study director and a record of QA inspections
- Describe the methods used and refer to any guidelines followed
- Describe the results, including an evaluation and discussion of any findings together with a conclusion
- Include individual data
- Give the location of any archive used for raw data

Changes to the report after final signature are made by means of a report amendment, in a broadly analogous way to protocol amendments.

OTHER ASPECTS OF GLP

There should be an archive where data may be securely stored and from which data may be removed only on signature of a designated management nominee. The data should be protected from theft, vermin, fire, and water. These criteria may be satisfied by a purpose-built building or by an appropriate cupboard, depending on the size of the test facility.

Regulation of GLPs is carried out by national authorities such as the FDA in the United States, the Japanese Ministry of Health and Welfare (JMHW) in Japan, or the UK GLP Monitoring Authority. There may be severe penalties for deliberately not following GLP, and the history of toxicological testing is dotted with fines and jail sentences of offenders.

SUMMARY

Toxicity testing, as opposed to research, is mostly conducted as part of safety evaluations and/or development programs for new chemicals, such as new drugs, pesticides, food additives, and cosmetics. Typically, the complexity and duration of the studies in animals increases as a program progresses.

- The regulatory context of development of the various classes of chemicals may differ according to expectation or differing circumstances of exposure, but the basic elements of a control group plus several treated groups is similar across all disciplines and experimental types, whether *in vitro* or *in vivo*.
- Every experiment should have a study plan with an objective, which should define the design of the study with respect to aspects such as duration, treatment, carrying system or solvents, investigations, reporting and archiving, and any guidelines with which the study is intended to be consistent, particularly if it is intended to be compliant with GLPs.
- Where appropriate, usually in animal studies, there should be demonstration of exposure through a toxicokinetic evaluation of the result of blood analyses from the test system.
- GLP is the regulatory framework that ensures the integrity of study conduct, and adherence to them is a fundamental requirement for the majority of studies submitted to authorities around the world for registration of all the main groups of chemicals.

REFERENCES

BVA/FRAME/RSPCA/UFAW. Joint Working Group: First report on refinement. Removal of blood from laboratory mammals and birds. *Lab Anim* 1993; 27: 1–22.

Cordier A, Amyes SJ, Woolley APAH. Correlation between fighting lesions and subcutaneous sarcoma in male B6C3F1 mice. Proceedings of the 29th Annual Meeting of the Society of Toxicology. Toxicologist 1990.

Department of Health. The United Kingdom Good Laboratory Practice Monitoring Authority: Guide to UK GLP Regulations, 1999. February 2000.

Derelanko MJ. *Toxicologist's Pocket Handbook*. Boca Raton, FL: CRC Press, 2000.

FDA: Code of Federal Regulations: Part 58 Good Laboratory Practice for Nonclinical Laboratory Studies. Title 21, Volume 1 (21CFR58.3) Revised 2007.

ICH Guidance. M3(R2) Nonclinical Safety Studies for the Conduct of Human Clinical Trials and Marketing Authorization for Pharmaceuticals, 2009.

OECD. OECD Principles on Good Laboratory Practice. Series on Principles of Good Laboratory Practice and Compliance Monitoring, Number 1 (as revised in 1997). January 1998.

4 Determination of Toxicity In Vitro *and Alternatives*

INTRODUCTION

There is considerable pressure to reduce and replace the use of animals in toxicological research or testing, for scientific, economic, and political reasons. This is driven by the three Rs put forward by Russell and Burch in 1959; these are replacement, reduction, and refinement, the aims of this being to replace animals in experiments, to reduce their use, and to refine experimental technique and protocol. The overall goal is to reduce the use of animals in a scientific manner, while ensuring product safety, i.e., to refine toxicological investigation so that human-relevant systems can be used for secure prediction of human hazard. There has been a large amount of success in these broad aims, and while some endpoints remain elusive, others have proved easier to be achieved *in vitro*.

The expansion of *in vitro* techniques continues as they become more refined and reliable and as understanding of their meaning and utility grows. Since the second edition of this book, the importance of alternative methods has grown hugely to the extent that strategies are becoming apparent by which they may be employed in safety evaluation as part of a three-part approach to the assessment of safety. A future in which toxicity is assessed *in silico*, *in vitro*, and finally *in vivo* is becoming more viable. However, it has to be said that regulatory acceptance is low at the moment; but it will grow. It may justly be said that the traditional approach to safety evaluation—the bread and butter of the practical toxicologist in industry—has been a shotgun fired in the hope that one of the pellets would hit a toxicologically significant endpoint and show toxicity that can be assessed. At least if toxicity in an endpoint has been shown in animals, it can be discussed and, with luck, dismissed as not human relevant; the absence of toxicity—illusory safety—cannot be dismissed in the same way.

In vitro toxicology, in the sense of replacing animals completely, is a Holy Grail for some people. Despite the considerable advances in technique and understanding in recent years, it is likely to remain so: an unattainably elusive goal for which the goal posts move as you get closer to them. *In vitro* literally means "in glass" (interestingly, my dictionary of toxicology does not attempt a definition, although the *Concise Oxford English Dictionary* indicates that it takes place in a test tube or culture flask, or outside a living organism). *In vivo* refers to the living, complete organism. These definitions are neat and offer black-and-white alternatives. The problem for the Grail hunters is that many *in vitro* techniques require the sacrifice of an animal to generate the test system; the numbers of animals used in this way can be considerable. A test involving primary cultures of hepatocytes or liver slices inevitably involves an animal to provide the cells or slices, although human lymphocytes can be obtained from volunteers, if need be on a regular basis. A refinement usually referred to as *ex vivo* involves treating a complete living animal waiting for an appropriate interval and then killing it and conducting the tests. This approach is used very successfully in tests such as the comet assay and unscheduled DNA synthesis (UDS); there are clear advantages in such an approach as the first part of the test is conducted in a whole animal with undisrupted metabolism, blood circulation, and dynamic relationships between organs and tissue systems. The major problem with *ex vivo* methods is that when they use an animal based-model rather than one that is human derived or humanized, there is still the gulf of extrapolation from an animal to a human situation, only it has been widened by the isolation of the cells from the complete animal. This complicates extrapolation to a complete human (or whatever target is being considered). The logical step would appear to be to develop test systems that are humanized to reduce the errors of extrapolation from animal-derived systems.

Methods traditionally thought of as alternative are those that have been developed to refine existing methods, reducing the number of animals used and replacing them where appropriate or possible. Probably the best example is the local lymph node assay (LLNA), which is taking over from sensitization testing in guinea pigs. This is based on the good predictivity with the new method and the fact that it is quantitative rather than subjective. While this still uses animals and it may not be seen as an alternative to animals, *in vitro* systems for the assessment of dermal and ocular irritation are truly alternative and have to overcome serious hurdles before creeping toward wider acceptance.

With the inception of the legislation relating to the European chemicals initiative Registration, Evaluation, Authorisation and restriction of Chemicals (REACH) and the 7th Amendment to the Cosmetics Directive, it was expected that the field of *in vitro* toxicology would be boosted as pressure was applied to reduce testing using animals. The various texts of guidance for REACH have made great play of demanding extensive toxicology data while repeating the point that animal testing should be minimized. While the guidance may indicate that use of animals is undesirable, actual practice has not necessarily had this effect. In the case of the Cosmetics Directive, the testing of completed products in animals and the use of new ingredients that have been tested in animals have been banned in Europe. As practically all ingredients accepted for use in cosmetics before the legislation have been tested in animals at some point, this situation may become uncomfortable for some in the future. Furthermore, it raises the issue of potential conflict between markets that require animal testing and those that have prohibited it.

The Organization for Economic Co-operation and Development (OECD) has published many safety test guidelines over the years and is a useful source of information on which tests have had guidelines published and, equally interesting, those that have been deleted, illustrating the development of *in vitro* testing and validation of new methods (Table 4.1).

TABLE 4.1

OECD Guidelines Not Using Animals—Current at 2016

Test Guideline	Title	Adopted	Revised
428	Skin Absorption: In Vitro Method	2004	
430	In Vitro Skin Corrosion: Transcutaneous Electrical Resistance Test (TER)	2004	2013
431	In Vitro Skin Corrosion: Human Skin Model Test	2004	2014
432	In Vitro 3T3 NRU Phototoxicity Test	2004	
435	In Vitro Membrane Barrier Test Method for Skin Corrosion	2006	
437	Bovine Corneal Opacity and Permeability Test Method for Identifying Ocular Corrosives and Severe Irritants	2009	2013
438	Isolated Chicken Eye Test Method for Identifying Ocular Corrosives and Severe Irritants	2009	2013
439	In Vitro Skin Irritation: Reconstructed Human Epidermis Test Method	2010	2013
455	Performance-Based Test Guideline for Stably Transfected Transactivation In Vitro Assays to Detect Estrogen Receptor Agonists	2009	2012
456	H295R Steroidogenesis Assay	2011	
457	BG1Luc Estrogen Receptor Transactivation In Vitro Assay to Detect Estrogen Receptor Agonists and Antagonists	2012	
460	Fluorescein Leakage Test Method for Identifying Ocular Corrosives and Severe Irritants	2012	
471	Bacterial Reverse Mutation Test	1983	1997
473	In Vitro Mammalian Chromosome Aberration Test	1983	2014
476	In Vitro Mammalian Cell Gene Mutation Test	1984	1997

Source: Adapted from OECD Guidelines for Testing of Chemicals: Full List of Test Guidelines, September 2014. www.oecd.org.

In addition, a draft guideline for the Syrian Hamster Embryo (SHE) Cell Transformation Assay was published in 2013. Deleted guidelines include the following:

- 401: Acute Oral Toxicity (deleted 2001)
- 472: Genetic Toxicology: *Escherichia coli*, Reverse Assay (merged with 471, 1997)
- 477: Genetic Toxicology: Sex-Linked Recessive Lethal Test in *Drosophila melanogaster* (deleted 2014)
- 479: Genetic Toxicology: In Vitro Sister Chromatid Exchange Assay in Mammalian Cells (deleted 2014)
- 480: Genetic Toxicology: *Saccharomyces cerevisiae*, Gene Mutation Assay (deleted 2014)

- 481: Genetic Toxicology: *Saccharomyces cerevisiae*, Mitotic Recombination Assay (deleted 2014)
- 482: Genetic Toxicology: DNA Damage and Repair/Unscheduled DNA Synthesis in Mammalian Cells In Vitro (deleted 2014)
- 484: Genetic Toxicology: Mouse Spot Test (deleted 2014)

The following is a broad review of *in vitro* technique, strengths, weaknesses, and future potential. Inevitably, many of the areas covered in this chapter are relevant to those in others, and there will be some overlap; in general, the more comprehensive coverage will be in the main chapter.

RATIONALE FOR *IN VITRO* TOXICOLOGY

Table 4.2 gives a comparison of *in vitro* versus *in vivo*. In essence, a single *in vitro* test will tend to answer a simple question. While a battery of appropriately chosen

---

TABLE 4.2
In Vivo* versus *In Vitro

In Vivo	*In Vitro*
Whole-body responses.	Responses of single tissue, cell, or cellular organelle.
Many endpoints in one test.	Usually only one endpoint.
Can answer complex questions.	Can only answer simple questions.
Metabolism is built in; separate studies on metabolites are usually not needed.	Metabolism systems have to be added and metabolites tested separately.
Flexibility of route of administration, dosing schedule, and duration.	Exposure via presence in culture medium (except in perfused organs). Limited life span for the test system.
Use of different species as needed.	Test system choice more limited.
Ease of interpretation as a result of knowledge of dose (mg/kg) and systemic exposure.	Interpretation is more difficult without information on expected or actual *in vivo* concentrations.
Comparability with man; biochemical and physiological processes are similar across species and similar mechanisms of toxicity.	Extrapolation to man from a single cell or subcellular organelle of a laboratory animal is much more difficult as the result must first be extrapolated to the complete animal.
Predictive; repeat-dose animal studies have been shown to predict 71% of human toxicities.	Patchy predictivity; lack of whole-body response limit use in mainstream safety evaluations but potentially good for single endpoints (e.g., QT interval prolongation).
Experience of use; huge repository of expertise background data and published papers.	Limited experience with some systems; possibly due to inconsistent protocols, test system derivation, etc.
Generally technically undemanding.	Can be very difficult technically; reproducibility questionable for unvalidated tests.

Source: Compiled with the assistance of Snodin DJ, An EU perspective on the use of in vitro methods in regulatory pharmaceutical toxicology, *Toxicol Lett*, 127:161–168, 2002.

tests can answer a broader range of questions, such questions may be answered more easily by a simple study in animals.

In vitro systems have many strengths but also weaknesses, which have hindered their acceptance by regulatory authorities—especially when used to answer general questions. Their strength lies in the simplicity of their endpoints and their consequent use in the evaluation of toxicological mechanisms. Their principal weakness has been that they are not readily able to respond to chronic exposure and they cannot easily replicate the interrelationships that exist in the body between the various organs and tissues. In particular, they cannot (currently) give any indication of an effect that will accumulate over prolonged administration, such as progressive renal failure or neurological and psychological changes. However, they have a considerable role to play in lead candidate selection and in the evaluation of substance groups, such as cosmetics, where *in vivo* techniques may not be used. With technologies such as combinatorial chemistry, there are increasingly large numbers of chemicals to screen for toxicity or efficacy to aid selection of lead candidates for development. Traditional methods are too slow and costly for this. For a series of compounds for which a particular mechanism of toxicity has been identified, this may be investigated *in vitro* and the least toxic compounds selected for development. The objectives of screening include the following:

- Ranking a series of compounds in terms of effect and selecting one for development
- Mechanistic investigations
- Examination for specific effects, for instance, assessment of gap junction patency
- The early discarding of compounds that would fail later in development

Screening of members of a compound series for specific activity or adverse effects is critically important in some cases. For instance, retinoids may be screened for embryotoxicity. The degree to which compounds in a series express an effect may be due to structure–activity relationships. In such cases, it may be possible to conduct an initial computer-based screen for the presence in the structures either of a specific set of parameters (for instance, bond angle or interatomic distances) or for chemical groups or structures that may be related to a specific effect. Such computer-assisted design techniques have utility in both toxicology and pharmacology.

In many areas of chemical development, especially for pharmaceuticals, failure of compounds may occur after considerable expenditure of animals, time, and money. Reasons for failure include formulation problems, toxicity, poor efficacy, and poor acceptability (dermal formulations should be neutral in color—not bright yellow). Clearly, some of these are not toxicological in nature, but commercial pressures to develop a compound may result in the best of a bad series being selected. Reasoned application of *in vitro* techniques can bring forward the stage at which compounds are rejected. However, it must be remembered that very few useful compounds are without some form of toxicity, and it is likely that the overeager application of such early-discard techniques or screens may result in the rejection of compounds that would have been successful.

HOW AND WHEN TO USE *IN VITRO*
OR ALTERNATIVE TECHNIQUES

As indicated in the previous sections, *in vitro* has enormous value as a screening process, which can usually be completed quickly and inexpensively. Although the subsequent testing of the chosen compound may have to be done in animals, the existence of *in vitro* screening programs diminishes their use, contributing to the achievement of the three Rs. On this basis, it may be helpful to consider that many *in vitro* techniques act as additional methods rather than complete replacements for the more traditional tests.

One of the themes of this book is the differences in philosophy, testing protocol, and regulation between the various classes of chemicals. For example, animals cannot be used for testing cosmetic ingredients or products, meaning that the only tests that can be performed are *in vitro* or in human volunteers. The other extreme is seen with pharmaceuticals, where the vast majority of testing, which is undertaken to support clinical studies in man, is conducted in animals. Some areas of safety pharmacology are conducted *in vitro* (for instance, the hERG assay for the prediction of QT interval prolongation), and the core battery of genotoxicity tests contains three *in vitro* tests and one *in vivo* (a major reduction in animal usage has been due to the increasingly widespread incorporation of the *in vivo* genotoxicity test into routine 28-day toxicity studies in rats). The development of the hERG assay is one example of how a better *in vitro* technique can replace another. Previously, the usual *in vitro* assay for assessment of the potential to prolong QT interval was the papillary muscle of the guinea pig or the Purkinje fiber of the dog; it is probably relevant to question the ethics of such tests given the increasing use of the hERG assay and of *in vivo* telemetry. The hERG assay uses a stable cell line, is reproducible, and, although technically demanding, can be conducted relatively inexpensively and quickly. *In vivo* telemetry allows the animal to be used again as it does not need to be killed to achieve the experimental objective.

For agrochemicals and chemicals that fall under the REACH legislation, the choice drivers will be different again.

Before embarking on a program of *in vitro* studies, it is important to define the objectives. Apart from anything else, is the decision to use *in vitro* an appropriate one?

Is the intention to use a single test routinely to screen individual compounds for a particular effect, or is it to look at a series of compounds with the objective of choosing which one to develop? Whole-embryo culture was popular at one time as a screen for potential teratogenic effects but has fallen out of favor. In some cases, it was used to bring forward the reproductive toxicity studies in a development program; however, in at least one company, after reviewing the rationale and what the reaction would be to the answer, it was concluded that the test did not make a useful contribution to the decision processes, and it was dropped. This comes back to the statement made in Chapter 1; if you do not know what you will do with the answer, do not ask the question. On the other hand, whole-embryo culture or related assays may well be useful in screening chemical series, such as retinoid derivatives, and ranking them for development.

It is probably more important to avoid any question where the answer is not understood; in this case, the question should be reframed so that a system can be devised that gives meaningful answers. An example of this type of test is the SHE Cell Transformation Assay, for which a new draft OECD guideline was produced in 2013. Historically, this test was known to give results that correlate with carcinogenicity, although the reasons for this were not fully understood. The guideline makes the point that the results of this assay should be used as part of a testing strategy, in a weight-of-evidence approach, and not as a stand-alone test.

CONSIDERATIONS IN SCREENING PROGRAM DESIGN

Given this important caveat about understanding the results of a test, a test or screening program should be designed with care; for programs where a sequence of tests is conducted, it may be useful to set up a flowchart to give a set of predefined limits to the process. It is a hazard of any screening process that good compounds will be rejected from development because of adverse results in ill-chosen or ill-applied tests that are not truly predictive of effects in the target species. Much comparative metabolism work *in vitro*, for instance, is conducted in isolated hepatocytes from various species, including humans, the objective being to select a second species for toxicity testing. The human hepatocytes are derived from cadavers or liver surgery, and are typically a mixture from several individuals, who may or may not be defined in terms of lifestyle (smoking or alcohol consumption), disease, or genetic metabolic polymorphism. It is unlikely that such a system will produce consistent results from one test to another; on this basis, there is considerable sense in developing a metabolically competent line of human hepatocytes in a secondary (immortal) culture. This would have the advantage of being defined metabolically and being reproducible. The development of several lines to cover the major polymorphisms would significantly extend the utility of such a test. In addition, when fully validated, it is likely that this would gradually become accepted by regulators by default.

Unless the endpoint of concern is absolutely critical to the success of a compound and it is well established that an *in vitro* assay is reliably predictive of that endpoint, it is unwise to rely on a single test to screen a series of compounds. Equally, failure in a single test should not necessarily bar a compound from further development; any flowchart should take account of this possibility. At the end of the process, which may include short *in vivo* tests for a selection of compounds from the series, a decision may be reached by weighing all the results and striking a balance between desired activity and predicted toxicity.

AREAS OF USE OF *IN VITRO* TOXICOLOGY

There is a vast range of different *in vitro* assays, and they are broadly divisible into those tests that are accepted by regulatory authorities and those that are not, except that this simplistic division is complicated by the acceptance of some tests by some regulators, but not by others. In general, the most difficult set of regulators to satisfy as to a test's acceptability are those overseeing pharmaceutical development and registration; this assessment is based on the relatively few *in vitro* tests that are accepted as part of a pharmaceutical safety package of data. Tests that have been accepted in pharmaceutical development, or whose use is mandated by authorities, include genotoxicity assays in bacteria and cultured mammalian cells, the hERG assay in safety pharmacology, and various isolated heart preparations, including perfused rabbit hearts, guinea pig papillary muscle, and Purkinje fibers from dogs.

The testing of cosmetic ingredients and products, at least for those to be marketed in the European Union, is now exclusively *in vitro* due to the 7th Amendment to the European Cosmetics Directive. Vinardell (2015) wrote a review on the use of *in vitro* techniques in submissions to the EU authorities, which provides a useful overview of the use of alternative methods. The testing of cosmetics is covered in more detail in Chapter 19.

The European chemicals legislation, REACH, was partly conceived in the expectation that sharing data would reduce the numbers of animals used in testing and that the use of animals should be avoided. However, this lofty principle has foundered on the innate conservatism of regulators, and the use of animals has not reduced and may actually have increased as new tests are demanded. Despite this, assays for endpoints such as skin corrosion and eye irritation are increasingly important. In addition, the use of *in silico* techniques and computer software for the prediction of toxicity will also increase.

CONSIDERATIONS IN THE DEVELOPMENT OF *IN VITRO* TESTS

The concept of using *in vitro* techniques rather than animals has many attractions, not least speed and cost. However, they also carry a number of significant disadvantages that limit their routine use—especially for complex endpoints such as repeat-dose toxicity and reproductive function and developmental toxicity. Current test systems are limited to single cell type or organ (tissue slice or whole-organ perfusion) and can be technically demanding and, as a result, difficult to reproduce between laboratories. Replication of *in vivo* exposure conditions may be difficult, especially for lipophilic or insoluble chemicals. Furthermore, because the systems have a limited life expectancy, they are not suitable for the examination of chronic effects, especially those that accumulate gradually over prolonged exposure. Many toxicities are multifactorial and dependent on interrelationships between tissues/organs that are not readily reproducible *in vitro*.

The pursuit of *in vitro* techniques should be directed at development of test methods that are acceptable to regulatory authorities as replacements for inferior tests, and to focus attention on identifying toxicities *in vitro* that may be expected *in vivo*. For example, an *in vitro* screen may suggest potential for renal toxicity indicating the need for specific examinations in animal studies and, perhaps, influencing dose selection so as to minimize any suffering in test animals. The use of *in vitro* techniques can—potentially—replace "severe" procedures such as eye irritation. Successful development of these methods will lead to reductions in animal use, through gradual regulatory acceptance and reduction of the numbers of compounds entering full development. Financial benefits should not be a sole reason for replacing animals but should not be ignored as a factor; rather, the impetus for this process should be to do it better.

It would be difficult to underestimate the influence of legislation of the development of *in vitro* techniques, the European Cosmetics Directive and REACH being prime examples. A contrary viewpoint is given in other areas of use, such as pharmaceutical development, where lack of regulatory acceptance and, it has to be said, satisfactory tests have delayed the wider adoption of some tests for certain endpoints, particularly those that are complex or multifactorial.

A further boost for development of new approaches to safety evaluation has been given by the European Partnership for Alternative Approaches to Animal Testing, which was launched in November 2005. This is a joint initiative from the European Commission and a number of companies and trade federations that are active in various industrial sectors to promote the development of new methods that uphold the three Rs, as modern alternative approaches to safety testing. The Partnership's work focuses on a number of areas, including existing research; development of new approaches and strategies; and promotion of communication, education, validation, and acceptance of alternative approaches.

Broadly, the criteria for development of an alternative or *in vitro* method include the following:

- The mechanism by which changes are induced should be well understood.
- The test system should be a single target organ and/or cell type.
- The compound should be water soluble, as lipophilic compounds may not gain access to the test system.
- There should be a large set of existing data that can be used in the interpretation and validation of the new test.

To be successful, a new toxicity screen should be

- Robust: The test should be relatively easy in technical terms; complication leads to error, and specialist equipment means expense. New animal models should not have overonerous husbandry requirements.
- Readily transferable between laboratories.
- Understood: There is little sense in producing data unless the mechanism of their generation and their significance is well understood.
- Reproducible: If not, its utility and relevance may be questioned. Baseline data for individual animals, plates, or replicates should not be so variable that change is indistinguishable from historical control data.
- Predictive: With good sensitivity and specificity.
- Quick: Lengthy experimental phases mean slowed development or lead candidate selection and additional expense.
- Cost effective: There is no future for any test if the costs outweigh the value of the results.

Solvents should be chosen carefully, taking into account their potential to react with the test chemicals. The best option is water, but this may not be practicable for lipid-soluble compounds.

The usual refuge in such cases, DMSO, has its own toxicities and should probably be used with caution as it has been shown to react with some classes of chemical, such as carboxylic acid halides. These were positive in bacterial reversion tests due to reaction with DMSO to form dimethylsulfide halides. These are alkylating agents and sufficiently stable to produce positive results; when tested in water, the results were negative (Amberg et al. 2015).

VALIDATION OF *IN VITRO* METHODS

Validation is a perennial problem with *in vitro* systems as it is often a lengthy process, the LLNA taking 16 years to validate. Having said that, many would say that animal tests have not been properly validated. However, they are at least well understood and produce data that may be extrapolated to humans or another target species, which cannot be said so readily for *in vitro* data. Validation seeks to answer questions relating to the reliability and relevance of the method under evaluation. The European Centre for the Validation of Alternative Methods (ECVAM) and, in the United States, the Johns Hopkins Center for Alternatives to Animal Testing (CAAT) and the Interagency Coordinating Committee on the Validation of Alternative Methods (ICCVAM) have done a lot of work on the inception and validation of new methods. Validation should seek to answer questions of the following type:

- Is the biochemistry and mechanism *in vivo* and in the proposed *in vitro* system fully understood?
- Is the *in vitro* method reliably predictive for the endpoint *in vivo*?
- Where a surrogate marker of effect is used, what is the reliability of this marker of effect *in vivo*?
- What are the robustness, reproducibility, and reliability of the method, including among laboratories?
- Is the method sensitive and selective?
- What is the correlation of the *in vivo* test results with those of *in vivo* methods?
- Do the data mean what we believe they mean?
- Have the limits of the method been investigated and defined?

Ultimately, validation should determine if the *in vitro* method is fit for the purpose. There is an understandable tendency to validate methods with known chemicals; nephrotoxicity models are often validated with mercuric chloride or similar classic toxic compounds. Probably, the best method of validation, following exploration of the method with known toxicants, is to use it in parallel with more conventional assays so that the data can be compared when both tests have been completed. Ease of validation is likely to decrease with increasing complexity of endpoint. Genotoxicity tests have been accepted because they investigate a relatively simple endpoint in a series of robust assays that can be readily transferred among laboratories and that are, for the most part, relatively simple to conduct. On the other hand, tests for immunological effect have been difficult to devise and validate, and the complexity of the endpoints studied must play a role in this.

Development of new *in vitro* methods is seen as cutting-edge research, with the bonus that it can be quite inexpensive to set up and run. The result has been a plethora of new methods proposed or new test systems examined, and it has been difficult to discern any focus in the field as a whole. In such a situation, it is essential that validation be carried out rigorously on tests that have emerged as potential replacements. As a result of this vast effort, new tests that are also viable have taken years to emerge, although some notable exceptions such as the LLNA have been promulgated and accepted in shorter periods. However, part of the key to the success of this method is that it is conducted in a whole animal; the advantage over prior methodology is that it is quantitative and uses fewer animals.

There is a vast range of references on validation of alternative methods, many of which may be obtained from sites like that of ECVAM or the journal *In Vitro Toxicology*. Zeiger (2003) points out that the validation process should be scientific, flexible, and transparent. Validation determines the reliability (reproducibility) and relevance (quality of measurement or prediction) of the proposed test method, where appropriate, by comparison with the method it may be replacing. Naturally, judgment is essential in this process as there can be no sensible way in which values for these parameters can be fixed. One of the goals of *in vitro* or alternative methods must be not only to replace animals, but also to make the predictions more relevant to the target species, usually humans. Thus, predictors of human effect should be validated against the human effect rather than that in a laboratory species or other surrogate.

Validation is usually carried out by a number of laboratories simultaneously, using blind-labeled compounds dispensed from a central source. The results can then be examined, when all laboratories have completed their separate experiments. The experience over the last 10 years or so has allowed some retrospective examination of failure (early in the process) and success (based on the experience gained in failure).

In 1991, a joint program was established by the European Commission and the European Cosmetics, Toiletry, and Perfumery Association (COLIPA) to develop and validate *in vitro* photoirritation tests. The first phase involved the evaluation of phototoxicity tests established in cosmetic industry laboratories together with a then-new assay, the *in vitro* 3T3 NRU phototoxicity test. This uses mouse fibroblast cell line 3T3 in a photocytotoxicity test, which has NRU as the endpoint for cytotoxicity.

A prevalidation study was conducted with 11 phototoxic and 9 nonphototoxic test chemicals; the 3T3 NRU PT test correctly identified all the chemicals, a result that was subsequently confirmed independently in Japan. In the second phase of the study, which was funded by ECVAM, this test was validated in 11 laboratories in which 30 test chemicals were tested blind. The results of this blind testing were reproducible, and there was very good correlation between *in vitro* and *in vivo* data. Although the study was considered to be ready for regulatory acceptance, concerns were raised that more ultraviolet (UV)-filter chemicals should have been tested. Subsequently, a further 20 chemicals, evenly divided between nonphototoxic and phototoxic, were successfully tested. Finally, in 2000, the test was officially accepted by the European Commission and published in Annexure V of Directive 67/548 EEC on the classification, packaging, and labeling of Dangerous Substances.

This lengthy process has the clear hallmarks of success: a carefully performed evaluation of candidate tests followed by a second blind testing by a number of laboratories, all coordinated centrally. When questions were raised about its regulatory acceptance, they were answered carefully and in a similar manner to the earlier validation. Progress in evaluation of alternatives in this area continues; Lelievre et al. (2007) reviewed favorably a reconstructed human epidermis test system for evaluation of potential systemic and topical phototoxicity.

TEST SYSTEMS AND ENDPOINTS

Test systems for *in vitro* testing range from purified enzymes and subcellular fractions such as microsomes used in metabolism studies, through single cells, isolated tissue components such as nephrons, tissue slices, and up to isolated organs such as perfused livers, kidneys, or hearts. A refinement is *ex vivo* testing, in which an *in vitro* preparation is made from the tissues of animals that have been treated in life. Table 4.3 summarizes the types of test systems that are available for *in vitro* experiment.

Subcellular test systems have quite specific endpoints that are usually relatively easy to measure. They have prominent use in areas such as metabolism where

---

TABLE 4.3
In Vitro **Test Systems**

Test System Type	Examples
DNA Probes	Microarrays
Isolated enzymes	Cytochrome P450s
Subcellular organelles	Microsomes
Single cells—primary culture	Hepatocytes and hepatocyte couplets
Single cells—cell lines	Tumor cell lines, Caco-2 cells, stem cell cultures
Stem cells	The great white hope
Tissue slices or organ components	Liver or kidney
	Isolated nephrons
Perfused intact organs	Liver or heart
Cultures reproducing tissue architecture and function; different tissue cultures interlinked to give a "circulation"	Liver bioreactors and similar systems
	Potential for use of stem cells in these systems
Embryo culture	Whole embryo or micromass
Fertilized eggs	Chicken eggs used for irritancy and corrosion prediction
Excised tissues	Bovine eyes used in ocular irritation studies
Ex vivo studies	Unscheduled DNA synthesis; comet assay
Studies on particular tissues	Comet assay in various cell types derived from toxicity studies

Source: Adapted from Woolley A, *A Guide to Practical Toxicology*, 1st ed., London: Taylor & Francis, 2003.

---

aspects such as rate and extent of metabolism can be assessed; this is also a useful and inexpensive test system in which to assess potential species differences. Human microsomes or purified human enzymes are often used.

Cultured cells are essentially available in two types: primary cultures, which are derived from a freshly killed animal (although they may be cryopreserved), and secondary cultures, which are immortal cell lines. Of the first, rat hepatocytes and human peripheral lymphocytes are frequently used. These cells have some limitations, in that they tend to deteriorate quickly; for instance, freshly isolated hepatocytes lose their metabolic capability over a few hours, limiting the time available for testing. Another disadvantage is that, for cells such as hepatocytes, the blood–bile duct polarity of the cell is lost in the process of producing a single-cell suspension. This important aspect of hepatocyte function can be maintained by the use of hepatocyte couplets, in which the biliary side of the cells is maintained in the middle space between two hepatocytes.

The use of culture methods to produce blocks of cells that have some of the characteristics of the original tissue is also a technique with considerable potential. Hepatocytes can be induced to maintain functionality for several weeks when mounted in an appropriate matrix of collagenous material. With the development of the concept of organ-on-a-chip, followed by the related human or body on a chip and the use of microtissue samples and microfluidics, the potential to explore the interrelationships of organs *in vitro* is becoming more practicable.

In contrast, the secondary cultures or cell lines have much longer use times but often lack the in-life characteristics of the cells from which they were derived. These cells are represented by Caco-2 cells used in *in vitro* assessment of absorption; however, selection of only one cell line from the hundreds available is essentially nonrepresentative and deceptive of the huge choice available. Cells may also be derived from animals in toxicity tests, for tests such as the comet assay for DNA damage. Recently, the ethics of the use of human-based cell lines has been questioned. This is particularly highlighted by the HeLa cell line derived from Henrietta Lacks, who died of cancer in 1951 and, prior to her death, unknowing and without consent, provided cancerous cervical cells, which became the first human immortal cell line. It was only in the early 1970s that Mrs. Lacks's family discovered that her cells had been taken, and then only because they were contacted by researchers looking for family genetic history.

The endpoints available for study in cells are wide ranging and include mitochondrial function, membrane integrity (assessed by leakage of markers into the culture medium), and effects on protein and DNA (omics), quite apart from growth (division rates), plating efficiency, assessment of gap junction patency, viability, and death. Specific cell lines have specific roles in tests used in regulatory assessments, for example, hERG cells in cardiovascular safety assessment of the potential for novel drugs to elicit cardiac effects in life and the use of Langerhans's cells in the assessment of immune effects.

In vitro experiments with cells are often associated with advanced technique beyond such simplistic endpoints as dye exclusion or enzyme leakage into the culture medium. Flow cytometry has been shown to be useful in short- and long-term experiments, in which various aspects of cellular function can be assessed, including mitochondrial function and cell cycle modulation. While this looks at cell populations, the characteristics of individual cells can be assessed in tests such as the hERG assay, in which the potential to affect the cardiac action potential is assessed. This is an example of a test accepted, in fact required, by regulatory authorities for drug development.

The subject of cells in culture should not be left without considering the potential impact of stem cells. Stem cells have two properties that are attractive to the toxicologists; they reproduce themselves identically, and they can give rise to different cell types, such as heart, liver, kidney, or pancreatic islet cells. There are two basic types of stem cell—embryonic stem cells and adult stem cells, which go by a variety of names. Davila et al. (2004) carried out a review of the use of stem cells in toxicology and concluded that they have considerable potential in toxicity testing. If a range of human-derived cell lines that are similar in every respect to their tissue analogs can be produced, and if these cells can be used in assays that are reproducible, their impact could be immense. In particular, these authors single out that hepatotoxicity and cardiac effects (QT prolongation) could be beneficiaries of the use of stem cells. Although such cells have huge potential in toxicity testing, it should be remembered that the cells that are derived should be reproducible. There is little sense in deriving hepatocytes from a stem cell line on several occasions if the cells that result each time are different in subtle ways. An immortal culture of human hepatocytes that maintained its metabolic capability without change is likely to be more useful than a multiple derivation of hepatocytes from stem cells that are not the same each time.

Precision-cut tissue slices have the advantage that tissue architecture is maintained, although exposure of the cells to the test item may be limited to the outer layers of the slice, and luminal spaces, such as renal tubules, may be collapsed. In addition, the surfaces represent areas of damage. Although their lifetime has been limited, this problem is being surmounted in culture systems, which can extend the life of the slices for several days, during which exposure can be continued. Liver and kidney are favorites for this type of test system, but other tissues such as lung can also be used. Liver slices can be maintained for up to a week in some systems, and the decline of biomarkers, such as enzymes, measured in the slices and/or in the culture medium, which is replaced regularly. It is possible to detect change in clinical chemistry parameters, such as marker enzymes, and in histology, including proliferative changes, or those associated with storage of substances such as glycogen. With such systems, it is relatively easy to carry out comparisons between species.

Isolated tissue components, for example, nephrons, pose more of a technical challenge and are not used to the same extent as other *in vitro* systems. While the renal nephron may be a technical challenge to isolate, other tissue components are used more frequently. These include use of papillary muscles from guinea pigs and Purkinje fibers from various species in cardiac safety pharmacology investigations. Cardiac testing also makes use of isolated perfused hearts, notably in Langendorff's preparation, which uses a rabbit heart. Other organs used regularly include the liver and kidney, but given sufficient skill to connect them to a perfusate, there is little reason why other organs should not be used in perfusion experiments.

At this point, we start to consider alternative systems that do not use vertebrates. Fertilized chicken eggs have been used in embryo toxicity screening and in assessment of irritation and corrosion. *Hydra* has also been used in reproductive toxicity screening for embryological effects, but it is fair to say that this has not been as widely adopted as the other reproductive screens. Looking at other invertebrates, it is clear that the pupa of most insects goes through a process of reorganization and organogenesis that is probably disruptable by reproductive toxicants; although there is some work on developmental toxicity in insects, it does not appear to have been used specifically to study disruption of morphogenesis in a way that may be of use in developing new methods for studying chemicals.

The final type of *in vitro* test is that which takes a whole animal, treats it with the test substance on one or more occasions, and then uses part of that animal in an *in vitro* test. These are often extensions of *in vitro* tests. Two such tests, the UDS and comet assays, can be performed on cells treated *in vitro*. However, there is some elegance in allowing the full organism access to the chemical for a period of hours and then isolating appropriate cells from its tissues and assessing effects that may have occurred.

TARGET ORGAN TOXICITY

The following sections look at areas of application of *in vitro* testing, firstly in terms of target organ toxicity and then by field of application. These areas have been chosen in the knowledge of the requirements of pharmaceutical development; however, they illustrate the broad potential for *in vitro* testing and research.

The early determination of target organ toxicity *in vitro* is an important step in the decision processes that determine which compounds should be developed and which should be dumped immediately. Put like that, it would seem that the detection of toxicity in a particular organ *in vitro* should be the death knell of any compound; however, this should be viewed dispassionately in the context that such decisions should be taken from a broad database and not on the basis of a single test. So, the aim should be to rank candidate compounds according to effect and to make decisions based on as wide a set of data as possible.

Expression of toxicity at unexpected target sites, or at unexpected intensities, by drugs or other chemicals has been a major factor in their withdrawal from the market. While clinical human toxicity at major target organs may be predicted from appropriate nonclinical studies, some target organs may be less easily predictable. For instance, cerivastatin (Baycol), a statin intended to reduce cholesterol levels, showed an unexpected risk of rhabdomyolysis (severe muscle damage) especially when used at a high dose or with gemfibrozil, another lipid-regulating agent. It is unlikely that such an effect would be predicted by any of the routinely used test systems, particularly those *in vitro*, as there is unlikely to be any routine test for such an endpoint. The diversity of toxicological endpoints will always be a massive hurdle to successful prediction of target organ effect *in vivo* from *in vitro* assays, covering single effects or mechanisms.

In the following discussion, particular attention is paid to the liver as an illustration of the type of approach that is possible in studying target organ toxicity *in vitro*.

LIVER

As a major site of metabolism and elimination of xenobiotics and, as a result, toxicity, this is probably the organ in which most effort has been invested in *in vitro* toxicological testing. This has been helped by its accessibility and availability from a number of species, including humans. In addition, it is a relatively large discrete organ unlike others, such as the adrenal (discrete but very small) or the immune system (extensive, complex, and diffuse). Physically, it is easy to work with; the cells may be readily dissociated into primary cultures of single cells, the organ may be sliced with precision and the slices used in experiments up to several days, and it is relatively easy to perfuse.

Given that the liver is such an important organ for toxicity, it is inevitable that this is one of the most dynamic areas of research for *in vitro* methodology. There are several areas of research, including development of new test systems, development of new markers of effect, and investigation of mechanisms. Two major reviews have been published that summarize progress in this important area: Godoy et al. (2013) and Soldatow et al. (2013).

Although the liver is composed of hepatocytes (ca. 80%) and other cells (ca. 20%), its processes are subject to highly complex regulation, and disturbance of the architecture readily disrupts this. In simple terms, the liver is organized into lobules, the classic lobule being hexagonal with six portal triads of artery, vein, and bile duct and a single central vein. The hepatocytes are arranged in layers radiating from the central vein; each has a side exposed to blood flow toward the central vein and the other being on the biliary side, with bile flowing toward the bile duct in the portal triad. Although this arrangement is relatively simple, it is easily disrupted as isolated hepatocytes lack the biliary–blood polarity. In addition, the 20% of nonhepatocyte cells have complex regulatory functions. In addition to loss of polarity, isolation of hepatocytes disturbs a complex net of regulation by nuclear receptors and signaling pathways, some being activated and some suppressed; this results in changes in expression of many genes. Appreciation and understanding of these changes is crucial in interpretation of the results of any experiment.

Of the systems available for studying effects in the liver, isolated single cells have been the most popular. However, as indicated above, they have disadvantages in that they lose architectural polarity and undergo significant changes in regulation. In addition, they rapidly lose metabolic competence so that their viability is limited. Although subcellular systems are used, including purified enzymes and microsomal preparations, as we will see in the section on metabolism below, hepatocytes have the advantage that they are a fully functional cell complete with membranes and intracellular relationships. Another subcellular system that is regularly used is S9 mix, derived from the livers of rats, treated with an enzyme inducer, and used in genotoxicity studies where metabolic activation of the test substance is necessary. Precision-cut tissue slices are increasingly used as an evolution of the initial more crudely cut slices; they are the basis from which 3-D models and bioreactors have evolved.

As indicated, one of the major disadvantages of primary cell cultures is that they swiftly lose their metabolic capacity, gene expression, and relevance. A further disadvantage is that the majority of such cellular systems are derived from laboratory animals rather than humans. Accordingly, tremendous efforts are being made to produce immortal cell lines that retain their metabolic capabilities indefinitely. The Holy Grail of being able to use a consistently responding cell line in studies of comparative metabolism is gradually being realized, including the more distant goal of having lines of human cells that are immortal.

Soldatow et al. (2013) summarize the available systems as including perfused livers, liver slices, primary hepatocytes, and isolated organelles, particularly microsomes, which have an important role in comparative studies of metabolism. Newer systems use 3-D constructs, bioartificial livers, and coculture of various cell types, some of which attempt to replicate the *in vivo* architecture by the use of microfluid flow and microcirculation.

Research into markers of effect has blossomed in recent years. In early experiments, effects were marked by relatively crude endpoints such as death, exclusion of dyes, or leakage of simple markers such as lactate dehydrogenase. However, there has been much evolution in the sophistication of markers of effect in recent years. Godoy et al. (2013) reported that microRNAs, which are RNA of 19 to 25 nucleotides, which do not code, have been shown to play a major role in regulating gene expression, and cell differentiation and replication. Aberrant expression of microRNAs has been associated with different cancers. In addition, they are tissue specific; for example, miR-122 accounts for approximately 70% of all hepatic miRNA and has wide-ranging effects, offering much utility as a marker of change in the liver.

Specificity for the liver has been one of the drivers for this as there is clear utility in being able to distinguish damage in a particular tissue. A problem with the generic markers of toxicity, such as lactate dehydrogenase, is that they are universally expressed and do not necessarily indicate the tissue or cell type affected. Kiaa et al. (2015) reviewed the utility of miR-122 as a marker of drug-induced cellular toxicity in hepatic cells and indicated its suitability as a marker of hepatic change. In addition, they suggested that, due to the universality of expression of miR-122, it has the potential to act as a bridge between *in vitro* and *in vivo* experiments, with particular utility in the study of human systems of drug-induced liver injury *in vitro*.

KIDNEY

The options for the kidney mirror those for the liver, with the additional possibility of isolating entire nephrons, glomeruli, or fragments of the proximal tubule. Functionally, the kidney is the point of excretion for many endogenous chemicals and for xenobiotics such as drugs and their metabolites. It is structurally complex and with more cellular diversity than the liver. A suspension of cells obtained by enzymatic dissociation from the kidney will not be homogenous. If cells are isolated from the proximal or distal tubules, they can be used to study transport systems that can be inhibited by substances such as probenecid or quinine. As with the liver, there is a range of derived cell lines that can be studied.

Nervous System

The nervous system, from the central nervous system or peripheral nervous system, presents more of a problem, given that its principal function is to transmit electrical impulses or transfer small amounts of quickly decaying chemicals at synapses between neurons or other receptors such as those on muscles. Some of the endpoints are the same as those for other single-cell systems, such as cytotoxicity, apoptosis, and proliferation. More specific endpoints include electrophysiological aspects such as ion channels, and enzyme studies can include acetyl cholinesterase and other markers of effect. Techniques such as patch-clamp electrophysiology and calcium imaging allow the user to understand, respectively, how single cells and groups of cells behave in response to certain stimuli. Advances in microscopy have allowed insight into how nervous tissues respond to electric fields. These cells may not necessarily represent the complex network of nervous tissue but give understanding into how single or groups of cells respond. Such methods tend to rely on primary cell cultures taken from animals, and although they may be three Rs compliant, animal use could still be high. On a larger scale, the brain is suitable for study in tissue slice preparations, and it is possible to isolate the various cell types for individual study. The drawback of these techniques, however, is that change in one aspect of this complex system does not necessarily directly correlate with effects in life. Brain slices can be studied with techniques such as electrophysiology, autoradiography, genetic analysis, and histopathological staining. They allow intact nervous systems to be investigated and experimented upon.

Immune System

The immune system is a complex, diffuse set of tissues and cell types that is distributed throughout the body. While certain tissues are clearly closely associated with it, such as the thymus and lymphoid tissues, generally, others do not have such obvious connections. The lung may not have an immune function as such, but it is home to a population of immunosurveillance macrophages. Other immune response systems are contained in the blood in the form of immunoglobulins and proteins of the complement cascade and the various populations of leukocytes, the numbers of which vary rapidly according to a variety of immune stimuli. The immune system plays a central role in the body's responses to foreign proteins or to proteins that have been linked covalently to smaller molecules to form haptens. These go on to be presented to the appropriate cell to elicit a response, which can be devastatingly quick and sometimes fatal. One particular problem is the dissimilarity between animal and human immune systems, making the prediction of whole-body human effects by extrapolation from parts of animal immune systems *in vitro* especially tricky.

One endpoint that is of clear interest for a viable, predictive (and preferably simple) *in vitro* assay is sensitization, which is discussed in more detail below. This is responsible for a great deal of occupational disease and ill-health, and given the history of less-than-perfect predictivity of animal models such as the guinea pig, the interest in an alternative animal-free model has been intense. This has been somewhat alleviated by the regulatory acceptance of the LLNA, but this is still an animal-based test, although it is considered to be objective and predictive.

OTHER ORGANS

As a general inconvenience, chemicals do not always target one of the major tissues reviewed briefly above. One of the results of enzyme induction in the rodent liver is thyroid follicular hypertrophy, which can lead to thyroid tumors, if allowed to persist. This may be characterized as an indirect effect due to liver change and has been generally regarded as not being relevant to humans. However, direct effects on the thyroid are possible, and alternative tests have been devised to assess this. One example has suggested the use of tail resorption in the tadpole of the frog, *Xenopus laevis*, a species that has also featured in reproductive toxicity testing (Degitz et al. 2005). In this work, treatment of tadpoles in the early stages of metamorphosis with inhibitors of thyroid hormone synthesis was associated with a concentration-related delay in development, and the authors suggest that this may be indicative of a viable test system for thyroid axis disruption.

Disruption of thyroid hormones has been identified as an important endpoint in the regulation of chemicals. Expanding on the use of *Xenopus* larval metamorphosis, guidelines have been established by the OECD and the US Environmental Protection Agency (EPA) for testing potential endocrine disrupters. These are based on evaluation of alteration in the hypothalamic–pituitary–thyroid axis, with thyroid gland histopathology as a primary endpoint (Miyata and Ose 2012).

The lung is another target organ, which can be studied *in vitro*, either by perfusion, by isolation of cell types, or as tissue slices.

While the isolated, perfused heart or related tissues, such as the dog or guinea pig papillary muscle, have been used in safety pharmacology studies, these methods still entail the use of an animal. Doherty et al. (2015) described structural and functional screening in human-induced pluripotent stem cell–derived cardiomyocytes as a means of investigating cardiotoxicity. They examined 24 drugs for effects, both structural and functional endpoints, including viability, reactive oxygen species generation, lipid formation, troponin secretion, and beating activity in the derived cardiomyocytes. There were no effects in drugs that were described as cardiac safe. Sixteen of 18 drugs with known cardiac effects showed changes in the derived cardiomyocytes in at least one method. Further classification of the effects as structural or functional or both was possible by taking C_{max} values into account. This study has potential lessons in this and other areas, indicating that a multi–endpoint approach is likely to be more informative in evaluating potential toxicity in a particular tissue.

INTEGRATED SYSTEMS IN DEVELOPMENT

These include so-called organ-on-a-chip and body-on-a-chip systems, which promise to circumvent the static nature of traditional *in vitro* test systems. As pointed out static, monolayer cultures do not have critical elements of the environment *in vivo*, including blood flow, mechanical stress, or the three-dimensional architecture that is critical to correct tissue function such with the liver or kidney. Of these factors, fluid flow and architecture have been identified as key attributes. Scaffolds such as collagen/hyaluronic acid or beads or fibers can be used in conjunction with microfluid flows to better mimic the situation *in vivo*. The natural evolution of this is the body on a chip, which is the subject of an EU project to develop a comprehensive model *in vitro* to identify multiorgan toxicity and the effects of metabolic activity on efficacy (Body-on-a-Chip [BoC], EU project reference 296257). The website for the project (http://cordis.europa.eu/project/rcn/104027_en.html) gives the central objective of developing "a versatile and reconfigurable pharmaceutical screening technology platform that relies on organotypic three-dimensional spherical microtissues." This will include human microtissues such as brain, liver, heart, tumors etc.

It is clear that there will be progress in the development of new techniques to examine toxicity to various tissues *in vitro* over the coming years, and that anything written now will soon be out of date. However, although it is now possible to examine the effects on individual tissues over periods up to several weeks, the next challenge will be to examine the relationship between these effects and other tissues, which are relevant to toxicity *in vivo*. Minor shifts in homeostasis in one organ can lead to major change in others, and it is these insidious and progressive toxicities that represent one of the biggest challenges in toxicity assessment. Initiatives such as body on a chip and the use of microfluid circulation will lead the way in this, together with the increasing use of human or humanized systems.

FUNCTIONAL TOXICITY TESTING

The following sets out to review tests for functional change; the order in which these have been reviewed is unashamedly derived from the order in which the various sections of a pharmaceutical registration package are presented. However, this is a handy method of illustrating the diversity of techniques and test systems that is available. For the most part, we have not attempted to describe methods, because this is intended to be no more than a brief review of each area; more details may be given in the main chapters that address them. The exception to this is the alternative methods such as those for skin and eye irritation, where a little more detail has been given.

PHARMACOLOGY AND SAFETY PHARMACOLOGY *IN VITRO*

Pharmacology, which is essentially a branch of toxicology (albeit generally conducted at lower doses or exposure concentrations), has immense potential in terms of *in vitro* experimentation. This is because there is usually only one endpoint or, at least, a series of closely linked endpoints manifested in a single-cell system, tissue slice, or perfused organ. In what might be called general pharmacology, as opposed to safety pharmacology, there is a range of routinely conducted assays, including studies of receptor inhibition, in which a known bank of receptors is incubated with the test compound to determine, for example, binding affinities or inhibition concentrations (IC_{50}). In safety pharmacology, the use of the hERG assay and various cardiac preparations has been reviewed above; however, this area is constrained by regulatory oversight, and innovation is perhaps not as quick as in other areas. While cardiac function is covered to some extent by the hERG assay and its relatives, there are currently no *in vitro* alternatives for renal, respiratory, or nervous function. This should not rule out the development of screening assays, which may become accepted eventually.

METABOLISM

The main areas where *in vitro* techniques in metabolism can contribute significantly to chemical toxicity assessment are drug interactions and comparative metabolism. Both these areas can be investigated in the same system types, namely, purified enzymes, microsomes, or hepatocytes. Tissue slices may also be used. In a number of recent cases, where a drug has been withdrawn from the market, the reason has been found to be due to inhibition or acceleration of the metabolism of one drug by another. It is especially useful to know about the potential inhibition of the cytochrome P450 enzymes, particularly CYP3A4. Ketoconazole is a potent inhibitor of CYP3A4 and interacts with terfenadine, which was withdrawn due to effects on the electrocardiogram, seen as QT prolongation resulting in torsades de pointes. Cisapride is another inhibitor of CYP3A4, which was withdrawn for similar reasons. It is clear that inhibition of these enzymes is a critical element in drug toxicity and safety assessment, and early detection is the best way of devoting precious resources to the best available molecules.

Comparative metabolism is undertaken relatively early in development of chemicals and is a useful method of assessing which would be the best second species to be used in a full program of toxicity studies. Hepatocytes or microsomes prepared, typically, from rats, dogs, monkeys, and humans are incubated with the test chemical to determine the rate of elimination of the parent and, with radiolabelled material, the appearance and identity of metabolites. Other species such as mice and minipigs can also be used. Studies of this type have helped to reduce animal use by cutting out studies in inappropriate species.

KINETICS

The overall kinetic behavior of a chemical in the body is the result of four basic processes, absorption, distribution, metabolism, and elimination (ADME). This can be more simply viewed as the interactive effects of absorption and clearance, which for convenience is usually expressed as clearance from the central compartment, i.e., the blood. Caco-2 cells, a cell line derived from human adenocarcinoma, have often been used in *in vitro* assessment of absorption, although there is no simple correlation between Caco-2 cell permeability and human gastric absorption. Some idea of their reliability may be obtained by reference to physicochemical parameters. Clearance can be assessed using hepatocytes in a targeted variation of metabolism studies, described above.

While the gut is a complex system for the study of absorption *in vitro*, the skin is potentially easier to work with, being somewhat simpler and, more importantly, readily available as robust preparations, either as constructed systems or as skin taken from animals or humans. Having said that, skin compromised by diseases such as dermatitis or psoriasis has different properties from normal skin. Davies et al. (2015)

developed a model for the study of absorption across compromised skin, using tape stripping of dermatomed pig skin. Endpoints and their relationships studied included transepidermal water loss (TEWL), electrical resistance (ER), and tritiated water flux (TWF), which are markers of skin barrier function. In any such investigation, it is important to establish normality in terms of the parameters studied; normal values for these three endpoints have been published for six species, including humans. It is notable that of the three, the most robust was ER, a parameter evaluated in some tests of skin corrosivity and irritation.

TOXICITY TESTING

In a limited sense, toxicity may be tested in cell lines using a number of endpoints as touched on above. These include dye uptake or exclusion (neutral red or trypan blue, respectively), growth, viability, enzyme leakage into the culture medium, and so on. While it has been a long-standing complaint that long-term exposure was not really possible *in vitro*, the advent of tissue slice techniques, which allow more prolonged treatment, is an important advance. The disadvantage will continue to be, however, that these single-tissue systems lack the relationship with other tissues that may mitigate (or increase) the toxic effects of the chemical under study.

One aspect of toxicity testing that should be remembered is the fact that very often, the concentrations that are tested are orders of magnitude higher than those that are likely *in vivo* and that interpretation of the results should take this into account.

CARCINOGENICITY

There is huge potential for evaluation of carcinogenic potential by *in vitro* testing. It is relatively simple to detect a genotoxic carcinogen; the challenge lies in assessing the potential for nongenotoxic carcinogenicity. Having said that, it is apparent that the lines between the two old certainties—genotoxic and nongenotoxic—are becoming increasingly blurred. This demonstrates the fact that if the choices appear to be black and white, it simply means that the gray nuances in between have not been discovered or understood yet.

There is a range of nongenotoxic mechanisms that should be susceptible to *in vitro* investigation, some of them relatively simple but fundamental. J. E. Trosko has long advocated a test in which gap junction patency is investigated, the basic theory being that a cell that does not communicate with its neighbors is likely to be shut off from regulatory processes, and thence, to become cancerous. Another relatively simple endpoint that is likely to repay regular investigation relates to epigenetic factors, such as the levels of methylation of DNA, which affects gene expression, which is, in itself, another area for exploration through techniques such as microarrays and genomics.

One problem for any form of carcinogenicity testing, including tests using animals, is that the actual impetus for the cancer may be a small perturbation of a physiological balance that is not clinically obvious but that has a long-lasting effect on the homeostatic status of tissues or individual cells. The detection of such small change is very difficult at the moment, for a variety of reasons, particularly background noise that obscures the treatment-related response (see Chapter 2); this is a potential growth area for *in vitro* testing. The future of carcinogenicity testing is considered in more detail in Chapter 8.

REPRODUCTIVE TOXICITY

While it is possible to culture cells from the testis or other reproductive organs, the main effort of research into *in vitro* techniques has concentrated on developmental toxicity as an endpoint, if such a complex phenomenon can be reduced to the singular. The techniques that have been developed include whole-embryo culture assay and the limb bud or micromass assays. Other systems have been tried at intervals, including *Hydra* and fertilized hens' eggs. In addition, other models may be useful, including invertebrates and fish. Embryonic stem cells also offer opportunities for test development. Further details of some of these tests are given in the chapter on reproductive toxicity testing (Chapter 7).

Säfholm (2014) investigated the developmental and reproductive toxicity of progestagens in a test system using *Xenopus (Silurana) tropicalis*. She found that larval exposure to levonorgestrel caused severe impairment of oviduct and ovary development, causing sterility, while no effects on testicular development, sperm count, or male fertility were found. The findings in this thesis indicate the potential for this test system for investigation of developmental toxicity. It should be pointed out that this is only an example of many systems being investigated.

SENSITIZATION

Sensitization, either respiratory or dermal, is a complex process, which is difficult to mimic *in vitro*. More than one cell type is involved, and the mechanism is complex and not well understood. As a further complication for *in vitro* testing, the chemicals that trigger sensitization are not always water soluble. In fact, they are usually lipophilic with reactive groups, which can form haptens with endogenous proteins. Sensitization is an essential step in developing allergic contact dermatitis, a condition that is very costly to industry; the pathways involved were described by the OECD (2012a,b) as an adverse outcome pathway (AOP).

Urbisch et al. (2015) point out that, as with cardiotoxicity above, a single method would not be enough to cover the full AOP for skin sensitization and that a battery of tests is likely to be needed. These authors indicated that the nonanimal test methods showed good predictivity compared to the murine (LLNA) and were better when compared with human data, when a two-out-of-three methods approach was used. This gave accuracies of 90% and 79% when compared to human and LLNA data, respectively.

Reisinger et al. (2015) carried out a systematic evaluation of 16 nonanimal test methods for skin sensitization safety assessment with a view to prioritizing methods for future development. The ultimate goal is to establish a data integration approach to skin sensitization safety assessment, using data for bioavailability and exposure and metabolism in the skin. As an example of tests reviewed, they described the use of human peripheral blood monocyte–derived dendritic cells for the *in vitro* detection of sensitizers. Subsequently, the results of a ring study using these cells were reviewed by Reuter et al. (2015). Five laboratories evaluated seven chemicals, including six known sensitizers (with one prohapten) and one nonsensitizer. The conclusion was that the test correctly assessed the sensitization potential of the chemicals and, as an important observation, could be added to a toolbox of *in vitro* methods for assessing sensitization potential.

IRRITATION AND CORROSION

This is an area of testing where there has been a huge effort in the development and validation of new methods, which have often included commercially available test systems. Irritation is a reversible, nonimmunological inflammatory response produced at the site of contact, while corrosion is the production of irreversible damage at the contact site as a result of chemical reaction. Given these relatively simple definitions, it might be supposed that development of reproducible human-relevant tests would be simple. Unfortunately, this has not been the case, in part due to the complexity of the tissues being investigated, especially the eye, for which irritation testing is still an important target.

Given the emotive nature of testing in the eyes of animals such as rabbits, the development of reliable test methods has been a priority, which is beginning to show success. For ocular irritation, various models are available, such as the Bovine Corneal Opacity and Permeability (BCOP, OECD 437) assay or the Isolated Chicken Eye Test (ICET, OECD 438). Other OECD test guidelines for irritation and/or corrosion are given in Table 4.1. In addition, models using fertilized hens' eggs, for example, the chorioallantoic membrane test, have been evaluated with some success. Testing for irritation and corrosion is looked at in more detail in Chapter 9.

PHOTOTOXICITY

This is a long-standing endpoint, which has been of interest to groups as diverse as celery pickers, psoriasis patients receiving psoralen plus UVA therapy, and users of aftershave. The validation of the 3T3 NRU PT (OECD 432) test has been described above. Phototoxic chemicals absorb light in the range of sunlight, and the 3T3 NRU PT test *in vitro* was shown to be predictive of acute phototoxicity *in vivo*. However, the guideline notes that the test does not predict other effects such as photogenotoxicity, photoallergy, or photocarcinogenicity and gives no indication of potency. Nor has it been designed to examine indirect mechanisms of phototoxicity, such as the influence of metabolites (the test does not include any system for metabolic activation). In addition, the test is broadly useful for water-soluble chemicals, and the use of solvents such as DMSO (with their own toxicity and reactivity issues) may be needed. Other limitations include killing of the fibroblasts with UVB and poor suitability of the test for assessing phototoxicity of dermal products.

ECOTOXICOLOGY

This is another area that is greatly influenced by the regulatory impact of REACH in Europe. As an illustration of the approaches used, Zurita et al. (2005) reported on the ecotoxicological evaluation of diethanolamine using a battery of microbiotests. Their test systems included bacteria for the inhibition of bioluminescence, algae for growth inhibition, and *Daphnia magna* for immobilization, while a hepatoma fish cell line was used for a variety of toxicological endpoints such as morphology, viability, and metabolic studies. Perhaps inevitably, the frog *Xenopus* has also been used in ecological risk assessment.

PITFALLS IN *IN VITRO* TOXICOLOGY

The basic challenge with any *in vitro* technique is interpretation of the data in relationship to the animal from which the system was derived and then, if needed, extrapolation to humans. The absence of interaction between organs or tissues and the static nature of many (older) systems mean that some effects are seen that would be absent in life. Complications become evident if simple systems are used *in vitro*, in isolation from modifying influences such as transporter systems, blood circulation, other cell organelles, or fractions that may be important *in vivo*. Concentrations tested should be relevant to those expected in life, e.g., in toxicity studies or clinically. Also, the transient nature of most *in vitro* systems means that chronic administration and the detection of progressive effect are not possible. These test systems are often technically demanding and difficult to reproduce from one laboratory to another. An additional factor to consider is the sheer volume of data that are produced by some of the new techniques and the consequent requirement for suitably validated pattern recognition or data-plotting software.

Where the software "cleans" a data set to simplify a data plot, there should be confidence that the correct data are being excluded; otherwise, useful information may be lost. As already pointed out above, the choice of time point can be critical in achieving data that can be analyzed to best advantage. In the absence of good validation work, the significance of the differences seen may be misinterpreted and erroneous conclusions drawn. If the mechanism is not understood correctly and the meaning of the data is not clear, there is no point in doing the test, except as part of a validation exercise. It is clear, in these cases, that these unproved systems would be totally unsuitable for toxicity prediction or safety evaluation.

Godoy et al. (2013) point out in their review of *in vitro* systems for liver toxicity that a vital message for the development and use of *in vitro* systems is that the situation *in vivo* should always be remembered.

OMICS

Much attention has been focused on genomics, toxicogenomics, and proteomics, which, respectively relate to the expression of genes in normality, gene expression following toxic exposure, and protein expression as a result of gene activation.

This follows the recognition that mechanisms of toxicity are reflected by the profiles of the genes expressed in response to the toxic insult; protein expression is a consequence of gene activation. The use of DNA microarrays in which thousands of DNA probes or synthetic oligonucleotides are mounted on a chip in a known order and then hybridized with target DNA is able to determine which genes are activated in target tissues as a result of a toxic exposure. The result is often presented at scientific meetings as a rectangle of several hundred colored dots, together with the confident assertion that "this shows expression of 'A' genes." This may be exploited in toxicity studies by taking tissue samples, particularly liver, to study differential gene expression in the presence or absence of toxicity.

It has been shown that the protein expression profile is associated with the effect and mechanism of toxicity and that these profiles are broadly similar within each group of compounds that, in life, have similar effects and mechanism. However, within each group, each compound has its own distinct profile of protein expression. By setting a desired profile or indicating thresholds for decision, compounds can be selected or rejected for further development.

The use of these techniques needs careful consideration of objectives before committing to a system that is likely to be expensive to buy and maintain, and may be difficult to alter. Genomics offers the possibility of use of open or closed systems. In the former, the endpoint is not specified, and all genes expressed can be highlighted as a pattern of spots on the array. In the latter, a specific number of genes, for instance, a few hundred, may be investigated, giving a pattern of effect in these target genes. In contrast to genomics, protein expression can (currently) only work as an open system in which all proteins are examined. Because of the vast amount of data that are generated using these techniques, pattern recognition software is important in interpretation of the data.

Protein and gene expression both change with time, and the number of genes expressed a few hours after exposure may be an order of magnitude greater than that at 1 hour. Both proteomics and genomics offer a snapshot of the effects of the compound at a particular time point and do not reliably show the past or the future. Choice and consistency of time point are therefore important in ensuring that the data generated contain enough information—but not too much—to achieve the experimental objective. The study of proteins expressed is a reflection of what is happening in the cell, whereas the pattern of gene expression is a reflection of potential. Both techniques are, however, necessary to give a full picture of what is happening.

Genomic and proteomic expression both produce patterns of effect that may be associated with particular toxicities. This principle is also exploited in the use of nuclear magnetic resonance (NMR) spectroscopy, which can be used to reveal the amounts of small molecules (up to a molecular weight of 600) in a biological fluid, especially urine. The readouts from this can carry 3000 vertical lines, each specific for a particular chemical, such as citrate or hippurate. The problem is that there will be simultaneous changes in the quantities of many endogenous metabolites, quite apart from those due to the test compound. Sorting these data and then using appropriate software to plot them is an important part of the process. One aspect of this technology, which has been around for a long time, albeit without the pattern recognition software to make it more universally useful, has been the name given to it—metabonomics. If there is not much debate about the utility of the technique, there is a fair amount of speculation on how to pronounce it correctly.

A significant strength of these pattern-based techniques is their potential, when used as open systems, to show changes that may indicate toxicities due to novel mechanisms. However, they cannot show what the mechanism of toxicity is, and further investigation, which could be prolonged, would be needed. The problem here is that compounds showing novel patterns of change are likely to be dropped from development and not investigated further, unless the pattern is common to the members of a promising series of compounds. This may represent a missed opportunity for examining new mechanisms of toxicity that could be of significance in humans. Many successful, but toxic, compounds with acknowledged benefits might have been dropped from development if such techniques had been used for them. Such knowledge might prevent some of the more unpleasant surprises of chemical development and marketing but, equally, is probably responsible for the rejection of compounds that may have benefit despite toxicity.

The basic tenet, that chemicals can induce a particular profile of protein expression, is a useful one. The profiles produced by investigational chemicals may be compared with those from a library of known toxins, as an indication of anticipated effect. The weakness of this, and similar systems, is that it is critically dependent on the size and content of the database against which the profiles are compared. Inherently, it is unlikely that new types of toxicity can be predicted by this type of database-dependent system, because their discovery is dependent on their presence in the database for previous compounds. Much trouble is caused by unexpected or novel effects, which become apparent following significant human exposure and which are not predicted by routine screening techniques. The possibilities for comparing members of a series is very good, however, if the effects of the first few members have been adequately characterized.

The interest in these technically demanding and expensive technologies is based on their perceived potential as methods of early rejection of compounds from development. The savings in time, laboratory space, animals, and expense could be significant. However, for medicines at least, it should be appreciated that many successful drugs have significant toxicities that were found in preclinical development. Insofar as drug development is an art, part of this art is to assess the relevance and impact of these effects and to judge if the drug can be used beneficially in appropriate patients. The basis on which compounds are rejected or selected must be chosen in advance; otherwise, indiscriminate rejection may limit the development of effective chemicals—medicines or pesticides—in the future.

As with many such systems, it is usually better to reach a decision on the basis of several strands of data. Kramer et al. (2004) examined the integration of genomics and metabonomics data by treating rats with single doses of five known drugs or vehicles. They looked at urine samples collected up to 168 hours after using NMR, and gene expression profiles were determined in livers on four occasions up to the same time. Traditional data were collected in the form of clinical pathology on urine and serum samples. There was good correlation between ketone bodies monitored in urine and expression of genes involved in ketogenesis, when a peroxisome proliferator-activated receptor (PPAR) agonist was tested. They also found that while one technique alone could not separate low dose from control, this could be achieved by using both genomics and metabonomics together. Given that one of the great challenges in toxicity testing is to determine the significance of small differences, this technique may have utility in the long term for application in toxicity studies.

FUTURE UTILITY

The future of *in vitro* toxicology was reviewed in 1997 by a working party of the British Toxicology Society (Fielder et al. 1997). Subsequently many reviews on the future of *in vitro* testing have been published, especially by ECVAM, including a major review edited by Worth and Balls (2002). The whole approach is put in perspective, however, by the fact that this "nonanimal"-based report recommends tests such as whole-embryo culture, for which freshly killed animals are required, albeit not treated *in vivo*.

One of the areas of particular interest with *in vitro* techniques is their potential to replace mammalian tests in areas such as pharmacology and quality testing. The prime example of this is the LAL test, which exploits the sensitivity of amebocytes from *Limulus*, the horseshoe crab, to detect pyrogens in solutions for infusion; it is significantly more sensitive than the rabbits it replaces. Pharmacological models increasingly use *in vitro* techniques, such as the use of human cloned potassium channels for assessment of the potential for prolongation of the QT interval of electrocardiograms.

Other applications of *in vitro* techniques are discussed under the relevant sections below, particularly reproductive toxicity, genotoxicity, and irritancy and corrosivity. The last two have frequently been associated, in the past, with unacceptable animal suffering, particularly in the eye irritation tests in rabbits. As a result, there has been considerable effort expended in finding viable alternatives, the challenge being to reproduce the complexity of the situation in the living eye in the context of *in vitro* simplicity.

The early work with *Xenopus* is illustrative of the imaginative thinking that is being applied to the development of alternative strategies for safety evaluation. The downside of this is that, no matter how imaginative the final battery of assays, covering the whole gamut of toxicological endpoints *in vitro* will be time consuming and probably expensive. In addition, due to the absence of organ interactions in current *in vitro* systems, the search for an all-embracing system of *in vitro* tests is likely to be unsuccessful for the foreseeable future. However, they can be used to ensure that the animal tests that are finally needed are performed according to the most efficient design possible with the most appropriate endpoints studied.

Regulatory acceptance is a critical area for success in this field. Unless those in regulatory authorities can be persuaded that a new technique is a good model for effects in man (or any other target species), they are quite right to demand more information. This has come to be more carefully supplied in the form of rigorous validation protocols, which have seen acceptance of an increasing number of techniques in recent years.

In the context of the three Rs of Russell and Burch, these techniques hold out an almost mystical promise of a world of safety evaluation and testing that does not depend on animals. The problem is that they are often considered in an emotional sense, which ignores the wider context of the purpose of testing. A dispassionate, scientific approach is more likely to lead to abandonment of experimental protocols such as the 2-year bioassay for carcinogenicity in rodents, which uses hundreds of animals and can take up to 36 months.

SUMMARY

In vitro techniques have enormous potential that will be realized with increasing effect, leading to gradual regulatory acceptance of validated tests that have been shown to have relevance to prediction of human hazard.

- Their basic weakness and strength is the limited number of endpoints that can be covered in a single system in comparison with a whole animal, in which all the interrelationships between tissues and organs are intact.
- *In vitro* methods are one of the three tools in the toxicologist's chest, together with *in silico* and *in vivo* tests, and will be used in a much more targeted manner in coordination with these in the future.
- The ethical pressures to reduce reliance on animals, together with the increasing constraints of cost and time, will drive development of these techniques forward.
- Successful validation of these tools is essential for regulatory acceptance.

Ultimately, such alternative methods will come to play a more central role in safety evaluation in conjunction with *in silico* and *in vivo* methods.

REFERENCES

Amberg A, Harvey J, Czich A, Spirkl H-P, Robinson S, White A, Elder, DP. Do carboxylic/sulfonic acid halides really present a mutagenic and carcinogenic risk as impurities in final drug products? *Org Process Res Dev* 2015; 19(11): 1495–1506.

Davies DJ, Heylings JR, McCarthy TJ, Catherine M. Correa CM: Development of an in vitro model for studying the penetration of chemicals through compromised skin. *Toxicol In Vitro* 2015; 29(1): 176–181.

Davila JC, Cezar GG, Thiede M, Strom S, Miki T, Trosko J. Use and application of stem cells in toxicology. *Toxicol Sci* 2004; 79: 214–223.

Degitz SJ, Holcombe GW, Flynn KM, Kosian PA, Korte JJ, Tietge JE. Progress towards development of an amphibian-based thyroid screening assay using *Xenopus laevis*. Organismal and thyroidal responses to the model compounds 6-propylthiouracil, methimazole, and thyroxine. *Toxicol Sci* 2005; 87(2): 353–364.

Doherty KR, Talbert DR, Trusk PB, Moran DM, Shell SA, Bacus S. Structural and functional screening in human induced-pluripotent stem cell–derived cardiomyocytes accurately identifies cardiotoxicity of multiple drug types. *Toxicol Appl Pharmacol* 2015; 285(1): 51–60.

Fielder RJ, Atterwill CK, Anderson D, Boobis AR, Botham P, Chamberlain M, Combes R et al. BTS working party report on *in vitro* toxicology. *Hum Exp Toxicol* 1997; 16(S1): S1–S40.

Godoy P, Hewitt NJ, Albrecht U, Andersen ME, Ansari N, Bhattacharya S et al. Recent advances in 2D and 3D in vitro systems using primary hepatocytes, alternative hepatocyte sources and non-parenchymal liver cells and their use in investigating mechanisms of hepatotoxicity, cell signaling and ADME. *Arch Toxicol* 2013; 87: 1315–1530.

Kiaa R, Kelly L, Sison-Young RL, Zhang F, Pridgeon CS, Heslop JA, Metcalfe P et al. MicroRNA-122: A novel hepatocyte-enriched *in vitro* marker of drug-induced cellular toxicity. *Toxicol Sci* 2015; 144(1): 173–185.

Kramer K, Patwardhan S, Patel KA, Estrem ST, Colet JM, Jolly RA, Ganji GS et al. Integration of genomics and metabonomics data with established toxicological endpoints. A systems biology approach. *Toxicologist* 2004; 78(1 S): 260–261.

Lelievre D, Justine P, Christiaens F, Bonaventure N, Coutet J, Marrot L, Cotovio J. The episkin phototoxicity assay (EPA): Development of an in vitro tiered strategy using 17 reference chemicals to predict phototoxic potency. *Toxicol In Vitro* 2007; 21(6): 977–995.

Miyata K, Ose K. Thyroid hormone-disrupting effects and the amphibian metamorphosis assay. *J Toxicol Pathol* 2012; 25(1): 1–9.

OECD. The Adverse Outcome Pathway for Skin Sensitisation Initiated by Covalent Binding to Proteins. Part 1: Scientific Evidence. OECD Environment, Health and Safety Publications Series on Testing and Assessment 2012a; 168: 1–59.

OECD. The Adverse Outcome Pathway for Skin Sensitisation Initiated by Covalent Binding to Proteins. Part 2: Use of the AOP to Develop Chemical Categories and Integrated Assessment and Testing Approaches. OECD Environment, Health and Safety Publications Series on Testing and Assessment 2012b; 168: 1–46.

Reisinger K, Hoffmann S, Alépéec N, Ashikaga T, Barroso J, Elcombe C, Gellatly N et al. Systematic evaluation of non-animal test methods for skin sensitisation safety assessment. *Toxicol In Vitro* 2015; 26(1): 259–270.

Reuter H, Gerlach S, Spieker J, Ryan C, Caroline Bauch C, Mangez C, Winkler P et al. Evaluation of an optimized protocol using human peripheral blood monocyte derived dendritic cells for the In vitro detection of sensitizers: Results of a ring study in five laboratories. *Toxicol In Vitro* 2015; 29(5): 976–986.

Russell W, Burch R. *The Principles of Humane Experimental Technique. Universities Federation for Animal Welfare (UFAW)*. London: Methuen, 1959. May 1992, new edition. Available at http://altweb.jhsph.edu/.

Säfholm M. Developmental and Reproductive Toxicity of Progestagens in the *Xenopus (Silurana) tropicalis* Test System. Digital Comprehensive Summaries of Uppsala Dissertations from the Faculty of Science and Technology 1099. 51 pp. Uppsala: Acta Universitatis Upsaliensis, 2014. ISBN 978-91-554-8812-3.

Snodin DJ. An EU perspective on the use of in vitro methods in regulatory pharmaceutical toxicology. *Toxicol Lett* 2002; 127: 161–168.

Soldatow VY, LeCluyse EL, Griffith LG, Rusyn I. In vitro models for liver toxicity testing. *Toxicol Res (Camb)* 2013; 2(1): 23–39.

Urbisch D, Mehling A, Guth K, Ramirez T, Honarvar N, Kolle S, Landsiedel R et al. Assessing skin sensitization hazard in mice and men using non-animal test methods. *Regul Toxicol Pharmacol* 2015; 71(2): 337–351.

Vinardell MP. The use of non-animal alternatives in the safety evaluations of cosmetics ingredients by the Scientific Committee on Consumer Safety (SCCS). *Regul Toxicol Pharmacol* 2015; 71(2): 198–204.

Woolley A. *A Guide to Practical Toxicology*, 1st ed. London: Taylor & Francis, 2003.

Worth AP, Balls M. eds. Alternative (non-animal) methods for chemicals testing: Current status and future prospects. *ATLA* 2002; 30(S1): 125.

Zeiger E. Validation and acceptance of new and revised tests: A flexible but transparent process. *Toxicol Lett* 2003; 140–141: 31–35.

Zurita JL, Repetto G, Jos A, Del Peso A, Salguero M, López-Artíguez M, Olano D, Cameán A. Ecotoxicological evaluation of diethanolamine using a battery of microbiotests. *Toxicol In Vitro* 2005; 19: 879–886.

5 Toxicology *In Silico*

INTRODUCTION

Computational or *in silico* toxicology, that is, using computer models to determine toxicological endpoints, is perhaps the ultimate extension of the three Rs approach. Such methods have the potential to be quicker and cheaper than *in vivo* and even *in vitro* techniques. Once a model is established using existing data, there is no need to synthesize a chemical of interest—all that is needed is the chemical's structure. That is not to suggest that they are devoid of disadvantage, as George Box said, "Remember that all models are wrong; the practical question is how wrong do they have to be to not be useful" (Box and Draper 1987). Determining the extent to which the chemical of interest (test chemical) fits the data set with which the model has been trained (training set) is a key part of *in silico* toxicology. The question should always be asked—is this model valid for my compound; is it accurate; is it fit for purpose? A valid prediction should have a basis in science and be appropriate for the test chemical.

Before the advent of computational toxicology, considerable human expertise was (and remains) available to predict toxicity by scrutiny of structures, and there is no doubt that this is a powerful source of knowledge and judgment for use in predictions of toxicity and hazard. However, this knowledge base is best considered as being volatile due to illness, retirement, resignation, or death, and as a result is not always available where and when it is wanted. Furthermore, such judgment may lack consistency. The intention of *in silico* toxicology is to bring all such expertise to one place where it may be accessed at any time by anyone.

Structural activity relationship (SAR) programs attempt to link a chemical's structure (and in some instances, calculated physical properties) to its potential biological activity. SARs that use a mathematical relationship to estimate an endpoint's occurrence are termed quantitative SARs or QSARs. Those that do not provide such numerical assurances and are only qualitative drop the Q. Collectively, both types are referred to as (Q)SARs.

There are now multitudes of (Q)SAR programs available to help predict all manner of endpoints ranging from acute toxicity to phospholipidosis and peroxisome proliferation, from mutagenicity to reproductive and developmental toxicity. Outside of toxicology, computers have been used to generate models of pharmacokinetic parameters (such as absorption, distribution, metabolism, and excretion) as well as pharmacology models (such as receptor occupancy) and physicochemical endpoints (such as octanol–water partition coefficient). As the number of endpoints grows, so does the potential number of uses. Lead candidate selection and investigation for mutagenicity of a potential intermediate or an impurity are just two ways (Q)SARs are being used now to aid with the determination of toxicity.

(Q)SAR METHODOLOGIES

All (Q)SAR models begin with data gathering; it is what is done to the data following this process that defines the model type. As inferred above, there are two types of systems: those that use mathematical algorithms to produce quantitative results [(Q)SAR] and those that use rules (SAR), which may only indicate the plausibility of effects and do not offer the possibly fallacious comfort of a dubiously derived number. These are known as statistical and rule based respectively. Hybrid systems that use both methodologies do exist but are not widespread.

Rule-based SARs use rules generated by experts (generally humans) from literature to determine the toxicity of a structure. These rules try to determine plausible and probable mechanisms and link them to specific structures within the training compounds. If the program detects the presence of a triggering moiety—sometimes termed a toxicophore—in a compound, a structural alert will be triggered. Such structural alerts can be associated with probability (either numerical or nonnumerical) assertions; it is dependent on the user to determine how valid the prediction is to the compound. The method by which these rules are accessed varies from program to program. For instance, some programs, such as ToxTree (developed by Ideaconsult Ltd), use a decision tree–type approach, while others, such as Derek Nexus (developed by Lhasa Ltd), use pattern recognition software to identify potential toxicophores within the structure.

Statistical (Q)SARs use large data sets coupled with mathematical modeling (such as linear regression) to "learn" the toxicity of certain structures. The (Q)SAR models analyze the training structures for recurrent structural moieties and physical properties and can be curated by the manufacturer to eliminate any obvious false positives. There are multiple mathematical models for (Q)SAR models, such as univariate regression, multiple linear regression, partial least squares, artificial neural nets, fuzzy clustering and regression, K-nearest neighbor clustering, and so on. Each software provider will have their own, closely guarded, flavor of mathematical model with its own particular variation.

In a 2010 review of (Q)SAR written for the European Food Safety Authority (EFSA), the advantages and disadvantages of the various (Q)SAR approaches were compared; this is summarized in Table 5.1 (EFSA 2010).

Both methodologies use databases of new or historical data to validate their predictions. Sources of these databases vary between providers; some use publically available data alone, while others use proprietary data provided by external sources in addition. Whatever the source, curation of such data sets is a vital component as it removes errors and allows the format to be tailored into one that can be read by the software. Internal validation statistics give an indication of the goodness of fit of the model to the data used to generate it by analyzing similar or the same structures. Such validation can reveal how scrupulous the developers of the software have been, as a model that outperforms the assay it is modeling may be suspect, such as model overfitting problems. Model overfitting is where a model contains more features than there need to be or is more complex than required, thereby violating Occam's razor or the principle of parsimony (Hawkins 2004).

External validations allow the user to view how accurate a model is when faced with previously unseen compounds and allow assessment of the model's performance.

TABLE 5.1
Comparison of *In Silico* Methodologies

Method	Advantages	Disadvantages
Rule based	• Mechanistically connected to the predicted endpoint • Provide reasoning for the predictions • In many cases support the prediction with literature references or expert knowledge	• Often restricted and/or ill-defined applicability domain • Usually cannot explain differences of the activity within a chemical class • Usually have lower accuracy of the prediction than statistical models
Statistical	• Usually have high accuracy of the predictions • Predictions can be used for preliminary research when mechanism of action is unknown	• Usually difficult to interpret the model predictions • Often do not provide mechanistic reasoning of the predictions • Often nontransparent to the end user

Source: EFSA: Applicability of QSAR analysis to the evaluation of the toxicological relevance of metabolites and degradates of pesticide active substances for dietary risk assessment. *EFSA Journal.* 2010. Copyright Wiley-VCH Verlag GmbH & Co. KGaA. Reproduced with permission.

The ability of a program to predict the validity of a result is measured in a number of ways:

• Accuracy (also known as concordance): ratio of true results to false results
• Sensitivity: ratio of true positives (TP) to false negatives (FN)
• Specificity: ratio of true negatives (TN) to false positives (FP)

These measures are calculated using the following formula:

$$\text{Accuracy} = \frac{TP + TN}{TP + FP + TN + FN} \quad \text{Sensitivity} = \frac{TP}{TP + FN} \quad \text{Specificity} = \frac{TN}{TN + FP}$$

Accuracy tells one how often the model correctly predicts the empirical result—low accuracy indicates that the model did not perform well with the data set and the model may not be applicable for use with such compounds. Sensitivity gives an idea of how able the model is to correctly identify empirically positive compounds (true positive)—from a regulatory point of view, this is a vital component of a (Q)SAR model. A high specificity indicates that a model is well able to predict the endpoint of concern and produces few false-negative results (i.e., those that are empirically negative but give positive results in the model), whereas a low sensitivity means that there are a high number of falsely negative results. In a similar fashion, specificity gives a measure of how able the model is to correctly identify empirically negative compounds (true negatives). A model with low specificity could be expected to produce more false-positive results, though of course, this depends on the data set.

Overall, it is considered more acceptable to have a high sensitivity (model is more likely to correctly predict true positives) than high specificity (model is more likely to correctly predict true negatives)—though in an ideal world, both would be high. A model that shows low sensitivity and high specificity is unlikely to be acceptable from a regulatory standpoint without adequate justification for its use, as it cannot correctly predict potentially toxic substances.

Due to the vast numbers of computational toxicology programs and endpoints available as well as continuous model improvements—a database that is not updated is useless—there are large swathes of research papers that have performed such validations and derived performance figures. Although it is tempting to report these figures, it is neither practical nor feasible to report on specific programs (or rather, specific snapshots of programs) because as soon as the data are generated, they may be superseded and outdated. That is not to say that such comparisons are not worthwhile; far from it. The increasing accuracy of these programs has led to acceptance by regulators of results from *in silico* predictions in place of an empirical assay. As is discussed in the International Conference on Harmonisation (ICH) M7 case study (Case Study 5.1), negative predictions of mutagenicity by such software have been accepted as indicating the absence of a structural alert and that no further toxicological investigation would be necessary. The combination of *in silico* tools with expert user knowledge and interpretation has been shown, in one study, to increase the mutagenicity negative predictivity value of the software alone from 94% to 99% (Dobo et al. 2012).

Performance figures, whether expressed as percentages or hit rates, are meaningless unless they are considered in the context of what the output will be used for. Opinions on the significance of an accuracy of 75% may vary. Such figures are produced from comparison of the computer result with those from actual tests, with all their procedural differences or irregularities compounded by debates on the interpretation of the results.

Outside of bacterial mutagenicity, if such systems are used as an adjunct to the process of hazard prediction and not as the sole means of assessment, they have considerable utility and cannot be ignored. There are some important caveats, however. The various systems available have significant differences in performance in different areas, and some are better with particular molecular classes than others. This means that, for given uses, some systems will be better than others, and this must be taken into account when choosing a system.

As implied earlier, it is better not to use individual systems in isolation, and there is a range of supporting tools that can be used as adjuncts to the prediction process. These include any system used in the design of the molecule, for instance, for pharmacological (Q)SAR screening. There are also systems that allow literature searches to be based on molecular structure or on parts of the molecule. It is also probable that physiologically based pharmacokinetic models will become of increasing importance, although they have a more traditional role in risk assessment.

Ultimately, in consideration of the performance of computer systems, the cynical view must be remembered, that the toxicity of any novel molecule cannot be accurately predicted until the data for that molecule are entered into the database. Equally, it must be asked at the beginning of the process if you actually need an accurate prediction or whether you simply need to rank compounds in terms of expected toxicity, so that the best candidate can be selected for development. In this case, the consistency and relevance of output will probably be more significant than ultimate accuracy.

MODEL VALIDITY

Whatever the endpoint, a (Q)SAR must be valid for the chemical of interest and fit for the intended purpose for which it is being employed. Furthermore, if the results of the analysis are to be used for regulatory submission, the (Q)SAR should be relevant and acceptable by the target authority. This can be displayed as a Venn diagram (Figure 5.1).

Further to this, the Organization for Economic Co-operation and Development (OECD) has (at the 37th Joint Meeting of Chemicals Committee and Working Party on Chemicals, Pesticides and Biotechnology) adopted five principles for establishing the validity of the (Q)SAR models for use in regulatory assessment of chemical safety (OECD 2007). These are as follows:

1. A defined endpoint

    Endpoint refers to any physicochemical property, biological effect (human health or ecological), or environmental fate parameter that can be measured and therefore modeled, ideally drawn from the same experimental protocol and conditions but this is not always possible.

2. An unambiguous algorithm

    To allow transparency, the QSAR model should be expressed in the form of an unambiguous algorithm.

3. A defined domain of applicability

    Allows expression of model limitations in terms of the types of chemical structures, physicochemical properties, and mechanisms of action for which the models can generate reliable predictions.

4. Appropriate measures of goodness of fit, robustness, and predictivity

    This principle expresses the need to provide two types of information: (a) the internal performance of a model (as represented by goodness of fit and robustness), determined by using a training set; and (b) the predictivity of a model, determined by using an appropriate test set.

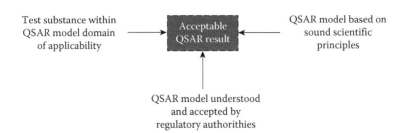

FIGURE 5.1 Factors contributing to an acceptable (Q)SAR model. (From European Chemicals Agency, http://echa.europa.eu/.)

5. A mechanistic interpretation, if possible

The intent of this principle is therefore to ensure that there is an assessment of the mechanistic associations between the descriptors used in a model and the endpoint being predicted, and that any association is documented. The "if possible" caveat recognizes that such interpretation is not always available.

One of the most important aspects of these principles is the domain of applicability or applicability domain. Each (Q)SAR is built from a specific set of empirical results, be it from mutagenicity, skin sensitization, or neurotoxicity, as a training set. If a test substance lies outside of (Q)SAR's applicability domain, then the results generated are of low reliability, and the model is not applicable to the test compound. Furthermore, such a prediction may not be accepted by the regulator.

REGULATIONS, QSARs, AND EXPERTS

As regulators become more familiar with *in silico* systems, so their results become more likely to be accepted alongside empirical data for read-across purposes or ultimately in place of *in vitro* or *in vivo* assays. The ICH M7 guideline on the assessment of potentially DNA-reactive pharmaceutical impurities allows for the replacement of an empirical assay for bacterial mutation with an *in silico* approach—this is expanded upon in Case Study 5.1. *In silico* tools can also be a useful adjunct when assessing the potential toxicity of extractable and leachable compounds in medical devices, pharmaceutical packaging, and even food contact products. Such evaluations coupled with targeted read-across are looked at in Chapter 17.

The European Commission's Joint Research Centre (JRC) has produced the QSAR Model Reporting Format (QMRF), which is a "harmonised template for summarising and reporting key information on (Q)SAR models, including the results of any validation studies." The QMRFs are structured according to the OECD QSAR validation principles and can be accessed via the JRC website or requested from the software's manufactures. In a similar vein, the QSAR Prediction Reporting Format (QPRF) provides a template on how to report substance-specific predictions generated by (Q)SARs.

The Registration, Evaluation, Authorisation and restriction of Chemicals (REACH) guidance document on QSARs and grouping of chemicals lists seven ways in which a QSAR can be employed:

1. Provide information for use in priority-setting procedures
2. Guide the experimental design of an experimental test or testing strategy
3. Improve the evaluation of existing test data
4. Provide mechanistic information (which could be used, for example, to support the grouping of chemicals into categories)
5. Fill a data gap needed for hazard and risk assessment
6. Fill a data gap needed for classification and labeling
7. Fill a data gap needed for persistent, bioaccumulative, and toxic (PBT); and any that are very persistent and very bioaccumulative (vPvB) assessment

In addition, ECHA has produced a practical guideline on "how to use and report (Q)SARs," which details many of the same things as in the above guidance document but offers practical examples of how to go about such analyses.

As a guide, an *in silico* report should be written in a similar manner to any standard toxicology report. The report should contain a title page, an abstract, an introduction, a methods section (with descriptions of software used and version numbers), a results section, a discussion section (which may include performance figures for the QSAR), and a conclusion. If possible and applicable, results (either full or abbreviated) should be included in the report. In many cases, it is not practical to include 200 pages of raw output for each structure, so a summary of the data should suffice. At the other end of the spectrum, many expert systems do not produce any output unless an alert has been triggered; in such cases, it is important to list which endpoints were not triggered.

Although this may be a somewhat biased opinion, expert interpretation is a key element in the use of (Q)SARs and can make the difference between a regulatory acceptance and the bitter taste of rejection. Under ICH M7, the use of an expert is a vital component of the *in silico* analysis. Expert opinion may dismiss a positive prediction, but equally, the expert may reverse the *in silico* consensus to dismiss a negative prediction. The dismissal of such false-positive and false-negative results is vital in an expert's analysis and is based mainly on his/her ability to interpret the output and see if the results apply to the test structure.

In assessing a molecule for potential adverse effects, a human expert looks at it in terms of molecular structure, size, constituent groups and elements, ionization potential and polarity, and probable metabolism. It should also be possible to say if one group will affect the influence of another nearby group, for instance, by electron withdrawal or steric hindrance. The molecular weight gives a rule-of-thumb guide to whether it will be excreted in the bile or urine; the physicochemical data will give some idea of absorption potential and corrosivity. The human expert should assess the three-dimensional structure of the molecule and whether it has any chiral centers of asymmetry. Previous knowledge and experience may be available to the expert to warn of possible effects associated with that class of chemical or with particular structural conformation. This may also be associated with lateral thinking that leads to literature searches for suspected relationships or contributing factors.

Amberg et al. (2016) have examined the various instances in ICH M7 where a clear answer from the two-model approach was not forthcoming, such as a single *in silico* positive, or where a structure is outside of the domain of applicability of the model. The paper details the various situations experts may find themselves in when performing an ICH M7–compliant review and is a good starting text for such analyses. For example, for statistical models, Amberg et al. suggest six points for consideration when assessing a result, namely, coincidental features (does the training set contain reactive features not found in the test compound?), mitigating features, limited training set examples, no significant positive model features, irrelevant training set examples, and incorrect/inadequate underlying data. Some parties, including the US Food and Drug Administration (FDA), advocate the use of a third QSAR system when investigating a disputed or out-of-domain result; however, there comes a point when one should recognize that perhaps no computational model is appropriate for the test structure and another route should be taken.

CASE STUDY 5.1   ICH M7

The first step to regulatory acceptance of (Q)SAR use in place of an *in vitro* assay came in June 2014 with the publication and adoption of the ICH M7 guideline by the US FDA, Europhean Medicine Agency (EMA), and Japanese Pharmaceuticals and Medical Devices Agency (PDMA) (ICH M7 2014). This guideline allows for the replacement of a bacterial mutagenicity assay, for pharmaceutical impurities, following an initial analysis of the available database and literature, with "two (Q)SAR prediction methodologies that complement each other." It goes on to indicate that "one methodology should be expert rule-based and the second methodology should be statistical-based" and that these (Q)SARs should follow the general validation principles laid down by the OECD.

If relevant carcinogenic or genotoxic/mutagenic data are found on the chemical of interest, it should be classified as class 1, 2, or 5 (see table at the end of this case study). In the absence of data, then the separate methodology (Q)SARs analysis is employed, which then leads to classification into class 3, 4, or 5. The crucial statement of this document is "the absence of structural alerts from two complementary (Q)SAR methodologies (expert rule-based and statistical) is sufficient to conclude that the impurity is of no mutagenic concern, and no further testing is recommended." It is also important to note that the ICH M7 guidelines also state that "if warranted, the outcome of any computer system–based analysis can be reviewed with the use of expert knowledge in order to provide additional supportive evidence on relevance of any positive, negative, conflicting or inconclusive prediction and provide a rationale to support the final conclusion."

If both software systems give valid, negative results, then the structure is classified as class 5, and no further testing is required. In cases where a positive result is observed—in the statistical and/or the rule-based (Q)SAR— expert judgment must be used to determine whether the prediction is valid. If the prediction is valid and the alerting structure is not shared with the parent, then the structure should be classified as class 3. If the prediction is valid and the alerting structure shared with the nonmutagenic parent, then the structure should be classified as class 4, and no further testing is required. The key point about the latter class is that the parent shares the structural alert in the "same position and chemical environment." This is quite an ambiguous statement and relatively open to interpretation; thus, it is often necessary to run the parent structure alongside the impurities to validate the predictions.

In some instances, the results of the (Q)SAR analysis may not be clear cut. A prediction that is not in domain or has equal weight of evidence for and against the result does not automatically allow one to classify the structure. In these instances, expert judgment must be used to determine if the substance is mutagenic or nonmutagenic. The arguments made should be based on rational scientific argument as well as being clear and easily understood. If no such argument can be made, based on the available data, then use of a third (Q)SAR program should be considered.

If a compound is categorized as class 3, there are two options. Firstly, the daily exposure of the patient to the compound can be controlled to below the Threshold for Toxicological Concern (TTC). The TTC for lifetime exposure (>10 years) is 1.5 µg/day and is based on an excess lifetime cancer risk of <1 in 100,000. If the duration of exposure is less than 10 years, the TTC value chosen may be adjusted. It should be noted that if multiple impurities are found to be mutagenic, the TTC value is 5 µg/day for a period >10 years. If the exposure to a substance cannot be controlled, a bacterial reverse mutation (Ames) assay can be conducted with the substance of concern; if it is mutagenic, then it is categorized as class 2; if it is nonmutagenic, then it is class 5. The guideline goes on to suggest that in the event of a positive Ames assay, further *in vivo* assays [such as the transgenic mutation assays, Pig-a assay (blood), micronucleus test, (blood or bone marrow), rat liver unscheduled DNA synthesis (UDS) test, and comet assay] may be conducted. The assay choice depends on the circumstances of the positive Ames test and is further expanded on in the guideline text.

Taken from ICH M7 guidelines:

Class	Definition	Proposed Action for Control
1	Known mutagenic carcinogens	Control at or below compound-specific acceptable limit
2	Known mutagens with unknown carcinogenic potential (bacterial mutagenicity positive,[a] no rodent carcinogenicity data)	Control at or below acceptable limits (appropriate TTC)
3	Alerting structure, unrelated to the structure of the drug substance; no mutagenicity data	Control at or below acceptable limits (appropriate TTC) or conduct bacterial mutagenicity assay: If nonmutagenic = class 5 If mutagenic = class 2
4	Alerting structure, same alert in drug substance or compounds related to the drug substance (e.g., process intermediates), which have been tested and are nonmutagenic	Treat as nonmutagenic impurity
5	No structural alerts, or alerting structure with sufficient data to demonstrate lack of mutagenicity or carcinogenicity	Treat as nonmutagenic impurity

[a] Or other relevant positive mutagenicity data indicative of DNA reactivity-related induction of gene mutations (e.g., positive findings in *in vivo* gene mutation studies).

In a draft addendum [ICH M7(R1)], released in June 2015, the ICH released details of how compound-specific acceptable intakes may be derived and gives examples of such calculations. It is a useful starting point if the analysis goes down this route (ICH M7 2015).

It is important to consider both the parent structure and its indication when performing an assessment under ICH M7. In some circumstances, such as where the parent is itself mutagenic or carcinogenic (i.e., ICH M7 class 1), a substance-specific value may be derived based on carcinogenic potency and other such considerations. Alternatively, if there is a known threshold and associated mechanism for the parent's genotoxicity, then such threshold may be applied with adjustment to the impurity.

Drug substances/products intended for the advanced cancer indications (as defined in ICH S9) are not included in the scope of ICH M7. Furthermore, ICH M7 is not applicable to the following drug substances/products: biological/biotechnological, peptide, oligonucleotide, radiopharmaceutical, fermentation products, herbal products, and crude products of animal or plant origin. Neither is the guideline intended for excipients used in existing marketed products, flavoring agents, colorants, and perfumes. However, even though it is not intended for such indications or products, it can act as a useful guide on how to conduct such computational studies.

No matter what the results of the (Q)SAR analysis, any argument made must be scientifically valid and based on the best knowledge of a structure's potential mode of action (if any). Finally, expert knowledge is subjective and may vary between experts in terms of content and quality. ICH M7 represents the (small) first step on the road to regulatory acceptance of (Q)SARs for further endpoints. It is doubtful, at least for the foreseeable future, that (Q)SARs will replace empirical assays; however, their use with targeted read-across is certainly a less resource-intensive way of determining potential toxicity.

CHOOSING A (Q)SAR

Selection of a computational toxicology program is largely a personal choice and very dependent on the circumstance requiring its use. The requirements for lead candidate selection in a very early stage of development are likely to be different from the needs for the assessment of a potentially mutagenic impurity. Furthermore, although it is possible, it is not advisable to build a computational toxicology model from scratch, unless there is a very specific goal in mind. Quite apart from having to master the dark art of computer programing, the builder would have to procure a large data set, and curate and then validate the model—quite ignoring the fact that such a model would be unlikely to be accepted by a regulator.

For these reasons, it is more usual to buy or locate a ready-made program that may allow adaptation (user trainable) if necessary. As a brief guide, a practical (Q)SAR program should

- Allow easy entry of molecular structures
- Contain the endpoint of interest
- Recognize structures associated with toxicity or unwanted pharmacology
- Predict interactions between different parts of the molecule, electron-withdrawing characteristics, etc.
- Be transparent as to how or why the prediction results were produced
- Allow prediction/calculation of physicochemical properties, such as log P and pKa, or give scope for data entry
- Be sensitive to the significance of chemically minor changes (such as substitution of S for N) that may be toxicologically significant
- Be easily updated with continuous development by the producers
- Be regulatory acceptable and comply with OECD principles on (Q)SAR validation

ENDPOINTS

Almost every possible empirical endpoint has a corresponding *in silico* model attached to it; thus, it is possible to find models on almost anything. Generally, simpler endpoints such as mutagenicity will be easier to predict than more complex ones such a behavioral change. It should also be considered that absence of an alert in a structure does not necessarily equate with a negative prediction, merely that the model does not contain sufficient evidence to link a moiety to a particular endpoint. The following is a small selection of possible endpoints available, though not every endpoint will be available in every *in silico* tool:

- Physiochemical properties: log P, water solubility, vapor pressure, boiling point, melting point
- Pharmacokinetics: exposure, absorption, distribution, metabolism, excretion
- Pharmacology: human Ether-à-go-go Related Gene (hERG) channel inhibition, anticholinesterase activity, receptor-binding efficacies, protein binding, DNA binding
- Single-dose toxicity: acute inhalation toxicity LC_{50}, acute oral toxicity LD_{50}, high acute toxicity
- Repeat-dose toxicity: organ toxicity (hepatotoxicity, nephrotoxicity, splenotoxicity, neurotoxicity, cardiotoxicity, ocular toxicity, bone marrow toxicity, etc.), maximum tolerated dose (MTD), maximum recommended daily dose, chronic lowest observable adverse effect level (LOAEL), gastric irritation, Cramer class
- Genotoxicity and carcinogenicity: bacterial mutagenicity, mammalian mutagenicity, *in vitro* clastogenicity (chromosome damage, sister chromatid exchange), *in vivo* clastogenicity (mouse micronucleus, *in vivo* chromosome damage), human carcinogenicity, rodent carcinogenicity

- Reproductive and developmental toxicity: developmental toxicity (including teratogenicity), developmental neurotoxicity, reproductive toxicity, fertility, testicular toxicity, estrogenicity
- Local tolerance: skin irritation, eye irritation, corrosivity, skin sensitization, photoallergenicity
- Other toxicities: α2u nephropathy, methemoglobinemia, phototoxicity, respiratory sensitization, lachrymation, anaphylaxis, mitochondrial dysfunction, methemoglobinemia, phospholipidosis
- Ecotoxicology: acute toxicity (fish, daphnia, etc.), chronic toxicity
- Environmental fate: bioaccumulation, biodegradation, soil absorption

COMMERCIAL VERSUS NONCOMMERCIAL

(Q)SARs can be found as open-source (i.e., free-to-use) programs or on a commercial license basis. In addition, they can also be developed in house using techniques such as support vector machine and comparative molecular field analysis. For most purposes, however, it is less time consuming and more practical to use an off-the-shelf product rather than taking the time to construct and validate one's own model (also, such in-house models may not have the same degree of regulatory acceptance). The question of whether to use the commercial or noncommercial software depends on the application; for investigatory work, an open-source program may be adequate, but for a regulatory submission, a commercial (Q)SAR may be more applicable. It should also be noted that at least three of the commercial systems (Derek Nexus, Leadscope Model Applier, and CASE Ultra) are the subject of research collaborations with the US FDA.

The main difference between commercial and noncommercial software is principally one of support and continuing development. Although some of the open-source programs have originated from organizations such as the OECD, US Environmental Protection Agency (EPA), and European Union, their development may not be as sustained as that of the commercial systems. Furthermore, users are generally left to their own devices when they find a prediction that cannot be explained.

As technology advances, open-source software is becoming increasingly prevalent and, in some cases, can give a good indication of toxicity. However, when using these programs, it should always be considered that a database that is not updated on a regular basis is potentially out of date, and hence potentially unreliable. This may be due to the development of new knowledge and understanding that has not been incorporated into the software. Open-source software offers a cheap method of initial assessment but does not currently offer the same regulatory credibility and acceptance as some commercial software.

Commercial software is not free to run and can be associated with relatively high setup and maintenance costs. When choosing a commercial (Q)SAR, the support provided by the issuing company and the regularity with which it is updated should be considered. Commercial software tends to have better user interfaces and generally has fewer bugs and glitches. It is usually updated regularly—once every

12 months is typical—and offers the confidence that it incorporates the latest thinking and understanding in the endpoints addressed. Finally, the commercial software companies have dedicated teams of technicians to help with little hiccups and should offer training on their tools.

PITFALLS IN *IN SILICO* TOXICOLOGY

One of the major pitfalls in these *in silico* technologies is the more complex the endpoint, the harder it is for the software to predict. Relatively simple endpoints, such as predicting whether or not a substance will be DNA reactive, are achievable and produce consistently reliable results. However, judging whether this substance will go on to produce carcinogenicity—based on a limited data set and only physiochemical parameters—can be extremely difficult. Toxicity that occurs through specific receptors, is idiosyncratic, or is immune based may not be predicted well by general toxicity QSARs.

Furthermore, QSARs are very much like sewers—you get out only what you put in. If bad or poorly curated data go into the training set of the QSAR, when a similar structure comes along, then it will be poorly predicted relative to empirical results. For instance, if an experimental Ames assay uses metabolic activation from maize in place of the more common rat liver, this would generally not be considered a valid test. However, the QSAR may not be aware of this when it is learning the data and can create false-positive results. Another example of this is the carboxylic acid halides, which were found to be positive when assayed for mutagenicity in the presence of DMSO (due to a solvent-created artifact) but negative when tested in acetone (Amberg et al. 2015). Thus, it is of vital importance to understand where the data have come from and how the predictions have been generated when evaluating their validity.

Databases are also affected by the tendency to place emphasis on the chemicals that have been shown to have the predicted toxicity over those that were negative or not toxic. Absence of toxicity is as important as presence in prediction systems. A correctly curated data set should be able to account for a difference in the number of positive and negative structures. Equally, there is a tendency for the toxic chemicals—the development mistakes—to be buried in company archives in confidential reports that are not allowed into the public domain. Where the data or rule bases are developed by the users in a cooperative manner, there is a corresponding increase in the strength of the system and, inherently, of its credibility and the confidence that can be placed in its predictions.

The final pitfall of *in silico* toxicology is one of lack of acceptance and understanding on the part of regulators or other interested parties. You may have selected the most appropriate programs and written the most elegant argument based on the results of similar structures, but it could be rejected if the argument is not considered valid. This, however, is true of any assay and any argument whether it be empirical, *in cerebro*, or *in silico*.

SUMMARY

Computational toxicology offers an inexpensive and quick method of rapidly assessing the potential hazards of a compound or group of compounds.

- From lead candidate selection to prediction of genotoxicity, the use of such programs is growing.
- (Q)SAR models are tools and should not be used alone as the final answer—interpretation is key.
- Not all (Q)SARs are created equal; selecting the correct model for the compound of interest is vital if the results are to be viewed as valid.

Regulatory acceptance is growing, and an *in silico* approach can be used in place of a bacterial reverse mutation assay for potentially DNA-reactive impurities; whether this will expand to other genotoxicity assays and other endpoints remains to be seen.

REFERENCES

Amberg A, Beilke L, Bercu J, Bower D, Brigo A, Cross KP, Custer L et al. Principles and procedures for implementation of ICH M7 recommended (Q)SAR analyses. *Regul Toxicol Pharmacol* 2016; 77: 13–24.
Amberg A, Harvey JS, Czich A, Spirkl H-P, Robinson S, White S, Elder DP. Do carboxylic/sulfonic acid halides really present a mutagenic and carcinogenic risk as impurities in final drug products? *Organic Process Res Dev* 2015; 19(11): 1495–1506.
Box GEP, Draper NR. *Empirical Model Building and Response Surfaces*. New York: John Wiley & Sons, 1987.
Dobo KL, Greene N, Fred C, Glowienke S, Harvey JS, Hasselgren C, Jolly R et al. In silico methods combined with expert knowledge rule out mutagenic potential of pharmaceutical impurities: An industry survey. *Regul Toxicol Pharmacol* 2012; 62(3): 449–455.
EFSA. Scientific Report submitted to EFSA. Applicability of QSAR analysis to the evaluation of the toxicological relevance of metabolites and degradates of pesticide active substances for dietary risk assessment. Question No EFSA-Q-2009-01076, 2010.
Hawkins DM. The problem of overfitting. *Journal of Chemical Information and Computer Sciences* 2004; 44(1): 1–2.
ICH M7. ICH M7—Assessment and Control of DNA Reactive (Mutagenic) Impurities in Pharmaceuticals to Limit Potential Carcinogenic Risk, 2014.
ICH M7. Addendum to ICH M7: Assessment and Control of DNA Reactive (Mutagenic) Impurities in Pharmaceuticals to Limit Potential Carcinogenic Risk: Application of the Principles of the ICH M7 Guideline to Calculation of Compound-Specific Acceptable Intakes M7 (R1), 2015.
OECD. Guidance Document on the Validation of (Q)SAR Models. ENV/JM/MONO(2007)2, 2007.

6 Safety Pharmacology

INTRODUCTION

The purpose of safety pharmacology studies, in the context of drug development, is to support first administration of novel drugs to man, usually healthy human volunteers. The main reason for their conduct is the avoidance of life-threatening side effects that can be elicited even by drugs that are on the market. Safety pharmacology, as a key discipline in pharmaceutical development, started emerging in the 1990s. In 1991, the Japanese Ministry of Health and Welfare issued a guideline for general pharmacology, which included studies to detect unexpected effects on the function of major organ systems. The intention was to encourage companies to carry out a series of tests from a first tier and then to use suggested tests from a second tier to investigate any effects. This chapter is a review of safety pharmacology as required for the development of most types of pharmaceutical under the guidelines promulgated by ICH—the International Conference on Harmonisation of Technical Requirements for Registration of Pharmaceuticals for Human Use. Although there is extensive reference in the text to the text of these guidelines, we have also attempted to put safety pharmacology in its proper context in terms of practice, interpretation, and the pitfalls that may be encountered. A review by Redfern et al. (2002) helps to put the subject more deeply into context and is a very useful text.

It is useful to draw a distinction between safety pharmacology and what may be characterized as the investigation of the intended pharmacology of the compound under test—in other words, an estimate of its potential clinical efficacy. In current parlance, *primary pharmacodynamics* refers to studies exploring the mode of action or effects of a substance relative to its intended therapeutic target, while studies of *secondary pharmacodynamics* examine modes of action that are not related to the intended target.

Historically, both these areas of experiment have been referred to as general pharmacology. Safety pharmacology is a refinement of secondary pharmacodynamics in that it looks typically at a defined set of organs (as detailed in the core battery) and tends to ask more general questions of test systems (although some are very specific, as we shall see). ICH guideline S7A (2000) defines safety pharmacology as "those studies that investigate the potential undesirable pharmacodynamic effects of a substance on physiological functions in relation to exposure in the therapeutic range and above." The overall intention is to identify effects of the kind that have led to the withdrawal of successful drugs from the market, sometimes after many years of consumer exposure. One of the challenges to this intention is that some of the more serious reactions to drugs are idiosyncratic and thus will not be identified by any of the nonclinical studies—including safety pharmacology. Safety pharmacology, therefore, is not a catchall discipline infallibly weeding out dangerous drugs; however, when used critically, it can be a useful tool in the risk assessment process that leads to the authorization to market a new drug. These studies may also highlight effects that may be expected in cases of human overdose.

Under normal circumstances, the core package of safety pharmacology studies should be completed before the first trial in man as recommended in ICH S7A. Follow-up studies may be conducted subsequently, but the normal regulatory requirement is to do the core package early in drug development.

GENERAL PRINCIPLES

In most safety pharmacology experiments, exposure of the test system is transient, either as the result of a single dose *in vivo* or through a defined incubation period *in vitro*, perhaps at rising concentrations. This may be modified when safety pharmacological investigations such as behavioral studies, electrocardiology, or respiratory evaluation are added to the investigations conducted in an otherwise standard toxicity study.

As noted, safety pharmacology studies are used to investigate undesirable effects on organ function that may have relevance to human safety. The key word here is *relevance*, and there has been an often acrimonious debate as to what exactly constitutes a relevant effect, or what is an acceptable safety margin between effect and anticipated human exposure levels. In most cases, the severity of the indication has to be taken into account, greater risk being acceptable with more severe indications such as cancer or other life-threatening conditions.

As always with guidelines, the dictum is to use appropriately selected test methods and systems according to the expected effects of the chemical under investigation, while at the same time, taking note of the guideline suggestions of appropriate organs for investigation. Thus, effects may be expected or reasonably anticipated due to the intended pharmacological target (for example, proconvulsive effects in anticonvulsive agents) or effects that are not due to the intended pharmacological action (for example, QT prolongation and cardiac effects in a variety of noncardiovascular drugs, such as the antihistamine terfenadine or the antipsychotic haloperidol). The evidence from secondary pharmacodynamic studies may be used in the selection of tests and test systems, and in the overall interpretation.

The guidelines give some indications of test species, saying that they should be chosen according to relevance to man, in terms of metabolism, pharmacological sensitivity to the class of compound, and so on. Although this would seem to give a broad choice, the palette of possibilities is in fact limited by convention and practicality. It may be considered that an exotic species of animal may be a good model for a particular effect or pharmacological target; however, proving that this is valid and scientifically justifiable is likely to be time consuming and expensive. In addition, regulatory authorities tend to be (rightly) conservative in these matters, meaning that test system choice is made mostly from a number of well-established *in vitro* systems and whole animals or excised organs from well-known laboratory species such as the rat or guinea pig and, more rarely, the dog or nonhuman primate.

TESTS TO BE CONDUCTED

ICH S7A states that the objectives of these studies are the identification of undesirable pharmacodynamic properties relevant to human safety, to evaluate effects seen in toxicity or clinical studies, and to elucidate the mechanism of such effects, whether expected or observed. The ICH S7A guideline is careful to say that it offers recommendations for safety pharmacology studies and, by implication, that the guidelines offered are not the equivalent of boxes that must be ticked. As ever, though, if there is not some sort of study of an indicated endpoint or organ system, a scientifically supportable reason should be given for its absence. While the guideline is at pains to say that a rational approach to selection and conduct of safety pharmacology studies should be used, the fact that it exists is the driver for much safety pharmacology. The unspoken philosophy is often, apparently, to fit the test program to the regulatory framework rather than to look for a testing paradigm that is relevant to the compound. Such an approach is shallow, and thought must be given to justify the proposed testing program. Suffice it to say, if your compound does not fit the pattern, the reasoning for selection of tests or their omission has to be scientifically supportable.

One useful principle to consider is not to do a test that makes no scientific sense. For example, there is likely to be little need to do a study on gastrointestinal (GI) transit time for a compound applied dermally for a skin condition, providing that absorption into the systemic circulation is known to be low. For compounds such as antibodies with very specific receptor targets, it may be enough to monitor pharmacology endpoints during the toxicity studies. However, the context of the test compound in relation to its chemical class and mode of action must be considered before deciding against safety pharmacology tests; if the compound is from a new class or has a novel mode of action, it is more likely to require testing than if it is from an established class with known pharmacological effects.

The comfort zone offered by convention and precedent is as strong in safety pharmacology as in other areas of toxicology, and the use of known methods and protocols will always be preferred to novel, unvalidated, and unfamiliar tests. Having said that, however, the guideline indicates that the use of new technologies and methods is encouraged provided that the principles are soundly based.

Most of the tests conducted routinely are defined by well-established protocols, and generally, there is little change to these, unless there is good reason to modify the design. The text in Focus Box 6.1 has been compiled using the ICH S7A guideline, which forms the basis for much of the routine testing. The implied nonroutine tests, whether nonroutine as a result of endpoint, design, or test system, are the subject of individual scientific justification and are not ruled out.

WHAT TO TEST?

The exact identity of the compound (or compounds) to be tested may not be as straightforward as first thought might suggest. While testing the parent compound is a given, there may be strong arguments for testing major human metabolites that are not seen in laboratory animals, an argument that is enhanced if the metabolites are known to be pharmacologically active. If the compound under development has enantiomers or isomers, consider testing these as well (especially as enantiomers can express notably different activity). For instance, the $S(+)$ enantiomer of vigabatrin is known to be the active enantiomer in epilepsy, while the $R(+)$ is inactive; likewise, the enantiomers of thalidomide have different activities. Additional testing may also

FOCUS BOX 6.1   THE CORE BATTERY OF TESTS AND FOLLOW-UP STUDIES WITH EXAMPLES OF INVESTIGATIONS

Core battery tests:
- Central nervous system (CNS)
  Motor activity, behavioral changes, coordination, sensory/motor reflex responses, and body temperature
- Cardiovascular system (CVS)
  Blood pressure, heart rate, and the electrocardiogram (ECG)
- Respiratory system
  Respiratory rate and other measures of respiratory function (e.g., tidal volume or hemoglobin–oxygen saturation)

Supplementary tests:
- CNS
  Behavioral pharmacology; learning and memory; ligand-specific binding; neurochemistry; and visual, auditory, and/or electro-physiology examinations
- CVS
  Cardiac output, ventricular contractility, vascular resistance, and the effects of endogenous and/or exogenous substances on the cardiovascular responses
- Respiratory system
  Airway resistance, compliance, pulmonary arterial pressure, blood gases, and blood pH
- Renal and urinary system
  Urinary volume; specific gravity; osmolality; pH; fluid–electrolyte balance; proteins; cytology; and blood chemistry determinations such as blood urea nitrogen, creatinine, and plasma
- GI tract
  Gastric secretion, GI injury potential, bile secretion, transit time *in vivo*, ileal contraction *in vitro*, and gastric pH
- Autonomic nervous system
  Binding to receptors, functional responses to agonists or antagonists, direct stimulation of autonomic nerves and measurement of cardiovascular responses, baroreflex testing, and heart rate variability

Source: Compiled from ICH Harmonised Tripartite Guideline:
Safety Pharmacology Studies for Human Pharmaceuticals
S7A, November 2000, available at http://www.ich.org.

be indicated if the formulation of a compound radically alters its pharmacodynamic properties, for example, if a liposomal form is developed.

Another group of compounds that pose problems are biotechnology products such as proteins or antisense nucleotides with very precise specificity for species or receptors [these are considered in ICH S6 (2011)]. If the target specificity of a protein means that there is no likely interaction with the receptors normally investigated in routine safety pharmacology studies, there may not be any scientific utility in performing routine studies. Another point to consider is that a humanized monoclonal antibody is unlikely to have any relevant effect in a wild-type rodent and, as for the toxicity program, the most relevant test species should be chosen. One expensive alternative is to develop a version of the molecule that is specific for your chosen test species, for instance, a mouse interferon or antibody. This does not mean that safety pharmacology for these compounds can be ignored; it has to be considered and a logical rationale put forward for the proposed testing program, even if it is essentially one without formal studies. For this type of compound, it is likely that some of the points covered in safety pharmacology studies can be looked for in the toxicity studies.

DESIGN

ICH S7A indicates that safety pharmacology studies should be designed to define the dose–response relationship of the adverse effect observed and to investigate the time course of any such effects.

The basics of study design for safety pharmacology are given in Focus Box 6.2. Doses or concentrations used should span the therapeutic and pharmacodynamic ranges; it is important to do this so as to take into account any differences among species in terms of sensitivity. Where no pharmacodynamic effect can be induced, it is acceptable to use a dose that produces moderate adverse effect in studies of similar duration or route of administration. However, it should be recognized that some effects produced may complicate interpretation of the data and may themselves set a limit on dose. Therefore, the high dose should produce some toxicity with the proviso that the toxicity should not affect the parameters measured. For instance, in a respiratory study, minor change in the plasma activity of hepatic marker enzymes or body weight loss may be acceptable, but emesis, hyperactivity, or muscular tremor would not be. Selection of the maximum dose possible is indicated because adverse pharmacological effect may be associated with receptor interactions where affinity is several orders of magnitude lower than that of the intended receptor. The low dose should not be lower than the primary pharmacological dose or the human clinical dose. Where other data are not available, it may be acceptable to use multiples of the pharmacological dose; where toxicity is not limiting, a margin of 100-fold the pharmacologically active dose should be considered. Although it is desirable to include several doses or concentrations, where no effect is produced on the endpoint for safety pharmacology, a single limit dose may be acceptable.

The timing of measurements is important, and the time points chosen need to be scientifically justifiable. Ideally, they should be timed according to pharmacodynamic or pharmacokinetic data and should include measurements at T_{max} and when there is no drug present in order to assess maximal effect and any reversibility or

FOCUS BOX 6.2   BASICS OF SAFETY PHARMACOLOGY—TEST DESIGN

- The objective is to define dose–response or concentration–response relationships and to investigate time course of effect.
- Doses should include and exceed the primary pharmacodynamic or therapeutic range. In the absence of effect, the dose should be high enough to produce moderate adverse effect in studies with similar route of administration and duration. Concentrations *in vitro* should be chosen to elicit effect.
- Normal design is control and three doses or concentrations.
- A single group tested at a limit dose may be enough, if no pharmacological effect is seen.
- Duration is usually single dose or short *in vitro* exposure (often to successively rising concentrations).
- Compounds to test: parent compound (and enantiomers or isomers) and any major metabolite expected to be present in man.
- Measurement at least at the time of maximal pharmacodynamic effect or plasma concentration (T_{max}) and when there is no drug present.
- In general, these studies should be designed and conducted according to Good Laboratory Practices.

delayed effects. The type of measurement may be indicated by compound class as well as by guidelines.

Although safety pharmacology studies are usually single dose, repeat-dose studies may be indicated by results from repeat-dose toxicity studies or human experience. The pitfalls of including safety pharmacology studies in routine toxicity studies are discussed later in this chapter.

TEST SYSTEMS FOR SAFETY PHARMACOLOGY

As with much of toxicology, the test systems used are often those in wide use that have been validated and accepted by the various regulatory authorities. Rodents are the main models to be used, either for *in vitro* applications or as complete living animals. The use of nonrodents is expensive, and they tend to be used much less frequently, although the use of *in vitro* preparations such as Purkinje fibers from dogs or sheep has been more frequent in the recent past, but this has now declined. The test systems most often used are indicated in Focus Box 6.3.

SAFETY PHARMACOLOGY IN TOXICITY STUDIES

It is logical, and appropriate under the three Rs to make as much use of animals in toxicity studies as possible, provided this can be done without compromising the

**FOCUS BOX 6.3   TEST SYSTEMS
FOR SAFETY PHARMACOLOGY**

- *In vitro* preparations

  Cells in culture [human Ether-á-go-go-Related Gene (hERG) assay]

  Isolated tissues—papillary muscle, Purkinje fibers

  Perfused organs—heart, kidney

- Rodents

  Mouse—modified Irwin test, GI tract transit time, renal function

  Rat—modified Irwin, renal function, respiratory, GI tract transit time; telemetric surgical implants may be used when appropriate

- Nonrodents

  Anesthetized animals for CVS and respiratory studies (but not for ventilation effects), sometimes combined with renal function

  Freely moving telemetry-implanted animals, for CVS and respiratory studies

integrity of the other investigations or vice versa. In view of this, it is an attractive option to "bolt on" some additional observations to the protocol for, for instance, a 4-week study in rats.

However, Redfern (2002) points out that this apparently attractive option has some disadvantages, namely, that repeated administration may mean that effects are examined rather than responses and that the results of the tests may be influenced by any organ impairment that has been produced by the test substance. In addition, the response to the drug may be affected by developing tolerance to repeated administration. The basic message is that such measurements should be made on day 1 of a study before such influences are manifested as effects due to repeated treatment rather than responses. The bonus of such an approach is that a later examination can be used to chart change from day 1. The drawback is that very often, other time-critical examinations are conducted on day 1, including toxicokinetic sampling, and the addition of other, complex, time-consuming tests can make day 1 a logistical nightmare, which could be a straightforward invitation for error and which may compromise the whole study.

Recent developments in collection of cardiological data have helped the process of this type of study. While the use of surgically implanted telemetric devices for remote collection of electrocardiogram (ECG) data is now routine in dog and non-human primate studies, it is prohibitively expensive for a standard 28-day study. However, the use of jacket external telemetry systems does not involve surgical implantation, merely the wearing of a jacket that protects sensors on the animal and carries a transmitter for the data to be collected remotely. These systems are inexpensive, reliable, and versatile and may be able to be used as a substitute for a formal implant telemetry study, according to circumstance and regulatory agreement.

TESTS AND THEIR CONDUCT

The following is a brief review of a few typical studies that are often performed for safety pharmacology programs. The core battery is looked at first, with particular emphasis on cardiovascular studies, followed by a few examples of supplementary studies.

Central Nervous System—Modified Irwin Screen

These are deceptively simple studies; animals (rats or mice) receive a single adminis-tration of the test compound and are then observed for up to 24 hours. In fact, if there are no visible effects after about 8 hours, the later examinations may be dropped completely. However, this apparent simplicity belies the complexity and number of the observational endpoints that are routinely assessed. The main design points of a typical study are as follows:

- Four groups of five animals, usually rodents (one sex is acceptable, if there is unlikely to be a difference between the sexes), assigned to control and three treatment levels.
- Single administration, usually oral or intravenous; the chosen route of administration should not impede the observations required, so continuous infusion and inhalation are not likely to be practicable.
- Assess behavioral changes before dosing and at appropriate intervals at for instance, 1, 2, 4, and 6 hours postdose; if effects are seen after 6 hours, fur-ther observations may be carried out after 24 hours, or longer, if indicated.
- The observations include, but are not limited to, cage-side assessment, handling and physical observations, and observation in a standard arena for open-field testing. Actual endpoints assessed may include alertness; locomotor and exploratory activity; grooming; tremors or muscle spasms; posture; gait; coat condition; respiration; aggression; skin color (e.g., for peripheral vasodilation); startle response; reflexes (including tail flick test for pain); examination of the eyes for miosis, mydriasis, corneal reflex; etc.

As with other tests, it is important that the laboratory conducting the test has a good background of experience with the test and the strain of animal used, as this is important in correct interpretation of the often small differences that are encountered. Every so often, there is a suggestion that this type of observational battery could be extended to larger laboratory species such as dogs or nonhuman primates. This is fea-sible, as has been demonstrated by some laboratories, but the number of parameters examined needs to be reduced. The results of this type of test may be supplemented by the observations from toxicity studies, particularly the single-dose studies.

Cardiovascular System

Useful background to this critical area of nonclinical safety evaluation is given in ICH guideline S7B (ICH 2005). This indicates that the objectives of these studies include identification of the potential of a test substance and its metabolites to delay ventricular repolarization, and to relate the extent of any effect to the concentrations to the test substance and/or its metabolites. It is pointed out that the results of these studies may indicate the mechanism of any effect and can be used in conjunction with other data to make an estimate of risk of cardiac effect in humans.

The endpoint that is critical here is the QT interval, or more correctly, the QTc or corrected QT interval, although other endpoints such as heart rate, blood pressure, and peripheral effects should not be forgotten or ignored. The potential for new medicines to affect repolarization of the heart has led to the removal of a number of noncardiac compounds from the market in recent years. This serious side effect has been associated with sudden cardiac death due to prolongation of the QT interval of the ECG. The QT interval has become a major focus of pharmacological investigation both in terms of the effects of potential new medicines and in terms of a major research effort to develop new models that are quick to implement, accurately reflect human potential for effect, and are acceptable to regulatory authorities.

The QT interval, as defined by Fermini and Fossa (2003), is the period between the beginning of the QRS complex and the end of the T wave, as shown in Figure 6.1, and is a measure of the duration of depolarization and repolarization of the ventricle. The major concern for QT interval prolongation is the increased risk of life-threatening ventricular tachyarrhythmia, including torsades de pointes, particularly

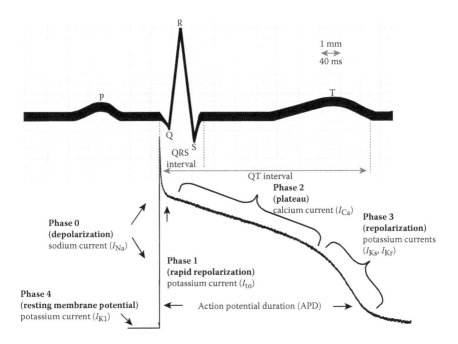

FIGURE 6.1 Temporal correlation between action potential duration and the QT interval on surface ECG. (Picture of a normal ECG beat with intervals and peaks indicated.) (Fermini B and Fossa AA, The impact of drug-induced QT interval prolongation on drug discovery and development, *Nature Revi Drug Discov*, 2:439–447, 2003.)

when combined with other risk factors (e.g., hypokalemia, structural heart disease, bradycardia). However, the relationship between QT interval and this severe effect is not simple, and compounds that prolong QT, such as the calcium channel blocker verapamil, do not necessarily cause torsades de pointes clinically. In fact, the effect appears to be a function of increase in the QT interval when corrected for heart rate—the value known as the QTc. To put this in perspective, Fermini and Fossa report that data on QTc intervals in cases of torsades de pointes suggest that a QTc of more than 500 milliseconds indicates a significant risk of cardiac arrhythmia, and for individuals, an increase in maximum QTc interval of 60 milliseconds over baseline is also indicative of risk.

One of the concerns, from the perspective of safety evaluation, is the fact that this propensity of noncardiac drugs to prolong QTc has been discovered some time after marketing of the drug started. For instance, the antihistamine terfenadine was marketed from 1985 and was associated with about 429 serious cardiovascular events and 98 deaths up to 1996. This is a relatively low incidence, quoted to be approximately 0.25 per million daily doses sold (Fermini and Fossa 2003), and underlines the difficulties faced in predicting serious adverse events from medicines; a small difference from background is very difficult to detect and, if detected, to interpret meaningfully.

The potential for effect on the QTc interval is assessed in a number of tests, of which the most popular is now the hERG inhibition assay. This may be supported by other *in vitro* assays but is almost always followed by a test in a dog or nonhuman primate that has a telemetry implant for recording heart rate, ECG, and blood pressure.

The hERG Inhibition Assay

This assay examines the blockade of K^+ channels expressed in human embryonic kidney cells stably transfected with the product of the hERG gene, a human ion channel responsible for the I_{Kr} repolarizing current. Inhibition of this current is associated with prolongation of the cardiac action potential, which is in turn associated with cardiac arrhythmias such as torsades de pointes. The technique is a highly specific, low-throughput assay, requiring considerable technical skill and specialist equipment. Changes in the ionic current are measured using a voltage clamp technique. Variations on this theme that have higher throughput (e.g., measurement of rubidium flux) have been developed. The following are the main points of the assay:

- Four or five concentrations of test substance and an appropriate positive control, such as terfenadine.
- Investigate in an appropriate number of cells according to whether the test is a screen or a definitive assay.
- Exposed by continuous perfusion or static bath.
- Voltage steps: −15, −5, 5, 15, 25 mV, for example.
- Determine the half maximal inhibitory concentration (IC_{50}), if appropriate.

The potential hazards of using such channel-based assays were illustrated by Abi-Gerges et al. (2011), who pointed out that most hERG assays had been conducted with heterologous systems expressing the hERG 1a subunit; however, both hERG

1a and 1b subunits contribute to the K^+ channels producing the repolarizing current I_{Kr}. They tested a diverse range of compounds to assess any differences in sensitivity between these channels. They found that most compounds had similar potency for the two channels. However, fluoxetine (Prozac) was a more potent inhibitor of hERG 1a/1b than 1a channels, which results in a lower safety margin. Conversely, other agents were more potent in blocking hERG 1a compared with 1a/1b channels, including dofetilide, a high-affinity blocker.

In Vivo Cardiovascular Studies

These tests use either anesthetized animals (usually without recovery) or conscious, freely moving animals fitted with telemetry devices. The use of freely moving animals, either with surgically implanted transmitters or external transmitters contained in external jackets, has huge advantages; the test is conducted in a whole, unrestrained, freely moving animal, without the potentially confounding effects of anesthesia or restraint by technicians. The clinical route of administration may be used, and importantly, the animals can be used a number of times before the surgical implants need to be replaced. A drawback, however, is that the surgical implant is expensive and has a time-consumingly low throughput. Typical species include dogs and nonhuman primates, but pigs may also be used.

Parameters recorded include ECG, heart rate, and blood pressure. In anesthetized models, parenteral administration is necessary, and many anesthetics interfere with cardiac function and add another layer of potential interaction with the test substance, making interpretation more difficult in some cases. A typical protocol for animals with telemetric implants is as follows:

- Four animals, usually of the same sex.
- Three doses given on one occasion each, plus control to all the animals, with an appropriate washout period between doses; a randomized block design may be used. The washout period should be consistent with the pharmacokinetics or pharmacodynamics of the test substance.
- Parameters measured include blood pressure, heart rate, ECG (at least lead II and PR, QT, QTc, and RR intervals, and QRS duration), and core body temperature.
- Body weights and food consumption recorded as appropriate.
- Respiratory parameters such as rate, tidal volume, and minute volume may also be examined to give information on another core area of safety pharmacology testing using separate instrumentation.
- Measurements carried out shortly before dosing and then continuously or at frequent intervals until 24 hours postdose.
- Records are made via receivers placed either in the dogs' home pens or in specially equipped pens.

The animals are freely moving and, crucially, at very low levels of stress, meaning that any differences seen, relative to baseline or controls, are more likely to be related to treatment.

Other Tests

Other tests that have been used include the use of isolated Purkinje fibers from dogs or sheep, allowing the effects of a drug on cardiac channels to be studied *in situ*.

This means that the effects on several ion channels can be detected. However, this assay is technically demanding, of low throughput, and a negative result does not exclude proarrhythmic tendencies and consequent risk of torsades de pointes in humans. It also requires the death of a whole animal when a telemetry study might give a more meaningful result.

The papillary muscle from the guinea pig and isolated rabbit heart (in, for example, Langendorff's preparation) may also be used to investigate the potential for effect on QT interval.

Another test is the guinea pig monophasic action potential (MAP), which uses an anesthetized guinea pig and has been reviewed by Hauser et al. (2005) and Marks et al. (2012).

Respiratory System

The respiratory system is subject to a diverse range of internal and external influences that result in complexity of control and response. The safety pharmacology studies normally conducted are those that test the effects of systemically administered test substances rather than those given by inhalation.

The number of drugs that can affect the respiratory system is surprisingly large. Murphy (2002), who published a very useful review of assessment of respiratory function in safety pharmacology, lists more than 60 drugs with the potential to affect bronchoconstriction or pulmonary injury, together with 25 agents and classes that can influence ventilatory control. An important consideration in rationalizing the need for respiratory assessment is that patients often have compromised respiratory function due to asthma, bronchitis, or emphysema. For example, nonselective β-adrenergic antagonists, given for glaucoma or cardiovascular disease, may be associated with life-threatening side effects in asthma patients. Mild suppression of respiration may be life threatening in patients with conditions such as sleep apnea, or if the drug is taken in conjunction with other drugs that suppress respiratory function, such as tranquilizers or sedatives.

A typical study may use rats to assess effects of a drug on parameters, such as respiration rate and tidal volume. For this, the typical protocol could include the following:

- Three treatment groups of between 5 and 10 rats, allocated to group randomly based on tidal volumes recorded during the acclimatization period.
- Two additional groups receive the vehicle control and a reference substance, such as morphine hydrochloride.
- Measurement of respiratory parameters in plethysmography chambers attached to appropriate electronic recording devices, to measure thoracic volume.

- Analysis of each parameter at intervals before dosing and at appropriate intervals afterward.
- Acclimatization to the recording chambers may be necessary before each measurement.

Studies should aim to examine the two basic functions of the respiratory system: pumping air and gas exchange. Parameters indicative of effect on the air pump function include respiratory rate, tidal volume, and minute volume. Respiratory rate and tidal volume are independently controlled and may be subject to selective alteration. Measurement of either of these alone cannot reliably indicate change in pulmonary ventilation. For example, theophylline increases minute volume by increasing tidal volume without affecting respiratory rate, and morphine depresses minute volume by reducing respiratory rate with no effect on tidal volume. Additional measurements may include flow of inspiration and expiration, detection of hypoventilation and hyperventilation, and distinguishing between central and peripheral nervous effects. Effects on gas exchange may be assessed by measurement of lung airflow and compliance.

The usual test systems are dogs and rats, although special compounds may require special animals, such as nonhuman primates or transgenic mice. Guinea pigs may be used for compounds that have activity on leukotrienes or histamine. It is important that evaluation of ventilatory effects be undertaken in conscious animals as most anesthetics alter ventilatory reflexes. It is essential that the animals be acclimatized to the apparatus before the experiment, to ensure that any change in pattern is distinguishable from baseline.

GI Tract

Rats or mice are used in studies on the GI tract, which typically involve administration of the test substance, followed by a bolus dose of charcoal and subsequent tracking of the meal through the intestine. This is usually achieved by killing the animals and removing the gut and measuring the distance the charcoal has traveled since administration. A protocol may include the following:

- Five to 10 rats or mice per group, dosed at three treatment levels and a control.
- Administration by oral gavage of a small amount of charcoal suspension (1 mL).
- Half to 1 hour after the charcoal dose, the animals are killed, and the distance the charcoal meal has traveled along the intestine from the stomach is measured together with the total length of the intestine.
- The stomach may be weighed to give an indication of gastric emptying.

Alternative approaches being developed include gamma ray counting of a labeled bolus that can thus be tracked in the living animal over a longer period than is possible using the more conventional study design described.

RENAL FUNCTION

This is usually a straightforward assessment of renal function in terms of concentrating ability and electrolyte composition of urine collected over a defined period, following administration of the test substance. The protocol main points are as follows:

- Rats are loaded with saline before administration of the test substance.
- Groups of 5 to 10 animals are treated at three dose levels and a vehicle control.
- The animals are placed individually in metabolism cages and urine collected at intervals such as 0 to 4, 4 to 8, and 8 to 24 hours postdose.
- Water and food are withheld over the first few hours and then returned for the later collection(s).
- The pH and volume of each urine sample are measured.
- Each sample is analyzed for sodium, potassium, and chloride concentrations, which are corrected to output, using the volume of urine excreted and the animal body weight (mmol/kg).
- It may also be sensible to examine the activities of urinary enzymes such as alkaline phosphatase or *N*-acetyl glucosaminidase.

PITFALLS OF SAFETY PHARMACOLOGY STUDIES

Inevitably, there are several aspects of these studies to be aware of, both in conduct and in interpretation. The presence of confounding factors in the design should be considered carefully. In studies with anesthetized animals, it is possible that the choice of anesthetic may have an influence on the parameters measured and so produce results that are skewed, giving a false impression of effect or absence of effect.

Toxicity may interfere with the responses of the animals, and so, for instance, animals may respond differently in an Irwin screen at a toxic dose as compared with a dose that is pharmacologically active. In some cases, parameters may be estimated rather than measured directly, and this can have unintended consequences. For example, in respiratory studies, estimation of tidal volume rather than direct measurement may give misleading results as tidal volume is dependent on other variables, such as breathing rate and pattern, temperature, and humidity. An increase in the breathing rate of a rat from 40 to 70 breaths per minute may cause a 30% underestimation of tidal volume (Murphy 2002). The bottom line here is that if the parameter is influenced by factors that are not properly controlled, direct measurement is preferred to estimation.

Lack of acclimatization to experimental circumstances may also produce a set of data that changes over repeated measurement as the animals become more familiar with the equipment or surroundings. In addition, because the tests are almost always single doses, they measure response to treatment in naive animals, and changes that accumulate over repeated dosing will be missed. However, this point may be mitigated by appropriate clinical observation in standard toxicity studies or by inclusion of safety pharmacology investigations. Having said that, the warnings given above about including safety pharmacology endpoints in routine toxicity studies have to be considered.

Many of the assays, both *in vivo* and *in vitro*, are technically demanding, and it is important that the laboratory conducting the test should have adequate experience and skill in conduct and interpretation. The possession of a good historical control database is important here since the group size is small and may lead to chance differences from control, which, in the absence of dose relationship, may not be treatment related.

For *in vitro* experiments, it is important that the tissues or cells are treated appropriately from the moment of removal from the culture or from the animal. If the handling is in any way below best practice, the end result may not be good enough for definitive interpretation.

As illustrated by the work of Abi-Gerges et al. (2011), it is important to note that initial understanding of the mechanisms or channels may be incomplete and that additional knowledge may well affect how extrapolation is carried out between test systems and humans.

SUMMARY

Safety pharmacology studies are conducted almost exclusively for pharmaceuticals, although there is no particular reason why judicious use for other chemical classes should not provide useful information, perhaps for exploration of mechanism.

- The conduct of safety pharmacology studies has been increasingly mandated by regulatory authorities as a response to clinical findings, sometimes after several years of successful marketing.
- Although such responses are often reflexive and excessive, we are now at a point where the tests requested are reasonable and, if justifiable, inappropriate tests can be omitted or different ones suggested, in order to address the core battery of tissues and organ systems. These consist of the central nervous, respiratory, and cardiovascular systems; supplementary tests may be appropriate to investigate the findings of the core battery or to examine an expected effect of the test compound.

Above all, these tests should not be conducted to tick boxes, even if that is often effectively what happens. It is important to consider each test suggested on its scientific merit and to be able to defend its conduct or omission, if and when required. The guidelines available, especially ICH S7A and ICH S7B, give a robust framework for designing these tests, but despite the temptation, they should not be treated as gospel to justify the conduct of meaningless tests.

REFERENCES

Abi-Gerges N, Holkham H, Jones EMC, CE Pollard CE, Valentin J-P, Robertson GA. hERG subunit composition determines differential drug sensitivity. *Br J Pharmacol* 2011; 164(2b): 419–432.

Fermini B, Fossa AA. The impact of drug-induced QT interval prolongation on drug discovery and development. *Nat Rev Drug Discov* 2003; 2: 439–447.

Hauser DS, Stade M, Schmidt A, Hanauer G. Cardiovascular parameters in anaesthetized guinea pigs: A safety pharmacology screening model. *J Pharmacol Toxicol Methods* 2005; 52: 106–114.

ICH Harmonised Tripartite Guideline: Safety Pharmacology Studies for Human Pharmaceuticals S7A. November 2000. Available at http://www.ich.org.

ICH Harmonised Tripartite Guideline: The Non-Clinical Evaluation of the Potential for Delayed Ventricular Repolarization (QT Interval Prolongation) by Human Pharmaceuticals S7b. May 2005. Available at http://www.ich.org.

ICH Harmonised Tripartite Guideline: Preclinical Safety Evaluation of Biotechnology Derived Pharmaceuticals S6(R1). June 2011. Available at http://www.ich.org.

Marks L, Borland S, Philp K, Ewart L, Lainée P, Skinner M, Kirk S, Valentin J-P. The role of the anaesthetised guinea-pig in the preclinical cardiac safety evaluation of drug candidate compounds. *Toxicol Appl Pharmacol* 2012; 263: 171–183.

Murphy DJ. Assessment of respiratory function in safety pharmacology. *Fundam Clin Pharmacol* 2002; 16(3): 183–196.

Redfern WS, Wakefield ID, Prior H, Pollard CE, Hammond TG, Valentin JP. Safety pharmacology—A progressive approach. *Fundam Clin Pharmacol* 2002; 16(3): 161–173.

7 Determination
General and Reproductive Toxicology

GENERAL TOXICOLOGY

In broad terms, general toxicology is something of the poor relation of toxicology. It can be seen as lacking the glamor or intellectual rigor of other areas of toxicological investigation because it sets out to be a catchall; to paraphrase Gerhard Zbinden, "It looks for everything but hopes for nothing." However, it is central to safety evaluation of novel chemicals, as effects that may be seen in other more specialized areas can also be detected or supported by well-designed general toxicity studies. For example, microscopic evaluation of testes in a general toxicity study may indicate the potential for effects in formal fertility studies in the program of reproductive toxicity studies.

TEST SYSTEMS FOR GENERAL TOXICOLOGY

The large majority of test systems for general toxicity are animal based, due to the need to demonstrate toxicity elicited after repeated administration over long periods, something that *in vitro* systems are only just beginning to do. A further factor militating against such systems is the multiplicity of endpoints examined in a classic animal study and the limited number of such possibilities in a cell culture that lacks the complex interactions between tissues seen in whole animals. For a complete picture, general toxicology also requires an estimation of the absorption, distribution, metabolism, and elimination (ADME) of a compound and of the pharmacokinetics following single and repeated administration. With pharmaceuticals particularly, it is important to choose one species that is as close to humans as possible in terms of ADME; two such similar species would be better, but as one is almost always the rat, this luxury is not always possible. The use of comparative *in vitro* metabolism data and preliminary *in vivo* data should allow a scientifically justifiable choice of test species to be made. In practice, the test species are chosen from a relatively limited palette of possibilities, restricted by toxicological conservatism and regulatory acceptance, as indicated in Table 7.1. The advantages and disadvantages of each system are summarized in Focus Box 7.1.

Of all the aforementioned test systems, NHPs have been the subject of most debate and ethical pressure to avoid or proscribe their use. Until about the 1990s, NHPs were wild-caught animals, which produced a variety of problems that were either not understood or simply ignored. The animals were generally of unknown age and origin, with unknown diseases or parasite profiles. In particular, they could—and did

---

TABLE 7.1

Test Systems for General Toxicology

Rodent	Nonrodent
Rat	Dog
Mouse	Nonhuman primate
Hamster	Minipig
	Rabbit

---

in some cases—transmit fatal zoonoses to humans, such as hepatitis B, and viruses such as rabies, Ebola, and Marburg. As a result of their inconsistent origins, ages, and histories, they gave inconsistent historical control data. These problems have been largely circumvented by captive breeding; however, control over viral status is dependent on source, as established breeders are able to provide certified virus-free animals.

Ferrets were suggested, at one time, as an alternative to dogs, because they have a similar gut microflora to humans. However, they offer more problems than solutions, and we have never seen a general toxicity study in ferrets. Rabbits have been the species of choice in short-term dermal toxicity studies but not for other routes of administration. The minipig is increasingly used in dermal studies due to the similarity of the skin to that in humans.

The age of the animals used should be considered. Young animals tend to metabolize chemicals somewhat differently from adults, and this can lead to unexpected results due to age-related differences in metabolic capabilities, especially in rodents. Similarly, with dogs, it is quite normal to use immature animals at about 5 months or 6 months old. The consequence of this is that reproductive toxicities, such as testicular atrophy, may not be apparent in shorter studies; differing stages of sexual maturity in a small group may also make interpretation of change much more difficult. For practical reasons of age and safety of handling staff, primate studies are usually conducted with immature animals of about 2 to 3 years old. It is possible to obtain mature monkeys, but they are expensive and difficult to handle safely, a factor of some importance, given the sometimes fatal diseases that they can pass on either through a bite or through feces or urine.

FOCUS BOX 7.1 CHARACTERISTICS OF TEST SYSTEMS FOR GENERAL TOXICOLOGY

The following attributes for the main test species are given in no particular order and without guarantee of completeness:

- *Rats/mice:* Easy to house; small, meaning that relatively little test substance is needed; well understood with ample historical control data; multitude of strains; short life spans; good regulatory acceptance;

traditional; few ethical problems associated with their use; genetic consistency; statistically robust designs are relatively easy to achieve at sensible cost; not necessarily good models for humans; metabolism tends to be rapid and systemic exposure lower than in humans; males have greater metabolic capacity than females, which often leads to sex-related differences in toxicokinetics and toxicity. Sexually mature at 5 to 7 weeks. Mice are easier than rats to manipulate genetically in the construction of transgenic models that may have greater human relevance than conventional strains—albeit at significant cost.

- *Hamsters:* Alternative to rat or mouse but rarely used except in specialist studies and some carcinogenicity bioassays; few historical control data; the species to use when all other rodent options are exhausted.

- *Beagle dogs:* Reasonable size to work with; good natured; well accepted and now the only dog available in the United Kingdom for laboratory experiments; well understood; ample historical control data; good regulatory acceptance; can react badly to compounds such as nonsteroidal anti-inflammatory drugs; low workplace handling risk; usually weigh between 10 and 15 kg, needing large amounts of test substance; large areas needed for stress-free housing and husbandry. Sexually mature at 7 to 12 months.

- *Minipigs:* Similarity of skin to humans makes them suitable for dermal studies; kidney structure similar to humans; increasing regulatory acceptance; omnivorous diet gives gastrointestinal similarity to humans; large: "mini" can mean up to 50 kg with a norm of about 20 kg and a consequent effect on test substance requirements; not as easy as dogs to dose or to take samples from; some metabolic peculiarities, particularly in sulfation. Sexual maturity at 4 to 5 months for sows and 3 to 4 months for boars. The use of the minipig in toxicology has been reviewed by Svendsen (2006).

- *Nonhuman primates (NHPs):* Species used at present normally macaques (cynomolgus or rhesus monkey); the marmoset has been used in some circumstances; complex to keep, group housing gives optimum results; supposedly closer to humans in terms of ADME but not always; good regulatory acceptance; size generally between 2 and 5 kg (cynomolgus monkey) or 250 and 600 g (marmoset); small size of marmosets means lower compound requirement (good for biotechnology products), but this alone cannot justify selection; more expensive and less available than dogs; small size of marmosets may necessitate use of satellite groups for pharmacokinetic determinations; marmosets are subject to stress and diet factors; intense ethical and government pressure against use of any NHP. Sexual maturity at 14 to 18 months (marmosets) or 4 to 5 years (macaques).

STUDY DESIGNS IN GENERAL TOXICOLOGY

Studies in general toxicity include the shortest and the longest studies in toxicological investigations, the only others of comparable length being the perinatal and postnatal development reproductive study and the 2-year carcinogenicity bioassay. The shortest and simplest study is the single-dose acute study, which is intended to characterize severe toxicity following a single large dose. The original objective was to calculate the LD_{50} or median lethal dose at which 50% of the treated animals died; this was established statistically from the results of several treatment groups of up to 10 animals of each sex. The results were not always reproducible as acute toxicity may be significantly affected by many factors including strain or supplier of animal, diet, or environmental factors. Although the LD_{50} test itself is no longer conducted, the concept is retained as a useful indicator of toxicity ranking—the figure can be estimated approximately from the data from the initial sighting studies, usually in terms of a dose greater than, for example, mg/kg. In an acute study, the animals are dosed once and observed for 14 days, which is both a strength and a weakness. Some toxicity expressed in the period immediately following administration, for example, liver toxicity seen with carbon tetrachloride, may be completely repaired by the end of the 14 days. However, 14 days may allow any slowly developing toxicity to be expressed; this can be seen with cytotoxic anticancer chemotherapies. The basic designs for general toxicity studies are summarized in Table 7.2 (see also section "Study Design Basics and Confounding Factors" in Chapter 3). The observations and measurements indicated in this table are discussed below.

DOSE SELECTION IN GENERAL TOXICOLOGY

For the initial studies, dose levels are selected based on knowledge of the compound or of similar compounds or on the effects in other species. The first studies are dose range finders and use small groups of animals to select a high dose level that will be usable in the main study. The designs of these studies tend not to be regulated in the same way as the main or pivotal studies, so they are more flexible in the approach that can be used. In each case, the objective is to select a high dose for the main study that can be given for the whole treatment period without excessive toxicity.

For rodents, groups of three males and three females (or just one sex in the first instance) receive different dose levels for a few days to assess toxicity and to select a high dose for the main study; if females only have been used in the first phase, a group of males is dosed at the selected dose to check for sex-related differences in effect. The chosen dose level may then be administered to a larger group for 7 or 14 days.

For nonrodents, the usual approach is to dose one male and one female for 1 or up to 3 days at gradually increasing dose levels so as to elicit an effect of treatment. There may be a washout period between dose levels. When toxicity has been seen, a second pair is dosed at the indicated dose level for 7 days or longer to check for toxicity in treatment-naive animals.

Dose range–finding studies can sometimes be found to be nonpredictive of effects because they are conducted in fewer animals than the main study. They are often

TABLE 7.2
Summary of Basic Designs for General Toxicology

Species		Acute (Single Dose)	2 or 4 Weeks	13 Weeks	26 Weeks
Group size[a]	Rodent	5 m + 5 f[b]	10 m + 10 f[c] (rev 5 m + 5 f in 2 groups)[d]	10 m + 10 f (rev 5 m + 5 f in 2 groups)	15 m + 15 f
	Nonrodent	Not used	3 m + 3 f (rev 2 m + 2 f in 2 groups)	4 m + 4 f (rev 2 m + 2 f in 2 groups)	4 m + 4 f
Clinical pathology	Rodent	Not done	Wk 4 + end rev	Wk 4 or 6, 13, rev	Wk 13, 26
	Nonrodent	Not done	Wk 0, 4 + end rev	Wk 0, 4 or 6, 13, rev	Wk 0, 13, 26, rev
Ophthalmoscopy	All	Not done	Wk 0, 4, end rev	Wk 0, 13, rev	Wk 0, 13, 26, rev
Necropsy	All	All groups	All groups	All groups	All groups
Histopathology[e]	Rodent	Gross lesions	Control, high dose and affected tissues at low- and mid-dose groups. Affected tissues only in reversibility groups.		
	Nonrodent	Not done[f]	All animals are examined.		

Source: Adapted from Woolley A, *A Guide to Practical Toxicology*, 2nd ed., London: Taylor & Francis, 2008.

Note: f, female; m, male; rev, reversibility studies; Wk, week.

a Except in acute studies, there are usually three treated groups and a control.

b Groups of 2 m + 2 f used for dose selection.

c This may be reduced to 5 m + 5 f in some cases.

d Reversibility studies usually controls and high dose for rodents, controls and high dose for dogs; examinations normally conducted in final week.

e In addition, organ weights are recorded on all except acute studies. In-life observations on all studies include clinical signs, body weight, and food consumption.

f Acceptable data on acute toxicity may be obtained from early dose range–finding studies.

conducted without some of the complex investigations used in main studies—for instance electrocardiography may not be done in nonrodents, or clinical pathology in rodents; histopathology is rarely done in any such study. These omissions may save money, but they do not allow a full overview of potential effect.

Another factor in failure of dose range–finding studies can be the age of animals used relative to those in the main study. If animals, especially nonrodents, are taken from stock, they are likely to be older than animals ordered for and treated in the main study; this can lead to expression of different toxicities in the two studies. It is also difficult to transfer the results of a dose range–finding study from one laboratory to another as some factors—especially husbandry—can lead to expression of more or less toxicity, making the dose range–finding data from the first laboratory irrelevant.

STUDY DURATION

The duration of studies in safety evaluation is largely fixed by toxicological convention and increases as the program progresses, from 14- or 28-day studies, to 13 weeks, to 26 weeks (rats) or 39 or 52 weeks in nonrodents. The basic design of all these studies is the same, namely, a control and three treatment groups. These receive dose levels that are based on a high dose expected to cause toxicity, a low dose calculated to be a high multiple of expected human exposure, and an intermediate dose level at an approximate geometric mean of the other two. One of the objectives is usually to determine a no-observed-effect level (NOEL) or a no-observed-adverse-effect level (NOAEL), from which safe exposure levels for humans may be estimated. Typical dose level choices could be 10, 30, and 100 mg/kg/day. With increasing study duration, the number of animals tends to increase from 5 or 10 per sex per group in rats in 14- or 28-day studies to 15 per sex per group or more in the longer ones. In addition, the longer studies will probably include animals allocated to recovery or reversibility studies to assess the regression of effects when treatment is withdrawn. In practical terms, this means that 18 weeks should be allowed for a 13-week study in rats, divided into 1 week for acclimatization to the study room, 13 weeks for treatment, and 4 weeks without treatment. For nonrodents, the study durations are the same, but animal numbers are lower for reasons of ethics, space, and cost. Typically, a 14- or 28-day study in dogs will be conducted with three dogs per sex per group and a 13-week study with four per sex per group, with additional animals allocated for reversibility studies (see section below).

In some cases, for pharmaceuticals, a so-called extended single-dose study is acceptable for single administration to human volunteers. For instance, they are accepted by the US Food and Drug Administration (FDA) and by the European Medicines Agency (EMA) as toxicity studies to support microdosing studies in man (in which small doses of radiolabeled material are administered to give an early indication of pharmacokinetics). Similar-sized groups of animals to those used in the 2- or 4-week study designs given above are used, with an interim kill 2 days after dosing and a second after 14 days. Although they have the advantage of lower test item requirements, they use more animals and take the same amount of time as a conventional study—which will have to be performed later in any case. As a result, they do not generally have any significant advantages over routine designs.

PARAMETERS MEASURED IN GENERAL TOXICOLOGY

General toxicity studies are relatively nonspecific screens for adverse effects that are not necessarily predictable, but that are likely to arise in known ways or manifestations. For this reason, the measurements that are conducted in these studies are very similar across programs and study types. Where toxicity of an unexpected or new type is seen, it is often investigated in specifically designed mechanistic studies. The normally conducted measurements and observations are discussed briefly below. They can be broadly divided into three categories—assessment of exposure (toxicokinetics); in-life observations and clinical pathology; and postmortem investigations.

Assessment of Exposure

This is an integral part of the majority of toxicity studies, including studies in reproductive toxicology. Without exposure, there is no effect; the course of exposure can determine the clinical response to treatment, both within the dosing interval (pharmacological actions) and with repeated treatment (development of tolerance or progression of toxicity). The basic parameters to be assessed include area under the concentration curve (AUC), half-life of elimination ($t_{1/2}$), and clearance and maximum concentration and time after dosing at which it occurs (C_{max} and T_{max} respectively). The usual practice is to take six samples post dose (e.g., 0, 0.5, 1, 2, 4, 8, and 24 hours post dose) from at least three animals per sex per time point.

Depending on the size of blood sample required, it may be necessary to add satellite animals to each group; this has been regularly needed for small species such as rodents and marmosets but is not needed for larger animals such as other NHPs, dogs or minipigs. This is one area where a significant increase in the sensitivity of analytical methods has led to a reduction in the numbers of animals used, fulfilling the reduction aspect of the three Rs. Sample sizes of 50 μL or less usually mean that satellite animals are not needed in rodent studies.

In-Life Observations

Of the investigations summarized in Focus Box 7.2, clinical observation and measurement of growth and food consumption are usually the most informative. Effects on the eyes are rare, but this is an examination that is common to all regulatory guidelines, in deference to the importance of ocular effects in humans. Although examination of the other senses would also seem sensible, it is difficult to achieve, the only other occasionally examined being hearing, usually by means of a whistle or other sudden noise. A deeper investigation of the nervous system can be achieved through neurological examination, a relatively simple estimate for neurotoxicological potential, which can be performed in most species. Other in-life examinations, relevant to the expected effects of the test substance, may include measurements of testicular size and semen sampling or examinations such as electroencephalograms or electroretinography. The last two are rarely used and are of questionable utility in general toxicology.

FOCUS BOX 7.2   IN-LIFE OBSERVATIONS AND MEASUREMENTS IN GENERAL TOXICOLOGY

* *Clinical observations:* Clinical observations following administration are the most basic investigation and give information on the effects of the compound that may be expected at high doses in humans. Subjective indications of ill-health, such as headache or nausea, are not readily assessable in animals; however, lack of activity or abnormal posture may be a consequence of these. Salivation at or immediately after dosing is seen frequently in oral toxicity studies and may simply reflect the expectation of dosing or taste of the test substance.
* *Food consumption and body weight:* These are nonspecific indicators of toxicity that may be affected by many factors such as general malaise, pharmacological action, sedation, or other neurological effect. They act as critical early indicators of effect before other examinations are employed to investigate further. If the dose volume is high, then it is possible that food consumption will reduce in the absence of actual toxicity. High dose volumes of vehicles such as corn oil may also cause a reduction in food consumption in all groups, including controls, and may be associated with increased weight gain.
* *Water consumption:* It can be measured if there is suspicion that kidney function is affected. It should be noted, however, that water consumption will tend to be lower if food consumption is also reduced.
* *Ophthalmoscopy:* It is performed before treatment and at the end of the treatment and, if appropriate, reversibility periods.
* *Electrocardiography:* This is useful for assessing unwanted or pharmacological effects on the heart and can be allied with blood pressure measurements. Although routine in nonrodents, it is only practicable in rats if sophisticated computerized systems are used. Blood pressure measurement is normally indirect by use of a pressure cuff on the tail but can be direct from an artery by use of a pressure transducer. These measurements can, and should, be achieved by the use of jacket telemetry systems—though this can prove trickier in nonhuman primates as they can take the jackets off with ease. Many laboratories are moving toward implantation of telemetry systems, but these are still in the development phase and can prove expensive as surgery is required.

Clinical Pathology

The next group of investigations is performed on blood, urine, or feces to assess the effects of treatment on the function and status of a number of major organ and tissue systems. They may give early warning during a long-term toxicity study that is not apparent from in-life observation or may support these findings; equally, they may be indicative of early toxicity that has resolved later in the study. Clinical pathology investigations are relatively simple and give quantitative data, which are amenable to statistical analysis. Interpretation of variation from controls or historical control data depends on the interrelationship of observed differences, the presence of dose relationship, and other changes in the study.

Hematology examines the numbers and morphology of the erythrocytes, platelets, and leukocytes in the peripheral circulation (Evans 2009a).

- *Erythrocyte parameters*: Hemoglobin, red cell count, hematocrit, absolute (calculated) indices (mean cell volume, mean cell hemoglobin, mean cell hemoglobin concentration), reticulocyte count, red cell distribution width. Cell morphology, assessed on a blood film, if indicated by other measurements.
- *Leukocyte parameters*: Total and differential white blood cell counts. Morphology.
- *Coagulation*: Prothrombin time, activated partial thromboplastin time, fibrinogen concentration, platelet count.

These measurements give insight into the condition of the bone marrow and the presence of peripherally induced anemias. The coagulation measurements give some indication of the condition of the liver, in that prolongation may mean that there is reduced synthesis of coagulation factors due to changes in hepatic synthetic capacity. These examinations may be extended by examination of smears to determine any effects on the bone marrow. Fibrinogen, as well as being a precursor of fibrin, is one of the acute-phase proteins, which vary in concentration according to conditions such as inflammation.

Clinical chemistry is intended to examine the function of several organ systems, particularly the liver and kidney, through determination of the activity of enzymes and of measurement of a number of analytes, such as urea, proteins, and electrolytes (Evans 2009b).

- *Enzymes*: Alkaline phosphatase, alanine, aspartate aminotransferase, lactate dehydrogenase, gamma-glutamyl transpeptidase, leucine aminopeptidase, creatine kinase; sorbitol dehydrogenase may also be examined as a test of liver function.
- *General analytes*: Urea, creatinine. Glucose, total protein and differential protein electrophoresis, albumin, albumin/globulin (A/G) ratio. Cholesterol, triglycerides, creatinine. Total bilirubin.
- *Electrolytes*: Sodium (Na), potassium (K), chloride (Cl), calcium (Ca), phosphate (PO_4).

Liver function is indicated by changes (usually increases) in the activities of several enzymes that are more or less specific for differing functional changes; the concentration of the various proteins is also useful in this respect. Kidney function is shown by the concentrations of urea, creatinine, and electrolytes. There is no reliable enzymatic indicator for kidney damage in the plasma, but several enzymes may be assayed in the urine, for instance, alkaline phosphatase and N-acetyl-β-glucosaminidase.

Other tissues or organs may be assessed through the activity of other enzymes such as creatine kinase for heart-related effects, and aspartate aminotransferase for changes in musculature (in the absence of change in alanine aminotransferase, as together, these two enzymes are markers of liver toxicity). Changes in alkaline phosphatase and lactate dehydrogenase may be further assessed through isoenzyme studies, to indicate if the liver or another tissue is the prime organ of effect. Some enzymes are more appropriate than others in the various species used. Marmosets have low peripheral activities for alanine aminotransferase, rats have low gamma-glutamyl transpeptidase, and in other NHPs, the plasma activity of alkaline phosphatase and lactate dehydrogenase tend to be more variable and consequently less useful than in other species.

Urinalysis is the main in-life window on kidney function, through examination of urinary electrolyte concentrations (used with volume to calculate total output), stick tests, and microscopic examination of sediment obtained after centrifugation. It is either quantitative or semiquantitative:

- *Quantitative*: Volume, osmolality, or specific gravity; pH; electrolyte concentrations (Na, Cl, PO_4, Ca); urinary enzymes; creatinine.
- *Semiquantitative*: Appearance/color. Stick tests for protein, glucose, ketones, bilirubin, and blood. Microscopy of the deposit left after centrifugation.

Although the stick tests are given here as being semiquantitative, instrumentation is now available to read them and obtain a quantitative value. Unlike blood, urine should be collected over a period of hours, preferably overnight, and so requires special collection cages in which the animals can be isolated. It is possible to allow access to water over this period, but one of the functions of urinalysis is to determine the ability of the kidneys to produce concentrated urine.

Postmortem Examinations

At the end of the study, the animals are killed humanely and subjected to a thorough postmortem (autopsy or necropsy), in which a range of organs are weighed and the tissues examined *in situ* and after removal. Up to 50 organs or tissues may be retained in fixative against histopathological processing and microscopic examination (Table 7.3). The purpose of these examinations is to detect morphological effects that may correlate with other changes seen in in-life or clinical pathology. Changes that may have long-term consequences for the animal should also be found, such as endocrine-induced hyperplasias that might develop into tumors in later life.

Organ weights indicate effects due to atrophy, for instance, in the testis, or of adaptive hypertrophy, which may be seen in the liver following administration of enzyme inducers or peroxisome proliferators such as diethylhexylphthalate. Increased weight may also reflect increased storage, for example, of lipid. Some organs are weighed routinely, but the data do not necessarily reveal much that is useful, due to variability of postmortem blood loss, lung weight being an example. The weight of the uterus is greatly affected by the stage of sexual cycle at the time of kill, and this should be taken into account when examining the weights. A further

TABLE 7.3

Organs and Tissues That May Be Retained at Necropsy

All gross lesions	Pancreas
Adrenals[a]	Pituitary[a]
Aorta	Prostate
Bone (sternum)	Rectum
Bone marrow smear	Salivary gland
Brain[a]	Sciatic nerve
Cecum	Seminal vesicles[a]
Cervix	Skeletal muscle
Colon	Skin
Duodenum	Spinal cord: cervical, thoracic, lumbar
Eyes/optic nerves	Spleen[a]
Heart[a]	Stomach
Ileum	Testes/epididymides[a]
Jejunum	Thymus[a]
Kidneys[a]	Thyroids/parathyroids[a]
Liver[a]	Tongue
Lungs (with main stem bronchi)[a]	Trachea
Lymph node: mesenteric, submandibular	Urinary bladder
Mammary gland or site	Uterus[a]
Esophagus	Vagina
Ovaries[a]	

Source: Adapted from Woolley A, *A Guide to Practical Toxicology*, 1st ed., London: Taylor & Francis, 2003.

[a] All these organs are weighed; paired organs should be weighed separately.

variable that must be considered is body weight, and this can be corrected for by expressing the organ weights as a percentage of body weight or brain weight. This becomes important when there is a significant difference in body weights between controls and treated animals. The weights of some organs follow body weight fairly closely (e.g., the liver); others tend to remain constant despite fluctuations in the animal's body weight, for example, the testes and brain.

Macroscopic appearance of the tissues as determined at necropsy is an important indication of effect and may be the only pathological evidence of change. Any abnormalities are noted and the tissue retained for microscopic examination. It is important that this examination is carried out by experienced technicians and that the information is accurately recorded. This is the link between the in-life observations, particularly information on the presence of tumors noted at clinical observations, and the pathological examination of the tissue sections.

Microscopic appearance assessed in stained sections cut from fixed tissue is the final examination of the study. Microscopic examination is used to detect any subtle or obvious differences between the control and treated animals, which may have arisen as a result of treatment with the test material. These changes may correlate with other evidence, for instance, from gross findings at postmortem or clinical pathology. The normal fixative is 10% neutral buffered formalin, although Davidson's fluid is used for the eyes and occasionally the testes. If it is intended to carry out testicular staging (i.e., assessment of all stages of spermatogenesis present), Bouin's fluid is often preferred. The normal stain used for the sections is hematoxylin and eosin, but specialist stains may also be employed, for instance, Oil-red-O (on frozen sections) for lipid or periodic acid Schiff's for glycogen. The use of electron microscopy is infrequent in routine toxicity studies, it being more applicable to mechanistic studies. The fixatives commonly used for electron micros-copy are glutaraldehyde and osmium tetroxide, and it is important that the samples are as fresh as possible. Although it is possible to carry out electron microscopy studies on formalin-fixed tissue, results are inferior, and they become less useful with increasing sample age.

Additional Examinations

Safety pharmacology: As discussed in Chapter 6, investigations relevant to safety pharmacology can be added to routine toxicity studies, for example, electrocardiog-raphy in studies with biological test items.

Micronucleus test: This is a good way of saving animal usage and implementing the three Rs. At the end of the treatment period, bone marrow samples are taken and assessed for the presence of micronuclei (see details of the *in vivo* micronucleus test in Chapter 8).

Reproductive endpoints: These may also be assessed in the course of general toxicity studies through records of estrus in nonrodents, plasma hormone analy-sis, organ weights, and histological examination. Testicular staging, in which the presence of the various stages of spermatogenesis is assessed, is a possible addi-tion to routine toxicity studies of 4 weeks or longer. Although many chemicals will show effect within 4 weeks, some may take longer, and treatment for at least one full spermatogenesis cycle is desirable. In practice, this means a 13-week study.

- *Male fertility*: Groups of untreated females may be added to a toxicity study in rats to assess male fertility after a set period of treatment, usually a full spermatogenic cycle of up to 70 days. After mating, the females are killed around day 15 of gestation and the uterine contents assessed for implanta-tions, classified as live, or early or late embryonic deaths. The number of corpora lutea graviditatis in each ovary is also recorded. The various indi-ces are then calculated in the same way as for a regular fertility study (see below).

- *Reproduction/Developmental Toxicity Screening Test and Combined Repeated Dose Toxicity Study with the Reproduction/Developmental Toxicity Screening Test [Organization for Economic Co-operation and Development (OECD) test guidelines 421 and 422 respectively]*: Male and female rats are treated in both these tests to provide initial information on toxicity and reproduction and/or development; both sexes are treated before mating (males for 14 or 28 days and females for 2 weeks) before and throughout mating and gestation and until day 4 of lactation. These studies give insight into toxicity and into effects on fertility and fetal development and postpartum viability. They are used principally for nonpharmaceutical chemicals.

REVERSIBILITY OR RECOVERY STUDIES

In studies of 4 weeks or longer, it is normal to include subgroups of animals that receive the same duration and dose levels of treatment as the other animals on the study, but are retained to assess the reversibility of any toxicity seen during the study. The usual length of such treatment-free periods is 4 weeks, which is normally enough to show the regression of treatment-related effects, either completely or in part. However, there may be reasons for using a longer period to achieve reversal of effects seen. These may include the toxicokinetics of the test substance or the type of lesion seen in the test animals. For chemicals with long elimination half-lives, such as some humanized antibodies, it is possible for clearance to be delayed for several weeks. As the continued presence of the test substance in the tissues or plasma may prolong the adverse effects of treatment, it is important to ensure that a period is allowed for recovery after complete elimination of test substance.

For certain types of lesion, usually those seen microscopically, a longer period without treatment is required simply because they take longer to regress. Among such changes are pigment depositions in the liver, e.g., hemosiderin or intracellular inclusions that have accumulated due to slow metabolism of their constituents. One example of the latter type is the accumulation in male rat kidneys of the complex of $\alpha 2u$-globulin with compounds such as trimethyl pentanol, the metabolite of tri-methyl pentane. Three months is normally the longest recovery period in routine use. Longer periods may be used but become increasingly difficult to justify; if a change is not reversible in 3 months, this may indicate the possibility of undesirable persistence of effect in humans.

The number of animals allocated to reversibility studies is also largely defined by convention. In studies using rodents, the usual number is five males and five females allocated to the controls and the high-dose group. It is now unusual to have reversibility studies at the intermediate dose levels, but this carries some risk if excessive toxicity is expressed at the high dose, leading to its premature termination. In studies with nonrodents, reversibility animals are typically included in the control and high-dose groups (the usual number is two males and two females, numbers being restricted for ethical reasons). In the absence of a formal reversibility study at intermediate dose levels, it is usually possible to make an estimate of the expected reversibility of effects seen, based on knowledge of type and extent of changes seen in the high-dose animals. Thus, adaptive change such as hepatocyte hypertrophy in the liver due to induction of hepatic enzymes is usually readily reversible, whereas other change may be expected to persist. Fibrosis, consequent upon extensive necrosis in the liver, would be expected to be irreversible, although function may not be seriously impaired if the lesions are not too extensive.

For nonrodents, it is possible, sometimes, to have reversibility animals only in the mid- and high-dose groups to indicate any dose response in recovery from toxic change, as the controls from the end of the treatment period may be sufficient to act as controls to the later sacrifice. However, this may not be sensible in shorter studies, for instance, in dogs, when a portion of the animals reach sexual maturity during the study; if the reversibility period is longer than 4 weeks, it may be advisable to include controls. To track reversibility in rats, animals for reversibility studies should be included in clinical pathology examinations at least at the end of the treatment period and at the end of the treatment-free period. This is not an issue with nonrodents.

The use of reversibility studies has been reviewed by Sewell et al. (2014) in a global cross-company data-sharing initiative on the incorporation of recovery-phase animals in safety assessment studies to support first-in-human clinical trials. They found a broad range of approaches to the inclusion of reversibility groups in this focused field of toxicity testing, finding several examples where the absence of reversibility groups did not affect the regulatory process. The group, which included people from regulatory agencies as well as industry, recommended that reversibility groups be included in studies for scientific reasons and should be considered in the context of the whole development program. The prospective need for reversibility groups may be assessed from the preliminary or early toxicity studies, and they could be included in later studies and in those with the most appropriate species. The numbers of animals should be minimized, as should the number of groups in which reversibility is studied. For nonrodents, it should be considered if animals for reversibility should be included in the control group; this recommendation is, perhaps, the most controversial of those made. However, depending on study design, it may be possible to assess reversibility in treated animals in the absence of controls; in dog studies where the end of the study and start of the treatment-free recovery period coincides with the onset of sexual maturity, it may be wiser to have controls. In studies in which controls are not needed for reversibility, it may be appropriate to have reversibility animals at the intermediate-dose group on the basis that reversibility at the high dose may not be apparent or, if the high dose has had to be terminated early, cannot be assessed.

EXAMINATIONS FOR SPECIFIC TOXICITIES

There are several areas of toxicity that do not merit their own special category of investigation, unlike genotoxicity or carcinogenicity, but that may be incorporated into general tests for toxicity. These include investigation of toxicities in the immune, respiratory, and nervous systems and in the skin. The problems inherent in these systems and investigations are sketched out below.

Immunotoxicity

As with other organ systems, the function of the immune system may be enhanced or suppressed by xenobiotic chemicals. Unlike most other organ systems, the immune system is not a discrete organ but an interrelated set of tissues distributed throughout the body. It includes the thymus, bone marrow, Peyer's patches, spleen, lymph nodes, and other lymphoid tissues. An effect on one part of this system may have contrary effects on other parts; consequently, interpretation of a small change in one area is made more complex by the difficulties of predicting the impact on other parts of the system. This complicates study of immune responses to xenobiotics with the added problem that, in general, animals are poor models for human immunotoxicity, particularly autoimmune reactions and hypersensitivity. With such a diffuse system, the best approach is to obtain a broad overview and then, if significant change is seen, to focus on the areas of interest in specific mechanistic studies. Accordingly, it is generally recommended that a tiered approach be adopted, the first tier being contained within the conventional toxicity tests. These examinations include differential leukocyte counts in peripheral blood, plasma protein fractions and the weights, and/or microscopic appearance of the lymphoid tissues. The distribution of lymphocyte subsets can also be examined by homogenization of tissues and flow cytometry. However, these investigations may not give a definitive answer as to whether there are changes that are truly indicative of a significant effect on immune function. The immune system is not static through the lifetime of an organism. The thymus involutes or atrophies with age, and this is quite normal; however, acceleration of involution relative to controls or expectation may well imply an immunotoxic effect. A further layer of complexity is added, when it is considered that such atrophy is also a response to stress, although this is usually accompanied by changes in the adrenal glands. The International Conference on Harmonisation (ICH) S8 guideline on immunotoxicity studies for human pharmaceuticals (Case Study 7.1) (ICH 2005) indicates that, in addition to findings in standard toxicity studies, additional prompts for immunotoxicity study may be the pharmacological action of the drug; the intended patient population; and factors such as similar structure to known immunotoxicants, drug disposition, and information from the clinic. Once change in the immune system has been identified, additional testing should be considered depending on the nature of the immunological changes observed, taking into account any concerns raised by the class of compound.

Immunotoxic investigations are additional to the normal assessment of skin sensitivity reactions, which are particularly useful for assessing workplace hazards and risk. Extended testing may include assessment of antibody responses, cytokine production, and susceptibility to infectious agents in mice, the intention being to define the cell population affected and any dose–response relationship. With a full set of data, an assessment of possible effect in humans may be made.

NEUROTOXICITY

The nervous system is toxicologically significant because of the far-reaching effects of change, which is often irreversible. While other tissues, such as the liver, have extensive repair capabilities following toxic insult, this is absent or very small in the nervous system. Also in contrast to other tissues, the nervous system has a more limited functional reserve, meaning that a 15% reduction in nervous function is likely to be much more significant than a similar reduction in renal or hepatic function. Detection of effects in the nervous system requires a range of special techniques that are often technically complex and require specialist interpretation. However, much can be done in a routine toxicity test as a first tier of neurological assessment. Clinical signs, combined where appropriate with a functional observation battery (see Chapter 6 for a description of this test), can lead to detailed neurological

CASE STUDY 7.1   ICH S8

The ICH S8 guideline on immunotoxicity studies for human pharmaceuticals details the steps that can be taken for investigation of unintended immunosuppression or enhancement (drug-induced hypersensitivity and autoimmunity are excluded). These steps take the form of further nonclinical testing and/or guidance on a weight-of-evidence decision-making approach for immunotoxicity testing. It does not apply to biotechnology-derived products covered by ICH S6 (see Chapter 19, Case Study 19.1) and other biologicals.

Additional immunotoxicity studies can be prompted by findings from standard toxicity studies, the known pharmacological properties of the drug, knowledge of the patient population, structural similarities to known immunomodulation, the disposition of the drug, and clinical information. Signs of potential immunotoxicity in standard studies include hematological changes, alterations in immune system organ weights, changes in serum globulins without plausible explanation, increases in infection incidence, and increases in tumors. This evidence is then analyzed, and it is determined, based on the weight of evidence, if additional studies are merited. The options for these studies are broadly as follows:

- T cell–dependent antibody response (TDAR)

    A TDAR study design is similar to a 29-day repeat-dose study with the inclusion of an immunization with a known T-cell dependent antigen such as sheep red blood cells (SRBC) or keyhole limpet hemocyanin (KLH) on day 24 of treatment to provoke a robust antibody response than can be compared across the controls and treatment groups; cyclophosphamide is used as a positive control.

- Immunophenotyping

    Immunophenotyping identifies and counts leukocyte subsets using antibodies and is usually carried out with flow cytometry analysis or immunohistochemistry.

- Natural killer cell activity assays

    Natural killer (NK) cell activity assays can be conducted if immunophenotyping studies demonstrate a change in number, if there is any indication of increased viral infection, or to investigate any other indication of effect. These tend to be *ex vivo* assays of tissues taken from treated animals and coincubated with target cells that have been labeled with ^{51}Cr.

- Host resistance studies

    In a host resistance study, groups of mice or rats treated with the test compound are challenged with varying concentrations of a pathogen (bacteria, fungal, viral, parasitic) or tumor cells. The infectivity of the pathogens or tumor burden is observed in vehicle and treated groups to determine the effect of the test item on host resistance.

- Macrophage/neutrophil function studies

  Macrophages/neutrophils are exposed to the test compound *in vitro* or taken from animals that have been exposed *in vivo*. Functional assays (phagocytosis, oxidative burst, chemotaxis, and cytolytic activity) are then performed on them.

- Assays to measure cell-mediated immunity

  These are *in vivo* assays where antigens are used for sensitization. The guideline makes the point that assays to measure cell-mediated immunity are not as well established as those used for the antibody response.

Once additional toxicity studies have been conducted, it is then determined if the information gained is sufficient for risk assessment and risk management purposes.

examination for reflexes, grip strength, coordination, gait, etc. Other tests such as the Morris water maze or open-field test may add further knowledge about specific brain area effects. Electroencephalography may prove useful, although the benefits over a thorough neurological examination conducted by a veterinary surgeon should be considered first. Similarly, ophthalmoscopic examinations may be supplemented by electroretinography, a rarely used method of assessing the electrical response of the retina to light impulses; it is time consuming and technically demanding, in both conduct and interpretation. Assessment of the senses is very limited in general toxicology. Hearing may be tested using a whistle or other noise, but the assessment is crude, as it is based on Preyer's reflex (the ears pricking forward). The loss of hair cells from the cochlear is associated with hearing loss, and to detect this, the use of scanning electron microscopy is recommended. The other senses, smell, taste, and touch, are not investigated routinely in toxicity testing, due to the difficulties in assessing these functions in laboratory animals.

In the blood, measurement of cholinesterases may indicate toxicity due to organophosphate or carbamate pesticides; however, after chronic administration, rats can show large decreases in activity without clinical evidence of effect. Organophosphates, which inhibit cholinesterases (as do carbamates), are classically associated with delayed-onset neuropathy, which has been tested routinely in chickens. This has been the species of choice for assessment of the target enzyme for this condition, neuropathy target esterase. Much emphasis is placed on histopathology, where the use of special fixatives and stains with appropriate microscopic technique can be very informative.

Respiratory Toxicology

Essentially, this is the field of toxicity resulting from inhalation of toxicants. Pulmonary toxicity as a result of systemic exposure, following administration by oral or parenteral routes, is not a common finding, paraquat being a prime example. The use of inhalation as a route of administration becomes important in assessing workplace hazards and, clearly, for medicines given by inhalation. Technically, inhalation is in a field of its own due to the problems of generating (and monitoring) the correct atmospheres, administering these safely to the animals, and thence wasting them to the outside through suitable filters.

The basic objective is to generate a respirable, uniform atmosphere from the test substance. It is important to determine the physical characteristics of the atmosphere generated and to calculate and sustain the correct rate of generation to achieve the desired dose or concentration. A relatively large proportion of inhaled material is eventually swallowed, giving a significant oral component to the toxicity elicited. Rats are usually exposed for up to 6 hours/day, restrained in tubes fixed onto a central cylindrical chamber so that only their noses protrude into the atmosphere that flows through the apparatus. Whole-body chambers can be used, although these use larger amounts of test material and result in dermal and oral exposure (in addition to that expected from clearance from the lungs). Dogs and NHPs are dosed through the use of masks.

Intranasal administration is relatively straightforward and can be performed using droplets or an aerosol of test solution. Vehicles should be chosen with care to avoid local irritation. A knowledge of the anatomical architecture of test species should be employed because, although the rat is used as the rodent species in intranasal studies, the nasal turbinates of NHPs have been generally considered to be a better model for humans than the dog. However, ethical pressures have tended to reduce the use of NHPs unless there is another, more pressing, justification, such as similarity of receptors or mechanism to those in humans.

Dermal Toxicity

Dermal administration is used less frequently than other routes but is relevant to the workplace and topical medicines. The main species used is the minipig, because the skin structure is close to that in humans. While studies may still be conducted for some chemical classes in rats and rabbits, they are no longer the default species for pharmaceuticals, for which a typical program for a topical treatment might include topical dosing studies in minipigs and subcutaneous dosing studies in rats. Careful choice of vehicle is essential as this has considerable influence on absorption of the test substance. In pharmaceutical toxicology, the formulation must be the same as or as close as possible to the clinical formulation to avoid any effects due to the vehicle. The potential toxicity of any excipients should be investigated by the use of sham-dosed controls in addition to a group that receives the vehicle only. Occlusion of the application site for several hours by wrapping the site in an impermeable dressing enhances absorption of the test substance and prevents ingestion. This is a feature of OECD test guidelines but may not be relevant for pharmaceuticals. Occlusion is normal practice in acute studies and up to 28 days but is not recommended in longer experiments. Due to the absorption characteristics of the skin, the concentration of the test substance and the area of the dosing site tend to be more important than the dose in mg/kg, especially as dermal toxicity studies are usually undertaken in part to assess local tolerance or irritation or other effects due to the test article–vehicle combination.

PITFALLS IN GENERAL TOXICOLOGY

The major pitfalls in study conduct, which result in spurious results, are related to the timing of the various examinations. Electrocardiograms (ECGs) can be expected to show two types of basic effect—pharmacological and toxicological. The pharmacological effects, wanted or unwanted, should be related to the presence of the test substance or an active metabolite and are generally seen in the few hours after administration. If effects are present 24 hours after dosing in treated animals, it may well be an indication of toxicity unless the elimination of the test substance is prolonged. The timing of ECG examinations is therefore important, based on what is required of the study. With continuing interest in QT prolongation, it makes sense to look at an ECG at the time of peak plasma concentration; examination after 24 hours should confirm the absence and transience of any effect. ECGs are usually only recorded in nonrodents that are not sedated. The process of restraint and application of electrodes is, at the first experience, a stressful process resulting in increased heart rate and blood pressure. It is useful to accustom the animals to the procedure by taking two or more recordings before the definitive measurements. Despite this, it is likely that heart rate and blood pressure will still be higher than normal, and this can mask effects of the test substance. To avoid this kind of error, the use of telemetric implants or jacket telemetry systems is recommended. With these internally implanted devices or attached devices, it is possible to record a number of parameters such as locomotor activity, ECG, heart rate, arterial blood pressure, respiration, and body temperature, although not all at the same time. Collection of these data from unrestrained animals gives a better indication of variation from normality than when in the presence of an observer or under restraint. In animals, it is possible for cardiotoxicity to develop in response to excessive pharmacology, and this must be taken into account in analysis of the data; toxicity without evident pharmacological cause needs careful interpretation.

In nonrodents, the timing of collection of blood samples for toxicokinetics relative to collection for clinical pathology is also critical; in rats, this is not such a problem, because separate animals are generally used to avoid collection of excessive volumes of blood from the same animals. Samples for clinical pathology should always be collected before those for toxicokinetics, if the same animals are to be used, or unless sample sizes are very small. The controls in rodent studies are not usually subjected to the same sampling regimen for toxicokinetics as the treated groups. In addition, it should be noted that there is increasing requirement for analysis of control samples in toxicokinetic studies, especially for pharmaceuticals. Varying sampling regimens and stress between controls and treated animals may introduce confounding factors. This could include a mild anemia, found at clinical pathology, that is not present in the controls but is not treatment related. In 2-week studies, with intensive toxicokinetic sampling in nonrodents, there may not be sufficient time between day 1 samples and clinical pathology, and day 14 toxicokinetic samples for complete recovery. If blood samples are taken immediately before the start of the treatment period from dogs or NHPs followed by day 1 toxicokinetic sampling, the sampling stress on the animals becomes significant and may complicate interpretation of the clinical pathology data.

Age and sexual maturity may present problems in the interpretation of general toxicity. For instance, beagle dogs become sexually mature between the ages of 7 and 12 months. In shorter studies conducted with sexually immature animals at the start of the treatment, this is likely to mean that there would be a range of sexual maturity when the study ends. As group size is small in shorter studies—usually only three per sex per group—it is possible for the distribution of sexual maturity to be uneven across the groups. In this case, a sexually mature control group may compare with sexually immature treatment groups, implying a treatment-related effect, which is entirely due to differences in maturity rather than an effect on testicular development. Spermatogenesis is very similar across the species, and effects are likely to be relevant to man unless a species-specific factor such as metabolism or pharmacokinetics is present. The implication is that short studies in dogs should be conducted with sexually mature animals, with testicular development being examined in longer studies conducted with animals that are sexually immature when the treatment period starts.

Small group size in nonrodents means that relatively infrequent effects may not be seen in the treated animals. If an effect is seen, for example, in the liver of one of a group of three at the end of the treatment period, its absence in two animals at the end of the recovery period does not necessarily mean that it is reversible. In this case, reversibility has to be judged from the nature and extent of the effect seen.

Although it is desirable to show toxicity, to give an estimation of the dose–response curve for a test substance, this is not always possible. In some cases, this is due to genuinely low toxicity even at high doses; here it is necessary to demonstrate absorption and adequate systemic exposure. Some drugs intended to have a local action in the lumen of the gastrointestinal tract may not be absorbed, but this is probably beneficial. For low-toxicity compounds that are not absorbed after oral administration, it may be necessary to use the intravenous route to elicit toxicity, if that is considered essential. In contrast, there are instances where the acute pharmacological action is so intense that it becomes a toxicological effect in its own right. In these cases, anesthetics and narcotic drugs being good examples, it may be impossible to demonstrate any toxicity apart from the excess pharmacological action.

Poor choice of test substance form or formulation can be a pitfall in any toxicity study. Particle size can be a limiting factor in absorption and thus in toxicity; micronizing a test material or changing the carrier system or vehicle can cause a radical increase in toxicity. Such changes should be avoided in the middle of a program unless some form of sighting or bridging study is conducted with the new form or formulation to ensure continued lack of effect.

Another factor to consider is the correct choice of examination for the test species being used. This is particularly true with clinical pathology where there are significant differences between species in the plasma activity of some enzymes. For instance, alkaline phosphatase is variable in cynomolgus monkeys. Leucine aminopeptidase has been suggested as an alternative (Evans 2009b) but is not widely used. Marmosets have very low activities for alanine aminotransferase and gamma-glutamyl transpeptidase activity is very low in rats.

In the final analysis, correct study design and interpretation will avoid the majority of these pitfalls and will facilitate interpretation of the whole data package. When reviewing a study report, it is important to understand where such problems can arise and to allow for them in your interpretation, bearing in mind that the pitfalls seen may not be included in the above analysis.

REPRODUCTIVE AND DEVELOPMENTAL TOXICOLOGY

GENERAL PRINCIPLES IN REPRODUCTIVE AND DEVELOPMENTAL TOXICOLOGY

The intention in studies of developmental and reproductive toxicity (DART), sometimes referred to collectively as reproductive toxicology, is to assess the potential for adverse reproductive effects in the target species (usually humans) due to exposure to chemicals, whether as a result of intentional (drugs and food additives) or unintentional exposure (pesticides and other chemicals). In contrast to general toxicology, which has a very broad approach to toxicity testing, the endpoints in reproductive toxicology are more defined, and there are specific stages to examine and evaluate. The reproductive process and its various stages are illustrated in Figure 7.1.

Despite this relatively simple definition of endpoints, reproduction is immensely complex and can be affected in many ways. Toxicity can occur during any part of the process, and the various tests are designed to examine every stage of the cycle in digestible chunks. However, there is enormous scope for different effects on

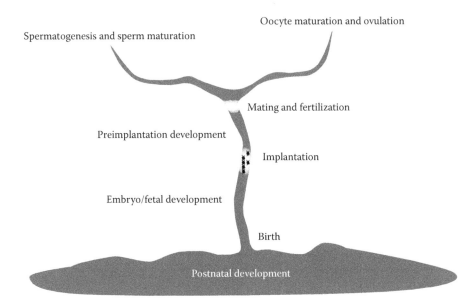

FIGURE 7.1 River of reproduction. (Courtesy of Newall, Derek, 1999. With permission.)

reproduction. There may be indirect or direct effects on the gonads, which have dual function as the source of the gametes and of sex hormones (the secretion of which is controlled by the pituitary). After gametogenesis, variations in behavior, fertilization, or effects on the processes of gestation can also influence the final outcome. The net result of this is that presence of an effect at one stage of the sequence does not necessarily pinpoint the origin of that effect; consequently, further investigation is needed to elucidate mechanisms and facilitate risk assessment. Where effects are seen, it may become necessary to break the process down further to determine the location of the effect, in terms of time and place. This element of timing is unique to reproductive toxicity, especially with respect to teratogenicity. For example, thalidomide was associated with reproductive effects in humans when given in week 4 or 5 of pregnancy. From the time of implantation until closure of the palate, the organs develop in the fetus according to a well-defined pattern and timing. Accordingly, treatment of a pregnant rat with a teratogen on day 8 of gestation will produce a different spectrum of effects in comparison with treatment on day 12 (Figure 7.2). Timing of treatment is also important in spermatogenesis, where single treatment may affect only one stage of the spermatic cycle. It is increasingly recognized that the visible processes of organogenesis are matched by biochemical changes that have profound influence on the toxicity of compounds in the fetus or neonate in comparison with an adolescent or an adult (Rasheed et al. 1997; Koukouritaki et al. 2002; McCarver and Hines 2002).

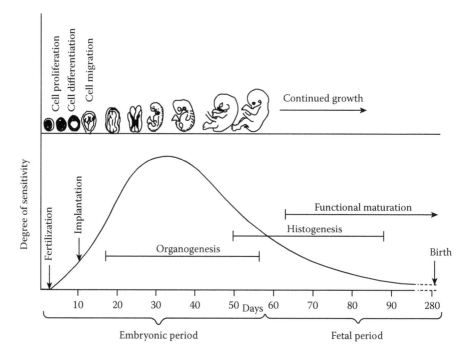

FIGURE 7.2 The stages of human embryogenesis. (Adapted from Timbrell J, *Principles of Biochemical Toxicology*, London: Taylor & Francis, 2000.)

Pregnancy is associated with a wide range of physiological changes that can affect the ADME of chemicals. Total body water and lipid are increased, associated in the former case with an increase in plasma volume. Because the total red cell population increases to a lesser extent than plasma volume, the effect is to reduce the red cell count to near-anemic levels. There is an increase in the extracellular space, which, with the increase in total body lipid, increases the available volume of distribution of chemicals. The body lipids accumulate over the early part of the gestation and are used in the latter part. Thus, where there is accumulation of lipid-soluble chemicals into the adipose tissue, rapid release late in gestation can lead to increased plasma concentrations, which have the potential for adverse effects in the mother and fetus. Plasma concentrations of albumin, important in reducing the free concentration of chemicals in the plasma by binding, are lower in pregnancy, partly due to the increase in plasma volume and partly due to decreased total body content.

Examples of reproductive effects with particular substances include testicular atrophy seen with the fumigant dibromochloropropane, teratogenicity with vitamin A or alcohol, transplacental carcinogenesis with diethylstilbestrol, and many more. Indirect effects through hormonal imbalance are also a frequent cause of reproductive toxicity, and specialist studies may be needed to investigate these. The presence of compounds that accumulate in animals is an environmental issue of some concern; for example, many organochlorines have been found to accumulate in marine mammals, although the effects of these have not necessarily been elucidated (Vos et al. 2003). Until recently, reproductive toxicity was relatively resistant to mechanistic explanation in contrast to other toxicities, although the number of elucidated mechanisms is increasing. Broadly, agents that affect cell division, apoptosis, membrane integrity, and other factors that are essential to organ differentiation and development are likely to have effects on the fetus. Hormonal disturbance may affect fertility, parturition, and lactation, and straightforward pharmacological action may impact on normal behavior at any point to disrupt normal reproductive processes.

JUVENILE TOXICITY STUDIES

These studies are typically conducted for pharmaceuticals to support clinical trials of pharmaceuticals in children; they have been extensively reviewed by authors such as De Schaepdrijver et al. (2008) and Bailey (see references for a selection). Bailey and Mariën (2009) make the point that these studies should be designed according to a scientific rationale, considering the toxicological information that may be obtained and its prospective utility—including ease of interpretation. They form part of a pediatric investigation plan (PIP), which should be discussed with regulators before being implemented. The objective of these studies is to investigate the potential for adverse effects on postnatal growth and development (Cappon et al. 2009). Such effects may not be apparent from the routine toxicity studies, which are usually conducted in animals that are older than the equivalent human stages of development.

While humans, minipigs, and NHPs may be born at approximately the same stage of development, rats and dogs are born less developed, so that by 1 week postpartum, rats and dogs have reached the human stage at birth. The age range in animals that is investigated is selected based on the equivalent stage of human development. While

TABLE 7.4
Age of Transition between Developmental Stages in Humans and Test Species

Species	Unit	Neonate	Infant/Toddler	Child	Adolescent
Rat	Days	<9 to 10	10 to 21	21 to 45	45 to 90
Minipig	Weeks	Birth to 2	2 to 4	4 to 14	14 to 26
Dog	Weeks	0.5 to 3	3 to 6	6 to 20	20 to 28
NHP	Months	Birth to 0.5	0.5 to 6	6 to 36	36 to 48
Human	Years	Birth to 0.08	0.08 to 2	2 to 12	12 to 16

Source: Adapted from Buelke-Sam, cited in Tassinari MS et al. Juvenile Animal Toxicity Studies: Regulatory Expectations, Decision Strategies and Role in Paediatric Drug Development. In Brock WJ et al. *Nonclinical Safety Assessment: A Guide to International Pharmaceutical Regulations.* 2013. Copyright Wiley-VCH Verlag GmbH & Co. KGaA. Reproduced with permission.

juvenile toxicity studies may be conducted in several species, the rat is typically used in preference to the dog, minipig or NHP. Table 7.4 gives the transition times between the various human stages of development.

These studies present considerable practical and technical challenges in the treatment of very young animals, especially rats. Their design is not prescribed by regulatory authorities in quite the same way as for other study types and should be considered carefully before being implemented; as with many development programs, it makes sense to discuss them with regulators before starting.

Test Systems for Reproductive Toxicology

The thalidomide tragedy had a huge influence on the choice of test species for assessment of reproductive toxicity, leading in particular to the use of the rabbit, which was sensitive to its effects. As indicated in Table 7.5, the rat, rabbit, and mouse are the principal test systems for examination of reproductive toxicity; the minipig is

TABLE 7.5
Stage of Reproductive Cycle and Preferred Test Systems

Stage of Cycle	Test System
Fertility and mating	Rat or mouse
Organogenesis—*in vivo*	Rat or rabbit; alternatives are mouse, minipig, or nonhuman primate
Organogenesis—*in vitro*	Whole-embryo culture, limb bud assay
Late gestation, parturition, and early development	Rat or mouse

Source: Adapted from Woolley A, *A Guide to Practical Toxicology*, 1st ed., London: Taylor & Francis, 2003.

increasingly used. Although NHPs are used in studies to evaluate effects during gestation (developmental toxicity studies), they are used only in special circumstances. In general, the placental structure of animals used in assessment of reproductive toxicity is not the same as in humans, although this may not be a factor that is always significant. A summary of the various test systems that are in reasonably regular use is shown in Focus Box 7.3.

Overall, there is no escaping the general acceptance, especially by regulators, of the rat, rabbit, and, to a lesser extent, the mouse as models for reproductive toxicity. They have the advantages of availability, size, length of reproductive cycle, and a level of understanding that is not so clear for the other systems. For all models, it is essential that there be a large amount of in-house historical control data to facilitate interpretation of the study data. In addition, this database has to be kept up-to-date by performance of new studies, in order to compensate for drift in the strain of animal used.

REPRODUCTIVE TOXICITY *IN VITRO*

The various alternative models indicated in Focus Box 7.3 are useful as screening methods in candidate selection. However, none of these *in vitro* systems has yet been accepted by regulatory authorities as a satisfactory alternative to whole-animal experiments. They lack the complex interrelationships that exist between the mother and the fetus and between the various tissues and dynamics in each. Other systems that might appear to have value in teratogenicity testing, such as metamorphosis in amphibians or invertebrate larval stages, have not been widely investigated, although the fruit fly, *Drosophila,* has been used in some experiments. Other possibilities include fish and chick embryos, the latter having been investigated as a screening system. Another possibility for the future is the nematode *Caenorhabditis elegans*; the complete cell lineage for this 900-cell organism has been elucidated. Care is needed in the collection of data and interpretation of these assays. There are also systems based on fragments of reproductive tissues (e.g., isolated seminiferous tubules), which can be used to investigate functional aspects of particular parts of the reproductive system, including the effects of one cell type on another. These specialized studies may be technically demanding and not readily transferable from one laboratory to another in a reproducible manner. These alternatives were reviewed in 1997 by a British Toxicology Society working party (BTS 1997). A positive result does not necessarily stop development of a chemical but merely emphasizes the possibility of reproductive effect, leading to an appropriate change in study timing.

Several *in vitro* assays for reproductive toxicity have been validated by European Centre for the Validation of Alternative Methods (ECVAM); these include the embryonic stem cell test, the micromass test, and whole-embryo culture.

The embryonic stem cell test takes advantage of the fact that embryonic stem cells differentiate in culture and studies the inhibition of differentiation. The test uses permanent mouse cell lines, embryonic stem cells to represent embryonic tissue, and for comparison, a line of fibroblasts to represent adult tissue responses to cytotoxicity. The test looks at three endpoints: inhibition of differentiation of the stem cells into cardiac myoblasts and inhibition of growth of both cell lines. Growth inhibition is

FOCUS BOX 7.3   CHARACTERISTICS OF TEST
SYSTEMS FOR REPRODUCTIVE TOXICOLOGY

- *Rat and mouse:* The reproductive cycle is completed relatively quickly; they do not hesitate to mate in laboratory conditions; relatively large litters are produced after a conveniently short gestation (about 20 days); inexpensive and easy to maintain; there is a wealth of historical control data. However, the rat produces a very large number of sperms in comparison with humans, and consequently, a relatively large reduction in sperm number or quality is necessary before effects are seen. Rats are the preferred rodent species for use in studies of embryotoxicity.

- *Rabbit:* The rabbit is a reasonable size and reproduces readily in laboratory conditions; gestation is relatively short at about 30 days; litter size is generally good; there are good historical control data for a number of strains. Disadvantages include their intolerance of compounds such as antibiotics or of chemicals that disturb their alimentary canals, including several vehicles, such as oils. They are not used in other reproductive studies, but they are ideal for longitudinal studies of the sperm cycle.

- *Minipig:* The use of the minipig is increasing in teratogenicity studies, and there may be a future for them in fertility studies. They have a clear advantage when the rabbit cannot be used due to rabbit-specific effects; gestation at about 115 days (3 months, 3 weeks, and 3 days) is long, and litter size is small (five or six). If they are used as the non-rodent species in the general toxicity studies, the reproductive studies can be done in the same species. However, they are large (approximately 35 kg) and so require large amounts of test material and more extensive housing. They require skill in dosing and in sample collection. They are particularly suited to dermal studies.

- *NHPs:* Used occasionally as an alternative nonrodent, but expense and availability limit their use. The usual species is the cynomolgus monkey; although marmosets have been investigated as an alternative, they are not used in regulatory toxicology. The rhesus monkey has been used but is a seasonal breeder, while the cynomolgus can breed at any time of year due to a menstrual period of about 30 days. Litter size is small, and gestation is long with a high miscarriage rate, meaning more animals are needed. The testicular physiology of the cynomolgus testis is said to be a good model for humans, and fetal malformation frequencies have been quoted at around 0.5%. Normal practice is to remove the fetuses via cesarean section and then to allow the mother to recover for reuse in another study. They cannot be used for fertility studies as such, but serial examination of parameters such as semen and sperm quality and hormone levels during a long-term study in mature males may give insight into effects on testicular function. This is based on the investigations used in infertile human males. Phylogenetic proximity does not mean necessarily that they are good models for humans in terms of ADME.

- *Alternatives:* There are a number of *in vitro* systems for assessment of teratogenicity, but none have been validated sufficiently for regulatory purposes. Whole-embryo culture is one of the best of these and has particular use when screening members of a series of compounds that are known to come from a teratogenic class, for instance, retinoids. In this case, the assay may be used to aid selection of lead candidates for development from a teratogenic series. The micromass assay is a variation on the whole-embryo culture theme, which uses primary cell cultures from the limb bud or brain. There is also an assay that uses *Hydra*, but use of this has not been pursued. Chick embryos have also been used occasionally (see Chapter 4).

studied in both cell lines as the embryonic stem cells have been shown to be more sensitive than adult cells.

The micromass test uses cultures of limb buds isolated from pregnant rats and tracks the inhibition of cell differentiation and growth. The test is considered to be useful in identifying the more potent embryotoxic chemicals and should not be viewed as a replacement method. The test is based on the tendency by undifferentiated mesenchyme cells in limb buds to form foci of differentiating chondrocytes. The inhibition of the formation of these foci is the test endpoint. The formation of cartilage by chondrocytes is an important process in the formation of the skeleton, and inhibition of cell proliferation and differentiation, intercellular communication, and interactions with extracellular matrix are thought to be parts of this process.

Whole-embryo culture has also been validated and is, intuitively, a more complete test system than the two preceding tests as it uses complete embryos instead of isolated cell lines or components of embryos. The assay uses embryos isolated from pregnant rats on day 9 or 10 of gestation. The isolated embryos are cultured for 48 hours during a period of major organogenesis and are then examined for heartbeat, yolk sac circulation and size, length of head, and crown to rump; the number of somites is scored morphologically, with assessment of any abnormalities that may be present. Although this test is more complete than the others described briefly here, it is still not seen as a replacement for conventional reproductive toxicity studies.

STUDY DESIGNS FOR REPRODUCTIVE TOXICOLOGY

As indicated, the process of reproductive toxicity assessment is broken up into manageable chunks, addressing fertility, embryonic development or fetal toxicity (teratogenicity), and perinatal and postnatal development including maternal function.

A full program in rats will cover the 63 days before mating in the males and may run into two generations, with continuous treatment of all animals up to weaning of the final litters (F2 generation). Such studies are lengthy and produce vast amounts of complex data.

BASIC DESIGN AND DOSE SELECTION

As with general toxicology, the standard design is for three treated groups with an untreated control. In the same way, doses are chosen after an appropriate set of sighting studies, in which treatment duration and timing mirror the study intent and duration relative to gestation and parturition as needed. (For the rat, the general toxicity studies are generally adequate for this, although it is usual to perform a sighting study in pregnant animals to confirm dose choice.) In rabbits, it is also usual to perform a small sighting study in nonpregnant animals and then to confirm in a few pregnant animals.

The chief design driver in reproductive studies is the stage of the cycle under examination. This determines length and timing of treatment and, to a large extent, the type of examination undertaken (Figure 7.2). Specialist design becomes necessary if, for any reason, it is not possible to carry out a normal study due to the expected effects of the test substance; in this case, the studies have to be broken up into individual stages. The designs of three main types of reproductive study are summarized in Focus Box 7.4. Although juvenile toxicity studies have been added to this, they do not fit readily into other categories of toxicity study and are generally started shortly after birth.

The duration and timing of the treatment period are relatively fixed, due to the time constraints of the processes examined. In rats, the spermatogenesis cycle is approximately 63 days. However, even with significant testicular toxicity, rats can be successful sires due to the large number of sperms produced. Histological examination is a good method for detecting effects in the testis; because the process of spermatogenesis is very similar between species, effects in one may be indicative of potential effect in humans. Differences in effects may, however, arise through differences in pharmacokinetics or metabolism.

Premating treatment in fertility studies is generally 28 days, with the possibility of increasing this to 63 days if effects are expected. In embryotoxicity studies, the treatment period is chosen to last from implantation until closure of the palate, which is approximately day 15 of gestation in the mouse, day 16 in the rat, and day 18 in the rabbit. However, it should be noted that the day of palate closure can vary slightly between strains, and this should be accounted for in the study design. Following closure of the palate, there is a treatment-free period until just before natural parturition, when the dams are killed and their uterine contents are examined. In minipigs, treatment is from day 11 to 35 with examination of uterine contents on day 110 of gestation. The treatment period in prenatal and postnatal development studies in rats is from day 6 of gestation to weaning of the litters; males are not treated.

In addition to these basic and traditional designs, the inception of Registration, Evaluation, Authorisation and restriction of Chemicals (REACH) in the European Union has produced an extended one-generation reproductive toxicity study OECD

FOCUS BOX 7.4   OUTLINE DESIGNS FOR SOME
EXAMPLE REPRODUCTIVE STUDIES

Note that these are examples only and should not be taken as the only option. Guidelines such as those for ICH and OECD (http://www.oecd.org) give preferred designs according to chemical class and use.

- *Rabbit (oral gavage) developmental toxicity study:* Four groups of 20 females. Time mated (day 0). Dosed from day 6 to 18 of pregnancy. Clinical observations daily. Body weights on days 0, 3–18, 22, 25, and 28 of pregnancy. Food consumption daily from day 3 to 6, then every two days. Postmortem examination on day 28 of pregnancy—parental females examined, gross abnormalities retained; fetuses—external, visceral, and skeletal examination and then retained.
- *Rat (oral gavage) fertility and embryonic development study:* Four groups of 25 males and 25 females. Dosed for (a) males—28 days premating, through mating and to necropsy; (b) females—14 days premating, through pregnancy to day 17 of gestation. Clinical observations daily. Body weights: males twice weekly, females twice weekly premating then daily to necropsy. Food consumption—males weekly, females weekly premating then at appropriate intervals during pregnancy. Postmortem examination on day 20 of gestation; parental animals—males' testes and epididymides retained; gross abnormalities retained; fetuses—external, visceral, and skeletal examination.
- *Rat perinatal and postnatal development study:* Four groups each of 25 mated females treated from day 6 of pregnancy to day 20 postpartum; 20 males and 20 females selected from F1 per group (not treated) and reared to sexual maturity and mated; necropsy on day 13 of gestation. Examinations are similar to those on other studies and may vary according to protocol.
- *Juvenile toxicity studies:* Group size is not readily defined but is selected to accord with intention, regulatory advice, and species. The studies are conducted in animals of an age comparable with the targeted developmental stage in humans. Effects on growth and development together with other standard endpoints are included, taking into account the results from studies in adult animals, with the objective of identifying toxicity related to age, such as developmental effects and any differences in sensitivity that may be age related. Neurological behavior and reproductive function may be examined.

test guideline 443, which, for this purpose, replaces the two-generation study. In addition, OECD test guidelines 421 (Reproduction/Developmental Toxicity Screening Test) and 422 (Combined Repeated Dose Toxicity Study with the Reproduction/Developmental Toxicity Screening Test) describe 28-day toxicity studies in which groups of 10 males are treated for 28 days (14 days prior to mating) and 10 females for up to 6 weeks, from before mating until day 4 or so of lactation.

MATING

There are three methods of obtaining pregnant animals for reproductive studies, natural mating, artificial insemination, or buying in time-mated animals from a supplier. Which method is used depends largely on the personal/corporate preferences of the laboratory performing the study. Natural mating, preferably using one male to one female, has the advantage that proven males can be used and that the sire of each litter is known; this means that particular abnormalities can be traced back to specific animals. Although successful and simple, natural mating requires the maintenance of an adequate stock of reproductively proven males. Artificial insemination uses pooled semen from several animals, so the sire cannot be identified for each litter.

Time-mated animals from suppliers are increasingly used and provide a good source of pregnant animals at reasonable reliability; these animals are mated naturally, offering the same advantages as in in-house mating without the need to maintain stud males. Sires can be traced and abnormalities ascribed to them, where appropriate.

GROUP SIZES

The ICH pharmaceutical guideline on detection of toxicity to reproduction says that there "is very little scientific basis underlying specified group sizes in past and existing guidelines nor in this one" (ICH 2005). Number of animals per study is chosen to give a satisfactory number of litters for evaluation and has been suggested as providing the best compromise between insensitivity in terms of detection of low-incidence effects and large number of animals, which may not increase the statistical sensitivity of the test. As always in toxicology, if the effect to be demonstrated is one with a high incidence, fewer animals are needed than for a rare event. It is, however, unusual to embark on a reproductive study with the sole intention of investigating a single effect. In an embryotoxicity study, the typical number of animals per group is 24 rats or 20 rabbits. Minipig studies inevitably use smaller numbers, for example, 12 females per group, which results in 9 or 10 litters of five to six. In rats, if the study includes investigation of more than one generation, it is likely that more animals will be needed to ensure that there are sufficient F1 litters at each treatment level from which to choose the males and females for mating to produce the F2 or subsequent generations.

PARAMETERS MEASURED IN REPRODUCTIVE TOXICOLOGY

Measurement of food consumption and body weight and recording of clinical signs is common to all study types *in vivo*. Record of litter size together with sex and

TABLE 7.6
Reproductive Parameters

Fertility	Embryonic Development	Prenatal and Postnatal Development
Both sexes	Females only	Dams
Time to mating	Litter size	Length of gestation
Females	Number and position classified as early resorptions, late resorptions, dead or live fetuses	Onset and duration of parturition
Litter size		Observe through lactation; necropsy at weaning
Fetal sex		
Number and position of implantations	Number of corpora lutea	Litters
Number of corpora lutea	Weight of gravid uterus and placentae	Number of pups, external malformation, body weights
	Fetal weights and sexes (live fetuses)	Survival
	Fetal abnormalities—external, visceral, and skeletal	Opening of eyes and pinnae, pupil and righting reflexes, startle response
		Learning test in swimming maze (postweaning)
		Ophthalmoscopy, Preyer's reflex, locomotor activity
		Sexual development and mating with necropsy of females on day 13 of gestation—numbers of corpora lutea and position and numbers of implantations

weight is also a feature of all reproductive studies. Clinical pathology is not normally performed, and histopathology may only be carried out on selected adults and offspring in multigeneration studies and in studies conducted according to OECD test guidelines 421 and 422 (see above). Each arm of the reproductive study program has its particular parameter measurements that may be loosely grouped based on fertility, embryonic development, and prenatal and postnatal development, which includes examination of maternal function up to weaning; these are listed in Table 7.6.

Fertility

The origin of the gametes is a factor in the sensitivity of the sexes to reproductive toxins. In the female, the ovarian germ cells are present before birth and decrease with age, being a pool of finite size that can be depleted, but not replenished. In the male, spermatogenesis is a process that is continuous from the point of sexual maturity, a factor that allows recovery according to the extent of the toxic insult. In a fertility study, mating behavior is assessed by recording the time taken to successful mating. Lack of mating can be investigated by pairing unmated females with successful males from the same group, while unmated males are paired with untreated females. The number and distribution of implantations in uterine horns, classified as early resorptions, late resorptions, or dead or live fetuses, are recorded to assess effects *in utero*, together with the numbers of corpora lutea in the ovary. Preimplantation losses are calculated by subtracting the number of implantations from the number of corpora lutea. For the males, assessment of sperm quality is increasingly recommended by sperm counts, motility, morphology, and quality. Testicular weight is a sensitive indicator of male effect and can be supported by histological processing and examination. Computer-assisted sperm assessment generates a large amount of data, which may not be fully understood, making interpretation difficult; it has been enthusiastically supported by regulatory authorities, presumably in the questionable belief that volume of data gives added security in assessment of prospective safety. However, the initial enthusiasm has not been matched by routine use; the use of this technique should be balanced against what can be revealed by alternatives such as testicular staging in routine toxicity studies (Creasy 1997).

Embryonic Development
Fetal weight gives an indication of maternal or placental function and of any retardation in development that has taken place. Smaller fetuses may have skeletal variations from controls that are a product of slower development rather than direct teratogenicity. The fetuses are assessed for abnormality by visceral or skeletal examination. Visceral development can be assessed by Wilson's sectioning of fetuses fixed in Bouin's fluid (not in rabbits), but this has increasingly been replaced by or combined with microdissection. Dissection of other fetuses may also be performed before the carcasses are cleared with potassium hydroxide and stained with alizarin red, which stains bone red for skeletal examination, and alcian blue for cartilage. Not all the parameters listed in Table 7.6 are universally applied; for instance, placental weight is rarely affected, although some classes of drugs have been known to produce differences from controls, notably some cardioactive substances. The weight of gravid uterus is useful for assessing effects on carcass weight, by subtraction of the uterus weight from the complete body weight immediately before necropsy.

Structural congenital abnormalities that potentially impair the survival or constitution of the fetus are classified as major abnormalities. Other defects are classified as minor abnormalities. Commonly observed variations in the degree of ossification from that expected of a day 20 gestation fetus, together with common variations in the extent of renal pelvic cavitation and ureter dilation, are recorded as variants. In some fetuses, an extra "wavy" rib may be seen; the significance of these has been widely debated down the years, but they are now considered to be without developmental significance. Embryofetal examinations in the rabbit are similar to those described above for the rat, with the exception that the head is treated and examined separately. All rabbit fetuses are dissected and cleared for skeletal examination.

Perinatal and Postnatal Development—Multigeneration Studies
This type of study is among the most complex toxicity studies conducted and can be made more complex by the addition of extra generations and longer treatment

periods. In the simpler of these studies, treatment ceases when the F0 females give birth. In multigeneration studies, usually conducted with agrochemicals or food additives, treatment may be continued throughout the study until termination of the F2 pups. It is possible to continue such studies into a carcinogenicity assessment by continued treatment of the F2 generation, although such studies are rare, due to complexity and expense. Mating performance is assessed in a similar manner to fertility studies. For the dams of the F0 and subsequent generations, the records of length of gestation and onset and duration of parturition are probably self-explanatory. The performance of the dams through lactation, coupled with the survival of the pups and their body weight gains during lactation, is also an indication of maternal function. In the studies where parturition is examined, the survival of the pups to day 4 postpartum is checked. At day 4, the litter size may be reduced, where necessary, to four males and four females to obviate effects on postnatal development that may be attributable to large or uneven litter sizes. The pups are examined for external abnormalities, and their development is charted according to the time of achievement of a series of physical, sexual, and sensory milestones—opening of eyes and detachment of pinnae, eruption of incisors, vaginal opening and balanopreputial separation, pupillary and righting reflexes, and startle response. Learning ability is usually tested in a simple Y- or E-shaped swimming maze after weaning. An open-field test is performed looking at general activity and exploratory behavior; locomotor activity is assessed by performance on a rotating rod. Reproductive function is assessed by mating of the F1 pups.

PITFALLS IN REPRODUCTIVE TOXICOLOGY

Although fertility studies are usually performed in rats, this should not be taken as the only method to assess male fertility. The rat produces approximately four times as many sperms per gram of testis as a man and is correspondingly less sensitive to effects on spermatogenesis. The mouse produces about three times more sperm than the rat. The use of data from routine toxicity studies in conjunction with those from reproductive studies gives an overview of testicular effects and so of the necessity for more specialized mechanistic studies. In using the results of testicular histopathology from routine toxicity studies, the effects of age at the start of treatment and uneven sexual maturity at the end of the study period, as referred to above for general toxicology, should be remembered.

The rat is a robust species in which most compounds can be investigated in a wide range of vehicles, including oils. The rabbit's gastrointestinal physiology, however, means that it is unable to cope with compounds or vehicles that disturb the physiological balance or osmotic environment in the gut. In essence, this means that antibiotics should be tested in other species such as mouse, minipig, or NHP and that lipid-soluble substances must be used as a suspension in an aqueous vehicle, with attendant problems of suboptimal absorption.

Compound type is also a consideration. As indicated, rabbits are sensitive to antibiotics, and hormonally active compounds may be inappropriate in rats or rabbits where the ovaries are responsible for maintaining hormonal control of gestation;

in humans and primates, this is carried out by the placenta. Selection of an inappropriate species may result in toxicity at doses so low that they approach expected human exposure or dosing levels or may, in rare cases, fall below it. This situation, which gives no margin of safety over human usage, is unsatisfactory and makes safety evaluation difficult; in these cases, the risk/benefit of the drug and indication for which it is intended should be considered. An alternative to these species is the cynomolgus monkey, although its use should be carefully considered against that of alternatives such as the minipig.

Various aspects of the reproductive process mean that it is subject to disruption by substances such as hormone derivatives or cytotoxic agents. With compounds such as these, it may become necessary to break the studies down into specific stages either to minimize the length of the treatment period or to investigate particular parts of the reproductive cycle. For instance, a fertility study usually requires mating of treated males with treated females; in some cases, it may be necessary to treat both sexes but to mate them with untreated partners. Effects on maternal function may be investigated by fostering the offspring of treated females on untreated females and vice versa. In dietary multigeneration studies, toxicity in the dams may be encountered during late lactation, when there is a marked increase in food consumption.

Transplacental carcinogenesis, as shown by diethylstilbestrol, is an uncommon effect, or at least has not been demonstrated to be detectable over normal background incidences of cancers, and is unlikely to be demonstrated by routine reproductive studies, as the effects do not become apparent until the offspring are adults. However, multigeneration studies go some way toward addressing this problem. Where there is some retardation of fetal development, for instance, due to lowered maternal food consumption or another indirect effect of treatment, there may be variations from control values, particularly in weight and/or skeletal development. These are not teratogenic effects but merely an indication of indirect toxicity. In similar ways, neurologically active compounds may affect maternal behavior or lactation and have indirect effects on pup survival through reduced maternal care. These compounds may also have indirect effects on fertility, if they affect mating behavior to the extent that mating is delayed or completely unsuccessful.

SUMMARY

Study designs in general and reproductive toxicity are constrained by conservative tradition and regulation. In addition,

- Dose selection for the first study in any species is potentially an imprecise process; for later studies, doses are selected based on effects in the most recent.
- Repeat-dose study duration is based on regulatory expectation and is often driven by traditional intervals: single dose (acute), 14 or 28 days (subacute), 90/91 days (subchronic), 6/9 months (chronic).

- Duration in reproductive studies is driven by the duration of the processes in life: spermatogenesis, gestation and organ development and perinatal and postnatal effects and development.
- Parameters measured are often extensive and have evolved over many years of experience. Specific observations and/or investigations may be added to account for prior expectation or for observations in earlier studies.
- Reversibility or recovery is an important point to consider in repeat-dose toxicity studies, although there is increasing pressure to discontinue the use of additional animals for this. The reversibility of any effect should be assessable from prior experience.

Finally, pitfalls in both general, repeat-dose and reproductive toxicity studies are often due to the impact of study procedures on animals, such as husbandry, technical effects due to restraint, and timing of sample collection.

REFERENCES

Bailey GP, Mariën D. What have we learned from pre-clinical juvenile toxicity studies? *Reproduct Toxicol* 2009; 28(2): 226–229.

British Toxicology Society (BTS) Working Party. Report on in vitro toxicology. *Hum Exp Toxicol* 1997; 16(S1).

Cappon GD, Bailey GP, Buschmann J, Feuston ME, Fisher JE, Hew K, Hoberman AM, Ooshima Y, Stump DG, Hurtt ME. Juvenile animal toxicity study designs to support pediatric drug development. *Birth Defects Res B Dev Reprod Toxicol* 2009; 86(6): 463–469.

Creasy DM. Evaluation of testicular toxicity in safety evaluation studies: The appropriate use of spermatic staging. *Toxicol Pathol* 1997; 25(2): 119–131.

De Schaepdrijver L, Rouan M-C, Raoof A, Bailey GP, De Zwart L, Monbaliu J, Coogan TP, Lammens L, Coussement W. Real life juvenile toxicity case studies: The good, the bad and the ugly. *Reproduct Toxicol* 2008; 26(1): 54–55.

Evans GO. *Animal Hematotoxicology: A Practical Guide for Toxicologists and Biomedical Researchers*. CRC Press, 2009a.

Evans GO, ed. *Animal Clinical Chemistry: A Practical Handbook for Toxicologists and Biomedical*. London: CRC Press, 2009b.

ICH. Guideline on Reproductive Toxicity. ICH Guideline 2005; S5(R2) http://www.ich.org.

ICH. Guideline. Immunotoxicity Studies for Human Pharmaceuticals. ICH Guideline 2005; S8 http://www.ich.org.

Koukouritaki SB, Simpson P, Yeung CK, Rettie AE, Hines RN. Human hepatic flavin-containing monooxygenase 1 (FMO1) and 3 (FMO3) developmental expression. *Pediatr Res* 2002; 51(2): 236–243.

McCarver DG, Hines RN. The ontogeny of human drug-metabolizing enzymes: Phase II conjugation enzymes and regulatory mechanisms. *J Pharmacol Exp Ther* 2002; 300(2): 361–366.

OECD. Guidelines for Testing of Chemicals. Available at http://www.oecd.org.

Rasheed A, Hines RN, McCarver-May DG. Variation in induction of human placental CYP2E1: Possible role in susceptibility to fetal alcohol syndrome. *Toxicol Appl Pharmacol* 1997; 144(2): 396–W0.

Sewell F, Chapman K, Baldrick P, Brewster D, Broadmeadow A, Brown P, Leigh Ann Burns-Naas LA et al. Recommendations from a global cross-company data sharing initiative on the incorporation of recovery phase animals in safety assessment studies to support first-in-human clinical trials. *Regulat Toxicol Pharmacol* 2014; 70: 413–429.
Svendsen O. The minipig in toxicology. *Exp Toxicol Pathol* 2006; 57: 335–339.
Tassinari MS, De Schaepdrijver L, Hurtt ME. Juvenile animal toxicity studies: Regulatory expectations, decision strategies and role in paediatric drug development. In Brock WJ, Hastings KL, McGown KM (Eds), *Nonclinical Safety Assessment: A Guide to International Pharmaceutical Regulations*. Wiley, 2013.
Timbrell J. *Principles of Biochemical Toxicology*. London: Taylor & Francis, 2000.
Vos JG, Bossart GD, Fournier M, O'shea T. Eds. *Toxicology of Marine Mammals*. London and New York: Taylor & Francis, 2003.
Woolley A. *A Guide to Practical Toxicology*, 1st ed. London: Taylor & Francis, 2003.
Woolley A. *A Guide to Practical Toxicology*, 2nd ed. London: Taylor & Francis, 2008.

8 Determination
Genotoxicity and Carcinogenicity

GENOTOXICITY

Genotoxicity—or genetic toxicity—is defined as the exploration of toxic action on DNA and the wider effects on genetic material and expression of genetic change, including genetic differences in absorption, distribution, metabolism, and elimination (ADME) and susceptibility or resistance to toxicants. Such genetic change can lead to a phenotypic change and conditions such as cancer or birth defects. It should be considered that exposure to a genotoxic substance may not necessarily lead directly to cancer or birth defects, but it increases the risk of these endpoints occurring. Mutagenicity, a form of genotoxicity, is more specific in that it refers specifically to direct change in DNA, seen as changes in nucleotides in DNA and subsequent changes in gene expression.

GENERAL PRINCIPLES IN GENOTOXICITY

The intention in testing for genotoxicity is to determine the potential for damage to genetic materials and thereby to highlight any effects that might, with administration or exposure, lead to phenotypic change such as an increased incidence of tumors or birth defects through heritable effects in the germ cells. In the latter context, it is worth considering that changes in chromosomal number are usually fatal in laboratory animals but not in humans where conditions, such as Down's syndrome, are associated with an extra chromosome but do not lead to abortion. The genetic changes associated with some cancers are given in Table 8.1.

Though individual genetic disorders are rare, collectively, they comprise over 15,500 recognized genetic abnormalities and affect approximately 13 million Americans. For instance, 3% to 5% of all births result in congenital malformations and 20% to 30% of all infant deaths are due to genetic disorders, while 11.1% of pediatric hospital admissions are for children with genetic disorders and 18.5% are children with other congenital malformations. In adults, 12% of hospital admissions are for genetic causes, and 50% of mental retardation has a genetic basis. Among chronic adult diseases, 15% of all cancers have an inherited susceptibility, and 10% of the chronic diseases (heart, diabetes, arthritis) that occur in the adult populations have a significant genetic component. Given this context, the assessment of the potential for new or existing chemicals to cause genetic damage is an important area of toxicological testing. This is the main area of toxicological investigation in which

TABLE 8.1

Genetic Associations with Cancers

Proto-Oncogene	Activation by	Chromosomal Change	Associated Cancer
c-*myc*	Genetic rearrangement	Translocation: 8–14, 8–2, or 8–22	Burkitt's lymphoma
c-*abl*	Genetic rearrangement	Translocation: 9–22	Chronic myeloid leukemia
c-H-*ras*	Point mutation		Bladder carcinoma
c-K-*ras*	Point mutation		Lung and colon carcinoma
N-*myc*	Gene amplification		Neuroblastoma

Source: Courtesy of Dr. Mike Kelly (personal communication). Adapted from Woolley A, *A Guide to Practical Toxicology*, 1st ed., London: Taylor & Francis, 2003.

in vitro testing has been accepted, principally, on the basis of the relatively simple endpoints that are examined in these tests.

Testing for genotoxicity or mutagenicity, in a regulatory context, became much more frequent in the early 1970s following the development of the bacterial reverse mutation assay (or Ames test) by Bruce Ames. This simple test, using specially derived strains of *Salmonella typhimurium* for which histidine is an essential amino acid, determines the ability of a chemical to produce mutations that allow the bacteria to grow in the absence of histidine. The basic hypothesis was that carcinogenesis originated through damage to DNA and chemicals that damage DNA are more likely to be carcinogenic than those that do not. The problem with this is that, while there is good correlation between mutagenesis and carcinogenesis, not all carcinogens damage DNA directly, and these are not readily detected by current methods that determine direct toxic effects on the DNA. The attraction of genotoxicity testing is that it offers a method of assessing carcinogenic potential that is quick, inexpensive, and usually *in vitro*, and can be performed early in the development of a chemical. This contrasts with the traditional approach, which is the use of long studies in rodents that can take 2 years to complete, are expensive, and are conducted later in development.

With the realization that the simplistic Ames test system examined only one endpoint in bacteria and resulted in a number of false positives, an increasingly large number of tests were developed to examine the effects on DNA in other ways. A more recent development has been that of determination of structure–activity relationships and the computerized prediction by expert systems of mutagenic potential. This is achieved by examination of the test structure for the presence of structural components or groups that have been associated with mutagenicity in other compounds. These are discussed in more detail later in the chapter on prediction (see Chapter 13).

Testing for genotoxicity acknowledges that there are basically two levels of effect—at the gene level and at the chromosome. At the former, mutations are sought that lead to localized changes at one or a few bases in the DNA, thereby changing the coding for the protein produced by the gene. A change from one base to another, or the misreading of a chemically altered base, may lead to a different amino acid being inserted into an otherwise normal protein; this is a point mutation. When a base or base pair is inserted or deleted, this is known as a frameshift mutation, as the reading frame of the code is changed, leading to an abnormal protein product. At the level of the chromosome, the changes are broadly in terms of structure or number; there may be changes in number due to effects on mitosis or meiosis and translocations, rearrangements, breaks, or gaps, which indicate an effect on the chromosomes themselves. DNA or chromosomal damage is detected directly or indirectly. Direct evidence comes from the induction of genetic change, such as the ability of Ames test bacteria to divide in the absence of a previously essential amino acid, or by examination of chromosomes in metaphase where breakages and abnormalities are evident under the microscope. Indirect evidence of genetic damage may be obtained by measurement of DNA repair in tissues. This is easier to detect in tissues that do not normally divide and is used in the assessment of unscheduled DNA synthesis (UDS) in hepatocytes.

The majority of chemicals are not directly genotoxic, and one disadvantage of a bacterial system *in vitro* is that the bacteria lack the enzymes that are responsible in mammals for the activation of chemicals to toxic metabolites. Consequently, a metabolizing system was devised that uses the microsomal fraction from homogenized rat liver, which contains the majority of xenobiotic metabolizing enzymes; this is known as S9 mix. For normal regulatory purposes, this is prepared from the livers of rats treated with an enzyme-inducing agent, such as repeated doses of β-naphthoflavone and/or sodium phenobarbitone or Aroclor 1254 (although this is used less often now). In some cases, S9 mix may be prepared from other tissues such as kidney or with the use of other inducing agents. One factor to consider is that S9 has its own intrinsic toxicity and that incubation of mammalian-derived cells should be limited to only a few hours. S9 mix is rich in the microsomal elements of metabolism typified by cytochrome P450, which carry out the initial reactions (phase 1) of metabolism. It has much less of the phase 2 metabolism systems, which conjugate the metabolites with endogenous molecules to make them more polar and thus easier to eliminate. This is unlikely to be a significant problem in most cases, as these conjugates are unlikely to be mutagenic in their own right. However, it means that the test system may be exposed for an unrealistic time to active metabolites that might otherwise be removed by conjugation.

Test Battery and Study Design

In assessing the genotoxicity of a new chemical, it is normal to use several tests that examine different endpoints or mechanisms of effect, in recognition of the limited scope of individual tests. The types of test used are typically a bacterial mutation test, usually the Ames test; an evaluation of chromosome damage in mammalian cells; an *in vivo* test for chromosomal damage, such as the micronucleus test in rodents; and possibly a test for gene mutation in mammalian cells. The general composition of the chosen test battery is largely determined by the type of chemical and the regulatory authorities at which it is aimed. The nature of the chemical may also influence the choice of tests; for instance, excessively bacteriotoxic materials, such as antibiotics, are not suitable for bacterial assays, although some guidelines may still require them. For an antibiotic, two tests using mammalian cells, one for gene mutation (e.g., mouse lymphoma L5178Y TK assay) and one for chromosomal damage [e.g., Chinese hamster ovary (CHO) cell assay] with a micronucleus test *in vivo*, might be recommended as a basic test battery. For a compound known to disrupt cell division, test design is critical, and harvest times and exposure concentrations must be carefully chosen; however, mammalian cell mutagenicity could still be possible. In every case, the choice of tests should be made on a rational and scientific basis, bearing in mind the regulatory guidelines for the class of chemical. In the event of a single positive result that is of borderline biological significance, the test battery may be expanded to include studies of DNA interaction, damage, or repair such as UDS or the comet assay. However, when responding to a positive result with additional testing, it should be noted that an increased number of ill-chosen tests may not clarify the picture. In fact, such an approach may simply serve to produce interpretative uncertainty.

The design of genotoxicity studies follows the broad pattern of other toxicity tests, using a control and several increasing concentrations of test material. In a typical Ames test, there could be a control and five concentrations of test material up to a maximum of 5 mg per test plate. There should be at least three plates for controls and at each test concentration. Positive controls, using known mutagens, should be run at the same time to ensure that the bacteria are responding as expected. The positive controls are chosen according to the strain of bacteria and whether they need metabolic activation through S9 mix or are directly mutagenic (Table 8.2) (Woolley 2003). Benzo(a)pyrene or 2-aminoanthracene are used with S9 mix to demonstrate sensitivity to metabolically activated mutagens but are not the only choices available.

As with other branches of toxicology, the choice of exposure concentration or dose level is of crucial importance. This is normally achieved using dose range–finding

TABLE 8.2
Direct-Acting Positive Controls Used in the Ames Test

Species/Strain	Direct-Acting Mutagens (No S9 Mix)
Salmonella typhimurium	
TA 1535, TA 100	Sodium azide
TA 97A	9-Aminoacridine
TA 98	2-Nitrofluorene
TA 1537	9-Aminoacridine
TA 1538, TA 98	2-Nitrofluorene
TA 102	Cumene hydroperoxide, mitomycin-C
Escherichia coli	
WP2, WP2 uvrA	4-Nitroquinoline-*N*-oxide

Source: Adapted from Woolley A, *A Guide to Practical Toxicology*, 1st ed., London: Taylor & Francis, 2003.

studies to assess toxicity. In the Ames test, toxicity is assessed by adding a small amount of histidine to the agar medium, allowing a small amount of growth, which is seen as "background lawn." At toxic concentrations of test substance, this background lawn is thinner than in the controls. This leaves a reduced number of bacteria available for mutation, and as a result, fewer mutant colonies are formed, which may give a false impression of nonmutagenic effect. In experiments with cultured mammalian cells, a degree of toxicity is considered desirable, as this demonstrates exposure of the cells. However, cytotoxicity itself can give rise to false-positive findings of genotoxicity, either due to the apparent chromosomal damage visible when there is a high proportion of dead or dying cells or due to chance clonal selection of mutant cells, when high levels of toxicity are used in a mammalian gene mutation assay. Positive results at toxic concentrations should be interpreted with caution. The maximum level of desirable toxicity at the highest concentration is around 50% in the chromosome aberration assay and around 80% in the mouse lymphoma L5178Y TK assay. It should also be noted that genotoxicity has not only been shown at concentrations where the test material is insoluble, but that dose responses have been observed past the concentration at which precipitation occurs. Accordingly, insolubility is not necessarily a valid criterion for choice of the highest concentration. If other criteria, such as the pH of the medium or osmolarity, do not limit the concentration of test substance, the usual maximum concentration is set at 5 mg/plate, 5 mg/mL, or 10 mM.

For genotoxicity tests in whole animals, doses are chosen according to the known acute toxicity of the test substance. The route of administration is chosen according to the expected route of exposure in humans but is normally oral or by intravenous or intraperitoneal injection. For *in vivo* tests, such as the micronucleus test in rodents, it is necessary to prove exposure of the target cells (normally bone marrow) to the test substance by analysis of plasma samples. Additional animals may be necessary for this. Although it may be reasonable to assume that intravenous injection is associated with target cell exposure, where compounds are precipitated or rapidly transformed in the plasma, this assumption may be misplaced. Exposure is less certain with intraperitoneal injection and even less so with oral dosing. However, it is generally assumed that the plasma concentrations of the test substance give a good indication of the concentrations to which the target cells in the bone marrow or liver are exposed, as these tissues have a good blood supply.

The duration of exposure is also a factor to consider in study design, although to a very large extent, this is indicated in the guidelines and literature. Due to its toxicity, exposure of mammalian cells with S9 mix is generally shorter than without it; in a human lymphocyte study, this can mean 3 hours instead of 24 hours. In whole-animal experiments, the number of doses is a factor to consider; generally, single administration of a high dose is used, although several doses may be used in some cases. The cells of interest may be sampled or harvested at different times after administration to take account of different times of onset of effect. Taking the micronucleus test as an example, it is normal to harvest bone marrow cells on at least two occasions, for example, 24 and 48 hours after dosing; an additional harvest at 72 hours is recommended in some guidelines. An older design involved giving two doses, 24 hours apart, and sampling on one occasion only.

Increasingly, micronucleus tests are conducted as part of a 28-day toxicity study, with evaluation in the fourth week of treatment; this has been credited with saving the use of many animals. In some designs, satellite animals, such as controls retained for toxicokinetic sampling, are dosed with a positive control such as cyclophosphamide; other designs rely on positive controls from previous studies.

TEST SYSTEMS AND TESTS

There is an extensive history of genotoxicity test systems, including the use of mice in the mouse coat color spot test, to assess mutation due to radiation, and the dominant lethal assay also in mice, both of which were developed in the late 1950s and 1960s. The use of the fruit fly, *Drosophila melanogaster*, has a longer history in mutation research but has fallen out of favor. These tests depended on anesthetizing the flies at intervals to check effects. The potential disadvantages of imperfect anesthesia and the escape of the flies into the open laboratory or their death, together with the technical demands of difference recognition, may have had some influence on the decline in their use. The stress associated with chasing an expensive experiment around a laboratory with a butterfly net could not be expected to increase its popularity with toxicologists.

Progress in the acceptance of new tests for genotoxicity is slowed by the multitude of test systems under development and by the consequent dilution of effort for really promising lines of research. The spectrum of validation for new tests or test systems is a constant problem, collaborative studies usually being conducted by a number of laboratories. These multicenter studies are expensive and cumbersome to organize, and may show up a lack of reproducibility in the more technically demanding assays. Following validation, there is the task of gaining regulatory acceptance and persuading companies developing new chemicals to use them. Sometimes, assays have gained credibility in industry through their use as screening assays before gaining acceptance from regulators.

Many genotoxicity assays are conducted *in vitro* using unicellular organisms or cell lines that have been produced with particular characteristics for the purposes of the test endpoint and that may be subjected to insidious genetic drift. As a result, test system characterization is an important factor in the conduct of these tests, in a way that is not seen *in vivo*. Whole animals have a much longer life span than microbial or cellular systems and therefore change more slowly, over a period of years rather than months. The rate of change in any animal species or strain is usually not large enough to cause problems, and a single change is unlikely to invalidate an experiment. With single-cell preparations, either bacterial or mammalian-derived, the generation times are quicker, and there is the possibility that the cells may lose the characteristics that are vital for correct performance of the test. For instance, the strains of *Salmonella* used in the Ames test have been modified in various ways to make them more sensitive to carcinogens. Modification has been performed to enhance absorption, via a rough coat, or to increase sensitivity to ultraviolet (UV) light or antibiotics. These characteristics are essential for the correct function of the tests, and because they are not immediately visible or verifiable by conventional biochemical testing, they must be checked in the stock cultures at regular intervals. The primary requirement

TABLE 8.3
Principal Genotoxicity Test Systems Used in Regulatory Toxicology

Mutation Event	Test Systems	Tests
Bacterial reverse mutation	*S. typhimurium*	Ames test for reversion to histidine independence
	E. coli	Ames test for reversion to tryptophan independence
Mammalian mutation *in vitro*	CHO or V79 Chinese Hamster cells, mouse lymphoma L5178Y cells	Mutation at HGPRT[a] locus Mutation at TK[b] locus
DNA damage *in vitro*	CHO or V79 cells, human peripheral lymphocytes	Chromosome aberration Chromosome aberration
	Primary cultures of rodent hepatocytes	Unscheduled DNA synthesis (UDS) or comet assay
DNA damage *in vivo* or *ex vivo*	Rat and mouse	Micronucleus test or UDS or comet assay
Mutagenicity	*In silico*: combination of expert (rule-based) and statistical software	Accepted for prediction of genotoxicity of pharmaceutical impurities and, by extension, in other applications.

Source: Adapted from Woolley A, *A Guide to Practical Toxicology*, 1st ed., London: Taylor & Francis, 2003.

[a] Hypoxanthine guanine phosphoribosyl transferase.

[b] Thymidine kinase.

for the Ames test is that the *Salmonella* or *Escherichia coli* strains used should not grow in the absence of histidine or tryptophan, respectively. The presence of rough coat can be ascertained through absorption of a high-molecular-weight dye and the consequent lethality. Sensitivity to UV light is checked by irradiation; in addition, the antibiotic resistance of some strains has been increased and is tested with the appropriate antibiotic. Finally, the relative sensitivity to known mutagens is checked in every study against expectation from laboratory background data ranges. Failure to complete these checks may produce unreliable results. This requirement may be less stringent for primary cultures of cells, such as hepatocytes, which are derived from animals of known strain and biochemical profile, which can themselves be characterized by conventional means.

Test systems for the evaluation of genotoxicity may be divided broadly into the categories in Table 8.3, which also lists the main tests in which they are used. Due to the multiplicity of test systems, only the major ones used in regulatory toxicology are discussed here, with references to tests or systems that may not be considered to be mainstream.

In Vitro Systems: Bacterial Cultures

Bacterial mutation assays were among the first *in vitro* toxicity tests to gain regulatory acceptance. The Ames test (increasingly referred to as the bacterial reversion assay) using strains of *S. typhimurium* has become the most widely conducted geno-toxicity assay, with the addition of *E. coli* in deference to Japanese wishes. The test is based on mutant *Salmonella* strains that cannot grow without histidine but that can be reverted to wild type by mutation, when they are able to synthesize their own histidine. After a period of incubation with the test material, with or without S9 mix, the colonies of revertant bacteria are counted. Each strain of bacteria used has a normal background incidence of mutation, and a dose-dependent increase from this is taken to be evidence of mutagenic effect. This effect should be reproducible in a second experiment, which is usually performed to a different protocol. Normally, four strains are used, selected from TA 100 and TA 1535, which detect base substitution; TA 98 and TA 1538, which detect frameshift mutations; and TA 97 and TA 1537, which detect single-frameshift mutations. TA 102 may also be used but is not generally required by regulatory guidelines. The same principles are applicable to *E. coli*, which is used to comply with Japanese guidelines and detects base substitutions; usually, a single strain such as WP2 uvrA is used. These strains are dependent on tryptophan, and mutations are revealed by the presence of colonies growing in the absence of tryptophan.

Salmonella and *E. coli* may also be used in forward mutation tests and in DNA repair tests, in which repair-deficient bacteria are mutated to repair-competent, which are able to form colonies that can be counted.

In Vitro Systems: Mammalian Cells in Culture

Mammalian cells may be used in mutation assays and in chromosome damage or aberration tests, also known as cytogenetic assays. The most commonly used mammalian cells are the Chinese hamster–derived cells (CHO and V79 lung-derived cell lines), the mouse lymphoma L5178Y cell, and primary cultures of rodent hepatocytes or human peripheral lymphocytes. The cell line cultures have advantages in that they are easy to culture consistently, but, in contrast to primary cultures of cells from tissues such as liver, they have little metabolic activity and tend to have abnormalities of chromosomal number (aneuploidy). The Chinese hamster cell lines tend to grow in sheets, which can make intercellular communication easier in some circumstances, leading to transfer of cellular components that may negate any mutation in the receiving cell. Therefore, the plating density of the cells needs to be controlled. These lines have particular use in chromosome aberration assays but may also be used for detection of mutations, for example, at the HGPRT locus. The mouse lymphoma L5178Y cell, which uses the TK locus, is more sensitive to mutagens than the Chinese hamster–derived cells; they can grow in suspension culture and thus do not have the problem of intercellular communication. In addition, it has been suggested that this cell line can be used in the detection of chromosomal abnormalities and mutations through differences in colony size, although the reproducibility of this has been questioned. This, using an appropriate protocol, potentially gives the cells a much broader scope than the Chinese hamster–derived lines for which mutation and chromosome damage are assessed in separate tests. If the TK assay is used, the size of colonies produced may indicate whether the damage is due to clastogenicity or mutation. There is the possibility that these cell lines may undergo some genetic drift in different laboratories, a factor that may ultimately lead to some inconsistency and irreproducibility of results.

Human peripheral lymphocytes are also used and have the advantage that they are from a relevant species and are also primary culture cells. They are used only for the assessment of chromosome aberration; these tests tend to be more expensive than those using cultured cell lines. It is important to ensure that the donors of the blood from which the lymphocytes are separated are free of viral infection, non-smokers, not too old, and not on medication that may be expected to affect the assay results. As stated earlier, it is important to demonstrate cytotoxicity in these assays, and positive controls are routinely used to demonstrate that the test system is valid. Once again, it is important to conduct a second experiment to confirm the results of the first.

The basic principle in mammalian cell mutation assays is to induce mutations that confer resistance to toxic nucleotide analogs. As with the Ames test, two independent experiments are conducted, preferably to slightly different protocols. Metabolic activity in these tests is provided by S9 mix. The cells are exposed for up to 24 hours without S9 or up to 6 hours with S9 mix, and are then cultured without the test substance to allow for phenotypic expression of mutation. Then they are cultured with the appropriate selective agent to check for the formation of colonies.

In chromosome aberration tests, the cells, which may be human peripheral lymphocytes, are exposed to the test substance for up to 24 hours and may have a treatment-free period. They are then treated with a spindle poison, which arrests the cell division in metaphase. Metabolic activation is again provided by S9 mix. The cells are taken onto microscope slides and stained. An appropriate number of metaphases, which may be 100 cells from two or three culture replicates at each of three treatment concentrations, are scored for the presence of chromosomal aberrations, which are seen as gaps, breaks or exchanges, and abnormalities of number. Although numerical abnormalities due to polyploidy and endoreduplication may be seen with other cell lines, aneuploidy is easier to detect in human cells. A chromosomal gap is an area in which the stain has not been taken up and where there is minimal misalignment of chromatid(s). A chromosomal break is defined as an unstained section accompanied by a clear misalignment of the chromatid(s). General opinion is that gaps are not as significant as breaks, but they are reported anyway, usually as separate totals from the other aberrations. More extreme disruption may be seen, and this is also reported. Cytotoxicity is determined by reductions in mitotic index for human lymphocytes. For cell lines, a variety of methods to assess cytotoxicity are available, including viable cell number, colony-forming ability, and MTT assessment of mitochondrial activity.

Hepatocytes isolated from rats may be used in a range of assays, such as UDS, which assesses repair that takes place following damage to DNA. The extent of DNA repair is assessed through the incorporation of tritiated thymidine into the nuclei of cells exposed to the test substance. The isolated hepatocytes are allowed to attach to glass microscope slide coverslips, where they are exposed to the test substance; they are then exposed to medium containing tritiated thymidine and, after fixing and drying, to photographic emulsion. The cells are stained, and the number of grains in the nucleus is assessed microscopically. There is also a method of measurement that uses liquid scintillation counting of the activity; however, this does not allow the exclusion of cytoplasmic grains from the total counted and so is less sensitive but also less time consuming. UDS may also be examined in an *ex vivo* form of the test.

Cultured cells can also be used to assess sister chromatid exchange (SCE), in which sections are exchanged between the chromatids of a chromosome pair; however, the *in vitro* SCE test suffers from a high background incidence, which limits its sensitivity. SCE correlates well with genotoxicity and carcinogenicity but is not fully understood. It can be assessed from cells obtained from cancer patients or workers exposed occupationally to chemicals, where it may indicate increased effects on the DNA.

Ex Vivo Systems

In these systems, an animal is treated, and after an appropriate interval, tissues such as the liver are removed for further treatment *in vitro* followed by examination. This approach is used in a refinement of the UDS assay and in the comet assay. The advantage of this approach is that the chemical is administered to a whole animal and is subject to the normal processes of metabolism and elimination before its effects are examined in the target tissues. In the *ex vivo* UDS assay, the livers are removed from the animals at a suitable time after dosing, and sections or slices are treated with tritiated thymidine. The slices are fixed and then treated in a similar way to the cells in the *in vitro* methods previously described.

The comet, or single-cell gel electrophoresis assay, is potentially a powerful means of detecting DNA damage in cells from animals that have been treated with suspected carcinogens. The basic principle is to electrophorese the DNA from a single-cell nucleus, damaged DNA having a greater spread of travel (tail) than control, or undamaged DNA, the shape of the electrophoresis pattern giving the assay its name. The assay is simple and can be performed rapidly but may not characterize the type of damage that has occurred. Unlike many genotoxicity tests, it can be applied to any tissue believed to be a target for the test substance. The assay can be carried out after a single administration or could be included in routine toxicity studies as an indicator of DNA change and, by implication, of potential carcinogenicity. However, it should be seen as one element of a set of data collected to examine genotoxicity or carcinogenic potential and not be taken, by itself, to be a clear indication of hazard. It is likely that regulatory acceptance of this test will increase.

In Vivo Systems

Mice and rats (and occasionally hamsters) are used for examining the potential of chemicals to cause chromosomal damage by examination of bone marrow cells either by scoring of metaphases or, more usually, micronuclei in erythrocytes. Standard strains can be used in these routine tests, which are performed by administration of a dose at or near the limit of tolerance. Bone marrow is harvested from the femur 24 and 48 hours after a single administration and, in some designs, also at 72 hours. Another design has two administrations 24 hours apart and one harvest 24 hours after the second administration. Chromosomal damage is assessed in bone marrow smears by the presence of micronuclei, which are fragments of damaged chromosomes or whole chromosomes left behind when the nucleus is extruded following the final cell division in normal erythrocyte maturation. Up to 4000 polychromatic erythrocytes [as indicated in Organization for Economic Co-operation and Development (OECD) guideline 474] from each animal are assessed for micronuclei. The test can assess chromosome damage and spindle defects. Toxicity may be indicated by the ratio of polychromatic (early-stage) erythrocytes to normochromatic (late-stage) erythrocytes (the PCE/NCE ratio). A decrease in the ratio may be due to either prevention of early-stage development or replacement of dead bone marrow from peripheral blood. Therefore, a decrease in the PCE/NCE ratio indicates bone marrow toxicity and exposure. Opinions and guidelines tend to differ in the choice of an all-male design or one that uses both sexes, and on the number of erythrocytes that should be scored. In the event of a negative result, it is important to demonstrate exposure to the test substance, especially if the route of administration is oral or intraperitoneal. Consequently, it may be better to build these examinations into the original experiment with the same batch of animals, rather than do a separate experiment at a later date, which might not be equivalent in every respect to the original test. However, this uses more animals and does not comply with the three Rs. As noted above, these tests may be incorporated into standard 28-day toxicity tests, resulting in significant savings in animals and costs.

In addition, mice have been used in SCE assays, the mouse spot test, and the dominant lethal test, all of which detect mutations. However, these tests are more extensive in terms of animal numbers, take longer to complete than other assays, and are not routinely used. In the mouse spot test, pregnant females are treated on day 10 of gestation, and the offspring are checked for the presence of relevant spots of color difference in the coat, which imply the presence of mutation in the coat color genes of pigment cells. In the dominant lethal test, the effect of a prospective mutagen on the germ cells is assessed by single or sometimes limited repeated administration to males, which are then mated with a fresh, untreated female each week for a complete spermatogenic cycle. After 2 weeks' gestation, the uterine contents are inspected for implantations and implantation losses and fertility index. The presence of increased implantation loss implies that a mutation has occurred, and the week in which the effect is noted indicates the stage of the spermatogenic cycle that is involved.

Transgenic mice are also used occasionally in mutation assays, and it is likely that this will increase as validation of the various models proceeds and regulatory acceptance increases. Such systems have the advantage that they are *in vivo* and the chemical is subject to the dynamics of tissue interrelationships, metabolism, and elimination, in contrast to the tests conducted *in vitro*. They have the disadvantage that the animals are expensive and may need specialist care.

The OECD have approved an *in vivo* test, the Transgenic Rodent Somatic and Germ Cell Gene Mutation Assay (test guideline 488). This test uses four (control and at least three treated) groups of transgenic rats or mice with multiple copies of chromosomally integrated plasmid or phage shuttle vectors. The transgenes contain reporter genes for the detection of various types of mutations. Following 28 days' treatment, there is a 3-day treatment-free period to fix any unrepaired DNA lesions into stable mutations; the animals are killed, and genomic DNA is isolated from the tissues of interest. Mutations are scored by recovering the transgene and analyzing the phenotype of the reporter gene in a bacterial host deficient for the reporter gene.

This is an example of the integration of genotoxicity assays into 28-day toxicity studies in rodents, as exemplified by inclusion of the micronucleus test (which has saved the use of animals that would have been used in conventional micronucleus assays). The comet assay is another test that can be added to standard toxicity studies but that has not been taken up to any great extent, so far.

These two tests are both conducted in tissues that have to be harvested after the death of the animal; a test that can be performed in living animals would have advantages. The Pig-a assay analyzes DNA damage at the X-linked *Pig-a* gene locus. The *Pig-a* gene codes for a glycosylphosphatidylinositol (GPI), which attaches CD59 on the cell surface of peripheral blood erythrocytes (red blood cells) and reticulocytes. Following treatment, flow cytometry measures the frequency of cells in a blood sample without CD59 (Pig-a mutant cells). This test has been evaluated by an international working party (Gollapudi et al. 2015). The Pig-a methodology has the advantages that only a small amount of blood (<100 µL) is needed and animals do not need to be killed for blood collection, and it can easily be added to a standard toxicity study. It is a sensitive assay as low levels of background mutation can be detected and Pig-a and micronuclei can easily be analyzed in the same blood sample and provide complementary data, which will provide gene mutation and clastogenic/aneugenic data (Dertinger et al. 2015). The disadvantages are small, in that blood samples must be processed within 48 hours and, of course, flow cytometry equipment is needed. It is possible that the level of background expression may increase with age and the test may not be applicable for test articles that do not have a secondary effect or ability to penetrate to the bone marrow as the test requires exposure of the bone marrow to the parent compound or metabolite (Gollapudi et al. 2015). In addition, repeated exposure may be required to enable a response to be measured.

There is considerable scope for developing this test further. The gene is conserved across species, potentially widening the scope and applicability of the test; a human Pig-a assay has recently been developed that may have potential use in clinical studies. In addition, an *in vitro* method is under development. (We are indebted to Annette Dalrymple at British American Tobacco for information on this test.)

In Silico Systems

These are discussed more fully in Chapter 5; in summary, the use of *in silico* software for the prediction of genotoxicity, specifically mutagenicity, has been accepted under International Conference on Harmonisation (ICH) guidelines (ICH M7 2014; Case Study 5.1, Chapter 5) for the evaluation of impurities in pharmaceutical products' potential genotoxicity. This guideline is the first regulatory acceptance of *in silico* methods for prediction of toxicity and, although promulgated specifically for pharmaceutical impurities, has utility in other areas where genotoxicity and mutagenicity are important, including agrochemicals. They have also been used to explore results from *in vitro* tests that were not consistent with reasonable expectation and can be used to isolate effects that may be due to experimental design, such as choice of solvent. One hazard to be aware of in interpreting the results of such analyses is that the predictions of expert systems are dependent on the entry to the database of real tests; if the design and execution of these tests are flawed—for instance, by choice of a solvent that reacts with the test chemical to produce a mutagenic reaction product—then the prediction based on these will also be flawed.

PITFALLS IN GENOTOXICITY

The absence of toxicity is a major concern in genotoxicity assays, as it is often taken to mean that the test system was not adequately exposed to the test substance. This may be due to inherent insolubility of the test substance, making high concentration exposure difficult or impossible. However, as genotoxicity has been shown in the insolubility ranges of some substances, restricting test concentrations on the basis of solubility is not usually a valid option. Equally, it should be pointed out that there should also be an absence of toxicity at lower concentrations as toxicity itself may produce positive results in some test systems. In the Ames test, toxicity may lead to reduced colony counts, through a reduction in the number of viable bacteria able to demonstrate a response. With substances that are particularly bacteriotoxic, low achievable concentration may make the Ames test inappropriate.

An absence of response in positive controls may indicate that the test system was not what it was supposed to be, and the characterization of the cell or bacterial line should be checked, together with the laboratory background data accumulated from previous experiments. Poor characterization of the test system is a factor to bear in mind in looking at any set of unusual test data.

In chromosome aberration tests in mammalian cells *in vitro*, damage seen only at high concentrations may indicate that the harvest times were inappropriate; different harvest times in the second experiment may help to clarify effects seen. In a similar way, the use of a preincubation assay in the bacterial reverse mutation (Ames) test can provide alternative metabolic conditions in the second experiment. The use of different conditions in the second experiment following a negative or equivocal first experiment provides a more robust study with less chance of a false-negative result. Osmotic pressure or pH outside normal limits is also a source of invalid data and should be considered in the design of studies. In some cases, biologically irrelevant effects can be produced by choice of an inappropriate test system.

In tests *in vivo*, there is a problem if the test material has marked pharmacological effects at low doses that preclude high-dose testing. In these cases, it may be impossible to produce a high-enough exposure at the target cells. A similar lack of exposure may be seen in substances that are absorbed to a negligible extent. In these cases, a parenteral route may help to increase target cell exposure; however, intraperitoneal injection may not be appropriate, and intravenous injection may be difficult due to low solubility. This becomes problematic when an *in vitro* test has indicated a positive result that cannot be verified *in vivo* due to toxicity, poor absorption, or poor solubility. One approach is to consider the use of additional tests such as UDS, but the best administration route for this test is oral, as this is the route most likely to be associated with the highest possible concentrations in the liver, where the target cells are present. In the final analysis, a negative result *in vivo* achieved as a result of low-level exposure does not offset a positive result *in vitro*.

The possibility of false-positives should be considered. This is the chief reason for conducting a second experiment for *in vitro* assays, the object being to confirm the reproducibility of the first set of data, a factor that is particularly important where a marginal effect is examined. In the mouse micronucleus test, excessive stress may lead to a small increase in micronuclei.

Kirkland et al. (2007) described a workshop that examined how to reduce false-positive results in *in vitro* genotoxicity testing and the avoidance of subsequent animal tests. There was general agreement that genotoxicity tests in mammalian cells *in vitro* produced an unacceptably high rate of false-positive results. Having said that, one of the comparators was with rodent carcinogenicity studies, themselves a test of doubtful utility and human relevance in the assessment of carcinogenic potential. Amongst the factors identified in contributing to the rate of false positives were the following:

- Lack of normal metabolism in cell lines necessitating the use of exogenous metabolic systems such as S9 mix
- Impaired *p53* function
- Altered capabilities for DNA
- High concentrations of test chemicals—up to 10 mM or 5000 µg/mL, depending on factors such as solubility or toxicity
- Extent of cytotoxicity demanded by guidelines

The point was made—which should be considered for all experiments *in vitro*—that it is not clearly rational to exceed the *innate abilities of the cells with respect to metabolism, activation, and defense.* One of the problems with testing in cells in culture is that the appropriate enzyme systems, which are critical to activation of a compound so that it becomes genotoxic, are missing, due to either culture history or species differences. The rationale, therefore, was to increase the concentrations of the test chemical in the forlorn hope that some other enzyme might produce relevant metabolites and so produce a relevant effect. Hope is not a good or rational basis for scientific success.

From this, there was agreement that test cells in culture should have *p53* and DNA repair proficiency and defined phase 1 and phase 2 metabolism (with a broad set of enzymes). In addition, it was suggested that guidelines for concentration and cytotoxicity should be appropriate and that these measures together might reduce the incidence of false positives. The report of the workshop indicated that there was some evidence that human lymphocytes gave less frequent false-positive results than cell lines derived from rodents, and other cell systems were beginning to show promise.

Even if the appropriate line of cells is selected, it is necessary to characterize them fully. For example, a clone of L5178Y cells was in use that had an additional copy of the chromosome bearing the TK gene, and these cells were more sensitive to some mutations than the correct clone (Kirkland, personal communication). L5178Y cells are known to be oversensitive, and this may be in part due to deficiency in *p53*.

One area of fundamental importance is the purity of the test substance, as impurities have been associated with genotoxicity. A positive result with an impure early production batch of chemical may not be relevant to the effects of future batches. Equally, it should be borne in mind that a change in synthetic pathway during development may introduce new impurities that have not been properly tested in previous genotoxicity assays.

Care should be taken, in responding to a positive result, that additional tests are chosen that will help to explain the data produced rather than simply add to them. In data sets relating to older chemicals, it is possible to see the large number of tests that have been conducted to investigate positive results in early testing; these effects have been known to disappear when there has been a change in production methods. The initial response to a positive result should be to ask if it is biologically relevant and how it has arisen. With this information, it is then possible to design a set of investigations that will explain the initial data set.

GENOTOXICITY TESTING *IN VITRO*—SENSITIVITY AND SPECIFICITY

There is a huge pressure to adopt *in vitro* test methods in many areas of toxicology, including some that are simply not suited (currently) to such an approach, such as chronic toxicity. Although *in vitro* tests may exist and be used, this does not necessarily mean that they are effective in identifying hazards that are truly relevant to man. It has become obvious over the years since the Ames test was first used that a positive result did not mean infallibly that the tested chemical was a human carcinogen or even a rodent carcinogen. Equally, a negative result did not exclude carcinogenicity by nonmutagenic or nongenotoxic mechanisms.

It is normal to use a battery of standard tests to assess genotoxicity, typically two *in vitro* and one *in vivo*. However, the *in vivo* test, usually the rodent micronucleus test, is often described as being too insensitive. This is despite the fact that it uses a complete animal test system that should be more relevant biologically than specially adapted bacteria or isolated cells derived from long-dead mice. The intention in using a battery of tests is to catch some of the false negatives or false positives and to investigate different mechanisms of genotoxicity, so that an overall interpretation of the data can be reached without reliance on a single flawed test. The unspoken belief is that if a single test is flawed, then it is an improvement to use several flawed tests in harness but, crucially, to understand their flaws and how to interpret them. The second level of understanding is to know what tests would be useful to further investigate the false-positive results that are often found.

One of the main issues surrounding the currently standard tests for genotoxicity is their sensitivity and specificity in the detection of carcinogens or noncarcinogens, or in other words, the incidence of false-positive and false-negative results. An additional limitation in the extensive discussion that has taken place on this issue is the general tendency to speak about rodent carcinogens rather than those that cause cancer in humans. The implication of this is that, if the genotoxicity test or battery has poor sensitivity with respect to rodent carcinogens, the sensitivity of such tests in identifying human carcinogens is much less. This is based on the basic fact that most rodent carcinogens are not human-relevant carcinogens, often because their carcinogenicity is expressed at high doses that have no relevance to real-world exposures. In addition, genotoxicity tests, by definition, detect genotoxic carcinogens and not those that act via nongenotoxic mechanisms.

Kirkland et al. (2005, 2006) reviewed the ability of a battery of three *in vitro* genotoxicity tests to discriminate rodent carcinogens and noncarcinogens. The tests examined in detail were the Ames test, the mouse lymphoma assay, and a test for clastogenicity—the *in vitro* micronucleus test. These reviews clearly identified the strengths and weaknesses of the current approach to *in vitro* identification of genotoxic chemicals. Of the 554 carcinogens evaluated, 93% had corresponding positive results in at least one of the three tests; i.e., they had good sensitivity, but they were not good for identifying noncarcinogens (poor specificity). This poor specificity was illustrated by the finding that more than 80% of the 183 compounds that were noncarcinogenic in male and female rats and mice had positive data in *in vitro* genotoxicity tests. Of the three tests, the Ames test showed 54% sensitivity (correct positive responses) but showed the best concordance with the rodent studies in terms of its ability to give positive results for carcinogens and negative results for noncarcinogens. Adding the two other tests to the battery produced a decrease in sensitivity because this increased the numbers of positive responses from both carcinogens and noncarcinogens. The mammalian cell test had poor specificity, producing too many false positives.

The first of these reviews clearly showed the interpretative hazards posed by genotoxicity test results. Although the Ames test came out relatively well, the following points illustrate the problems:

- Of 206 carcinogens tested in these assays, only 19 gave consistently negative results in the full battery of three tests. Most of these were carcinogenic through a nongenotoxic mechanism or were very weak genotoxins.
- Genotoxicity data were found for 177 of 183 noncarcinogens in rodents, which showed that the Ames test was reasonably specific (73.9%), but the mammalian cell tests have specificity below 45% (i.e., a high incidence of false positives).
- Where all three tests had been performed, false-positive results were found for between 75% and 95% of noncarcinogens.

The authors indicated that if a chemical gave positive results in all three tests, it was three times more likely to be a carcinogen in rodents than not. Equally, a negative result in all three tests was associated with a twofold likelihood that the chemical would be a noncarcinogen.

The three tests reviewed by Kirkland et al. are not the only *in vitro* options. Sasaki et al. (2000) reviewed the utility of the comet assay by comparing the results with eight mouse tissues with the carcinogenicity data from 208 chemicals chosen from International Agency for Research on Cancer (IARC) monographs. Chemicals such as alkylating agents, azo compounds, and hydrazines were highly positive in this assay, reflecting the comet test's ability to show fragmentation of DNA molecules. However, the tissues that showed increased DNA damage were not necessarily those in which tumors developed. On the other hand, tissues that did express tumors usually showed DNA damage, indicating that organ-specific genotoxicity was a prerequisite but not necessarily predictive for carcinogenicity. This review indicated that the comet assay had a high-positive response for genotoxic rodent carcinogens (110 of 117 were positive) and a high-negative response for rodent genotoxic noncarcinogens, which suggests that the comet assay may be useful to examine *in vivo* the results of *in vitro* genotoxicity tests. This is supported by the observation that 49 of 54 rodent carcinogens that were negative in the *in vivo* mouse micronucleus test were positive in the comet assay.

It was acknowledged, in the report of a European Centre for the Validation of Alternative Methods (ECVAM) workshop (Kirkland et al. 2007) that *in vitro* genotoxicity tests, in mammalian cells, produce a high number of false-positive results. The concern is that these require considerable resources to investigate properly and result in increased use of animals. A number of problems with these tests were suggested. The tests rely on externally added S9 mix that is nearly always from induced rodent liver and is unlikely to produce metabolites that are relevant to humans. In addition, the cells used have impaired function of p53, and their DNA repair capability is not normal with respect to normal cells. The high concentrations used routinely, up to 5000 µg/mL, are predicated on the possible absence of relevant metabolizing enzymes. The hope (although this word was not actually used) is that use of high concentrations may elicit production of relevant metabolites through the activity of less prominent enzyme pathways. It has to be said that hope is a poor basis for scientific progress, and while pathways of further testing were suggested in the workshop report, the field is still in a state of flux.

In a further workshop report, Thybaud et al. (2007) sought to make recommendations for interpretation of common regulatory genotoxicity test batteries and to suggest strategies for follow-up tests. The high number of false positives was again noted. Although the results of the test battery may be negative, further testing may be considered necessary if carcinogenicity was seen in animal tests, if structural considerations indicated potential genotoxicity, or if significant human metabolites had not been tested. Any follow-up tests should be carefully selected based on mechanistic understanding or to elucidate mechanisms of action. Genotoxicity may arise through actions not related to direct reaction with DNA, and these may not be linear or have a threshold. Overall, the concentration at which the effects are seen is an important consideration as high concentrations are unlikely to be relevant to humans.

The various evaluations discussed above are, principally, to detect rodent carcinogens, and it is well known that these are often not relevant to human exposure, mechanism, or epidemiological experience. As with most biological systems, the black and white of the extremes merely point to the hazards of the gray areas in between; in this case, the two extremes are not entirely black and white themselves. It is not unusual for a test substance to display both positive and negative results across different genotoxicity assay types and sometimes even within the same assay. In some cases, it is possible to find an established mechanism of action that is associated with known positive results in a particular assay but negative results in almost all other assays. Where appropriate, extrapolation of such mechanisms, coupled with negative genotoxicity assays and carcinogenicity data, can be used to dismiss such positive results. Expert or rule-based structural activity relationship programs are good at identifying known toxicophores.

It is clear that interpretation of genotoxicity data needs a great deal of care. It is also abundantly clear that there is a need for a genotoxicity testing strategy that uses tests that are relevant to man and not to rodents.

CARCINOGENICITY

GENERAL PRINCIPLES IN CARCINOGENICITY

Cancer is usually a degenerative condition of old age, which is seen at high incidence in animals and in approximately 30% of the human population. Furthermore, it is a tremendously diverse condition, affecting practically every tissue in the body and, in terms of individual tumors, occurring at widely differing rates. For example, in humans, lung and breast cancer are common, but hemangiosarcoma is rare. As life expectancy increases, the background incidence of cancer will also tend to increase. The principle in carcinogenicity assessment is to screen chemicals for the potential they might have to cause, or be associated with, an increased incidence of cancer in humans. This process, which is broader than any single study, looks for structural similarities between the chemical and known carcinogens, and examines all the data from genotoxicity tests, metabolism, and pharmacokinetics, and the data from long-term testing in animals, usually rats and mice. The intention in these latter tests is to look for relevant tumor increases in animals or to look for mechanisms that may be expected to result in human tumors at a significant rate. These assessment methods are obligatory before a chemical is allowed to come into regular contact with humans through marketing. For existing compounds, natural or synthetic, there is the possibility of epidemiological study to elucidate relationships between observed tumors and human exposure.

Most known human carcinogens are genotoxic, and it is reasonable to assume that a chemical that is found to be clearly genotoxic in appropriate tests, including those conducted *in vivo*, is likely to be carcinogenic in humans. The problem with testing for carcinogenicity in rodents is that this tends to show whether the test material is or is not carcinogenic in rodents, and a careful extrapolation to the human situation is necessary before the risk of human carcinogenicity can be properly assessed. This extrapolation requires the careful assessment of all the available data and, possibly, the performance of additional mechanistic studies to explain any effects seen in animals or *in vitro* tests. The absence of carcinogenic effect in rodent tests should not be taken as definitive proof that a chemical will not be carcinogenic in humans. Equally, the presence of an effect in rodents, for a nongenotoxic chemical, is often taken as evidence that there will not be a similar effect in humans. The contradictory nature of these two positions calls into question the utility of the carcinogenicity bioassay, and there is, in fact, a growing acknowledgement that this is an unsatisfactory form of test, which will in time be replaced when satisfactory alternatives have evolved.

Carcinogenicity may be simply defined as the process of conversion of normal cells, so that they can form abnormal growths or tumors. However, this simple definition masks the complexity of the process, which is multistage and multifactorial. At its most basic level, genotoxic carcinogenesis has been described as a three-stage process—initiation, promotion, and progression. Initiation is where the initial change in DNA takes place and is fixed, promotion is the initial division of these cells to form a focus of less differentiated cells, and progression is where the focus of cells grows to become a tumor. Each of these stages is itself subject to a wide range of influences, which makes testing for the individual stages extremely difficult, if not actually impossible. The picture is made more complex when the intricate mechanisms of nongenotoxic carcinogenesis are considered. Where genotoxic carcinogenesis is the end result of direct effects on the genetic material (in terms of quality or quantity), nongenotoxic carcinogens act by producing changes in the expression of the genetic information. For instance, changes in the basic mechanisms of cellular control of programmed cell death (apoptosis) and division or simply increasing cell turnover can lead to cancer, without an initial direct effect on DNA. The extent of DNA methylation—epigenetic change—is also known to be an important factor in gene expression and cellular control as is the necessity for unhindered communication between cells via gap junctions. A further distinction between genotoxic and nongenotoxic effect is that the former, once fixed by cell division, is irreversible, whereas the latter can be reversed by withdrawal of the stimulus. The fact that there has been damage to DNA does not mean that cancer will develop; if the cells do not divide or are removed through natural processes of cell death or sloughing, cancer will not occur. Furthermore, carcinogens in humans can rarely be said to be acting alone. Human DNA is subject to a high background of "normal" damage due to environmental influences, independent of any specific xenobiotic chemical. When an additional potent carcinogen is added to this, the effects may be more than simply additive; chemicals to which we are exposed routinely in the course of everyday existence may serve to promote the effects of chemicals encountered at work or elsewhere. The carcinogenicity of mixtures, such as cigarette smoke, is bound up in the world of these interactions. In fact, smoking has a marked upward effect on the risk of cancer in workers who were employed in the asbestos industry or uranium mines. Conversely, a reduced incidence of cancer is associated with high levels of antioxidant or other protective chemicals, typically contained in a diet rich in fruit and vegetables.

In regulatory toxicology, the life span study in rodents has been the gold standard of assessment for many years, although this is changing gradually. The objective of these studies is to detect increased incidences of tumors in the treated groups that can be ascribed to the test substance. Although tumors may be caused by chronic inflammation or physical mechanisms such as implants, radiation, or fibers, the main emphasis here is on assessment of chemical carcinogenesis. However, the whole area of carcinogenicity assessment is under review as the relevance and utility of data provided by a classic 2-year study are increasingly questioned. This section sets out to provide a review of current methods and those that may supersede them.

TEST SYSTEMS FOR CARCINOGENICITY

The standard species in which life span carcinogenicity is assessed are the rat, mouse, and rarely, the hamster. The hamster is used very little in these experiments, particularly because of the lack of background data. Also, where temperature control was less than perfect, the prospect of an entire study going into hibernation in cold weather was less than ideal. Other species may be more appropriate than these rodents but are usually ruled out by long life span and, consequently, increased study length, housing requirements, and expense.

The usual approach has been to conduct bioassays over a 24- or 30-month period in rats and mice. The mouse has a long history in carcinogenicity testing; it was skin-painting experiments in mice that demonstrated the tumor promotion properties of phorbol esters, and they have also been used in photocarcinogenicity testing.

Historically, several strains of rodent have been favored, but each has its pros and cons. The choice of strain was greatly influenced by the US National Toxicology Program (NTP), which tended to use F344 rats and B6C3F1 mice. The former has a high incidence of testicular tumors and leukemia, and the B6C3F1 mouse is associated with a high incidence of liver tumors. Other strains have been used successfully, notably the Sprague Dawley CD and Han Wistar rats and CD-1 mouse, each with its own tumor profile.

Growth, survival, and tumor profile are inextricably entwined and have caused problems in the past. It has been noted, especially in studies where an unpalatable test substance is mixed with the diet, that lower food consumption in the treated groups is associated with lower tumor burden and longer survival, when compared with contemporary controls. The Sprague Dawley–derived CD rat was used extensively, until it was found that it was becoming increasingly overweight with a consequent reduction in life span (see Chapter 2 for further information) so that fewer than 50% of animals survived until the end of the treatment period. As this is one of the criteria of a successful carcinogenicity bioassay, there was a move toward other strains. One of the advantages of the F344 rat was that it was somewhat smaller than the CD, with better survival, eating less food, and requiring less test material.

The chief disadvantage of using a strain with a high incidence of a particular tumor is that it is difficult to show a small increase in tumors in the affected tissue, especially as the normal incidences can vary significantly between studies. In the final analysis, the choice of strain should be influenced by strains usually used in the laboratory that is expected to carry out the tests. This is a pragmatic decision based on the fact that the historical control data at the laboratory are important in the interpretation of the data; an apparently significant but small increase in testicular tumors may be dismissed as being within historical ranges. Such dismissal is even more authoritative if no dose relationship is present. Although there is good sense in using the strain of rat that was used in the general toxicity testing, this is not always possible, and is, in any case, not usually possible with the mouse, which is not often used in general toxicity.

It has been suggested that several strains should be used in a single study, which could be expected to address differences in response among strains. The problem with this is that the number of animals needed to show a weak carcinogenic response is large for statistical reasons. Hence, in order to detect a weak effect in only one strain, a large number of animals would be necessary in each strain, increasing the size of the study beyond practicable means. Furthermore, if a chemical is carcinogenic in only one strain of several tested, it becomes necessary to question the relevance of the result to the human situation. Given these considerations, if there was a reason to expect significant metabolic differences between strains, it would probably be better to choose a strain specifically for the carcinogenicity studies, based on closeness of metabolic relevance to humans. By the time it is necessary to perform the carcinogenicity studies, the information necessary for this choice should be largely present. However, such deliberation is rare, and more pressing concerns are the more obvious characteristics of the chosen strain, such as survival and tumor profile.

Test systems that allow demonstration of a carcinogenic response in a shorter time than the standard 2-year bioassay would appear to be attractive. The transgenic option, which can be completed in 6 months, are accepted by US regulatory authorities but are still viewed with skepticism in both the United States and Europe—although this is changing as more data become available. In these studies, the importance of various genes in carcinogenesis, for example, the *p53* or *H-ras* gene, is exploited by using strains of mice that are partially or wholly deficient in the gene of interest. These models appear to have some utility in assessing carcinogenic potential, but the same drawback as with the life span study exists, that the assay may produce responses that are irrelevant to humans. One aspect to be wary of in these assays is the potential for all animals, including those in the control group, to eventually have a particular tumor. This very much reduces the utility of the assay, as it reduces comparators to tumor number or size in treated animals in comparison with the controls or to time of observation of the first tumors. Overall, there are probably better ways of assessing carcinogenic potential, and these will become more important as the mechanistic bases of nongenotoxic carcinogenicity are elucidated.

STUDY DESIGN AND METHODS OF ASSESSMENT

The basic design of the classic rodent bioassay is another toxicological constant, defined by years of practice and regulatory acceptance (as well as rigidly conservative traditionalism). The norm is to treat three groups each of a minimum of 50 males and 50 females for at least 2 years, a treatment period that may be extended to 30 months if survival indicates that this is necessary. There are often two separate control groups, giving 100 males and 100 females.

The route of administration is normally oral by intubation or admixture with the diet (see Chapter 3 for more extensive comment); the latter has the advantage of being simple and cost effective. Oral intubation or gavage has the advantage for pharmaceuticals that this is most likely to be the route of administration in patients. However, where oral intubation gives poor systemic exposure due to rapid clearance, this may be improved by dietary administration where the animals eat over an extended period. For agrochemicals or food additives, the most appropriate route is usually in the diet. Poor palatability of diet offered can reduce food consumption and consequently affects tumor profile and survival. There are occasional studies that are carried out by administration in the drinking water. However, it is extremely difficult to estimate spillage, making calculation of exposure very inaccurate. Drinking water administration is an inexpensive route of exposure and is thus often favored in academic studies. However, this route can make them irretrievably flawed before the first administration, let alone by the uncertainties of final interpretation and risk assessment. Other routes of administration include dermal or inhalation. Dermal administration is relatively simple, while exposure via the inhalation route is highly complex and, due to the amounts of high-cost equipment required, extremely expensive.

Although the classic bioassay approach uses both sexes of two species, there have been various attempts to get acceptance of a reduced protocol that uses male rats and female mice. Clearly, this is useful for picking up male rat–specific carcinogens, for instance, those acting via α2u-globulin. However, the ability to pick up such specific mechanisms does not necessarily make the assay results relevant to humans.

One type of carcinogenicity assay that does not fit well with more normal designs is the photocarcinogenicity study, a fairly straightforward concept, which is not at all straightforward in its execution. The object of these studies is to determine the potential of the test substance to cause cancer in the presence of sunlight. Generally, they involve dermal dosing of mice followed by exposure to UV radiation for known durations and known intensities. Light intensity is difficult to monitor, as the light sources tend to degrade with use. The studies can last between 6 and 12 months. Problems arise when all the animals, including controls, show skin tumors, which reduces the useful data to time of onset and individual burden of tumors rather than incidence. Also, there has been a lack of consistency in design, strain of rodent, number of animals, and the UV exposure system used, meaning that comparison between protocols is extremely difficult and that the results are less reproducible.

CHOICE OF DOSE LEVELS

Correct dose level is critical in these studies, especially for regulatory acceptance. Generally, it is required that the high dose be chosen as a maximum tolerated dose (MTD) that is responsible for toxicity that will not shorten an animal's survival other than by carcinogenicity. It is important that exposure should be for the lifetime of the animals; reduced survival due to toxicity reduces time of exposure and so lessens the opportunity for tumor formation. However, it should also be recognized that increased survival could be associated with higher tumor burdens due to the natural incidences being higher in old age—tumor profile may also be affected. A 10% reduction in body weight gain is considered to be acceptable evidence of toxicity, but care should be taken to ensure that this reduction is not simply due to indirect factors such as poor palatability of diet offered and, consequently, lower food consumption. The lowest dose level is chosen as a suitable multiple of expected human exposure, based on anticipated pharmacokinetics or expected daily intakes either as food additives or as residual pesticides on foods.

The use of the MTD has been widely criticized, especially on the basis that the doses thus selected are often unrealistically high. Pharmacokinetics and metabolism at high doses are frequently unrepresentative of those at lower doses; in addition, a general relationship between toxicity and carcinogenicity cannot be drawn for all classes of chemicals. A further consideration is that most human carcinogens, which are mostly genotoxic, are carcinogenic at less than the MTD. Other criteria for dose choice have been suggested, such as pharmacokinetics and systemic exposure [Area Under the concentration Curve (AUC)] or metabolism.

PARAMETERS MEASURED

Measurement of food consumption and body weight gain should always be carried out and are, obviously, critical in dietary studies for calculation of achieved dose levels (see Chapter 3 for further information). In studies with administration of constant concentrations, typical in agrochemical studies, the achieved dose will fall as the study progresses due to the animals' growth and the fact that food consumption will tend to remain similar throughout the study. In the estimation of dose levels in these studies, it is important to reduce scatter of food as far as possible or to be able to make a reasonably accurate estimate of this, as this has a significant impact on the accuracy of the dose calculations. Young animals tend to play more than old animals (and scatter more food), and unpalatable diet will be scattered more as the animals dig into it looking for something better; this is usually more of a problem in the early weeks of a study.

Clinical observations, especially for palpable swellings, which give an indication of the time of onset of tumors and their location, are routine. From these data and those collected at necropsy, the tumor burden for each animal can be assessed, as it is possible for treatment to produce a greater number of tumors in individual treated animals than in the controls. Skin tumors would be a good example of this type of effect, as they are easily seen clinically or at necropsy and each would be sampled and examined. In some cases, the onset of a tumor type may be accelerated by treatment, although the overall incidence of tumors may remain very similar to that in the control group. This is particularly important with tumors that are present in the test strain at high incidences, such as mammary tumors in Sprague Dawley rats.

Histological processing of a wide range of tissues and their examination is the primary endpoint of a carcinogenicity study; this would normally include examination of, at least, a blood smear and, more usually, hematological processing of a blood sample just before the end of the treatment period. It is important to ensure that the pathologist has experience of reading these studies and is using terminology that is consistent with that used by other pathologists. In contrast to other types of toxicological data, which may be graded for severity, a tumor is either present or absent, and there can be heated debate among pathologists over the diagnosis of a tumor or group of tumors. At such times, reliable, independent peer review of the sections is vital, although this does not always solve the problem. Unlike numerical data, which are wholly objective and can be examined according to whether they were obtained with correct technique, pathologist opinion may be partly subjective and is very much dependent on factors such as skill of histological processing and sectioning and on the experience of the pathologist. Also, unlike a set of numbers, which can be accepted or transformed for analysis, the same set of slides may be examined by several pathologists, each of whom can express subtly different opinions on them. The problem with this is that while there may be one favorable opinion pointing to an absence of effect, there may be two others—one noncommittal and one indicating carcinogenic effect. It is not possible to ignore the unfavorable opinions, and all must be reported.

The time of exposure of the individual animals to the test substance may differ significantly due to the fact that animals can die at different times during the study—typically, mortality will increase in the last 26 weeks of treatment. Also, the animals may die early as a result of toxicity or causes unrelated to treatment, reducing the time of exposure to the test substance, and therefore potentially reducing the final incidences of tumors that might have formed later. Reduced survival, for instance, due to nephrotoxicity, may be associated with a similar or lower incidence of a tumor type than in the controls, which may give a false negative with respect to carcinogenicity. For this reason, the data resulting from the microscopic examination are processed to give an age-related adjustment to tumor incidences.

OTHER SYSTEMS FOR CARCINOGENICITY ASSESSMENT

The process of carcinogenicity assessment is progressive and not solely reliant on the results of the bioassay studies. As discussed, the causes of cancer are multifactorial, and the results of a rodent bioassay may not be relevant to humans, particularly if they indicate a nongenotoxic mechanism. Equally, testing a confirmed genotoxic chemical in a 2-year bioassay is an irrelevant waste of animals. In view of this, the assessment process to determine carcinogenic potential should itself look at as many different aspects as possible of the test material and its effects. This is reviewed in Focus Box 8.1.

Beyond the extension of investigations in routine toxicity studies, there are a multitude of proposals that are based on accelerated protocols to study tumor incidences. These are generally based on a faster time to tumor and may involve transgenic animals or surgical techniques such as partial hepatectomy. With the latter technique, the theory is that the fast reparative proliferation in the liver would provide an environment that favors the early emergence of tumors. The less expensive option of causing hepatic damage with carbon tetrachloride did not take off in any significant way. The use of transgenic animals is examined in Focus Box 8.2 and Table 8.4.

The question should be asked whether these studies are simply a shorter method of producing tumors that have as much relevance to humans as those produced in a full-length bioassay. Results are clearly dependent on model choice.

The evidence shows that model choice is critical and that there is no guarantee that selection of a model according to class of chemical will be viable as you could, theoretically, choose the model according to the result you want.

It is apparent from the data in Table 8.4 that the performance of these assays was patchy at best, although they seem to have some utility in assessment of genotoxic compounds. The negative results for phenacetin have been ascribed to genotoxicity that was considered only to be weak.

The complexities of designing new carcinogenicity assays were put into perspective by a review by Jacobson-Kram et al. (2004) from the US Food and Drug Administration (FDA), who pointed out that determination of carcinogenic potential is an exercise that is "complex and imperfect." The disadvantages of the current approach include the duration (at least 3 years including a 13-week range-finding study and postmortem histopathology) and expense. The authors also acknowledge that the current system is imperfect for hazard assessment due to the number of false positives, leading to unrealistic risk assessment when the extrapolation is made to human exposure at relevant dose levels. In addition, the current assays require a large number of animals and provide little information on mechanism of action. These authors suggest that a perfect carcinogenicity assay (in the event that such a beast can be designed) would identify all chemicals that could be potential carcinogens in humans at exposures relevant to humans; would have no false negatives or false positives (100% sensitivity and specificity, respectively); could be used to rank

FOCUS BOX 8.1   ADDITIONAL MEANS
OF ASSESSING CARCINOGENIC POTENTIAL

The following are used in addition to the data from classic life span bioassays:

- *Molecular structure:* This can be computer driven (see Chapter 5 for further information) and is known as quantitative structure–activity relationship (QSAR) or, without the quantitative aspect, simply as SAR. It is well established that certain molecular groups or structures are associated with carcinogenicity and their presence acts as an early indication of carcinogenic potential.
- *Genotoxicity studies:* Genotoxicity studies, as discussed above, also indicate if the substance is likely to interact with DNA and so be associated with increased cancer incidence.
- *Routine toxicity studies:* Data from routine toxicity studies should be reviewed. At the simplest level, the presence in the liver of foci of altered uptake of either hematoxylin or eosin, the stains routinely used in histological processing and examination, can indicate the presence of altered cells, which may be the precursors to tumor development. These foci can be investigated by the use of techniques to visualize the presence of various enzymes such as gamma-glutamyl transferase or the placental form of glutathione transferase, both of which may also be indicative of tumorigenic foci. In addition, simple hyperplastic change or the presence of chronic inflammation may be indicative of carcinogenic potential.
- *Hormonal levels:* Examination of hormonal levels in the plasma may also indicate changes that may lead to increased tumor incidences. This type of effect can also be assessed by microscopic examination of the various endocrine organs such as the pituitary, thyroid, or adrenal glands.
- *Immunosuppression:* An assessment of immunosuppression, as this has been shown to be associated with carcinogenicity.
- *Other investigations:* Other investigations of tissues and data from routine toxicity studies should be considered. The comet assay may give useful data when conducted at the end of studies to assess DNA damage in target tissues, such as the liver or gastrointestinal tract. Other tests that could be performed include proteomic investigations, to examine the levels of proteins that are expressed due to genetic changes, for instance, in the *p53* gene, deficiency of which is seen in many human tumors. DNA adduct studies could also be used to indicate effects on the DNA that might suggest a degree of carcinogenic potential. The extent of methylation of DNA is also an important factor in cellular control.

FOCUS BOX 8.2   TRANSGENIC ANIMALS
IN CARCINOGENICITY ASSESSMENT

- The use of transgenic animals in carcinogenicity has been popular in the United States but less so in Europe, and their use has been examined in a major study coordinated by International Life Sciences Institute (ILSI). [Data relevant to this study (presented by R. W. Tennant at a Satellite Symposium at Eurotox 2000) are available at http://dir.niehs.nih.gov/dirlecm/]. These models have been extensively reviewed by Tennant et al. (1998). Treatment periods are 26 weeks or longer, using groups of 15 to 25 males and 15 to 25 females, with significant effects on the statistical power. Design may be affected by the strain of mouse used.
- The following is an assessment of the data presented by Cohen et al. (2001), using seven models and 21 compounds, summarized in Table 8.4.
  - Nongenotoxic noncarcinogens were all negative in these models.
  - All known nongenotoxic rodent carcinogens, dismissed as not relevant to humans by mechanism or human data, were negative.
  - Peroxisome proliferators did not give consistent results.
  - Genotoxic carcinogens were positive or gave equivocal results. Phenacetin is a weak mutagen with a possible mechanism of carcinogenicity in humans associated with nongenotoxic effects, leading to cell proliferation, to which weak mutagenicity may contribute.
  - Hormonal carcinogens gave mixed results.
  - Peroxisome proliferators gave mixed results but were mostly negative.

carcinogens according to potency; would identify target organs or tissues and predict the types of tumors expected; would be rapid to conduct and inexpensive; and would be indicative of mechanism. The likelihood of such an assay being found is fairly small; obtaining consensual agreement to it would be an even higher hurdle.

The use of transgenic models was reviewed by MacDonald et al. (2004) in an assessment of the utility of genetically modified mouse assays for identifying human carcinogens. The principal emphasis of the review was on their use as tools in pharmaceutical development, but the comments made illustrate both the prospective utility and the doubts surrounding the use of these models.

As indicated in Focus Box 8.2, the choice of model is critical and is a function of the test material as well as regulatory acceptance of the model and suggested protocol. The availability (and cost) of the selected strain of mouse is a potentially limiting subset of these problems. The genotoxicity of the test material is clearly a critical consideration, and there may be a degree of uncertainty about model selection. This review suggested that the Tg.rasH2 model is the preferred model for nongenotoxic

TABLE 8.4
Comparative Data from Transgenic Models

	rasH2	Tg.AC Dermal	Tg.AC Oral	p53+/-	XPA-/-	XPA-/- p53+/-	Neonatal Mouse
Genotoxic Human Carcinogens							
Cyclophosphamide	E	E	+	+	+	+	+
Mephalan	E	E	+	+	+	+	+
Phenacetin	+	N	N	N	N	N	N
Immunosuppressants							
Cyclosporin	E	+	E	+	+	+	N
Hormonal Carcinogens							
DES	+	+	N	+	+	+	N
Estradiol	N	+	N	E	E	+	2N & 1+
Rodent Nongenotoxic Carcinogens—Human Noncarcinogens (Based on Human Data)							
Phenobarbital	N	N	N	N	N	N	N
Clofibrate (perox pro)	+	+					
Reserpine	N	N	N	N	N	N	N
Dieldrin	N	N	N	N	N	N	N
Methapyrilene	N	N	N	N	N	N	N
Rodent Nongenotoxic Carcinogens—Human Noncarcinogens (Based on Mechanism)							
Haloperidol	N	N	N	N	N	N	N
Chlorpromazine	N	N	N	N	N	N	N
Chloroform	N	N	N	E	N	N	N
Metaproterenol	N	N	N	N	N	N	N
WY-14643 (perox pro)			E				
DEHP (perox pro)				E			
Sulfamethoxazole	N	N	N	N	N	N	N
Nongenotoxic/Noncarcinogens							
Ampicillin	N	N	N	N	N	N	N
D-mannitol	N	N	N	N	N	N	N
Sulfisoxazole	N	N	N	N	N	N	N

Source: Adapted form Woolley A, *A Guide to Practical Toxicology*, 1st ed., London: Taylor & Francis, 2003.

Notes: +, positive; DEHP, Diethyhexylphthalate; DES, Diethylstilbestrol; E, equivocal; N, negative; perox pro, peroxisome proliferator.

test materials and is responsive to genotoxic compounds as well. In looking at the models reviewed, the following comments were made:

- *p53⁺ᐟ⁻:* The European authorities consider that this model should be acceptable for use in pharmaceutical submissions and, in contrast to the American authorities, would not limit its use to compounds that are genotoxic.
- *Tg.rasH2 model:* This was considered to be appropriate for genotoxic and nongenotoxic compounds by both European and American authorities.
- *Tg.AC model:* This was considered by both sets of authorities as a suitable model for dermally administered pharmaceuticals, but doubts were expressed by a US authority about the phenotypic stability of this model.
- *XPA⁻ᐟ⁻ and XPA⁻ᐟ⁻ or p53⁺ᐟ⁻ models:* While the European regulators thought that these models were promising, their conclusion was that further development was necessary. There was limited US experience with these models.
- *Neonatal mouse model:* This model has been accepted by European authorities and is considered, in the United States, to be appropriate in some circumstances for genotoxic compounds.

At the time of this review, the p53⁺ᐟ⁻, Tg.AC, and Tg.rasH2 assays, which are the most characterized of these models, were used most frequently in pharmaceutical development. There was evidently some debate about the duration of the p53⁺ᐟ⁻ assay with the possibility that this should be increased from 6 to 9 months to increase its utility. While the number of animals in early protocols was 15 per sex per group, it has been concluded that 25 males and 25 females offers a more powerful design.

The conclusions of this review were that the assays have value in identification of carcinogens and can act as an alternative to the 2-year mouse study in a carcinogenicity testing program. The emphasis was that these assays should not be considered on their own, but that they simply provide one strand of evidence that needs to be considered.

The fact that these two reviews were aimed particularly at pharmaceutical development is also significant. The new models are expensive, and while pharmaceutical companies may have the funding for such tests, other industries assessing chemical toxicity may not. The evolution of new, complex, and expensive assays is unlikely to be immediately welcomed in industries where profit margins are low and cost constraints are rigorously applied. Although safety should not be compromised by cost, pragmatically it is an important consideration; if the new assays are also as irrelevant as the ones they are replacing, in terms of human risk assessment, they will be even less welcome.

While the Europeans have been skeptical of these models, toxicologists in the United States have been more enthusiastic. There has been some agreement that these models have potential utility in the assessment of carcinogenic risk, but experience in use seems to be diluting the initial enthusiasm. It has been suggested (in a personal communication) that the p53⁺ᐟ⁻ did not give positive results with a number of genotoxic chemicals; however, this lack of positive response may mean simply that *in vivo* sensitivity to genotoxicity is less than sensitivity *in vitro*. This lack of concordance between *in vitro* and *in vivo* models may simply indicate that *in vitro* models are oversensitive. In any case, this suggestion casts some doubt on the utility of this model and perhaps (by implication) on the utility of rodents as experimental models for carcinogenesis. The authors of the review indicate that "general thinking has advanced beyond the notion that the traditional standard approach involving two species of rodent of both sexes exposed over their lifetimes is the only way to assess the carcinogenic potential of compounds *in vivo*." This realization may be important, but a cynic would point out that the next leap forward in thinking—that long-term studies in rodents are a highly questionable methodology for assessing carcinogenic risk that is relevant to humans—has not been taken. There has been some evolution of opinion since publication of the second edition of this book, but the basic situation is relatively unchanged. Like a man on a ledge, the concept of the leap is there, but no one has had the courage to take it.

Carcinogenic assessment is one of the areas in toxicology with the greatest scope for change in the way it is carried out and, furthermore, is a prime area in which the three Rs of Russell and Burch can be profitably applied. As the mechanisms of cancer generation become clearer in both general and specific senses, more methods of examining for these mechanisms will become apparent. Although identification of relevant mechanisms that can be reliably investigated will continue to be slow, it may be expected that there will be gradual acceptance of new protocols and investigations. This aspect of carcinogenicity is discussed below.

PITFALLS IN CARCINOGENICITY STUDIES

Inevitably, with these studies, there is considerable potential for pitfalls that have great significance for individual studies and the future of chemical safety evaluation. These can occur in any aspect of the study, starting with design and finishing with the conclusion.

The design of the classic bioassay is much dictated by tradition and regulatory preference, and with careful consultation, it should be relatively easy to avoid mistakes in this area. The possible exception to this is housing. Authorities in the United States have tended to prefer single housing for rats and mice, whereas Europeans tend to house them in groups of up to five of the same sex. Rats are social animals and are less stressed when housed in together. However, male mice tend to fight, and the injuries can reduce survival and compromise the validity of the study, especially if the test substance increases aggression. For this reason, male mice are housed singly. Recently, efforts have been made in some laboratories to investigate group housing of male mice from birth; however, such housing paradigms are not widespread nor widely accepted. The design of cages and the use, or not, of bedding also provide some dilemmas. In studies where the test substance is mixed with the food, the amount of food discarded is a useful indication of palatability and is a critical factor in maximizing the accuracy of calculation of achieved dose levels. It is difficult to produce a sensible estimate of food scatter where the animals are housed in solid-bottom cages with sawdust bedding. Such estimates are much more secure when the cages have mesh floors suspended over absorbent paper (although these cages are not generally used for mice). It is now considered that sawdust bedding is better, as it is not associated with granulomatous lesions on the feet, which can lead to early sacrifice of the animals.

Dose level choice is critical in these studies, and correct design of the dose range–finding studies is vital, as is the careful interpretation of their results. Poor palatability of food in dietary studies can lead to lower body weight gain in comparison with the controls, but it is doubtful that a 10% decrease in body weight due to this would be accepted as evidence of toxicity. This could mean that the MTD was not reached and that, accordingly, the study objective was not achieved. Although the MTD is accepted as a method of dose level choice, it is better to have other support for this, for instance, pharmacokinetics. Equally, it should be reiterated that poor survival, due to excess toxicity or to characteristics of the chosen test strain, may also produce an invalid study. In any case, the toxicological relevance of an MTD achieved at unrealistically high exposure should be questioned in any risk assessment.

Once the study is designed and any controversy produced is overcome, the most contentious issue is the way in which the data are evaluated and interpreted. Faulty collection of the data will confound accurate interpretation. This includes incorrect estimation of food consumption and poor recording of clinical observations, particularly those relating to palpable masses, which affects the estimation of time of onset of tumor formation. This can be important where the tumor concerned is seen at high incidences and earlier onset may indicate treatment-related tumorigenesis. Inevitably, this leads us to the conduct of the necropsies and accuracy of recording of existing masses or tumors and their relationship to the clinical record. Once all the tissues have been sectioned and slides prepared, their evaluation is possibly the cause of more controversy and debate than any other part of the study. While toxicological pathology is clearly a science, it is a science with a high "art" content. Terminology can differ among pathologists, and interpretation of the sections can differ widely. For age-related analysis of the data, it is important that correct decisions are made as to whether a tumor was fatal, probably fatal, probably incidental, or incidental. Skewing these decisions can produce different interpretations of the data. One way around this is to ensure that the peer review of the sections is without reproach. In the event of disagreement, particularly in studies that are contracted out, a second or third pathologist opinion may not help the overall conclusion, as the original report will always stand as a valid alternative opinion.

One histopathological trap, especially where there are treatment-related increases in necropsy findings, is the tendency to examine more sections from treated animals than from the controls. The tissue typically affected in this respect is the liver, for which it is normal to examine two sections from different lobes in every animal. If necropsy shows a lesion in another lobe from those sampled routinely, that lesion is sampled in addition to the scheduled sections; discovery of a tumor in the additional section will increase the tumor incidence in that particular animal and in the treatment group as a whole. In strains where there is a high incidence of liver tumors, this has the effect of biasing the incidences upward and can suggest a treatment-related increase in tumor incidence where there is none.

In addition to these factors, there are others that have less impact but that can still be significant. The presence of high incidences of common tumors in control animals will tend to blunt the analysis of the data. Accordingly, the presence of good background data, or at least two control groups, is crucial in evaluation of the results. Complications are also introduced when the mechanism of toxicity is not present in humans, as is usually the case with nongenotoxic carcinogenesis in rodents.

In shorter studies in rodents or in studies *in vitro*, the same basic precautions in study design, dose level choice, and evaluation have to be observed, with the added complication that the technical conduct of the study has to be consistent with practice in other laboratories. This is particularly the case with the more complex *in vitro* assays such as the comet assay. However, perhaps the biggest pitfall in this type of study is in the understanding of the changes seen and their interpretation. If the origins of the data and the mechanisms of their generation are not understood, it is not possible to draw a supportable conclusion.

OVERVIEW OF THE FUTURE OF CARCINOGENICITY ASSESSMENT

In looking at the future of carcinogenicity assessment, it is probably worth taking a step back and surveying the field as it stands at the moment. Alan Boobis has pointed out (in a presentation to the British Toxicology Society Continuing Education Programme in 2006) that in the current test paradigm, compounds are tested at high doses in lifetime studies in rodents, in which the background incidence of some tumor types is very high (for instance, testicular tumors in Fisher F344 rats). The basis of the risk assessment is tumor incidence, but the relevance to man of such tumors produced at high doses is highly questionable. In addition, carcinogenicity may be secondary to toxicity expressed at these high doses. Quite apart from this sort of basic analysis, there have been a number of reviews examining the utility of the two-rodent bioassay test program in pharmaceutical and agrochemical development (Gaylor 2005; Doe et al. 2006).

Gaylor (2005) carried out an analysis of carcinogenicity studies under the auspices of the US NTP and suggested that almost all of the chemicals selected would have brought about a statistically significant increase in tumors at the MTD, if a larger sample size had been used (more animals, up to 200 per group, which is clearly unrealistic). On this basis, the bioassay based on the MTD is not distinguishing between carcinogens and noncarcinogens but simply not detecting weak carcinogens. In other words, it is simply a screen for potent cytotoxins at the MTD. Gaylor suggested that a bioassay should investigate the relationship between dose and cytotoxicity or other mechanisms that could result in an excess tumor burden, rather than whether a chemical is a carcinogen.

If it is accepted that the results of the classic rodent bioassay are of dubious relevance to humans, it becomes necessary to examine other methods of assessment of carcinogenic potential. Overall, it is relatively simple to detect genotoxic chemicals using established methods without the use of a full-length rodent bioassay. Equally, it has been shown that detection of nongenotoxic carcinogenesis in rodents is relatively easy. The challenge is to detect nongenotoxic carcinogenesis that is relevant to humans. Having said that, it is increasingly apparent that there are elements of promotion in cancer that are due to nongenotoxic chemicals or mechanisms, even with chemicals that are strongly genotoxic. It is also apparent that there is no simple battery of tests currently in existence that will reliably predict human carcinogenicity.

It is usual to consider the results of a range of tests in order to assess carcinogenic potential; however, it is clear that there is considerable scope to develop new tests that examine mechanisms of carcinogenesis that are not currently covered.

To answer the question about how to test for carcinogenic potential, it is worthwhile considering the origins of cancer as a multistep process. Typically, mutations in several genes are necessary, such as conversion of proto-oncogenes to oncogenes and inactivation of tumor suppressor genes. Mutations may result from direct interaction of a chemical or its metabolites with DNA or indirectly through the generation of reactive oxygen species, which are also present endogenously. If an endogenous process leads to excess production of such reactive species and there is, for whatever reason, a deficit in their removal, cancer may result. Normal cellular replication is inherently error prone, and the number of errors is likely to rise when replication is stimulated. However, it should be remembered that unrepaired DNA does not by itself mean that cancer is inevitable. If there is a deficiency in DNA repair, cancer becomes more likely, but an initiated cell (DNA-damaged or altered) will not give rise to a cancer unless it is stimulated to divide and allowed or encouraged to proliferate. A further factor here is that the body's natural defenses, principally the immune system, have to ignore the non-self-replication that is taking place. The stimulus for growth may be endogenous or exogenous. Overall, Alan Boobis highlights that increased DNA damage is procarcinogenic if the cell survives and that unprogrammed increases in cell proliferation or cell survival are also both procarcinogenic. These effects may be secondary to toxicity and are often subject to a threshold dose or concentration, below which carcinogenicity will not be seen. He suggests that the threshold for toxicity is the same or lower than the threshold for carcinogenicity.

These basic considerations may be used to inform a test strategy for carcinogenicity that does not use a life span study in rodents, although it has to be acknowledged that rodents will probably still have a role to play in these assessments.

One of the first techniques that can be applied to assess carcinogenic potential is the use of an expert system to examine structure–activity relationships in the molecule; however, it should be noted that they are principally of use in detection of structural groups associated with genotoxic carcinogenesis (Chapter 5). This is due to the inherent reactivity of these agents or their metabolites and the presence of DNA as a common target. In the case of nongenotoxic carcinogens, attribution of effect to structural aspects of molecules is more complex, due to the very wide range of mechanisms through which such carcinogenicity can be expressed. Where a particular structural group is associated with a particular nongenotoxic effect, for instance, peroxisome proliferation or nephropathy due to α2u-microgloubin, there is a possibility that this may be entered into computer databases for detection in future structures. However, the extent of the problem is underlined by the structural diversity of compounds that are associated with peroxisome proliferation.

Table 8.5 sets out a number of factors or effects that are important in nongenotoxic carcinogenesis and looks at how these effects may be detected. Additional mechanisms that can be investigated include changes in receptor interactions, which may be assessed through changes in the activity of tyrosine kinases. A proportion of these events can be covered in routine toxicity studies, either within the current

TABLE 8.5
Processes or Mechanisms in Carcinogenicity and Markers of Effect

Mechanism or Endpoint	Marker	Where or How Assessed?
Immune suppression	Leukocyte differential counts and plasma immunoglobulins; histology of lymphoid organs, T-cell activity. Loss of host resistance.	In routine toxicity or specific studies.
Chronic cell damage; increased oxidative damage	Histopathological change. Lipid peroxidation, decreased glutathione concentrations; lipid breakdown products.	Metabolite studies, *in vitro* studies, antioxidant concentrations. Routine toxicity studies. Proteomics or genomics.
Changes in intercellular communication	Test for gap junction patency.	*In vitro* cellular systems. Staining for connexin shows gap junctions.
Inhibition of tubulin polymerization	Function of spindle formation in mitosis or meiosis—aneugenesis.	Tests under development.
Cell proliferation	Hyperplastic foci. Proliferating cell nuclear antigen.	In routine toxicity studies and in proliferation responsive cell lines.
Hormonal disturbance	Hormone levels, e.g., thyroid hormones or estrogens.	In routine toxicity studies.
Chronic inflammation	Histological examination, sometimes backed up by clinical observations.	In routine toxicity studies.
Faulty DNA repair	Altered function.	*In vitro* in specific bacterial assays.
Alterations in apoptosis, especially inhibition	Histological examination. Overexpression of *p53* gene.	In routine toxicity studies, *in vitro* tests, and genomics. TUNEL assay.
Promotion	Alterations in gene expression and precancerous lesions or foci.	In routine toxicity studies or in specific promotion studies. Genomics.
DNA damage *in vivo*	DNA adducts, DNA synthesis.	Urine. Comet assay. Unscheduled DNA synthesis.
Changes in gene expression	Protein levels.	Proteomics *in vivo* or *in vitro*.
Epigenetic changes in DNA	Methylation levels.	Routine toxicity or specialist studies.

Source: Adapted from Woolley A, *A Guide to Practical Toxicology*, 1st ed., London: Taylor & Francis, 2003.

set of examinations or by extending those examinations to take in new endpoints. Examples include immune suppression, as seen with cyclosporine, hormonal imbalance, and using the comet assay to assess DNA damage in target tissues. Extension of routine histological processing and examination to include immunocytochemistry for specific markers of precancerous change is a relatively simple and cost-effective method of increasing the database for assessment of carcinogenic potential.

With increasing understanding of the factors and mechanisms relevant to carcinogenicity, the spectrum of options by which to investigate carcinogenic potential is growing all the time. For example, the link between epigenetic change and cancer is becoming more apparent, and this is clearly an avenue that should be explored.

J. E. Trosko has been at the forefront of the field in advancing the claims of epigenetic change as a crucial element in carcinogenesis. He and others have challenged the dogma that mutagenesis equals carcinogenesis and indicated forcefully that this is not the simple picture that purists (if there are any left) would like to believe. Trosko and Upham (2005) offer a broad definition of an epigenetic change as being one that changes genomic expression at the level of transcription, translation, or post-translation. This is the study of heritable changes in gene expression that happens without changes in the sequence of the DNA.

In a 2004 paper, Moggs et al. (2004) reviewed the implications of epigenetics and the relationship to cancer in the context of pharmaceutical development. The paradigm has been that cancer is due to damage to DNA or to alterations in cell growth, probably through changes in gene expression driven by receptor-mediated events. However, the authors indicated increasing evidence that other factors can affect gene expression, including changes in DNA methylation and in the chromatin, and in the function of molecules at the cell surface. The genetic code may be altered by DNA methyl-transferases together with proteins associated with the chromatin such as enzymes that modify histones. These changes contribute to the establishment and maintenance of altered genetic states; this affects cancer but is also important in other toxicities.

This theme was further developed by Serman et al. (2006), in a paper on DNA methylation as a regulatory mechanism for gene expression in mammals. These authors indicate three distinct epigenetic mechanisms that give an extra level of control to transcription and so regulate gene expression. These are RNA-associated silencing, DNA methylation, and histone modification, which are critical in the normal development and growth of cells. Methylation of DNA is involved in silencing genes at transcription and regulates imprinted genes and several tumor suppressor genes. DNA methylation has a role in normal embryonic development as well as in carcinogenesis.

Trosko and Upham (2005) examined concepts that have been ignored in carcinogenesis and particularly challenged the concept that a carcinogen is inevitably a mutagen and that the rodent bioassay is useful and relevant in the prediction of risks for human cancer. Taking a step back, the authors point out that a chemical that enters the body is distributed to tissues with three different types of cell: a few adult stem cells; progenitor cells, which would be expected to divide to form new cells (as in the bone marrow); and terminally differentiated cells. These cells interact through intercellular gap junctions and extracellular communication mechanisms. Although all may be damaged by chemicals, they may not be damaged to the same extent or in the same way.

In designing new tests for carcinogenic potential, it should be remembered that carcinogenesis occurs in an environment that is complex and where the target cell is one cell in a tissue that has a range of intercellular interactions. Many chemicals that contribute to the carcinogenic process do so without causing mutations or necrotic cell death. While it is clear that epigenetic mechanisms play a crucial role in carcinogenesis, devising a test or tests for them is difficult. In view of the complexity of the *in vivo* environment during carcinogenesis, devising a simple *in vitro* assay is likely to be a long-term project. However, it seems likely that a battery of carefully chosen tests may be able, eventually, to indicate carcinogenic potential. It is likely that animals will continue to play a role in such experiments.

Work reported by Yaxiong Xie et al. (2007) is an illustration of the type of research in progress. They found that exposure of pregnant mice to arsenic in the drinking water during organogenesis was associated with altered DNA methylation and aberrant gene expression in the livers of the newborn mice. The liver is a target for arsenic carcinogenesis, and this work is indicative of the type of change that may be looked for *in vivo*.

While it is tempting to rush into developing new tests for mechanisms relevant to carcinogenesis, it would be better to take a more detached view at first and then to devise a strategy from which new tests can be designed and validated. A number of toxicologists, including S. M. Cohen and Alan Boobis, have been putting forward the arguments for a change in the carcinogenic assessment process. Cohen (2004) pointed out that the underlying assumptions of the current approach are that the results in animals are relevant to man (species-to-species extrapolation) and that exposures achieved in animals at high doses are relevant to man (dose-to-dose extrapolation). However, these comfortable assumptions are increasingly, albeit slowly, undermined as experience and understanding grow. Cohen placed an emphasis on evaluating reactivity of the chemical with DNA and increase in cell proliferation that may be due to the test chemical. Both Cohen and Boobis emphasize the need to establish the mode of action or mechanism by which the chemical has its effects. The relevance of the findings to humans can be assessed by answering the following three questions:

1. Has the mode of action been established in animals?
2. Is this mechanism plausible in humans?
3. When pharmacokinetics and dynamics are considered, is the mechanism still relevant to humans?

The concept is that a combination of computerized models, existing genotoxicity tests, and other *in vitro* assays, together with studies up to 13 weeks in rodents, should provide the required mechanistic understanding. Clearly, some pharmacokinetic data will be required from humans. This is not a problem for pharmaceuticals but raises ethical questions when chemicals from other classes, such as pesticides, are being evaluated; in these cases, it is possible that microdosing techniques (administration of radiolabeled compound at doses of 10 to 100 µg, which probably gives a realistic human dose) may well have potential in this contentious area.

It is misleading in some ways to divide xenobiotics into classes such as pharmaceuticals, agrochemicals, and plant protection products. They are all foreign chemicals, and while the regulations for the tests addressed are different, the approach should be similar. It is clear that a stepwise or tiered approach to carcinogenicity testing would be appropriate, and this has been put forward by Doe et al. (2006), with respect to safety assessment of agrochemicals. These authors have suggested that the emphasis should be on producing data that are relevant to shorter periods of human exposure; they place less emphasis on long-term toxicity studies, and they do not recommend a mouse carcinogenicity study. All the data are considered in deciding a testing strategy. The end result should be the use of fewer animals to produce more relevant data.

There is still the desire to stay loyal to the paradigm of testing for carcinogenicity in rodents, although this is changing gradually. While Jacobson-Kram (2004) suggested that the transgenic models available now are not perfect tests for carcinogenicity, he also said that a chronic rat study in combination with a transgenic mouse may be a useful approach. However, he also said that the use of transgenic models may be an interim measure in the development of better tests. The use of toxicogenomics and proteomics will, eventually, aid the process of carcinogenic evaluation, although it is clear that the data will require careful interpretation, especially as it would be expected to be voluminous and complex.

Numerous *in vitro* systems have been investigated for detection of nongenotoxic carcinogens, such as the Syrian hamster embryo cell transformation assay, though it should be noted that this assay is not recommended for regulatory screening for carcinogenicity. The usual potential limitations of differences in metabolism and elimination *in vitro* exist in these assays. However, there is considerable scope for incorporation of oncogenes or inactive tumor suppressor genes, and these tests will develop further. An examination of changes in the thyroid, kidney, and liver of rodents following exposure to nongenotoxic carcinogens by Elcombe et al. (2002) concluded that there was no specific single alert for carcinogenesis in these organs but that careful choice of a range of markers could prove to be predictive, if time of evaluation and class of chemical were taken into account. The relevance of new assays to human carcinogenesis must be established for them to have their own credibility, rather than relying on the possibly flawed data from rodent carcinogens.

There does not appear to be any sensible way at the moment of monitoring the early stages of progression, other than by the appearance of tumors. It is possible that protein analyses, particularly in urine using nuclear magnetic resonance spectra, may be of use here.

Proteomics and genomics will become more powerful as their science develops. This can be used to assess the concentrations or levels of different protein products of genes, particularly oncogenes and proto-oncogenes, or of the genes themselves. The study of protein expression and the association of particular proteins with carcinogenesis would be a useful investigation to be added to standard toxicity studies. This sort of investigation may also have some relevance to examination of human samples in clinical trials of pharmaceuticals or in surveillance of patients after marketing is authorized.

There are also factors in carcinogenicity that are not readily tested in toxicity studies, as they are of relevance to the individual rather than to the population as a whole. These include the roles in carcinogenesis of viruses, diet and caloric intake, and genetic susceptibilities such as those predisposing individuals to breast cancer or to the skin cancer xeroderma pigmentosum (due to faulty DNA repair following UV irradiation). Ultimately, genomics will become crucial in assessing an individual's chance of developing cancer, particularly where there is a family history of a particular type of cancer.

In the future, it is probable that there will be an overall assessment of the results of a battery of tests, which should be chosen according to the class of chemical being investigated. This assessment will begin with a computer-based assessment of the structure of the chemical and its possible or probable metabolites, which will also be predicted. Although pharmacokinetic prediction is not particularly reliable at the moment, when based solely on structure and calculated partition coefficient, the role of physiologically based pharmacokinetic models will grow. From these initial assessments, the design of the toxicity studies may be adjusted to test the early predictions, and as the database develops, the study designs can be further refined. Specific *in vitro* tests will then be conducted to examine for common carcinogenic mechanisms or for those considered relevant to the test chemical. It is also probable that this battery will not include 2-year bioassays in rodents, except in special circumstances. The mechanisms of toxicity, which can be investigated in a focused manner *in vitro*, will become more important a priori rather than being examined after the toxicities have been expressed. The overall effect of this would be expected to reduce the numbers of animals used in box-checking studies and to increase the relevance and focus of information in the database, allowing an assessment of carcinogenic potential in the target species, usually humans. An attractive side effect of this increased focus is likely to be a reduction in the development times of chemicals by at least 2 years. The flowchart for such a test program is indicated in Figure 8.1, adapted from Cohen (2004).

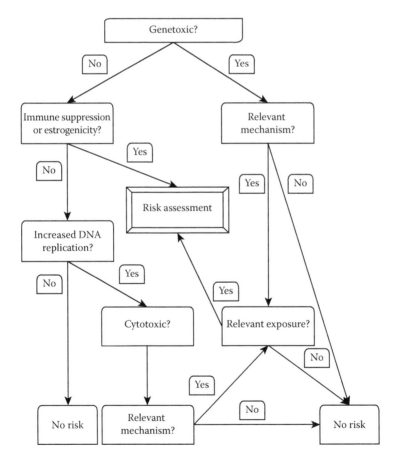

FIGURE 8.1   A flowchart for carcinogenicity risk assessment. (Adapted from Cohen SM, Human carcinogenic risk evaluation: An alternative approach to the 2-year rodent bioassay, *Toxicol Sci*, 80(2): 225–229, 2004.)

SUMMARY

The assessment of genotoxicity and carcinogenicity is evolving more quickly than other areas of toxicological evaluation. This is a function both of development of new endpoints and tests (e.g., the Pig-a assay) and of understanding (e.g., that a carcinogen in rodents is not automatically a carcinogen in humans). Furthermore,

- The essential principles of toxicology are the same in these related disciplines as in other toxicity testing: exploration of dose response and the consequent requirement for a number of concentrations or doses.
- Evaluation of genotoxicity was the first endpoint for which the results of tests *in vitro* were accepted by regulatory agencies; similarly, it has been the first discipline for which the results of *in silico* analyses have been accepted for prediction of mutagenicity, under the ICH M7 guideline. Both areas can be expected to expand and to be refined in the coming years.
- Although the presence of a positive signal *in vitro* has been taken to be an indicator of genotoxic potential, so labeling any such chemical as a potential mutagen, there is increasing realization that the concept of dose response and thresholds may be applicable to genotoxicity.
- New tests, such as the Pig-a assay, have the potential for exploring dose response in animals, without the need to kill them.
- The integration of genotoxicity tests, such as the micronucleus test, the comet assay, and the Pig-a assay, into repeat-dose toxicity tests has huge potential in furthering the cause of the three Rs of Russell and Burch by eliminating the need for separate studies.

In addition, the increasing understanding of the fallibility of rodent carcinogenicity bioassays is leading to a reevaluation of how carcinogenic potential is assessed. It is very likely that this will be assessed in the future from the results of *in silico* analyses, experiments *in vitro*, and repeat-dose toxicity studies of up to 90 days or 6 months. The elimination of the need for carcinogenicity studies in rodents has huge implications for the three Rs campaign and for costs; currently, two carcinogenicity studies in rats and mice use between 800 and 1000 animals and cost millions of dollars. There would also be a concomitant decrease in the time taken to develop a new pharmaceutical of up to 3 years.

REFERENCES

Cohen SM. Human carcinogenic risk evaluation: An alternative approach to the two-year rodent bioassay. *Toxicol Sci* 2004; 80(2): 225–229.

Cohen SM, Robinson D, MacDonald J. Alternative models for carcinogenicity testing. *Toxicol Sci* 2001; 64: 14–19.

Dertinger SD, Avlasevich SL, Bemis JC, Chen Y, MacGregor JT. Human erythrocyte PIG-A assay: An easily monitored index of gene mutation requiring low volume blood samples. *Environ Molec Mutagen* 2015; 56: 366–377.

Doe JE, Boobis AR, Blacker A, Dellarco V, Doerrer NG, Franklin C, Goodman JI et al. A tiered approach to systemic toxicity testing for agricultural chemical safety assessment. *Crit Rev Toxicol* 2006; 36(1): 37–68.

Elcombe CR, Odum J, Foster JR, Stone S, Hasmall S, Soames AR, Kimber I, Ashby J. Prediction of rodent nongenotoxic carcinogenesis, evaluation of biochemical and tissue changes in rodents following exposure to nine nongenotoxic NTP carcinogens'. *Environ Health Perspect* 2002; 110(4): 363–375.

Gaylor DW. Are tumor incidence rates from chronic bioassays telling us what we need to know about carcinogens? *Regul Toxicol Pharmacol* 2005; 41(2): 128–133.

Gollapudi BH, Lynch AM, Heflich RH, Dertinger SD, Dobrovolsky VN, Froetschl R, Horibata K et al. The in vivo Pig-a assay: A report of the International Workshop on Genotoxicity Testing (IWGT) Workgroup. *Mutat Res Genet Toxicol Environ Mutagen* 2015; 26;783: 23–35.

Jacobson-Kram D, Sistare FD, Jacobs AC. Use of transgenic mice in carcinogenicity hazard assessment. *Toxicol Pathol* 2004; 32(S1): 49–52.

Kirkland D, Aardema M, Henderson L, Müller L. Evaluation of the ability of a battery of three in vitro genotoxicity tests to discriminate rodent carcinogens and non-carcinogens I. Sensitivity, specificity, and relative predictivity. *Mutat Res* 2005; 584(1–2): 1–256.

Kirkland D, Aardema M, Müller L, Makoto H. Evaluation of the ability of a battery of three in vitro genotoxicity tests to discriminate rodent carcinogens and non-carcinogens II. Further analysis of mammalian cell results, relative predictivity, and tumor profiles. *Mutat Res* 2006; 608(1): 29–42.

Kirkland D, Pfuhler S, Tweats D, Aardema M, Corvi R, Darroudi F, Elhajouji A et al. How to reduce false positive results when undertaking in vitro genotoxicity testing and thus avoid unnecessary follow-up animal tests: Report of an ECVAM Workshop. *Mutat Res/Genet Toxicol Environ Mutagen* 2007; 628(1): 31–55.

MacDonald J, French JE, Gerson RJ, Goodman J, Inoue T, Jacobs A, Kasper P et al. The utility of genetically modified mouse assays for identifying human carcinogens: A basic understanding and path forward. The Alternatives to Carcinogenicity Testing Committee ILSI HESI. *Toxicol Sci* 2004; 77(2): 188–194.

Moggs JG, Goodman JI, Trosko JE, Roberts RA. Epigenetics and cancer: Implications for drug discovery and safety assessment. *Toxicol Appl Pharmacol* 2004; 196(3): 422–430.

Sasaki YF, Sekihashi K, Izumiyama F, Nishidate E, Saga A, Ishida K, Tsuda S. The comet assay with multiple mouse organs: Comparison of comet assay results and carcinogenicity with 208 chemicals selected from the IARC monographs and U.S. NTP Carcinogenicity Database. *Crit Rev Toxicol* 2000; 30(6): 629–799.

Serman A, Vlahovic M, Serman L, Bulić-Jakus F. DNA methylation as a regulatory mechanism for gene expression in mammals. *Coll Antropol* 2006 Sep; 30(3): 665–671.

Tennant RW. *Application of Transgenic Models for Toxicological Characterization. Crisp Data Base.* Bethesda, MD: National Institutes of Health.

Tennant RW. *Transgenic Mouse Models for Identifying and Characterizing Carcinogens. Crisp Data Base.* Bethesda, MD: National Institutes of Health.

Tennant RW. Evaluation and validation issues in the development of transgenic mouse carcinogenicity bioassays. *Environ Health Perspect* 1998; 106(suppl 2): 473–476.

Thybaud V, Aardema M, Clements J, Dearfield K, Galloway S, Hayashi M, Jacobson-Kram D et al. Strategy for genotoxicity testing: Hazard identification and risk assessment in relation to in vitro testing. *Mutat Res* 2007; 627: 41–58.

Trosko JE, Upham BL. The emperor wears no clothes in the field of carcinogen risk assessment: Ignored concepts in cancer risk assessment. *Mutagenesis* 2005; 20(2): 81–92.

Woolley A. *A Guide to Practical Toxicology.* 1st ed. London: Taylor & Francis, 2003.

Xie Y, Liu J, Benbrahim-Tallaa L, Ward JM, Logsdon D, Diwan BA, Waalkes MP. Aberrant DNA methylation and gene expression in livers of newborn mice transplacentally exposed to a hepatocarcinogenic dose of inorganic arsenic. *Toxicology* 2007; 236(1–2): 7–15.

9 Determination

Dermal Toxicity— Sensitization, Irritation, and Corrosion

INTRODUCTION

The three main aspects of dermal effects that are examined in this chapter and in regulatory toxicology are sensitization, irritation, and corrosion. Another factor to consider is phototoxicity, although that is, fortunately, not common. This chapter does not seek to cover general, reproductive toxicity or carcinogenicity studies using the dermal route of exposure, which may be needed in the development of a pharmaceutical or agrochemical and which are very similar to other studies of that type, the only real difference being the route of administration. This chapter deals with study types that are most relevant to occupational health, which can have unsuspected complexities of successful study conduct and interpretation.

Before looking at the testing needed to evaluate these three endpoints, it is worth considering their characteristics:

- Irritation is a reversible nonimmunological response at the site of contact, which may be seen as erythema (reddening of the skin) or edema (thickening due to accumulation of fluid under the skin).
- Corrosion is not reversible and is characterized by the production of irreversible damage at the site of contact as a result of chemical reaction. Both irritation and corrosion are dose-dependent reactions.
- Sensitization is an immunological reaction that, while it has dose dependency in the induction stage, is not clearly dose dependent following induction. Furthermore, the amounts of chemical needed to elicit an allergic reaction are much smaller than those needed during the induction phase.

From this, it is evident that the processes involved in sensitization and expression of allergy are more complex than for the former two endpoints, making assessment of sensitization *in vitro* more difficult. Although the emphasis of this chapter is on the skin, the examination of ocular irritation is also considered briefly.

GENERAL PRINCIPLES OF DERMAL TOXICOLOGY

The skin is the largest organ in body, up to 2 m², and forms one of the three most significant routes of exposure, along with oral and respiratory. Exposure via the skin is especially relevant in a domestic or occupational setting. The agents involved come from numerous sources and can be industrial chemicals, pharmaceuticals or agrochemicals and their intermediates, or chemicals encountered in routine domestic existence, among which cosmetics are a significant inclusion. The most typical dermal reaction is inflammation, characterized by redness, swelling, and heat, in response to irritants, sensitizing agents, or phototoxic substances. Corrosive agents can produce disfiguring burns, and excessive ultraviolet (UV) irradiation is associated with skin cancer.

In testing programs, dermal irritation and sensitization are often considered together as they are not usually central to a development plan but are clearly of interest in an occupational or domestic context. Statements on irritation and sensitization are essential in Material Safety Data Sheets that accompany chemicals sold to industry. There is some lack of consistency in nomenclature for the allergic form of skin irritation, contact hypersensitivity, or allergic contact dermatitis being used according to source.

There is a clear distinction between dermatitis due to irritation (contact dermatitis or irritant contact dermatitis) and allergic contact dermatitis. The former is associated with a dose- or concentration-dependent reaction that is due to direct interaction of the chemical or mixture with local skin constituents; the response is usually immediate and localized and requires similar concentrations in subsequent exposures to elicit a similar effect. Irritant contact dermatitis is distinct from the burns that are due to corrosive chemicals such as strong acids or bases. In cases of allergic contact dermatitis, there is a period during which relatively large amounts of the chemical may be tolerated without any obvious effects; this reaction-free period (induction) can last for years. After this initial period, sensitivity develops (in susceptible individuals), which may produce severe reactions triggered by minute amounts of the substance. There is usually a clear difference in effective concentrations between a pure irritant and one that has induced contact sensitivity.

Whereas a relatively simple set of chemical reactions determines irritation, allergic contact dermatitis is driven by a complex set of interlinked processes. These start, typically, with the passage of a hapten (a complex of a small exogenous molecule—the potential sensitizer—with an endogenous protein) into a Langerhans's cell in the epidermis, where it is processed and passed to a regional lymph node for presentation to T lymphocytes (known as T cells). Interleukin-1, produced by the Langerhans's cell, stimulates the T cell to produce further cytokines, which cause sensitized T cells to proliferate and act in the production of the clinical signs of sensitization at the site of exposure. These clinical signs in laboratory animals may be expressed as only erythema and edema. However, in humans, a wider range of symptoms, including pruritus, erythema, edema, papules, or vesicles, may be observed.

Contact dermatitis is a major cause of occupational ill-health and is an area of concern to toxicologists assessing exposure limits for the workplace; allergic contact dermatitis is less common but is, perhaps, of greater significance because of the very low exposures that are needed for a significant response. In a similar context, respiratory sensitization is also a significant occupational hazard in some settings, for instance, to animals in toxicology test facilities.

Phototoxicity is also a potential problem; for instance, the presence in celery of psoralens has been associated in celery pickers with extensive skin reactions brought about by sunlight-induced reaction of the psoralens with DNA, inhibiting DNA repair. Treatment of psoriasis with so-called psoralen and UVA (PUVA) therapy—8-methoxypsoralen and UV light—was associated with the induction of cancer. Certain cosmetic ingredients have also been associated with phototoxicity and with more mundane forms of dermatitis. It is worth pointing out that there is a distinction to be drawn between phototoxicity and photoallergy and vice versa. Phototoxicity is likely to occur on first exposure; it is dose related and may be seen after systemic or topical exposure. The mechanism is usually one of photoexcitation, leading to the generation of oxygen or other free radicals. Psoralens intercalate with DNA-producing adducts and inhibition of DNA synthesis. Photoallergy is a delayed type IV hypersensitivity reaction, which—as with other routes of sensitization—requires prior sensitization. Following induction, small amounts of exposure can lead to a reaction. Photoallergy induced by topical exposure is known as photocontact dermatitis and by systemic exposure as systemic photoallergy.

FACTORS IN DERMAL TOXICITY

Toxicity in the skin is affected by factors (summarized in Focus Box 9.1) such as local humidity/moisture, temperature, local injury, exposure to light, local concentration and location, and area exposed. Increased temperature and moisture, as found when the treatment site is occluded by a bandage, act to increase the local reaction. The location of exposure on the body is significant because the skin differs in thickness from one area to another and this affects the reactions seen, the head and neck being more sensitive than the palms of the hands or soles of the feet.

While it is probably not entirely useful to single out any one of these as being the most critical to take into account in designing or assessing dermal toxicity studies, it is possible to make some general remarks.

The form or formulation in which the chemical reaches the skin is a critical factor, as this can have a huge influence on transdermal absorption. This effect is so significant that nonclinical toxicity testing of topical pharmaceuticals is undertaken with a formulation that is as close as possible—preferably identical—to that to be used clinically. Deviation from this formulation, especially the addition, exclusion, or substitution of excipients, may require extensive retesting. For occupational toxicity, particularly irritation studies, the materials are often used as supplied, albeit moistened with water and applied to a moistened site.

The concentration of a chemical that can be applied in repeat-dose toxicity studies may be limited by the local effects seen and is often more important than the dose expressed in mg/kg of body weight. Local concentration also determines extent of sensitization but not necessarily of subsequent allergic reactions. Dermal application is not a normal route for investigation of systemic toxicity; in fact, the dermal route can be used to demonstrate the absence of systemic toxicity, which can be important in several fields of use, such as pharmaceuticals and agrochemicals.

Although it is possible to produce dermal toxicity as a result of systemic exposure, this is relatively unusual, and the potential for phototoxicity should be borne in mind.

FOCUS BOX 9.1   FACTORS TO CONSIDER
IN DERMAL TOXICOLOGY

The following is a guide to the factors that should be considered in the conduct of dermal toxicity studies, whether for occupational safety or for pharmaceutical development.

- *Form or formulation:* This has far-reaching effects on absorption into or through the skin. Many vehicles or solvents enhance absorption; the toxicity of a chemical used "as supplied" can be transformed by the addition of an appropriate carrier solvent.
- *Location of exposure:* The thickness of skin varies across the body. Although this is important in humans, in animals, the site of application is usually on the shaved or clipped back.
- *Area exposed and local dose:* The local concentration of the chemical and the size of the application site (cm^2) are usually more important in determining local effects than the dose in mg/kg.
- *Skin conditions at the site of exposure:* Warmth, humidity, skin damage, and local vasodilation enhance absorption and local effects. These are clearly relevant in animals' studies in which the application site is occluded.
- *The local concentration and the area of exposure:* These are more important in determining the extent of local reaction than the dose expressed in terms of mg/kg—generally given in mg/cm^2 or as a percentage.
- *Dermal metabolism:* The skin has significant metabolic capability, which can enhance local toxicity or produce reactive haptens, leading eventually to sensitization.
- *Local effects may limit investigation of systemic toxicity:* Excessive reaction at the application site may lead to the termination of repeat-dose toxicity studies.
- *Physicochemical properties:* These include polarity (lipophilicity) of the chemical, molecular weight, and pK_a. Chemical reactivity may lead to complex formation with macromolecules or results in irritation. Small lipophilic molecules are better absorbed than large polar ones.

The skin also has significant metabolizing capability that can result in the production of sensitizing or photoreactive molecules.

The condition of the skin at the site of exposure is also important as abrasion or other damage may enhance absorption. Transdermal absorption may also be increased by local vasodilation, which can maintain a diffusion gradient by removal of chemical from the dermis into the blood stream. In human terms, the presence of skin disease, such as psoriasis, is also likely to affect local absorption and reaction.

Dermal toxicity studies may be conducted with some form of occlusion of the application site; this involves application of a gauze and a waterproof layer over the site and application of a wrapping to keep the dressing in place for a defined period (usually six hours in repeat-dose studies). This is a feature of several Organization for Economic Co-operation and Development (OECD) test guidelines, which have recommended that this be done for studies up to 90 days' duration, but in practice, the 28-day study is the longest in which the animals are wrapped up on a daily basis.

Physicochemical factors that affect dermal toxicity include polarity (lipophilicity, indicated by the partition coefficient), molecular weight, pK_a, the pH of any solution, and the ease with which it reacts with local proteins to form haptens that could result in allergic sensitization. Small lipophilic molecules are more likely to be absorbed than large lipophobes, and reactive molecules are more likely to be associated with hapten formation than unreactive ones.

TEST SYSTEMS

Test systems for irritation and sensitization can be divided between the traditional (rabbit and guinea pig) and the new (*in vitro* systems for irritance and corrosivity studies and mice for assessment of allergic sensitization). The rabbit has been used for many years in the assessment of dermal and ocular irritations, although it is generally considered to be a more sensitive model in comparison with human skin. The rabbit *in vivo* assay is now being replaced by *in vitro* systems; negative results in such a test can then be confirmed with a small *in vivo* study. If the result of these *in vitro* assays is positive, it is likely that testing *in vivo* would not be required, if only on the ethical grounds of animal welfare (see Focus Box 9.2 for further details). The processes of dermal (and ocular) irritation in humans are complex and are therefore difficult to reproduce *in vitro*. There are a number of skin equivalents using dermal keratinocytes, which can be used to assess some aspects of ocular damage. The assessment of corrosivity is theoretically easier *in vitro* in view of the simpler nature of the damage caused (i.e., direct reaction with skin components), and a number of *in vitro* systems have been developed to assess this endpoint. One such test measures the reduction in electrical resistance across a sample of skin in response to exposure to corrosive substances. The leakage of enzymes such as lactate dehydrogenase into the incubation medium can also be used as a marker of toxic effect.

The guinea pig has been the traditional choice for conduct of dermal sensitization studies but, with increasing acceptance of the murine local lymph node assay (LLNA), is set for replacement by mice (assays described below in Case Studies 9.1 and 9.2 respectively). There is currently no generally accepted *in vitro* method of assessing hypersensitivity reactions, as with carcinogenicity, it seems probable that no single *in vitro* test will give a reliable assessment of sensitization and the complex processes involved.

For ocular irritation studies, the rabbit is still the *in vivo* model of choice, although systems such as the chick chorioallantoic membrane (CAM) test or systems using enucleated bovine or chicken eyes can be used to identify severe irritants. As with dermal irritation, a positive test *in vitro* can be used as justification of not doing a test *in vivo*.

FOCUS BOX 9.2   A STRATEGY TO DETERMINE IRRITATION AND/OR CORROSION

The following should be considered before embarking on a test program—the goal being to avoid unwarranted use of animals:

- Any existing human or animal data relating to the chemical or related compounds or mixtures.
- The presence of structures within the molecule associated with irritation.
- Physicochemical properties and reactivity.
- pH: extremes of pH—<2.0 or >11.5—may lead to severe local effects. This should include a review of buffering capacity.
- High toxicity following topical application may mean that it is impracticable to examine irritation.
- If dermal tests (using an appropriate species) up to the limit dose of 2000 mg/kg have been completed without local effect, specific testing for irritation may not be necessary.
- Results of validated *in vitro* or *ex vivo* tests for irritation or corrosion; if these are positive, it may be assumed that the substance is irritant or corrosive, and *in vivo* tests are not necessary.
- The final step is a progressive test in rabbits, in which a single animal is tested and observed, and, if no reaction is seen, further animals are tested up to a maximum of three. If the first animal shows evidence of irritation or corrosion, further testing is not required. The guideline indicates that substances that are known to be irritant or corrosive substances and those that are clearly not corrosive or irritant need not be tested *in vivo* (OECD 2016).

For cosmetics or their ingredients, the development of *in vitro* methods is a priority because, in Europe, testing these substances in animals is no longer allowed, thus making the evaluation of new ingredients by traditional methods illegal. This is contrasted by the approach used in other regulatory areas where *in vivo* study is still required by law.

STUDY DESIGN AND PARAMETERS MEASURED

DERMAL IRRITATION AND CORROSION

Ethically, dermal irritation and corrosion is a contentious area of testing, which has been associated with the general perception of suffering in experimental animals, particularly in the Draize ocular irritation test in rabbits. While it is always unwise to say that past problems are just that, much progress has been made in terms of experimental design and control. The strategy for irritation and corrosion testing has been refined such that these endpoints can be confirmed *in vitro* and tests in living animals are restricted. This strategic approach is described in the OECD guideline for acute dermal irritation and corrosion (no. 404, adopted in 2015), which seeks to avoid severe reactions in animals (Focus Box 9.2).

There is now a good range of *in vitro* tests for irritation and corrosion. For example, OECD guideline 439 (OECD 2015) describes an *in vitro* test based on reconstructed human epidermis to identify irritant chemicals or mixtures in accordance with the UN Globally Harmonized System of Classification and Labelling (GHS). These have been validated by organizations such as the European Union Reference Laboratory for alternatives to animal testing (EURL-ECVAM) and include assays described in OECD guidelines. Test examples include the transcutaneous electrical resistance assay (OECD 430 2015), the human skin model test (OECD 431 2016), and the *in vitro* membrane barrier test method for skin corrosion (OECD 435 2015). There is an extensive literature on these assays, and progress in their development has been rapid. Given the extensive and freely available description of these methods, details will not be given here, and the reader is referred to OECD guidelines, reviews, and various reports by EURL-ECVAM.

In animals, dermal tests are based on application of the test material to the shaved backs of rabbits for up to 24 hours under an occlusive dressing to prevent ingestion and to maximize the response. Following removal of the dressing, the response is graded at intervals up to 72 hours or until no further response is seen. Erythema, edema, skin thickening, exfoliation, cracking or fissuring, necrosis, and ulceration are assessed and scored relative to control values. Nonirritant substances may then be assessed in an eye irritation test (see "Ocular Irritation" section).

One of the difficulties with animal tests is that assessment of the reactions is subjective, although systems to assess reaction by means of reflectance colorimetry or spectroscopy are being introduced. The extent of erythema and edema are scored separately from 0 (no reaction) to 4 (severe reaction) and the combined scores used to calculate a primary irritation index. There are several formulae for this, but they are generally based on adding the scores together and dividing the totals by the number of test sites and scoring time points. In the European Union, the scores for erythema and edema are treated separately, while in other jurisdictions, they may be combined. In the European Union, the scores are used to produce a classification as follows:

Primary Irritation Index	Classification
0	Nonirritant
<2	Slightly irritating
2 to 5	Moderately irritating
>5	Severely irritating/corrosive

The various procedures are well described and compared in Derelanko and Auletta (2014). OECD guideline 405 gives details of test methods for both dermal and ocular irritation and corrosion.

Ocular Irritation

Following dermal and/or *in vitro* evaluation, nonirritant substances may be assessed in an eye irritation test, in which 100 mg or 100 µL of the test substance is instilled into the conjunctival sac of one eye of one or more rabbits (the other eye acting as a control). Reaction to treatment is observed and scored at appropriate intervals; if intense irritation is seen following dosing, the eyes are immediately irrigated to remove the chemical. As for the dermal assessments, only one animal is exposed at a time, and reactions are assessed before the next is treated.

Ocular irritation studies have been a regular source of controversy over many years and are moving slowly toward the use of *in vitro* systems. The potential for ocular toxicity should be assessed from physicochemical data relating to the chemical and from the results of *in vitro* screens before animals are used to confirm lack of irritation. It is generally accepted that if a chemical is shown to be a dermal irritant, then exposure of the eye *in vivo* is unnecessary.

The CAM test has been used as a screen to identify severe irritants before the use of animals. In this test, the response of the blood vessels in the CAM of fertilized chicken eggs is assessed after a brief application of the test chemical. Vasodilation and hemorrhage are scored, in comparison with controls and known irritants, at intervals after washing off the chemical. Scoring is based on the intensity of reaction over time following treatment; this period may be as short as five minutes. Other approaches have used excised bovine eyes, quantitative structure–activity relationship (QSAR) assessment, and various proprietary systems.

Irritation by Other Routes

It should be remembered that irritation is not simply a dermal or ocular phenomenon and that it can play a major role in toxicity studies using other routes, including oral and parenteral. Local irritation of the forestomach has been associated with carcinogenicity in long-term rodent studies, and irritation of the gastric mucosa may trigger emesis in dogs or nonhuman primates. Venous irritation at the site of administration—either intravenous or perivenous—may limit the dose or duration in parenteral toxicity studies. Inhalation studies in rodents may be associated with irritation in the larynx, which is not necessarily indicative of similar effects in humans. In studies by intranasal instillation, the presence in the formulation of absorption enhancers has been associated with microscopic changes in the nasal epithelium; these may represent a direct effect of the excipient or an exacerbation by treatment of such effects.

SENSITIZATION

In contrast to irritation and corrosion, less progress has been made in devising credible *in vitro* tests for sensitization, due to the complexity of the processes involved in induction and the subsequent reactions. Although animals have not been replaced, the introduction of the murine LLNA has led to a reduction in the numbers of animals used because previous protocols used large numbers of guinea pigs, whereas an LLNA uses 20 female mice.

ALLERGIC SENSITIZATION IN GUINEA PIGS

There are a large number of protocols using guinea pigs, based essentially on two designs: the Buehler test and the maximization test of Magnusson and Kligman. Both involve dermal treatment of up to 30 guinea pigs at relatively high, irritant doses to achieve sensitization, followed after 14 to 28 days by a challenge at a different site using a nonirritant dose to assess any allergic reaction. The maximization test is a more aggressive test than the Buehler test because the immune response, and the likelihood of sensitization, is enhanced by intradermal injection of the test substance with and without Freund's adjuvant. In addition, irritants such as sodium lauryl sulfate may be used before the challenge dose in order to induce mild inflammation. The main points of the maximization test are summarized in Case Study 9.1.

CASE STUDY 9.1   SENSITIZATION IN GUINEA PIGS—THE MAXIMIZATION TEST

OECD guideline 406 gives the main points of this test method as follows:

- At least 5 control and 10 treated guinea pigs; up to 10 and 20 may be needed if sensitization cannot be proved.
- The dose for induction should be mildly to moderately irritant but not systemically toxic. The challenge dose should not be irritant.

Treatment: induction (intradermal injection and/or topical application).

- Day 0: each treated animal receives pairs of intradermal injections each of 0.1 mL, given to the clipped shoulder region of as follows: (1) a 1:1 mixture of adjuvant and physiological saline, (2) test chemical alone, and (3) a 1:1 mixture of test chemical and adjuvant. Controls receive three pairs of injections as follows: (1) adjuvant and physiological saline, mixed 1:1; (2) undiluted vehicle; and (3) vehicle and adjuvant mixed 1:1.
- Topical administration is given on day 7 using a filter paper soaked with test chemical in the vehicle or vehicle alone, for treated and control animals respectively. The paper is applied to the clipped test area and held in place with an occlusive patch for 48 hours. If the chemical is not irritant, the test area is prepared by painting with sodium lauryl sulfate in Vaseline to induce local irritation.

Challenge:

- Day 21, the flanks of all animals are clipped and the test chemical applied to one flank using a patch or test chamber; the other flank is treated with vehicle only. The patches are left in place for 24 hours, and the reactions are scored.

The basic Buehler test uses dermal application of a mildly irritant dose to one flank on two occasions in the sensitization phase followed by a dermal challenge on the other flank up to four weeks later. The application site may be occluded by wrapping the animal in a dressing that keeps the site warm and moist. In both tests, response to treatment is assessed by scoring for erythema and edema 24 and 48 hours following the challenge dose. According to the OECD guideline, a mild-to-moderate sensitizer should give a response of at least 30% in a maximization test or 15% in the Buehler test. This response rate should be checked regularly, using known sensitizers such as hexyl cinnamic aldehyde, mercaptobenzothiazole, or benzocaine. OECD test guideline 406 gives further details of these methods.

ALLERGIC SENSITIZATION IN MICE (LOCAL LYMPH NODE ASSAY)

In the LLNA, sensitization is assessed by incorporation of tritiated thymidine into proliferating lymphocytes in the lymph nodes draining the site of topical application. The LLNA has a number of advantages over the guinea pig protocols, particularly

CASE STUDY 9.2   SENSITIZATION IN MICE—THE LLNA

The main points of this test are as follows:

- Four groups of five female CBA/Ca or CBA/J mice.
- Control (vehicle) and three concentrations of the test chemical to allow assessment of any dose response.
- Topical application to the dorsum of the ears over 3 consecutive days.
- Vehicles are listed in the guideline in order of preference: acetone/olive oil (4:1 v/v), dimethylformamide, methyl ethyl ketone, propylene glycol, and dimethyl sulfoxide. Vehicle choice is critical (see text for discussion of this).
- On day 6, each animal receives an intravenous dose of tritiated thymidine.
- Five hours later, the animals are killed; the draining auricular lymph nodes are removed; and the lymph node cells are isolated, washed, and subjected to scintillation counting as a measure of incorporation of tritiated thymidine.
- Results are expressed as a stimulation index derived by dividing the counts from test animals by those of the controls. A stimulation index of three or more, in the presence of a dose response, is indicative of a positive result.
- An EC3 value can be derived from the data by linear interpolation; this is the concentration of the test substance that is calculated to give a stimulation index (SI) value of 3 and gives an indication of potency when EC3 values are compared across different substances.

Source: Adapted from OECD guideline 429, 2010

with respect to numbers of animals used and welfare (e.g., no adjuvant is used). The test takes advantage of the primary proliferation of lymphocytes induced in the auricular lymph node following topical application of a sensitizer to the ear; the induced proliferation is proportional to the dose applied and to the potency of the sensitizer. The main points of the LLNA are given in Case Study 9.2.

While the test is a considerable improvement over the older protocols using guinea pigs (it gives a quantitative and dose-related response), it is not necessarily seen as a replacement, although that may effectively be the case. It is not infallible, and it is possible to induce false positives, which are discussed in the following section.

PITFALLS IN IRRITATION AND SENSITIZATION

Observation of the reaction to treatment in the guinea pig sensitization and rabbit dermal irritation tests is not fully quantitative and requires a degree of subjective judgment and, by implication, skill and experience in scoring the results. With the gradual adoption of quantitative methods such as the LLNA, this problem is being overcome for sensitization. However, this test is not infallible, and false positives may be produced by irritant chemicals, although this is a question of continuing debate. Assessment of irritation may be a problem where the test compound is opaque or colored and obscures local reaction.

With the dermal and ocular irritation tests in rabbits, the visible clinical signs are readily scored, but clinically relevant signs such as itching, pain, and other invisible effects cannot be assessed other than by inference from the behavior of the animal.

McGarry (2007) reviewed the LLNA with respect it its use for Registration, Evaluation, Authorisation and restriction of Chemicals (REACH) and reported concerns about the influence of the vehicle on the proliferation of the lymph node cells, increasing or decreasing it. In addition, she said that concerns have been expressed that the test has not been validated for formulations, such as emulsions, suspensions, and mixtures. She also raises the false positives given by some irritant substances. It is quite clear from the data reported that vehicle can have a profound effect on the results of the LLNA, but McGarry points out that this can also be a problem with other test systems for sensitization and in human patch tests. Given the importance of vehicle choice in dermal toxicology, it is not surprising that this should be a significant factor in the conduct of the LLNA. However, she concludes that the vehicle may influence the apparent potency of the sensitizer—that is, the concentration at which effects are expressed may be higher or lower depending on the vehicle used—but the identification of hazard is not generally affected. In an analogous way, it is possible that the components of formulations may influence the responses to sensitizers in the mixture. A further problem with mixtures or formulations is that they are often wholly aqueous and so are not so suitable for use in the LLNA. Irritant substances can produce false positives, and this is consistent with the use of the irritant sodium lauryl sulfate to enhance sensitization in the Magnusson–Kligman test. In fact, irritants can produce false positives in guinea pig assays as well as with the LLNA.

McGarry's excellent review highlights a number of aspects of the introduction of new methods. With the benefits of hindsight, it is evident that the LLNA is going through similar stages of development and understanding as the Ames test for bacterial mutagenicity. At first, the new method is seen as a major step forward; this is usually associated with enthusiasm, but experience in the early stages is limited by the inevitably short history of use. The next stage is a cooling period while experience with the new method develops and some shortcomings become evident; the inevitable consequence is that the method is compared unfavorably to the previous techniques, which become a gold standard, even if they are associated with the same problems. All methods, new or old, are tools to be used in safety evaluation, and like any tool, they have strengths and weaknesses; misuse is likely to give a skewed or erroneous result. True utility comes with understanding. The problems encountered with the LLNA are not specific to it but are found with any such test system. The conclusion is that the LLNA is, overall, an improvement on previous systems for sensitization and, in the absence of viable *in vitro* tests systems of sensitization, will remain the method of choice for the foreseeable future.

Although no single *in vitro* test is likely to replace animal studies, their use will inevitably increase, particularly by the use of batteries of tests to assess the individual processes that make up each overall human endpoint. It should be noted that if the data from a test are poorly understood, supplementing them with more uninterpretable data will not enhance the overall assessment. A plethora of mechanistic work from poorly chosen tests will merely confuse.

SUMMARY

Although often placed under the same umbrella of local tolerance, irritation and corrosion have different mechanisms and sensitivities to sensitization. However, for both forms, a number of points should be considered:

- Knowledge of the form and formulation of applied product coupled with the location of application are vital for understanding potential risks.
- Concentration is as important as the total dose; a product thinly applied over a large area may show different effects from one applied generously at a single small area.
- The condition of the skin at the site of application may alter how the compound is absorbed and how the skin reacts.
- *In vitro* assays as an alternative to *in vivo* ones are progressing well, but full regulatory acceptance may still require a small *in vivo* study to demonstrate the predictions.

REFERENCES

Derelanko MJ, Auletta CS. *Handbook of Toxicology*. 3rd ed. Boca Raton, FL: CRC Press, 2014.

McGarry HF. The murine local lymph node assay: Regulatory and potency considerations under REACH. *Toxicology* 2007; 238: 71–89.

OECD. Guidelines for the Testing of Chemicals. Test 404: Acute Dermal Irritation/Corrosion. 2015. Available at http://www.oecd.org.

OECD. Guidelines for the Testing of Chemicals. Test 405: Acute Eye Irritation/Corrosion. 2012. Available at http://www.oecd.org.

OECD. Guidelines for the Testing of Chemicals. Test 406: Skin Sensitisation. 1992. Available at http://www.oecd.org.

OECD. Guidelines for the Testing of Chemicals. Test 429: Skin Sensitisation: Local Lymph Node Assay. 2010. Available at http://www.oecd.org.

OECD. Guidelines for the Testing of Chemicals. Test 430: In Vitro Skin Corrosion: Transcutaneous Electrical Resistance Test. 2015. Available at http://www.oecd.org.

OECD. Guidelines for the Testing of Chemicals. Test 431: In Vitro Skin Corrosion: Human Skin Model Test. 2016. Available at http://www.oecd.org.

OECD. Guidelines for the Testing of Chemicals. Test 435: In Vitro Membrane Barrier Test Method for Skin Corrosion. 2015. Available at http://www.oecd.org.

OECD. Guidelines for the Testing of Chemicals. Test 439: In Vitro Skin Irritation: Reconstructed Human Epidermis Test Method. 2015. Available at http://www.oecd.org.

10 Determination
Environmental Toxicology and Ecotoxicology

INTRODUCTION

The environment may be loosely defined as the surroundings and conditions in which we live. Environmental toxicology is the study of toxic chemicals within that environment and the effects that they have on humans and populations. Ecotoxicology is specifically the study of environmental toxins on the flora and fauna that make up an ecosystem of a particular environment. The former has an implied human slant, while the latter is oriented more toward the effects of chemicals on the natural ecosystem studied as a whole, which, of course, includes humans.

Evaluation of the impact of individual chemicals on the environment is becoming increasingly important in regulatory terms. While the principal class of chemicals assessed for environmental toxicity has traditionally been agrochemicals (or plant protection products, as they are called in Europe), there has been a gradual increase in the awareness of the entry of pharmaceuticals into the environment as a result of treating both farm animals and human patients. This expansion of interest has increased with the implementation of the Registration, Evaluation, Authorisation and restriction of Chemicals (REACH) European initiative on chemicals. This vast piece of legislation lays down data requirements for chemicals, such that the extent of environmental assessment increases as the annual production (or import into the European Union) increases. The data requirements for REACH are covered in Chapter 19 (Focus Boxes 19.1 and 19.2).

Environment assessment is effectively relevant to all chemical products, whether they are agrochemicals, human or veterinary pharmaceuticals, or industrial chemicals. The only group of chemicals or products not caught up in this all-embracing net is cosmetics; however, it is likely that the major ingredients will be registered under REACH and covered there.

Furthermore, while the point of entry of a chemical to the environment may be relatively well defined—for instance, through spraying a field, treating a herd of cattle, or prescribing oral contraceptives to millions of people (leading to discharge into the sewage systems of the world)—the subsequent distribution of that chemical may be effectively worldwide. Broadly, the more stable a chemical is, the longer it will persist, thus facilitating wider distribution potentially to all parts of the world. For instance, bisphenol A and perfluorooctanoate have long half-lives and are found worldwide. The problem is that these stable, often lipophilic, molecules can bioaccumulate up the food chain, resulting in disproportionately high concentrations in groups such as marine predators. These high levels have been shown to have

239

significant toxicological effects on aspects of the marine mammalian life cycle, particularly reproduction (Vos et al. 2003).

From this discussion, it becomes evident that some environmental exposures arise intentionally, and some are unintentional. The impact may be local or, ultimately, global; the effects of release may be expected or unforeseen. For example, while it may have been predicted that treating cattle with the antiparasitic ivermectin might lead to release of unchanged drug to the local environment, it was not foreseen that it would affect the longevity of cowpats by affecting the viability of the insects that normally remove the dung. Also, the use of diclofenac in cattle in India was not predicted to result in a catastrophic decline in the population of vultures that, feeding on treated carcasses, ingested substantial doses of the drug and so died. Some releases of chemicals to the environment have become apparent after relatively poor regulation of their use combined with a lack of understanding of their potential impact. Thus, the insecticide DDT affects the thickness of eggshells in predatory, birds and organotin compounds are associated with imposex in dog whelks along shorelines (see Pollution, Routes of Entry, and Environmental Absorption, Distribution, Metabolism, and Elimination seciton).

RELEVANCE OF ENVIRONMENTAL ASSESSMENT TO CHEMICAL DEVELOPMENT

There is no escape from the relevance of environmental toxicology to chemical development in the twenty-first century. In fact, as discussed, it is likely to become more important, as REACH has extensive requirements for environmental assessment for existing chemicals, which need to be addressed in order to achieve registration. It has to be said, however, that the requirement for assessment may not translate into a need to test, provided adequate justification for not doing so can be made. The data to support a nontesting strategy may include proof of lack of discharge to the environment and physicochemical data, supported by a thorough literature search. New chemicals will face similar hurdles.

While agrochemicals have been subject to environmental assessments for many years, they are relatively novel for pharmaceuticals but are required in both the United States and the European Union. There is increasing awareness of the presence of drugs in the environment and the possible effects that they may have. For example, Zuccato et al. (2006) reviewed the presence of pharmaceuticals in the environment in Italy, examining the reasons for their presence and their effects. From the literature reviewed, it was clear that environmental contamination by pharmaceuticals is widespread. However, the distribution of individual pharmaceuticals in the environment is influenced by factors such as differences in the disease prevalence, prescribing habits, or market forces. The main sources of contamination are patients or treated animals. To a certain extent, environmental load can be predicted from sales figures and the rates of metabolism in the target species. Pharmaceuticals are intended to have defined pharmacological actions at low concentrations and so pose potential ecotoxicological threats to the environment and, ultimately, to humans. This area of toxicological endeavor, although comparatively new, is evolving rapidly. As with persistent pesticides, which were misused in the 1960s and 1970s, there has been recent realization of this potential problem, the extent and significance of which are gradually becoming apparent.

For household products, it is inevitable that many will find their way into the sewage system and thence into the environment when the sewage has been treated. For some industrial chemicals, deliberate environmental release may be unlikely, but it should be borne in mind that accidents cannot be planned and catastrophic releases occur every so often. In some cases, the impending pollution can be predicted, for instance, when a retaining system for water discharged from a mine begins to break down and threaten a major river system.

There are few chemical groups for which environmental assessment is not relevant; for most, it is becoming increasingly important.

GENERAL PRINCIPLES OF ENVIRONMENTAL TOXICOLOGY AND ECOTOXICOLOGY

One of the keys to environmental toxicology is the behavior of chemicals in the environment, apart from any inherent toxicity. There are similarities between the pharmacokinetic behavior of compounds in ecosystems and their behavior in individual organisms; both are governed by the interaction of the physicochemical characteristics of the chemical with the subject. The overall response is affected by the extent and duration of exposure and by the way the chemical is handled in the subject—in this case, the ecosystem. In ecotoxicology, the main subjects for studying effects are populations (individual species), biocenoses (communities of associated species), and whole ecosystems comprising a larger number of species, habitats, and functional features. In contrast to toxicology, where variability is limited to differences within one species or controlled differences in experimental technique, ecotoxicology deals with much greater diversity due to the presence of many species interacting in an essentially uncontrolled manner. These interactions eventually achieve equilibrium in a delicate balance, which is a function of interdependencies between the different components of the ecosystem. Effects on one species or group of species (e.g., insects) can have significant effects on a whole ecosystem. Changes in the ecosystem lead to adjustment and re-equilibration, which may have far-reaching effects out of proportion to the size of the original change. In the Gaian concept of James Lovelock, the Earth is seen as a single organism that regulates (and heals) itself. There may be truth in this, but on current performance, we are outdistancing the reparative processes and are in danger of compromising our own existence, of course at which point, the healing process may begin.

POLLUTION, ROUTES OF ENTRY, AND ENVIRONMENTAL ABSORPTION, DISTRIBUTION, METABOLISM, AND ELIMINATION

There is much heated debate about pollution and what constitutes pollution or a pollutant. A potentially useful definition of a pollutant is that of a chemical that has exceeded normal background levels and that has the potential to cause harm—always remembering that the potential for harm increases with concentration. Definitions of pollution refer to noxious chemicals discharged into the environment, and it has been suggested by some that pollution started when primitive humans lit the first fires. It is probably better to consider pollution as an excessive discharge into the environment that persists or accumulates to the extent that it causes harm. This may be considered on a local or national level as appropriate. Pollution as a result of fires, particularly coal fires, was noted before AD 1500 in London, and efforts were made to restrict the use of coal at intervals subsequently, culminating in the clean air acts in the United Kingdom in the second half of the twentieth century. For all the furore at the time and subsequently, it is unlikely that these localized and transient episodes of coal fire–induced smog have had any long-term adverse effects on the environment as a whole. Discharge into the environment as a result of industrial activities is easier to define as harmful in the long term. This would include unintentional discharge from mining reservoirs of water with high levels of heavy metals or other pollutants. These elements are present naturally in the environment but at lower, generally nontoxic concentrations; sudden high levels carry significant risk for water supplies, fisheries, and ecosystem well-being. In environmental toxicology or ecotoxicology, there are two major aspects for investigation: the environmental fate of a substance, i.e., what happens to a substance once it is introduced into the environment, and the ecological effects on the environment or ecosystem that follow its discharge.

In common with general toxicology, there are several routes by which a chemical enters an ecosystem, and there are different compartments analogous to the organs in the body of an animal, into which the chemical may be distributed. Following entry to the environment, the fate of a chemical can be described by the processes of absorption, distribution, metabolism, and elimination (ADME) in a broad analogy to similar processes in individual animals. As in animal pharmacokinetics, the chemical may be sequestered into individual compartments; in animals, this might be the bone or adipose tissue; in the environment, it might be a clay soil. In either case, sudden acceleration of release can result in harmful concentrations.

The major compartments of the environment may be summarized as water, air, soils, and flora and fauna (wild and domestic). Chemicals enter these via many diverse routes, including (intentionally) agrochemical spraying or illegal discharge, or (unintentionally) as air pollution from industrial fires, smokestacks, or vehicles or as run off into waterways from industrial sites or intensive farms. Many routes are not considered at the evaluation stage and can have had unexpected effects. One of the earliest noted examples of environmental effect resulting from agricultural use of a drug was ivermectin, which is excreted in the feces and increases cowpat life by killing the insects responsible for their degradation (Wall and Strong 1987; Madsen et al. 1990).

A more recent example is given by diclofenac, a nonsteroidal anti-inflammatory drug given to cattle in India. This has been blamed for a huge decline in the population of vultures, which has resulted in other effects in the ecosystem. Because cows in India are sacred, they are not eaten, and dead cattle were left for the vultures, which thus carried out an important role in public sanitation by removing carcasses. Reduced numbers of vultures due to the use of diclofenac resulted in increased numbers of rats and wild dogs; this has been associated with increased disease and a public health crisis as the differences between vulture and canine physiology mean that the dogs become carriers of disease. Dogs are a favorite prey of leopards, and the increased number of wild dogs has been associated with an increase in the leopard populations, and then increased leopard attacks on children.

The system into which a chemical is discharged—air, soil, or water—is important in determining the significance of the discharge and the extent to which it can be distributed through the ecosystem. In terms of distance of transport, the greatest distances are found with air, while water has the greatest capacity for movement in terms of volumes. Discharge into soil will ensure the lowest distance and the lowest volume of transport. While both soil and water have great potential as sinks for pollution, water pollution has the greatest potential to threaten populations due to the ease with which substances are transported. This threat can become global because of the effects of bioaccumulative toxicants on marine mammals (Vos et al. 2003).

As indicated above, clay soils have a high capacity for adsorption of some chemicals, which become tightly bound; as a result, their adverse actions are attenuated. Of course, in time, they will be slowly released from the clays into the rest of the ecosystem, giving a prolonged low-level exposure of organisms or a prolonged opportunity for degradation. Peat soils, on the other hand, do not have such adsorptive capacity, a contrast that was noted after the Chernobyl accident. The differences in binding of cesium-137 between the clay soils in the lowlands and the acid peat soils in the hills affected the amounts that were available, and this was reflected in the radioactive content of crops and livestock. In contrast to organic chemicals, metals—particularly heavy metals—are not degraded, and detoxification is dependent on their removal, irreversible binding, or dilution. Complex molecules may be broken down and eliminated from the ecosystem at greater or lesser rates according to chemical class. Simple carbon compounds are easily biodegradable, but halogenation may well prolong this process into years, as seen with molecules such as the dioxins and organochlorines such as DDT.

One important factor to consider in ecotoxicology is the ability of some chemicals to concentrate as they progress up the food chain until concentrations at the higher levels become toxic. This is the effect seen with the organochlorine DDT, long banned from the "developed world." The long half-life of DDT and its metabolite DDE, due to high lipid solubility and slow metabolism, results in increasing concentrations up the food chain until there is a clear effect, most easily seen in carnivorous birds. In the peregrine falcon, for example, eggshell thickness was diminished with increasing DDT exposure to the extent that breakages in the nest increased and the population declined.

The plight of marine mammals highlights the processes and effects of bioaccumulation and the difficulties of studying its effects. Marine mammalian toxicology is a discipline that has to cope with a number of challenges that make laboratory-based toxicology in rodents look easy. The subject species range in weight from a few kilos to tens of tons, are widely dispersed, are often rare, and are found in the largest continuous ecosystem on the planet. This area of toxicology has emerged from its early beginnings of simple analysis of tissues for contaminants into an era where detailed investigation of the effects of the contaminants is being addressed in an increasingly multidisciplinary manner. However, increasing fragmentation into disciplines may mean that it becomes more difficult to achieve a holistic view of marine toxicology (as it does with any other area of toxicology). This is true of marine mammal toxicology per se and of the marine ecosystem as a whole.

Marine mammals are a special case in toxicological terms, because they are at the top of the food chain. In addition, they have large blubber reserves, giving them the unlooked-for ability to accumulate lipophilic, persistent compounds such as polychlorinated biphenyls. Consequently, their offspring are subjected to high levels of these compounds from birth through their lipid-rich milk and are possibly predisposed to immunological deficiencies, with increased susceptibility to infection and tumors. Another aspect of this area of investigation is the blurring of definitions that sometimes occurs. For instance, it is quite easy to define whales as marine mammals as they never come ashore and also some species of seal as marine mammals as they come ashore only to breed. While these groups fit easily into the definition of *marine mammal*, as does the sea otter, other mammals such as the polar bear and more obviously littoral land-based animals may not be so easily categorized, though they are still subject to a marine environment and its contaminants. Vos et al. (2003), whose seminal book formed the basis of these two paragraphs, suggested 20 recommendations as to how this field of toxicology should be advanced. These include—as a nearly random choice—the integration of multiple approaches, compilation and dissemination of information, use of surrogate animal models (although if we have trouble justifying the use of a rat to evaluate safety for humans, the same process for safety evaluations for whales is likely to be even more fraught), understanding processes linking exposure to effects, and "understanding blubber physiology and estimating total body burdens of lipophilic contaminants."

Although pollution has traditionally been associated with molecular chemistry—solutions, emulsions, or aerosols, for example—or with microscopic particles, it is increasingly evident that larger pieces of waste may be significant too. There has been an apparently tacit distinction between litter and pollution. However, with the emergence of plastic as a major source of chemicals such as bisphenol A leaking into the environment from plastic waste, it is clear that pollution is not simply at the level of molecules, emulsions, or microscopic particles. For example, albatrosses pick up plastic during their foraging expeditions and regurgitate it into their chicks when they feed them, and it is possible for swallowed plastic waste to have fatal consequences in the guts of mammals.

FACTORS IN TESTING FOR ENVIRONMENTAL EFFECT

With the realization that environmental release of chemicals can have far-reaching effects, whether of unusually large amounts of endogenous (natural) substances or synthetic chemicals, has come the acceptance of the need for testing for potential adverse effects on the environment. The emphasis on such testing is inevitably on compounds intended for agricultural use as pesticides, but there have been initiatives to test pharmaceuticals for their environmental impact, and there is a continuing debate about the environmental effects of estrogenic compounds.

The problem of testing for ecotoxicity is that the scope for subtle change is much greater than in a single organism or test species and it is impossible to test every aspect of an ecosystem except in very large and complex experiments ("mesocosms" or field trials). The objective of ecotoxicity testing is to predict the behavior of a chemical in the ecosystem and to assess the potential for adverse effects in the situations under which it will be released. The major difficulty with this objective is the enormous diversity of the environment and the selection of representative test systems. Inherently, one species of fish cannot be considered to be completely representative of all other fish. Equally, an aquatic herbivorous invertebrate cannot be representative of an aquatic herbivorous mammal, although both are at the same trophic level in ecological terms. Because bees are not harmed by an agrochemical, it should not be assumed that it will be nontoxic to other less obviously beneficial insects (however one defines *beneficial*).

It seems probable that in the longer term, the ecotoxicological impact of genetically modified crops could be more significant than their immediate adverse effects on consumers. At this point, the extent of or scope for interaction of chemicals and the natural world, of which we are a part, becomes a topic of concern. The precise definition of ecotoxicity therefore becomes important. Clearly, if there are widespread effects on beneficial (or desirable) insect populations due to insecticidal gene expression in crops, this is a toxic manifestation of the crop and can be classified as ecotoxicity. Loss of a species is a clear-cut event with imponderable impact; if an effect is limited to a shift in populations of plants or animals due to cross-breeding, it may be more difficult to describe it as toxicity, although such an event may indeed be entirely adverse environmentally.

In assessing the potential for environmental effects, there are two roughly definable areas of investigation: those that are dependent on the physicochemical properties of the material that determine environmental fate and those that examine the potential for ecological effects. The first has to consider

- Physicochemical characteristics—partition coefficient (water and oil solubility), adsorption and desorption characteristics, volatility
- Fate and behavior—relative persistence, liability to abiotic degradation, final fate, rate and route of elimination

The potential for ecological effect is investigated via

- Effects on bacteria and other degrading organisms including assessment of biological oxygen demand
- Effects on higher organisms, such as bees, earthworms, fish, and birds, with extrapolation from laboratory species to environmentally relevant organisms

Although such assessments are made before the release of novel chemicals, there is the continuing need for monitoring after sales of the chemical have started. Such studies are the ecological equivalent of epidemiology and have similar weaknesses and uncertainties, unless the effects are unusual and clearly attributable to exposure to the suspect chemical. Thus, the thinning of eggshells in birds of prey was attributed to organochlorine pesticides through a series of field and mechanistic studies that together produced a body of evidence that was incontrovertible. The presence in the environment of synthetic estrogens is much more difficult to link to decreased sperm counts in men due to the inherent variability of the data and the different interpretations that are possible.

TEST SYSTEMS AND STUDY TYPES FOR ECOTOXICOLOGY

Test systems for assessment of ecotoxicology have been chosen on pragmatic grounds in helpless acknowledgment that assessment in every relevant species would be impossible. The following descriptions of test species and study types have been put together with reference to the Organization for Economic Co-operation and Development (OECD) guidelines and to *Principles of Ecotoxicology* by Walker et al. (2001). The intention is to give a flavor of the test species and studies conducted, rather than to attempt a definitive description. Furthermore, the various tests on algal growth and bacterial degradation are not considered here.

In many single-species studies, the objective is to determine the LC_{50} and a no-observed-effect concentration (NOEC). The LC_{50}—the median lethal concentration—is equivalent to the LD_{50}, seeking to determine the concentration at which 50% of the test system is killed. The concentration concerned may be that in water or in a diet. The values for these measures, particularly with tests conducted in water, are greatly influenced by the conditions under which the experiments are conducted. There seems to be a lack of standardization of some aspects of these tests—for instance, algae used as feed or in the characterization of important test components such as artificial soils. These may lead to deficiencies in trace elements or to other test parameters that have an unsuspected influence on the test data. It is probable that this situation will improve over time with development of knowledge in these areas, but in the meantime, it is equally good to be aware of the possibility of these problems and the difficulty caused in data interpretation, particularly when comparing data between laboratories. The majority of tests are single-species experiments, conducted in isolation; there is a brief discussion on mesocosm studies at the end of this section. Further details on study designs may be found in the appropriate guideline.

ECOTOXICOLOGY *IN VITRO*

As would be expected, there is a continuous effort to develop *in vitro* tests that have relevance to ecotoxicology. The following selection is intended to illustrate the breadth of test systems being employed. As with other areas of *in vitro* toxicology, this field is dogged by the obvious differences between the responses in the laboratory of a cell line derived from a target species and those of the complete organism in the environment.

Segner (2004) reviewed the use of cytotoxicity assays with fish cells as an alternative to acute lethality tests with fish and indicated that the concentration of chemical that would result in 50% mortality in fish in 96 hours (the LC_{50} value) *in vivo* could not be predicted from the values determined *in vitro*. The use of cell lines from relevant organisms is a recurrent theme but suffers the same disadvantages as the use of mammalian cell lines in other branches of toxicological assessment. The range of endpoints studied has similarities to those of other *in vitro* tests too, with parameters such as neutral red exclusion and enzyme leakage into the culture medium playing a significant role.

A promising approach might be to use a battery of tests to assess the potential for ecotoxicity, as used by Zurita et al. (2005) in an evaluation of diethanolamine, which is widely used as an intermediate and as a surfactant in cosmetics, pharmaceuticals, and agrochemicals. This investigation used systems representative of four trophic levels and included bacterial bioluminescence in *Vibrio fischeri*, algal growth inhibition in *Chlorella vulgaris,* and immobilization of *Daphnia magna*. A hepatoma fish cell line was used to study a range of endpoints. The fish cell line was the least sensitive of these systems, while *D. magna* and *V. fischeri* were the most sensitive. The authors concluded that diethanolamine was not expected to produce acute toxic effects in the aquatic environment. This seems to be backed up by acute toxicity data in fish and invertebrates (albeit somewhat elderly and variable) listed in the International Uniform Chemical Information Database (IUCLID) chemical data sheet. However, the authors hedged their bets by suggesting that chronic or synergistic effects with other chemicals were possible. This caution neatly encapsulates the ecotoxicological dilemma of testing a single species in the presence of single chemical, while isolating both from environmental reality.

Unsurprisingly, the most complete organism, *D. magna*, was the most successful in these tests, underlying the general rule of thumb that the more complex the test system, the more likely is it to be successful as a predictive tool. In this vein, the frog embryo teratogenesis assay using *Xenopus* has long been used in reproductive studies, though not at a regulatory level. Hoke and Ankley (2005) looked at the utility of this assay in ecological risk assessments. However, they indicated that this assay was relatively insensitive in comparison with acute toxicity data from tests with traditional aquatic test systems or with other amphibians.

INVERTEBRATES

D. magna are tested to assess effects on mobility and reproduction. In the immobilization test, the percentage of *Daphnia* that are not swimming after 24 or 48 hours is assessed for each concentration of test chemical. For the reproductive test, young *Daphnia*, less than 24 hours old, are used, and the total number of offspring produced by each animal that survives the test is assessed against the controls. The clone of *Daphnia* that is used is important as there are differences in sensitivity, which make comparison between experiments difficult. The algae used to feed the *Daphnia* can have an important effect on the test results, and in the absence of standardization, it may be difficult to compare results between laboratories.

Earthworms are studied by exposure to test chemicals in containers of artificial soil, with an assessment of mortality 7 and 14 days after application; at least two concentrations—one with mortality and one without—are examined, with appropriate controls. Experiments may include an assessment of reproduction, which has been found to be a more sensitive marker of effect in some cases. Another experimental procedure exposes the worms to the test material on moist filter paper.

Bees (used for agrochemicals only) are subject to acute oral and contact tests. Oral toxicity is assessed by feeding the bees with different concentrations of the test chemical, with mortality checks up to 48 hours. Contact toxicity is assessed by direct application to the thorax, which can also be used with other insects. Other invertebrates that can be used in test programs include wood lice, springtails, and marine arthropods from sediments, such as those found in estuaries and other areas with high pollution loading.

VERTEBRATES

Various species of fish are used, including rainbow trout, fathead minnow, zebra fish, and bluegill sunfish [for the US Environmental Protection Agency (EPA)]. Tests may be static (where the water is unchanged for the duration of the test) or semistatic (the water is changed at intervals), or flow through where the water is changed constantly. The duration of exposure is generally for up to 14 days, although shorter exposures are used to determine the LC_{50}.

Birds such as quail, mallard duck, pheasant, or partridge are used in a variety of tests, including dietary tests, which may use five test diets with increasing test substance concentration over a 5-day period. A 3-day off-treatment period follows.

The majority of these test systems are used in single-species studies with the classic toxicological design where controls and several treatment groups are examined, usually with the added dimension of time as a factor.

MESOCOSMS AND FIELD TESTS AND STUDIES

Single-species testing is, in some ways, analogous to *in vitro* toxicity test systems in that only a part of the ecosystem or animal is being examined and interrelationships between species or organs cannot be easily predicted. The problem of single-species testing can be partially circumvented by the use of mesocosms or field tests. While a mesocosm is an artificial ecosystem of a manageable and controllable size, field tests use preexisting areas in the environment for studies with chemicals such as pesticides or to determine the causes of observed environmental effects. The latter is analogous to an epidemiological study in humans. As always with toxicological investigation, size and complexity are associated with significant cost, and these experiments or investigations are inevitably expensive and time consuming.

Mesocosms are large-scale experiments that attempt to reproduce a section of the ecosystem in miniature, usually including a pond or water system such as an artificial stream. The use of "miniature" in this context is deceptive, however, because these may have a volume of 50 m³ or greater, with a surface area of up to 25 m². The advantage of both these test types is that they have a number of different species, which can interact in a way similar to that in the real world.

A mesocosm is constructed, in an appropriate container, according to the experimental duration and objective; longer experiments need larger systems. The components of the system and its origins and quantities are defined in guidelines (e.g., those from the OECD). All components, such as the sediment (with indigenous fauna and flora), fish species, plankton, and plants, are carefully characterized and sourced so as to be free from confounding contaminants. Before addition of the test chemical, the system is allowed to equilibrate and mature, the duration of this being proportional to the size of system. Experimental duration is influenced by the type of chemical being tested, persistent chemicals requiring longer examination than those that are readily eliminated by biotransformation or degradation. Several mesocosms may be set up to examine different doses of the test chemical, in which case the reproducibility of the system becomes critical to interpretation of the data.

Field tests, by definition, do not use constructed locations, but the experimental parameters are still carefully defined before the test is undertaken. The areas covered by field tests may be substantially larger than with mesocosm experiments and are typically performed for pesticides, which may be applied at doses expected to be toxic. Measurements made are dependent on the chemical class, habitat, type of agricultural system, and application method. They include determination of persistence of the chemical in soils, water, and the flora and fauna, including an estimation of any bioaccumulation risk. Study of population change in response to the application is an important aspect of these trials.

Changes in population that are noted independently of field tests are the trigger for field studies, the difference between a study and a test being that no chemical is deliberately applied in a study. As with an epidemiological study, the object is to determine the cause of an observed difference from expectation. Such studies depend on the initial observation—development of male sex organs in female dog whelks or declining reproductive performance in seals—and the painstaking investigations that follow. These include precise definition of the problem and analysis to determine the presence or not of abnormal chemical residues such as organic tin compounds or polychlorinated biphenyls, either in the affected species or in their environment. The relationship between the effect and the proposed cause is usually only accepted on provision of a credible toxicological mechanism of effect or an incontrovertible association that is not present in other locations. Frequently, as in many other walks of life, a strong or circumstantial association between a chemical and an effect is not enough to offer proof to authorities, especially if money is involved in rectification, either directly in cleanup costs or in increased costs for a profitable industry.

There is evidence that morphological change, resulting from pollution, may be counteracted by natural selective forces. Thus, populations of the peppered moth, *Biston betularia*, responded to carbon deposits on trees by increased proportions of a darker variant, the incidence of which has declined with declining carbon-based pollution. Similarly, there has been evidence that the development of male sex organs in female dog whelks (imposex, which hinders reproduction) is being circumvented through selective pressures. These population responses are apparently based on existing genetic diversity in the normal population, and it seems unlikely that this type of adaptive response to morphological change would be readily duplicated in the case of biochemical effects on basic molecular function.

An important aspect of field studies is the use of biochemical or morphological markers of effect to assess exposure. These may be easy to assess, as in the presence of imposex in female dog whelks exposed to organic tin compounds, or more challenging, as in the analysis of carcasses for chemical residues. Classic markers have included the thinning of eggshells in peregrine falcons, which was the mechanistic response to exposure to DDT and its major metabolite DDE. The routine monitoring of marker species can also be used in the assessment or development of local pollution. Increased metabolic capacity in the livers of river trout may imply exposure to excess concentrations of xenobiotics. Such hypotheses may be confirmed by analysis, and this could be extended to the carcasses of predatory birds or mammals such as herons or seals. These are markers of effect, and the distinction must be drawn between the presence of a chemical and its effects on individual species and its impact on the ecosystem as whole. To determine the impact of a chemical, it is necessary to carry out detailed population studies. Crucially, it is important to know what the population distribution was *before* the pollution occurred or to know the situation in an *identical* area in which no pollution has (yet) taken place.

ENVIRONMENTAL ASSESSMENT OF AGROCHEMICALS

Although environmental studies are now a part of pharmaceutical development, they were first conceived for agrochemicals and have reached a state of considerable refinement. While pharmaceuticals may be expected to reach the wider environment indirectly through the sewage system or, occasionally, by accidental spillage into a river or watercourse, pesticides are deliberately applied to large areas of the outdoors and so have much wider environmental access and potential ecotoxicological effects.

The studies conducted (often termed fate and behavior studies) are aimed at determining the fate of a chemical in the environment in terms of distribution, degradation (and mechanisms), and elimination from the ecosystem; this process is broadly analogous to the ADME studies conducted for pharmaceuticals. Any indication that a chemical will persist unduly in the environment is a flag for more extensive (and expensive) studies and more difficult justification of its use. Predicted environmental concentrations (PECs) are calculated and persistence is assessed; degradation products are assessed to ensure that they do not have any adverse effects that add to those of the parent compound. The PECs for parent and degradation products are used to assess exposure of nontarget species in soil and water, potential contamination of drinking water or groundwater, and potential effects in crops, which follow on from the treated crop.

The environmental distribution and breakdown of pesticides are dependent on factors such as the physicochemical characteristics of the chemical, the climate and weather conditions at the time of and following its use, and how it is used. As for a drug, the degradation can be described by a half-life, dependent on adsorption to soil, solubility, and breakdown by organisms such as bacteria. An indication of the mobility of a pesticide—how easy it is to elute it from soil—can be gained from the adsorption coefficient (K_{OC}) value, which gives a measure of adsorption affinity to soil. Mobility and degradation of pesticides differ between soils and are influenced by temperature and the amount of water in the soil. The concentration of a pesticide in the environment is dependent on the rate of application, the frequency of use, and the pattern of usage, and these have to be taken into account in the overall assessment.

Potential toxicity to wildlife is assessed by standardized laboratory tests using nontarget organisms such as birds, bees and other insects, and fish and aquatic invertebrates; effects on environmental bacteria are also assessed. Values for LD_{50} and LC_{50} are derived together with No Observed Effect Level (NOELs) and NOECs, and these are compared with the PECs. A predicted no-effect concentration (PNEC) is derived from the PEC to arrive at a tolerable concentration that should be associated with no effects. The overall goal is an indication of the overall toxicity of the material compared with the PECs to get an estimate of toxicity set against likely exposure levels. Internationally agreed-upon trigger values are used by the European Commission to decide whether the risk is acceptable or not.

While some of the studies are laboratory based and relatively easy to control, some are much larger and based outside in prepared containers or in the field. The container studies include microcosm and mesocosm studies; other studies may make use of artificial streams.

Ultimately, one of the species that could be exposed to pesticides or other agrochemicals is man, and it would seem sensible to obtain some information on the ADME of these substances in human volunteers. There has been much debate about the ethics of human studies with agrochemicals, and there has been considerable resistance to this, even to the extent of not using data when it has been generated. This does not seem to be entirely sensible. However, the recent advent of microdosing studies used for pharmaceuticals, where very small doses of radiolabeled compound are given to volunteers, may be relevant to agrochemical development. The use of small doses is consistent with normal expected exposure to pesticides, and it seems likely that these studies with their complex and expensive analytical techniques will prove to be more easily justifiable for low doses of pesticides than for pharmaceuticals, which are usually given at much higher doses than those studied in such experiments.

ENVIRONMENTAL ASSESSMENT OF PHARMACEUTICALS

Depending on the region of interest and the type of substance, there may also be a requirement to evaluate the environmental impact of pharmaceuticals. Certain classes of compound are exempted from this, including vitamins, peptides or proteins, carbohydrates, vaccines, and herbal products, on the basis that they are unlikely to pose any significant environmental risk. In Europe, this is a two-phase procedure, in which the first estimates the environmental exposure to the drug and the second assesses fate and effects in the environment. The estimation of environmental exposure undertaken in phase I is based entirely on the drug itself rather than on any metabolites or taking route of administration into account; it is also assumed that the major route of entry to surface water will be via the sewage system. Data relating to the dose per patient, the percent market penetration (to give an idea of how many people will use it), the amount of wastewater per person, and the dilution are used to produce a PEC for surface water. If this falls below $0.01\ \mu g/L$ for surface water and there are no other environmental concerns, it is assumed that there will be no risk to the environment if the drug is prescribed as expected. Substances that are potential endocrine disrupters and are persistent or highly lipophilic may need to be assessed in any case.

The second phase of the assessment is started if the PEC for surface water is more than $0.01\ \mu g/L$. This phase is itself in two tiers, A and B, in which a first base set of studies is conducted to assess aquatic toxicology and fate and, if indicated, a second tier in which more detailed study of emission, fate, and effects is conducted. The first part of tier A is to look at the fate and physicochemical properties of the drug; this includes an assessment of biodegradability and the sorption behavior of the drug, which is described by the K_{OC}, defined as the ratio between the concentration of the substance in sewage sludge or sediment and the concentration in the aqueous phase at equilibrium. A substance with a high K_{OC}, retained in a sewage treatment plant, may reach the terrestrial compartment via spreading of sewage sludge.

The aquatic effect studies of tier A include long-term toxicity in *Daphnia* sp., fish, and algae to predict a concentration at which effects are not expected; this is the PNEC, which is derived from NOECs determined in the various studies. The ratio between the PEC and the predicted NOEC is evaluated, and if this is less than 1, further testing in the aquatic compartment is not necessary. If this ratio is above 1, further testing in tier B is needed. This phase includes investigation of sediment effects and effects on microorganisms. The concentration of the drug in the terrestrial compartment is calculated unless the K_{OC} is greater than 10,000 L/kg.

For veterinary pharmaceuticals, the guideline places emphasis on veterinary medicinal products that will be used in food-producing animals that may not be individual treatments but may, for example, be used for treating a whole herd or flock. A tacit assumption is made that a substance that is extensively metabolized will not enter the environment. Separate consideration is given to substances used in the aquatic environment, which may enter the wider aquatic environment and those in terrestrial situations. Questions asked in the guideline include one about antiparasitic compounds, which may be a reaction in part to the environmental effects of ivermectin. Antiparasitic agents—but not those acting against protozoans—advance automatically to phase II. If the concentration at which the product enters the aquatic environment is calculated to be less than 1 µg/L or the PEC_{soil} is expected to be less than 100 µg/kg, environmental evaluation of the product may stop at phase I.

Phase II provides recommendations for standard data sets and conditions for determining whether more information should be generated for a given veterinary pharmaceutical. The tests are broadly similar to those indicated for human pharmaceuticals with appropriate adjustment for aquatic and terrestrial compartments. Animals that are reared in intensive conditions and those on pasture are given separate consideration, as are aquatic animals. The end process is calculation of the appropriate PECs and PNECs followed by a risk assessment of the environmental impact.

PITFALLS IN ENVIRONMENTAL TOXICOLOGY

The principal problem of ecotoxicology is the simplicity of the test systems relative to the complexity of the ecosystems and the multifactorial nature of many of the possible adverse variations that may occur. Although the test systems may be good models for individual components of the ecosystem, the specific tests may not be predictive for ecological effects in the target species or groups when they are removed from the relative simplicity of a laboratory environment. Furthermore, it is extremely difficult to assess the significance of change seen in a laboratory environment and to predict effects in the whole ecosystem.

The most complex ecotoxicological experiments attempt to reproduce entire ecosystems in miniature and to examine the reactions of components of this artificial system to the controlled introduction of the chemical. The principal difficulty with this type of test, apart from the eye-watering expense, is that with increasing experimental complexity, it becomes much more difficult to control the many variables. Although an artificial stream is probably a good reproduction of an ecosystem in miniature, it cannot reproduce the wider picture of the whole ecosystem.

A further factor is the likelihood of effects that are attributable to unconsidered relationships, as with the discussed example of the effect of diclofenac on vulture populations in India. Although this type of effect might be predicted by lateral thinking, rigidly regulated testing and data assessment do not readily lend themselves to such thought processes. The interrelationships and codependencies inherent in the ecosystem are not easily assessed a priori but, with the benefits of hindsight, become painfully predictable when the effects are first noticed. Extrapolating laboratory change (will an effect on one species significantly affect the whole ecosystem?) is fraught with difficulty.

SUMMARY

Loosely, the environment is the surroundings and conditions in which we live; environmental toxicology studies the effects of chemicals within that environment; ecotoxicology is the study of chemicals on the flora and fauna that make up an ecosystem of a particular environment. The environment has similarities to an individual animal in that a chemical that is introduced into it (administered) is taken up, distributed, degraded, and eventually, eliminated (ADME). Particular attention is paid to chemicals that may accumulate in the environment, such as polyhalogenated aromatic structures, and reach toxic concentrations in target organisms and that may increase in concentration up the food chain.

- The need for environmental testing has been underlined by the history of chemicals in the environment, such as DDT, ivermectin, diclofenac, and bisphenol A.
- Although the environment may have similarities to a single animal, testing chemicals for environmental effect is more complex than for toxicity in animals; taking the analogy with animals further, the individual species and organism types in the environment might be seen as equivalent to individual tissues in a test animal. The problem is that the diversity of species and organism types is so much greater than the number of tissues in an animal, that testing for effects in all is not practicable except by large and complex studies.
- Much environmental toxicity testing is based on tests in a few classes and species of organism that are assumed to be predictive of wider safety or effect. In addition, the use of *in vitro* and *in silico* methods should be considered.

As with all toxicity testing, the results of environmental tests should be interpreted as predictions and should be seen in the light of the behavior of similar chemicals. The principal pitfall is that the tests are extremely simple relative to the environment itself and it is not always possible to predict effects; the effects of diclofenac given to cattle in India on the vulture population is a prime example.

REFERENCES

Hoke RA, Ankley GT. Application of frog embryo teratogenesis assay-Xenopus to ecological risk assessment. *Environ Toxicol Chem* 2005; 24(10): 2677–2690.

Madsen M, Nielsen BO, Holter P, Pedersen OC, Brøchner Jespersen J, Vagn Jensen K-M, Nansen P, Grønvold J. Treating cattle with ivermectin: Effects on the fauna and decomposition of dung pats. *J Appl Ecol* 1990; 27(1): 1–15.

Segner H. Cytotoxicity assays with fish cells as an alternative to the acute lethality test with fish. *Altern Lab Anim* 2004; 32(4): 375–382.

Vos JG, Bossart GD, Fournier M, O'Shea T. *Toxicology of Marine Mammals*. London: Taylor & Francis, 2003.

Walker CH, Hopkins SP, Silby RM, Peakall DB. *Principles of Ecotoxicology*, 2nd ed. London: Taylor & Francis, 2001.

Wall R, Strong L. Environmental consequences of treating cattle with the antiparasitic drug ivermectin. *Nature (London)* 1987; 327(6121): 418–421.

Zuccato E, Castiglioni S, Fanelli R, Reitano G, Bagnati R, Chiabrando C, Pomati F, Rossetti C, Calamari D. Pharmaceuticals in the environment in Italy: Causes, occurrence, effects and control. *Environ Sci Pollut Res Int* 2006; 13(1): 15–21.

Zurita JL, Repetto G, Jos A, Del Peso A, Salguero M, López-Artíguez M, Olano D, Cameán A. Ecotoxicological evaluation of diethanolamine using a battery of microbiotests. *Toxicol In Vitro* 2005; 19(7): 879–886.

11 Interpretation
Basic Principles

INTRODUCTION

Interpretation can sometimes be seen as a dark art; the intention of this chapter is to give guidelines on the basic principles of interpretation of toxicological data and to indicate an overall philosophy. Interpretation is distinct from prediction. For example, in attribution of cause/effect before versus after the event, "She smokes a lot, so she may get lung cancer"—prediction versus the finding that she has lung cancer. "Did she smoke?" "Yes, 20 a day"; we have a possible contributing factor for this cancer.

The most basic object of interpretation is to assess the significance of difference, once it has been established that there is a difference to explain. The questions to be asked include the following: Is there a difference, and if there is, how has it arisen? Does an observed difference mean toxicity? Equally, if no difference is discernible, where one was expected, why has it not been detected? Were the methods used sensitive enough to show difference? Has exposure been achieved, or is the lack of effect due to a true lack of toxicity? Is the lack of toxicity relevant to other species? When the data are clear in showing an effect of exposure, there is usually little debate about the results, and with appropriate supporting studies, the mechanism is also generally accepted. However, the differences, especially in epidemiological studies, are often small, confined to one group or species, and have no clear origin or mechanism. In these circumstances, it is quite possible for conflicting interpretations to be put forward for the same set of data. Furthermore, additional studies may serve simply to produce contradictory results and are often performed to protocols that are not directly comparable with earlier work. The net result is that the data from the various studies cannot simply be combined to give a larger population size (and so more statistical power). The outcome is a body of data that is almost impossible to negotiate without falling foul of one group or another, and conclusions are left hanging.

THE INTERPRETATION CHALLENGE

The main challenges for toxicological interpretation [which are of clear relevance to the public (the ultimate customer of toxicological investigation)] include the causes of cancer (threat to the individual), reproductive effects (threat to the children), and general disease and debilitation, which can result in loss of quality of life or shortened life span. The public perception of risk and its assessment are dealt with in Chapter 14, but it is clear that public interest (and associated politicians) can put enormous pressure on the process of interpretation and may exert an undue influence on the end result. This is a particular problem when any degree of urgency exists, especially if more studies have to be conducted. It is all too easy in these circumstances to arrive at a conclusion, based on insufficient data, which is at best misleading or simply wrong.

Frequently, the task of interpretation is made more difficult by the lack of clarity of cause and effect. For any finding that has a significant normal incidence in the general population, asthma, for example, attribution of a set of cases to a specific cause can be tenuous. Unequivocal demonstration of cause and effect is possible only if there is a clear relationship between exposure to an agent and a condition present at a significantly higher incidence than normal. For this reason, minor increases in conditions that may be due to toxicity are extremely difficult to ascribe with certainty to individual chemicals or classes of chemicals. This leads to contradictory epidemiological studies that cause opinion to veer from one side to the other, in a manner that does nothing to help scientific credibility.

THE SCOPE OF INTERPRETATION

Data presented for toxicological interpretation range from the results of individual toxicity tests or whole data packages, or large epidemiological investigations, to a single data point from an occupational monitoring scheme. The complexity of interpretation increases as the number of measurements and the amount of data increase. One of the interpretative tricks with large multi–endpoint data sets is to group the data together into easily definable sets so that the conclusions can be better focused on the mechanism in operation. The complexity of a full clinical pathology data set for an individual in a health screen (perhaps 20 to 30 parameters) may be contrasted with the same type of data for a study in animals where there may be 40 to 50 parameters for 30 or more animals. With both sets of data, it is not sensible to try to interpret each parameter separately, because links between functional groups of analytes or hematological cell counts may be lost in the maze of increases and decreases and uncertain abnormalities. In these circumstances, one data point that is seen to be abnormal may be supported by other abnormalities or may be dismissed because there are no other supporting variations from normality. Equally, the changes seen at various exposure levels may be contrasted with those seen in other groups or in other members of the same group.

INTERPRETATION AS A DYNAMIC PROCESS

Interpretation is not a static process, and it is quite likely that new data will at least influence previous perceptions, if only to confirm them. As a toxicological program of testing or research develops, it should be possible to build up a picture from individual studies and to define extra studies to be undertaken in the light of these data. From the conclusions of the individual studies, the wider picture of the effects and mechanisms emerges, allowing overall conclusions on the activity, mechanism, and hazard posed by the test chemical. This in turn facilitates assessment of workplace risk, clinical dosing information, clinical treatment of overdose, acceptable daily intakes, or harvesting intervals (time to harvest from last spraying or treatment). Appropriate interpretation may also suggest better practice, for example, improvements in food storage in the light of conditions that may be caused by growth of fungi (e.g., aflatoxin produced by *Aspergillus* on peanuts) or molds.

There is often pressure to attempt interpretation of part data sets, for instance, partway through a long study or in the middle of an ongoing program of investigation. This should be performed with caution and in the clear understanding that data that follow on may invalidate the interim assessment. It is in this type of circumstance that interpretation of environmental disasters often comes adrift, leading initially to the wrong conclusions or inappropriate investigative studies.

STEPS IN INTERPRETATION

It is not sensible to set hard and fast rules for interpretation, as these are too readily disproved by exceptions; however, as the chapter develops, the general principles that should be applied to interpretation will emerge. These include the following, whatever the size of the data set:

- Assess the validity of the data.
- Look at all the validated evidence.
- Define the controls or baselines.
- Decide what evidence of exposure is available.
- Examine the mechanism proposed in support of the attribution.

When these have been adequately addressed, it may be possible to draw a conclusion as to cause and effect. Failing this, it should be possible to define further studies that should be conducted to elucidate the effects seen.

STUDY DESIGN

The first step is to assess the study design to ensure consistency with the study objectives and good practice. Part of this should be to look for procedural oddities (or deviations from protocol) that might influence the data. For instance, food consumption can be distorted by difficulties in recording discarded food or the practice of giving supplements, which may not be recorded quantitatively. From this base, the credibility of the data has to be assessed by review of the methods for factors such as sampling error, faulty procedure, or design. One such bias is found when more samples are assessed from treated groups in comparison with the controls, leading to an apparent treatment-related difference that is purely a product of sampling frequency.

A critical aspect of study design is the choice of dose or inclusion levels, as overload may lead to unrepresentative toxicity. This is particularly true in studies where the chemical is mixed with the diet, as high inclusion levels may have an effect on the nutritional value of the food offered. The comparators used also need to be examined, as the use of the wrong ones will invalidate the study. Thus, when trying to demonstrate similarity, old should not be compared with young, smokers with nonsmokers, uranium miners with office workers, etc.

CONTROLS AND EXPECTATION

One of the precepts of toxicology is the detection of adverse change from normality, which is defined by an appropriate control group(s). The controls in any experiment give a baseline of experimental normality against which all the treated groups or individual experimental units are assessed. It is therefore critical to be assured that the controls are, in fact, normal (as described in Chapter 2). Given that the numbers in any toxicity study will be merely a sample of a much larger population, it is inevitable that there will be a degree of normal biological variation between control groups in different experiments. Some of these variations will be extreme, and this can lead to an apparent difference that is not biologically real.

When confronted with an apparent treatment-related difference from controls, the assessment should seek to indicate if the treated-group values are simply higher than the controls or they have been increased as a direct influence of the test substance. Focus Box 11.1 summarizes the questions that need to be asked.

The presence of a dose response is a particularly important criterion in assigning a difference to treatment, while the presence of differences in associated parameters

FOCUS BOX 11.1   CONFIRMATION OF THE VALIDITY OF THE CONTROL DATA

The validity of the data from the controls in any experiment should be critically examined to confirm that they represent expectation or normality. Invalid controls call the whole experiment into question. The following should be considered:

- Were the controls experimentally appropriate and within the limits of expectation?
- If there were negative and positive controls, were they appropriate, and did they perform according to expectation?
- If there is more than one control group, are the data consistent between them?
- In studies with pretreatment data, were there any differences before treatment that might influence interpretation of difference later in the study?
- Have the data been distorted by procedurally related stress or, in animals, by the presence of an observer?
- Is only one parameter affected?
- How large is the difference?
- Is the difference reproducible or consistently present in other data or studies?
- Has the difference arisen through the way the data have been processed?
- When the validity of control data is checked against historical controls, is the comparison valid?

also lends weight to the argument for causal relationship with the test substance. To a degree, the size of a difference determines its reproducibility, as small differences seen in small studies are notoriously difficult to reproduce. The presence of the effect in similar studies or mechanistic evidence from related data would also support a relationship to treatment. The influence of data treatment procedures on the perception of difference cannot be ignored, especially when the only difference is statistical; appropriate data treatment may eliminate difference.

The critical question relates to the appropriateness of the controls and whether they were within expectation for the parameter under analysis; in other words, were they normal? Choice of appropriate controls is particularly critical in epidemiological studies, where confounding factors or poor differential diagnosis can invalidate a study. In toxicological studies, the choice of controls is easier as the experimental population is usually supplied as a uniform set of individuals that can be randomly separated into control and treated groups. In this case, you can be confident that the controls and treated groups have similar starting baselines. However, due to the presence of normal biological variation, especially with small group sizes, differences between the groups can be reasonably expected before treatment starts. At this point, it becomes useful, essential in some cases, to have historical control data to hand to assess where the control and treatment group values lie in relation to expectation.

USE OF BACKGROUND DATA IN INTERPRETATION

There will come a point in the examination of toxicological data when it must be decided whether an unexpected observation is natural or an unexpected difference is a change from normality. The contemporary study control should always be the first and chief comparator in any toxicity study. However, as indicated above, there is a role to be played in interpretation of toxicological data by focused use of historical control data. These data should be used to indicate if the controls have strayed from expectation and to back up the concept of normality; they should not replace the contemporary control. Only when the intention of an experiment, often an early or sighting study, is to look for gross differences from normal, should historical data be used to indicate normality in the absence of study controls?

Provided that the controls are selected from the same population as the treated groups, it is possible to be confident that they are truly comparable with them. With historical control data, care has to be exercised that this is true. The greater the similarity of the historical control individuals with those in the study with which they are being compared, the greater will be the confidence that can be placed in using them to support interpretation. The criteria that should be checked before historical control data are used in a particular study include (see Chapter 2 for further discussion of these points):

- Strain
- Route of exposure
- Age of test system
- Media or vehicle
- Supplier

- Husbandry or storage
- Study procedure
- Contemporaneity of data

The greater the deviation of the historical control parameters from the study test system, the less relevant they will be to the interpretation of that experiment. Of the points given, perhaps the most invidious is contemporaneity; control data tend to drift with time, and if this drift is not accounted for, differences may be perceived that are not in fact real.

Inappropriate use of historical control data is one of the easiest errors to make, especially if there appears to be no reasonable alternative. In fact, if there are no comparable historical control data, it is probably better to avoid their use entirely. It may be possible to use historical data from other laboratories, but this carries risks, which should not be ignored. Although the strain and age may be similar, the care of the test system and other factors such as environment, instrument settings, and so on may be sufficiently different to produce data that are not directly comparable. Such data may be used as a guide in the initial setting up of an assay but are of dubious use thereafter. This is illustrated by historical tumor incidences in rodents, which are available from suppliers. The problem here is that the data are compiled from studies conducted in different laboratories under undefined husbandry conditions. Differences in diagnosis and nomenclature used by the individual pathologists are also a confounding factor. It is therefore not possible to place much confidence in these data, but they are better than nothing and may be useful in discussing the incidences of rare tumors. In general, the less reproducible the test conditions, the less useful will be the historical control data from other laboratories.

There are some cases where use of historic control data is routine. Positive control data from previous studies are used routinely in the local lymph node assay, for instance. For these studies, it is accepted that repetition of a positive control with every study is not necessary and that inclusion of a positive control group every few months gives adequate assurance that the test system is responding as expected. These historic data validate the test system and response.

STATISTICS AND SIGNIFICANCE IN TOXICOLOGY

Statistics has come in for much criticism over the years, starting with Disraeli, supposedly quoting Mark Twain: "There are three kinds of lies; lies, damned lies, and statistics." Mark Twain was also supposed to have said, "First, get the facts, then you can distort them at your leisure" and "Facts are stubborn things, but statistics are pliable." Churchill added to this by saying, "The only statistics you can trust are those you falsified yourself" and "Statistics is the art of never having to say you're wrong." Ernest Rutherford's comment, "If your experiment needs a statistician, you need a better experiment," was made from the experimental perspective of a physicist for whom variation within an experiment would have been minimal compared with that seen in biology.

Statistics is revered among some toxicologists and some regulators, who see it as the final arbiter of difference and, by implication, biological significance, ignoring the fact that it is a tool that can be (and often is) misused. Statistics is too often used as an unstoppable force that drives interpretation, rather than as an assistant to this delicate process. Having said that, knowledge of statistical methods and their application is integral to toxicology.

Statistical analysis is routine in sufficiently sized toxicity studies, and the results can be slavishly reported to the general detriment of credibility. It is important, however, to remember that statistics is a fallible tool. A useful analogy is comparison of the use of statistics by toxicologists to the use of a lamppost by a drunk; they should be a source of illumination, not of support. It is easy to misuse them and to draw incorrect conclusions based solely on the presence of statistically significant differences.

In simplistic terms, there are three levels of significance that are important in toxicology: statistical, biological, and toxicological (or clinical), in increasing order of importance. Data should be analyzed with these significance levels in mind, taking into consideration any dose response (or its absence), the inherent variability or variance of the data being examined, and the sample size. Variance is a function of the range of the values (minimum to maximum) and of the deviation of the individual values from the mean and indicates the extent to which the values are distributed about the mean. Variance increases when the data include outliers, data points that are radically different from the majority of the group. Remember that high variation in a treatment group may be due to differences in response to treatment among individuals and not to normal biological variation. It is normal statistical practice to exclude outlying results from analyses. This is usually acceptable in control groups but may be more difficult in a treated group. If there is a single high value for a single parameter in a high-dose animal that is not supported by other results, it is probable that the single data point is an outlier and can be excluded. If a whole range of parameters is distorted from normality, another cause should be sought— for instance, extreme change induced before death. These data may be excluded from statistical analyses but may still be related to treatment.

Statistical significance means simply that the test group is different from controls in a numerical sense and that the difference in means is large enough for the effects of variance to be overcome. This can be numerically ridiculous and a disaster in presentation terms; computer programs often work on unrounded figures but report to one or two decimal places. This means that it is possible for a table to contain four group means for one parameter from different treatment groups, all with a value of 1.1 but with significant differences flagged for one or more; this is also often a reflection of the differences in variance in the data for each group. Equally, it must be pointed out that a difference that is not statistically significant may still be of biological significance.

Likewise, a statistically significant difference is not always of biological significance, when it relates to a change that may be important for the animal but that is not necessarily adverse and is probably reversible. Examples of this would include normal hepatic adaptation to treatment, possibly seen as a minimal hepatocellular hypertrophy or an increase in urine volume due to increased water intake after administration of a foul-tasting substance. Transient diuresis of pharmacological origin without other change would also be included, if it was not seen to excess.

Cessation of treatment or exposure is associated with a speedy return to normality. Biological significance does not equate to toxicological significance, especially as most data are representative of a single time point and do not analyze a continuum, which might show an increasing difference attributable to treatment. However, it should be noted that if a biological difference is allowed to persist, it may result in toxicity; a difference of biological significance, seen in short studies, may progress to toxicity in longer studies or over the course of a development program.

Toxicological significance denotes change, which, if allowed to persist, may impact the survival or well-being of the exposed population or test system. Although reversibility may mean that an adverse change is not of toxicological significance, the degree of change is important; for example, administration of carbon tetrachloride to rats can result in extensive liver damage, which is clearly the result of toxicity. This damage is clearly evident in the first few days following treatment, but due to the liver's powers of recuperation, there may be no difference from normality after 14 days.

Tumor data provide examples of the distinction between these levels of significance. A doubling of the incidence of a rare tumor over control incidences, if seen at the highest dose level, may not be statistically significant but would probably be considered to be of biological significance, if not of toxicological significance, depending on context. Equally, a 25% difference from control in testicular tumor count in some strains of rat may be flagged as statistically significant but is unlikely to be of biological or toxicological significance, especially if significant difference is established against only one control group or at the low- or intermediate-dose level. In some cases, an overall threshold of difference set arbitrarily at 10% has been used, as in "the difference was less than 10% from controls and was not toxicologically significant." This is nothing more than a numerical comfort blanket, with little scientific basis. A 10% increase in plasma activity for an aminotransferase is unlikely to be of any biological significance, whereas a 10% difference in plasma sodium concentration could be seriously unwelcome. This distinction is due to the low physiological impact of variability in enzyme activities versus the more precise homeostatic requirement for electrolyte concentrations.

At the simplest level, the use of statistics merely examines the differences between control and treated groups and gives a probability that the two groups represent different populations. In other words, they test that there is no difference between control and the treated group(s)—this is termed the null hypothesis. The P value gives a measure of the probability of the likelihood of the results occurring through chance alone; a P value of 0.05 equates to a 1 in 20 chance, while a P value of 0.01 equates to a 1 in 100 chance that the observed results are chance alone. If the calculated P value is below the chosen threshold of significance, the null hypothesis is rejected, and the result may be termed *statistically significant*. The results of statistical analysis can therefore be used to answer the question, "Is the difference from controls caused by treatment or exposure to the suspected factor?" Note that it generally only indicates the answer and does not provide it, unequivocally, in every case. All too often, a single statistically significant difference will not be enough to prove the wider hypothesis.

TABLE 11.1

The Effect of Variability in Data on Summary Statistics

Data Point	"Normal" Data	Variable Data	With an Outlier
1	31	36	31
2	29	24	29
3	27	25	27
4	34	36	34
5	32	42	32
6	29	19	63
7	28	36	28
8	31	23	31
9	33	33	33
10	30	30	30
Range	27–34	19–42	27–63
Coefficient of variation	7.3%	24.7%	30.9%
Mean	30	30	34
SD	2.2	7.4	10.5

Source: Adapted from Woolley A, *A Guide to Practical Toxicology*, 1st ed., London: Taylor & Francis, 2003.

Note: SD, standard deviation.

Statistical significance is driven by two fundamental factors:

- n: the number of data points in the group (group size)
- Variance: the difference of each data point from the group mean.

The smaller the sample size and the greater the variance in the data, the more unreliable will be the statistical values that result from any analysis; the effects of variability in data are illustrated in Table 11.1. For sample sizes of less than 10 or where the variance is large, a statistical significance is only a pointer to a difference that may be of biological or toxicological significance. It is the responsibility of the toxicologist to interpret the data to indicate the real significance of the difference, in biological or toxicological terms. At all times, it should be remembered that statistics are blind in that they are solely a numerical tool and can tell you nothing about the quality of the numbers or their origins; if they are used without discrimination, they are a blunt tool that can be a source of misinformation and erroneous conclusions.

STATISTICAL PROCESS

In analyzing data, it is important to use the statistical tests that are appropriate to the data type being examined, whether it is for a continuous variable or presence/absence data. The following is intended to show the approach normally taken in analysis of these data types and is intended simply as a guide to statistical method. For more technical explanation, one of the standard texts on statistics, such as Gad (2006) should be consulted, but having said that, it is extremely difficult to find an explanation of statistical method that is accessible to the mathematically challenged.

For a continuous variable, the first line of examination is at the level of group means and, usually, the standard deviation, which, in conjunction with the number of data points, n, gives a first indication of the variance of the data. Table 11.1 illustrates the effects of variability on the summary statistics (mean and standard deviation) for a representative set of data. The "normal" set has been constructed to represent typical values for a plasma enzyme such as alanine aminotransferase. The second—variable—set is a reworking of the first to introduce greater variability, while the third set illustrates the effect of a single outlying value.

The summary data give a first, crude indication of difference, assessed from the control and test group means and the overlap of their standard deviations. This has been described as the "very obvious test"; it has a pleasing simplicity, which is a bonus to many but may be frowned on by professional statisticians. There are two approaches to the analysis of continuously variable data, namely, parametric and nonparametric methods, the latter generally having less power than the former. Parametric analysis is the method of choice, but for this, the data should be normally distributed and have homogeneous variance.

The first step in statistical analysis of a data set is to confirm that the variance is homogeneous and, if so, to proceed to analysis of variance and other parametric methods. If the variance is possibly affected by the presence of outlying data points, it may be useful to perform the analysis with and without these values, although exclusion of individual data points should be done with caution. Analysis of variance uses the data from all groups and seeks to establish that the null hypothesis is true—that there are no differences among the groups—or that one or more groups are different. Although much used in the past, Student's t test is now acknowledged to be unsuitable where there is more than one group.

Where it has been decided that parametric analysis is not appropriate, nonparametric methods offer an alternative, although they are not easily applicable to complex data sets. They are mostly based on ranking the data and are particularly good when there is obvious deviation from the normal distribution but become more difficult when there are a number of tied values. The Wilcoxon rank sum or Mann–Whitney test is the simplest of these methods and is based on assessment of the ranks of the individual values, not on the original data themselves. The Kruskal–Wallis test is the equivalent of analysis of variance, used when there are more than two groups for comparison. This process is summarized in Figure 11.1.

Data that describe presence or absence are generally assessed using chi-squared or Fisher's exact test with more complex analysis being undertaken with tests for positive trend. The chi-squared test is appropriate for high-frequency findings and compares the observed with the expected frequencies, the latter being derived from all the data for the groups being tested. For data with lower incidences, Fisher's exact test is normally used, comparing the numbers of animals in each group with the lesion and those without it. In carcinogenicity bioassays, where analysis of tumor incidence is a vital component of the interpretation of the results, tests for positive trend are used. Cancer is more prevalent in older animals, and as a result of early death due to toxicity, the animals at the high dose will not have the opportunity

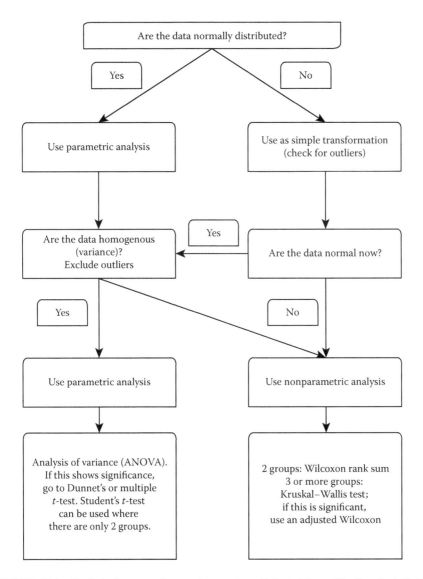

FIGURE 11.1 Statistical process for continuous data. (Adapted from Woolley A, *A Guide to Practical Toxicology*, 1st ed., London: Taylor & Francis, 2003; and Dickens A, Robinson J. "Statistical approaches". In Evans GO, ed. *Animal Clinical Chemistry—A Primer for Toxicologists*. London: Taylor & Francis, 1996.)

to express the same numbers of tumors as the controls. Where treatment causes an increased incidence of a tumor in animals that survive to old age, this will not be apparent if there is significant early mortality. Analysis is therefore conducted using the data from all the dose groups and takes account of the numbers of animals that die during the study. In addition, the tumors are categorized for each animal as fatal, probably fatal, probably incidental, or incidental; an additional category of uncertain may be added.

DATA TREATMENT AND TRANSFORMATION

In analyzing data, it is often useful to treat it in various ways in order to make differences easier to discern. One approach to this is to examine change rather than the original value. The classic example is analysis of body weight gain rather than the simple group mean data for the absolute weights. This can also be applied to parameters such as alkaline phosphatase, which should decrease with age; the absence of decrease may indicate treatment-related effect. In young rodents, food consumption may be readily correlated with growth by the calculation of food conversion ratios—in effect, the amount of body mass produced for each gram of food consumed. This figure declines as the growth phase is completed and becomes meaningless after that, as food consumption is maintained at the same values without significant gain in body weight.

For parametric analysis, it is necessary to have normally distributed data. With skewed data, it may be possible to achieve this by transforming the data, for instance, by using the log or square root of each data point, which can also be used where variability increases with the mean. More esoteric procedures, such as use of reciprocals or trigonometric functions, start to show a degree of numerical desperation, indicating that the use of nonparametric methods would probably be better.

ASSESSING EXPOSURE

Exposure (or lack thereof) is the ultimate arbiter of toxicity. Toxicological effect is always related, directly or indirectly, to exposure to an active molecule or to an agent such as radiation. Assessment of exposure is therefore essential to interpretation of toxicological data because, although the presence of an effect may indicate exposure, exposure cannot be assumed in the absence of effect. In addition, the presence of an effect does not necessarily mean that it can be attributed to the chemical under investigation. Focus Box 11.2 looks at some of the factors to be considered in assessing exposure.

The normal sequence of known chemical exposure resulting in an effect may be turned on its head when an effect is observed without an obvious explanation. In this case, the interpretation effort relies on finding common exposure factors and confirming that these are consistent with the observations. There have been many incidents or findings that have provoked such epidemiological study, including scrotal cancer in chimney sweeps due to soot, lung cancer, and smoking.

Evidence of exposure to the test substance is almost always achieved by analysis of samples taken from the medium *in vitro* or of blood or urine samples taken from animals. Occasionally, it is possible to point to effects seen and conclude that the test system must have been exposed, but this does not characterize the concentrations associated with the effect or if the parent molecule was present. Toxicokinetic analysis of blood samples gives an indication of basic kinetic parameters including half-life (time needed for the plasma concentration of the chemical to decrease by 50%) and area under the concentration curve (AUC).

FOCUS BOX 11.2   QUESTIONS IN ASSESSMENT OF EXPOSURE

Exposure, and its significance, should be assessed via the following considerations:

- Was the test system exposed to the test material or a metabolite? A particulate material may not be available to cells *in vitro*, and an oral dose may not be absorbed.
- Was toxicity due to a metabolite or an indirect effect (e.g., hormonal imbalance)?
- Did this exposure extend to the target tissue?
- Was the target tissue exposed to greater concentrations than elsewhere in the test system?
- In a life of mixtures, if there was exposure, was this the cause of the effect?
- Was the observed exposure sufficient to cause the observed effects? *Sufficient* may be defined from no-effect levels in previous studies; also consider any interspecies differences. *Sufficient* includes duration of exposure (area under the curve as well as treatment period).
- In absence of toxicity, was the exposure medium or vehicle appropriate to achieve exposure of the test system?
- Is the analytical method sufficiently sensitive and specific to detect the test substance or its derivatives?
- If a marker is used as a surrogate for the test chemical, is this specific and reproducible?
- Was there any cross-contamination of the controls that might invalidate the data?

INTEGRATION OF EXPOSURE INFORMATION

For the reasons given above, integration of exposure data into toxicological assessments is essential for a meaningful interpretation of the results. In assessing the likelihood of exposure, the physicochemical properties of the molecule should be considered, including partition coefficient, solubility, and absorption, distribution, metabolism, and elimination (ADME) (Table 11.2), as these will have a profound influence on the extent of exposure. The speed and extent of absorption and subsequent distribution into the test system are also critical. These factors determine the maximum concentration of the test chemical at the site where toxicity may be expressed, whether that is a protein or cellular organelle in a cell culture or in a tissue in a whole animal. Toxicity is usually seen once a threshold concentration or level of exposure to the test chemical or a metabolite has been exceeded. Systemically, this threshold may be associated with a particular level of exposure, defined by C_{max} (maximum plasma concentration) or AUC. With some toxicities, especially those relating to transient excess pharmacological action, it is possible to relate the onset,

TABLE 11.2
Selected ADME Factors and Their Impact on Toxicity

Factor	Impact
Long half-life	Longer systemic exposure and possibility of accumulation with repeated dosing.
Short half-life	Transient peaks of high concentration that may not elicit chronic toxicity.
High binding to plasma proteins	Low free concentration of active substance in plasma. Small changes in binding site availability may lead to large percentage changes in free chemical and so to toxicity, e.g., warfarin.
Tissue binding	Sequestration into a tissue compartment, such as bone or lipid, reduces the amount of chemical available to express toxicity. Sudden release later on may have serious consequences.
Metabolism	Can increase or decrease toxicity. Inhibition or induction of the enzymes or metabolism can have a marked effect on the toxicity of chemicals; simultaneous exposure to two or more chemicals may therefore have a much greater effect than an equivalent dose of either chemical alone.
First-pass effect	Significant metabolism of a chemical as it passes through the liver for the first time after absorption from the gut results in low systemic availability, reducing toxicity in more distant tissues.
Enterohepatic recirculation	Breakdown in the gut of conjugate metabolites excreted in the bile can lead to reabsorption, effectively increasing half-life and area under the curve.
Excretion failure or decline	Age-related decline in renal or hepatic function can lead to increased systemic exposure and hence to toxicity, e.g., benoxaprofen in elderly patients.
Concentration in tissues of elimination	Concentrations of chemicals in tissues responsible for their excretion can result in local toxicity, seen particularly in the distal tube of the renal nephron and in the bladder.

Source: Adapted from Woolley A, *A Guide to Practical Toxicology*, 1st ed., London: Taylor & Francis, 2003.

duration, and severity of effect to C_{max}. Where there is a long half-life of elimination, such that significant concentrations remain at the time of the next dose administration, it is likely that the chemical or its metabolites will accumulate. This can result in the appearance of toxicity at a late stage in the evaluation program, due to accumulation of effect. For instance, a 4-week study may show no effects, whereas the same dose levels in a 13-week study may be associated with minimal onset of toxicity just outside normal biological limits. In longer studies, this may progress to the extent that development of the compound should be stopped.

Although the plasma concentrations (usually equated with systemic exposure) of a parent compound are useful as a general indicator of exposure, they do not necessarily equate to concentration of the active molecule at the site of action. Brief excursions into toxic concentrations may only be associated with transient effects, such as those associated with excess or undesirable pharmacological action. The expression of chronic or persistent change is probably due to accumulation of effect (or deficient repair) following brief toxic concentrations or to sustained exposure to a concentration at which adverse change becomes apparent more slowly. This type of toxicity was apparent with the retinal effects associated with chloroquine, which has a high affinity for melanin in the retina. These irreversible effects are dependent not only on daily dose and duration of treatment but also on the total dose taken; toxicity may be expressed after withdrawal of the drug as the drug persists long after therapy is ceased.

However, toxic effect is not always directly attributable to the chemical that was added to the culture or given to the animals. Following administration, and simultaneously with absorption and distribution, the processes of elimination begin. These encompass metabolism and excretion, which are generally expected to result in the removal of toxic entities but which can increase toxicity. One route of removal from the plasma, which is properly part of distribution, is the sequestering of chemicals into tissues, where they are retained, effectively inactive, until released. For instance, highly lipophilic compounds, such as organochlorine pesticides, PCBs, and dioxins, accumulate in adipose tissue, from which they may be released to produce toxicity long after exposure has ceased. This can be a problem in pregnancy when significant lipid mobilization occurs during the third trimester. Another example is the binding of heavy metals, like cadmium, to metalloproteins and the subsequent toxicity when a storage threshold is exceeded.

Although, having demonstrated exposure of the test system, it is relatively simple to correlate change with the presence of the chemical or a metabolite or an indirect effect, the absence of effect needs considerable care in interpretation. Before a chemical can be truly said to be nontoxic, it is necessary to show that it was available to the test system and that significant concentrations at potential target sites were achieved. Incubation at high concentrations or oral administration of large doses does not mean that exposure was achieved. Poor absorption following oral administration (low bioavailability) is a frequent finding, resulting in low systemic concentrations; as a result, the toxicity of a parenterally administered chemical may be much greater than expected when effects are extrapolated from an oral dose. Poor availability may be due simply to the medium or vehicle in which the chemical was offered to the test system, and a change in this can result in significantly greater toxicity.

The assessment of exposure is dependent on the sensitivity and specificity of the methods used to detect the test substance. For small molecules, this is usually not a problem; for larger molecules such as peptides or proteins, the analytical challenges become more exacting, especially if the half-life is short or the concentrations very low. This leaves the problem of how to interpret the presence of effect in the absence of measurable exposure. There are a number of possible explanations for this, including analytical methods that are not sufficiently sensitive. Another possibility might be that the correct matrix is not being analyzed and that the correct place of analysis is the target site of activity. There is also the possibility that the pharmacological effect persists for longer than indicated by plasma concentrations due to persistent binding at a receptor.

Although it might be assumed that parenteral administration, in humans or animals, would result in rapid exposure to 100% of the given dose, this may not always be the case. Intravenous administration of an inappropriate formulation may lead to temporary deposition near the site of injection, perhaps as a result of local irritation or other damage. Equally, intramuscular dosing of a poorly isotonic formulation may result in slow release from the site of administration. These uncertainties pale into insignificance when the dynamics of oral dosing are considered. Before the compound can get into the systemic circulation, it has to cross the gut wall and pass through the liver without significant first-pass metabolism taking place. Excretion of conjugated metabolites in the bile can be associated with enterohepatic recirculation, where the conjugate is broken down in the gut and the active molecule is reabsorbed.

Although the processes of ADME are clearly important in animals, it should be borne in mind that their absence *in vitro* may have an adverse effect on the results, leading to false negatives or false positives. Although this criticism is partly met by the use of S9 mix, it may be necessary to use a preparation from a relevant tissue, such as the kidney. The toxicity of S9 to cell cultures is well known and should also be considered.

TOXICOKINETICS

In the context of toxicity studies undertaken for registration of chemicals, toxicokinetics is an integral part of the process, in terms of confirmation of exposure and of interpretation of the data. The following can only scratch the surface of this fascinating area of toxicology. The parameters calculated from the results of the bioanalysis are given in Table 11.3.

Clearance is a key concept; if hepatic clearance of a chemical is 60 L/hour and hepatic blood flow is 90 L/hour, the implication is that two-thirds of the chemical is removed by hepatic metabolism in one pass—an example of first-pass metabolism for chemicals given orally. Although half-life has been used traditionally as a key pharmacokinetic measure, partly because it is simple in concept, it needs to be used with some understanding of its derivation. As it is calculated from the volume of distribution and the clearance, it is dependent on these two parameters. However, it is still a useful concept; in general, a long half-life is likely to be associated with accumulation of the test chemical on repeated administration.

Accumulation of test chemical over the period of a repeat-dose toxicity study is quite common, and this may be accompanied by progressively accumulating toxic effect. This accumulation may be dose related, given that pharmacokinetic behavior at high doses may well be different from that at low or clinical doses. At high doses, absorption from the gut may become saturated or be limited by formulation; in addition, metabolism pathways may be saturated, resulting in higher AUC, slower clearance, and longer half-life. Pharmacokinetics at low doses relevant to human exposure may be significantly different from that seen at high doses, and it is possible that toxicity expressed at high dose would be irrelevant to humans, due to nonrepresentative pharmacokinetics, in relation to those seen at low doses.

There are numerous factors that influence the kinetics of xenobiotics and their ADME. Formulation can be critical, as indicated above; simply micronizing an insoluble chemical given orally may increase absorption. The type of molecule is also important, for instance, antisense oligonucleotides may be very rapidly

TABLE 11.3
Toxicokinetic Parameters

Parameter	Comments
Compartment	A hypothetical volume or space in which test chemical may be distributed or retained. Typically, the blood is one compartment; other tissues may form other discrete compartments—bone, adipose tissue, liver, etc. Distribution and retention in a particular compartment may affect toxicity; sudden release from adipose tissue in which long-term storage of the chemical has taken place may elicit toxicity.
Bioavailability	The percentage of a chemical that is available to the systemic circulation following oral dosing, usually in comparison with plasma concentrations following intravenous dosing. Calculated from the dose given and the AUCs found after oral and intravenous doses.
C_{max}	The maximum concentration of the test chemical or a metabolite, usually in plasma, after dosing.
T_{max}	The time at which C_{max} is reached.
AUC	Area under the concentration curve, usually $AUC_{(0-24)}$ or $AUC_{(0-\infty)}$; a measure of systemic exposure related to half-life and plasma concentration. Increasing the AUC, e.g., through formulation changes, may result in greater toxicity.
Volume of distribution (V_d)	The apparent volume of the body occupied by the chemical. The total amount of chemical in the body is divided by the plasma concentration at C_{max}. For chemicals that are quickly distributed to other tissues, it is possible for the volume of distribution to exceed the body volume. A low volume of distribution may imply little distribution outside the blood.
Clearance (Cl)	This describes the efficiency of elimination of a compound from the blood or a tissue compartment or the body as a whole. Defined as the volume of blood cleared of chemical in liters per hour or milliliters per minute. Total clearance is the sum of the clearance values from all the compartments in the body.
Half-life	The half-life of elimination $(t_{1/2})$ is the time needed for the plasma concentration of the chemical to decrease by 50%; it is dependent on volume of distribution and clearance and is calculated as $t_{1/2} = (0.693 \times V_d)$ divided by clearance.

distributed to the tissues, meaning that their systemic presence is brief and at barely detectable levels. Slowly metabolized compounds, such as dioxins or organochlorine insecticides like DDT, may accumulate in lipid tissues and have half-lives in years. In these circumstances, the traditional method of assessing exposure in the plasma is not useful; the unchanged chemical may well be present in the target tissues at toxic concentrations but cannot be detected by the usual methods. A biomarker of effect may be useful in these cases, or a biopsy and analysis of an appropriate tissue.

For some chemicals, metabolism is rapid and may result in metabolites that are pharmacologically active or toxic. These metabolites can have different pharmacokinetics from the parent molecule. Add to this the potential for some chemicals to induce (or inhibit) their own metabolism, and it becomes evident that interpretation of toxicokinetics should not be attempted in isolation from other information on the chemical, including data from specialist ADME studies, histopathology (induction of metabolism in the liver is often accompanied by hypertrophy of the hepatocytes), and any available information on human pharmacokinetics.

THE REALITY OF DIFFERENCE—THE INTERPRETATION OF SMALL DIFFERENCES

Toxicology for regulatory purposes is largely about the desire to demonstrate the presence or (preferably) absence of difference from normality, and all interpretational effort is directed at this deceptively simple objective. It is relatively easy to spot differences so large that they are barn-door obvious—a 10-fold increase in colony count or enzyme activity or an unusual, rarely observed, pathological lesion. Differences of medium size are also relatively simple; they are consistent or outside the normal limits, or there is a clear dose response. The real challenge is provided by the small differences, often at the lowest dose level. Small differences may be hugely significant, not for their short-term effects but because they may perturb normal physiology or homeostasis by a small amount that has an increasing effect with its continued presence. Hormonal change is a classic example of this; it is interesting to note that these are not routinely investigated in toxicity studies.

One of the reasons for looking for difference is the perceived obligation to show toxicity at one dose and thereby imply safety at a lower one. This leads to pressures to assign significance or otherwise to trivial differences; we must show toxicity, so this is a significant toxicological change. Or we must show a no-observed-effect level (NOEL), so this small difference at the bottom end of the dose–response curve is irrelevant. Although it has been pointed out that, to demonstrate a NOEL, there must be effects at higher doses, this edges toward irrelevance when the doses are vastly higher than those expected in humans. Where there is a small difference at the lowest dose level, which is supported by increasing differences at higher doses, it cannot be escaped that treatment has probably had an effect (or at least an influence) at the lowest dose. The significance of this difference is where interpretation becomes more complex. The no-adverse-effect level (NOAEL) is a useful concept because it acknowledges the presence of treatment-related change while putting it into perspective. The problem is that difference from controls, which is inevitable when using biological systems, is open to misinterpretation unless it is barn-door obvious, as shown by the very obvious test referred to above. For small differences, it is difficult to assign significance.

Having noted a difference from controls, the first question to ask is whether there is a dose response. For example, if treated animals are different from controls, is this within background data ranges? It is noticeable that increases or decreases are often present in treated animals and are dismissed as being "within normal ranges." It is quite possible for a 5% difference in a biochemical parameter to fall inside normal limits and still be treatment induced, especially if it comes at the bottom of a dose–response curve where the parameter is progressively and clearly affected at higher doses. Such differences show an influence by treatment but are often not biologically significant at the time of observation.

A degree of difference is inevitable as a result of normal biological variation within the limits of normality defined by contemporary controls or similar, independent studies. Any pressure to downgrade a difference by defining limits after the event should be resisted. Although differences due to normal variation are expected, it is reasonable to expect some change in the test system when administering pharmacologically active chemicals at relatively high doses.

Where there is a classic dose–response curve, interpretation is relatively simple. With U-shaped dose responses, it is not so easy, the first problem being to demonstrate that the curve is in fact abnormal, with maximal or minimal effect at intermediate doses rather than at the extremes. Vitamin A shows such a curve—toxicity at low doses due to absence, benefits at normal levels of exposure, and toxicity at high doses. Having said that, where there is a difference from controls that is not obviously on a dose–response curve, it is usually easy to dismiss it as being due to chance variation within normal limits. Other reasons for dismissing a difference are that it is present in only one sex (although in rodents, this is often not sensible, due to metabolic differences between males and females) or within background data, or inconsistent between examinations. The statement that it is of "equivocal significance" simply means that it may be related to treatment but is not understood. Sometimes the significance of minor differences seen in early studies becomes apparent with prolonged treatment, when lack of biological significance can be replaced by clear toxicological effect. A 10% deficit or increase may not be significant at 4 weeks but may become fatal if age-related decline in function is accelerated during prolonged treatment.

Differences become impossible to interpret satisfactorily when the data, or the mechanism that generates them, are not understood completely. There is, apparently, a touching, unstated, belief that more data on more parameters will mean better safety evaluation; this is a fallacy. A single difference from controls does not necessarily mean that the function of the tissue or organ system is impaired in proportion to the difference, as compensating mechanisms exist that cope with change in one direction by regulating in another. The overall goal of evaluating a range of parameters is to look at the function, which is the product of many processes that work together. Thus, liver and kidney functions are examined in a range of tests in the course of routine toxicity studies, and changes in individual parameters are assessed against the data for related measures of organ toxicity. With increasing severity of effect on, say, the liver, the number of parameters affected and the size of difference from controls increase, usually with dose. At the low end of the dose–response curve, at or near the NOAEL or NOEL, increasing the number of parameters to be examined may actually confuse the situation because normal biological variation will ensure a selection of differences in both directions, making secure interpretation nearly impossible.

There is a hint of this in the increasing emphasis on new areas in pharmaceutical development and the enthusiasm for technically demanding methods that generate numbers in the absence of clear understanding of their biological significance. As an example of a single-parameter test, sister chromatid exchange went through a phase of popularity. It was agreed that there was a clear correlation between positive results and mutagenicity, as indicated by other tests, but the mechanism and significance of the effect were not understood, and it did not become a standard test for regulatory purposes. As a single endpoint, it was never interpreted in isolation from other genotoxicity data. There is also a tendency to react to human toxicity with new test requirements. One example of this is the severe cardiac events linked to QT interval prolongation in people taking drugs such as cisapride. Although this type of effect may halt development of a promising drug, it should be noted that there are drugs on the market that are known to cause QT prolongation but are not withdrawn. Once again the question must be asked where the threshold of difference for rejection lies. A more general, philosophical question could ask if a one-size-fits-all approach to this type of problem is scientifically valid.

THE REPRODUCIBILITY OF DIFFERENCE

Ultimately, there is only one way of confirming the significance of a small difference, and that is to see if it is reproducible, either in a second experiment (as for *in vitro* tests) or in the next, usually longer, toxicity study. Furthermore, if the second experiment is performed using slightly different methods, the reproducibility or otherwise of the difference becomes much more significant. For experiments that are inherently weak statistically—those with small group sizes or with incomplete data sets—it is not unknown for a second test to show up a different set of statistically significant differences from controls. In this case, it is easy to write off the differences as being due to normal biological variation; this illustrates neatly the importance of considering statistical versus biological or toxicological significance. Small group sizes, combined with measurement of parameters that have large inherent variances, will tend to throw up statistically significant differences that disappear on repetition.

While it is easy to live with nonreproducibility in small differences, it is more complex when larger; apparently toxic differences are not reproduced in successive studies. The potential reasons for this include those in Focus Box 3.2, Chapter 3, to which should be added any changes in study design that may have taken place. Another consideration is that toxicity may be expressed early in a study but, due to development of tolerance or adaptation to treatment, will not be evident later on.

SUMMARY

In attempting interpretation of any data set, the following questions are among those that should be answered:

- Are the data a true reflection of the methods used, taking the test system characteristics into account?
- Has there been any effect on the parameters examined that may be due to procedures [e.g., increased heart rate as a result of restraint in recording electrocardiograms (ECGs)]?
- Have the controls or other baseline comparators performed as expected?
- What evidence of exposure is available?
- Are the data more variable than usual, particularly in treated groups?
- Have the data been affected by any processing?
- Were the statistical analyses appropriate and decided before the study was started?

- Statistical analysis is a blind tool that says nothing about the quality of the data or their origins; they should be used with care and discrimination.
- Is there a dose response?
- Is there a NOEL or a NOAEL?
- Is there a plausible mechanism that could explain the differences?

All the data should be considered in reaching a conclusion; there is little sense in reviewing one parameter in isolation as true treatment-related difference is usually supported by change in several parameters. This concept of linkage is discussed in the next chapter.

REFERENCES

Dickens A, Robinson J. Statistical approaches. In. Evans GO (Ed), *Animal Clinical Chemistry—A Primer for Toxicologists*. London: Taylor & Francis, 1996.

Gad SC. *Statistics and Experimental Design for Toxicologists and Pharmacologists*, 4th ed. Taylor & Francis, 2006.

Woolley A. *A Guide to Practical Toxicology*, 1st ed. London: Taylor & Francis, 2003.

12 Interpretation
Different Data Types

There is no such thing as a right answer for interpretation; it is impossible to cover every eventuality. The following sections give an overview of the types of toxicological data and attempt a basic guide on how to approach each type.

INDIVIDUAL DATA SETS

The simplest type of data that may be considered to be toxicological relates to a single parameter for one individual, such as a marker for occupational exposure. This may be presented as a single time point or a series of time points, which give a chronological profile of exposure. With this type of data, it is important to be sure that the marker is either a direct marker of exposure or a surrogate marker such as an easily measured effect. In general, the more remote the analyte or effect from the parent compound or the greater the natural background incidence or concentration, the more difficult it is to draw supportable conclusions from the data unless they are clearly at the extremities of or outside the normal range. The best marker of exposure is the parent compound, a known metabolite or by-product. However, use of these is often not possible, and an indirect marker of effect such as inhibition of cholinesterase activity in the plasma following exposure to organophosphate insecticides can be used. Although it is possible to analyze urine or plasma for DNA adducts, these are not necessarily specific and may reflect lifestyle or other exposures. A more general approach is to look at clinical pathology data, which rarely look at a single parameter in isolation, as a guide to abnormality that may be work related. With any such data set, which is unlikely to have contemporary controls unless part of a full epidemiological study, it is important to have access to robust, trustworthy historical data ranges. If there is any sample analysis for the individual from before the start of exposure, this is clearly a significant advantage, although the date and circumstances in which the sample was taken should be considered. It is also probable that such data will be available for other workers and can be combined to give an overview of the exposed population. Within this data set, it may also be possible to identify subsets of individuals who have been subject to greater or lesser exposure depending on their workstation.

For chemicals that are accumulated into tissues and released slowly, assessment of exposure may be difficult during the early stages, as the effects only become apparent when a concentration threshold is crossed. For instance, cadmium has a half-life measured in decades and accumulates in the kidney until a critical concentration of around 200 µg/g is reached. At this point, cadmium toxicity becomes apparent through increased urinary excretion of proteins, and the cadmium concentration in the urine rises as a late herald of renal toxicity.

SAFETY PHARMACOLOGY

For much safety pharmacology, as with other areas of toxicology, interpretation is often dependent on the experience (and historical control data) of the laboratory performing the tests. On many occasions, the interpretative effort is focused on a small difference in one or two parameters in a single test, perhaps on one occasion at the high dose. It is very tempting to overinterpret such minor differences and to assume that there is a treatment-related effect when it is probable that there is not. Criteria that need to be applied include dose relationship at the time point considered and the character of the data at other time points.

Another factor is the presence or absence of supporting data from other parameters in the test or from other studies, including toxicity studies if appropriate. As a last resort, historical control data from the laboratory should be consulted, providing that it is reasonable to relate it closely to the study under consideration. For instance, route of administration, strain, and age should be the same or very similar; ideally, the vehicle should also be the same or at least have a similar degree of toxicity. A vehicle that affects transit time through the gastrointestinal (GI) tract would not be an appropriate comparator for an oral study conducted with a simple aqueous solution. One area where historical control data are useful is when a positive control, for instance, furosemide, has not induced the expected effects in a renal function study in rats. The positive control results serve to confirm that the test system has performed as expected; if the positive control does not produce the expected effect, the validity of the test has to be questioned very carefully.

In interpretation, the presence of confounding factors in the study design or execution should be considered. The presence of excessive toxicity leading to reduced activity is likely to affect interpretation of an Irwin screen. As mentioned in Chapter 5, anesthesia can have profound effects on parameters measured in cardiovascular and respiratory studies. Where a set of measurements may be influenced by an animal's reaction to its surroundings, such as in a restraint tube for respiratory measurement, it is important that a suitable period of acclimatization is allowed.

When you are satisfied that the data are valid, the doses at which the adverse effect is seen should be compared to the doses eliciting the primary pharmacodynamic effect in the test species or the proposed therapeutic effect in humans, if feasible. It should be remembered that there are species differences in pharmacological sensitivity (ICH 2000).

Redfern et al. (2003) wrote a comprehensive survey on the relationship between preclinical electrophysiology, QT interval prolongation in the clinic, and torsades de pointes, with particular reference to the hERG assay, which examines blockade of K^+ channels expressed in stably transfected human embryonic kidney cells. They came to the conclusion that the data set they analyzed confirmed that most drugs associated with torsades de pointes in humans also block the hERG K^+ channel at concentrations similar to the free plasma concentration found in clinical use. They also suggested that a 30-fold margin between the concentration at which hERG channel block is seen and the clinical plasma concentration represents an adequate margin of safety for compounds that are positive in this assay. There are some caveats, however; verapamil has a twofold safety margin between its effective plasma concentration and its IC_{50} in the hERG assay but has not been associated with torsades de pointes. At the other extreme, the margins for some drugs associated with torsades de pointes are very much larger than 30-fold. It is clear, as with everything else, that a judgment has to be made in each case and in the light of results from more than one test.

The extent of the interpretative problem is put into perspective by the statement by Redfern et al. that over a hundred drugs prolong QT intervals in man but that many of these have a history of safe clinical use. The retrospective discovery of torsades de pointes potential in a hitherto safe drug does not mean that it is suddenly unsafe; the margin of difference between the concentration at which it is clinically effective and that at which it has effects on the hERG channel should be taken into account, quite apart from the clinical history of use.

Overall, the interpretation of safety pharmacology data has to take all the available data into account. For assessment of the risk of QT prolongation in man, the data from *in silico* models, *in vitro* hERG channel assays, and studies using telemetric measurements from dogs together with blood concentrations necessary for therapeutic effect should be considered (Pollard et al. 2010). As an additional complexity, it has emerged that (Abi-Gerges et al. 2011) that the hERG channel consists of at least two subunits, which confer different sensitivities for different drugs. For example, fluoxetine was more potent at blocking hERG 1a/1b than 1a channels, while E-4031 showed the reverse.

GENERAL TOXICOLOGY

In some ways, general toxicology is the least precisely defined and the broadest of all the branches of toxicological investigation, due to the number and variety of endpoints examined. A typical program of toxicity studies includes studies from single dose up to 12 or 24 months in length. This breadth of investigation poses a number of challenges in interpretation, which are best approached by taking an overview of the data, rather than trying to interpret change in each parameter in isolation from the others. In many ways, the presence of different parameters acting as markers of change in different organs or tissues gives greater security of interpretation, as change in one parameter may be supported by change or lack of change in another. In addition, the findings in one study should be reproducible in succeeding studies, giving confidence that marginal effects are treatment related or spurious. There are classic associations that are useful to remember in everyday situations, particularly in liver and kidney toxicities, which are the most frequent target organs; some of these are listed in Table 12.1.

As can be seen from Table 12.1, changes in some parameters can often be tied in to pinpoint change in a particular tissue or organ system. Some of the associations are unexpected. Thyroid change may be associated with a marginal anemia due to variations in the plasma concentrations of thyroid hormones. Antibiotic administration in rats can be associated with greatly increased cecum size but also with a

TABLE 12.1
Classic Associations in Toxicology

Liver Toxicity	Renal Toxicity
Increased plasma activity of liver marker enzymes, e.g., ALT and AST	Increased water consumption and urine volume. Urine parameters may change, e.g., enzymes and cellular debris.
Decreased plasma total protein concentration	Increased plasma concentrations of urea and creatinine. Proteinuria.
Increased coagulation times due to decreased synthesis of coagulation factors	Severe renal toxicity may lead to decreased erythrocyte parameters due to effects on erythropoietin synthesis.
Increased liver weight due to enzyme induction or accumulation of lipid or glycogen	Increased kidney weight.
Change in color or size at necropsy	Change in color or size at necropsy.
Histological findings such as necrosis or centrilobular hypertrophy due to enzyme induction	Histological change, e.g., basophilic tubules or necrosis, papillary necrosis, or glomerular changes.

Source: Reprinted from Woolley A, *A Guide to Practical Toxicology*, 1st ed., London: Taylor & Francis, 2003.

decrease in the peripheral neutrophil count, both as a result of a decrease in the intestinal burden of bacteria. Perhaps the most difficult changes to interpret are those that have a multitude of different causes. Reduced growth and food consumption may be due to sedation, true appetite suppression (as opposed to poor palatability of the food offered), abdominal discomfort, or other less specific and less easily identified causes. Reduced growth in the absence of reduced food consumption may indicate an effect on the GI tract or, in some cases, an effect on basal metabolic rate, as seen with decoupling of cellular respiration. In rats, either of these will be associated with reduced efficiency of food conversion, a measure of the amount of weight gained per gram of food consumed; however, this is no use outside the growth phase—generally around 15 to 20 weeks of age in rats. Pituitary tumors in rats are linked with a range of clinical signs such as hunched posture and torticollis but particularly with weight loss and lowered food consumption.

One of the most common effects seen in toxicity studies is induction of metabolism enzymes in the liver as a response to treatment with a xenobiotic. This is usually reflected in the liver by increased weight and hepatocyte hypertrophy in the central area of the liver lobule, around the central vein. Under the light microscope, the cytoplasm in the affected cells may have the appearance of ground glass. Under the electron microscope, a large increase in the endoplasmic reticulum is evident—the endoplasmic reticulum being the site of many drug-metabolizing enzymes, particularly the cytochrome P450 family.

In rats, this hepatic effect is often associated with changes in the thyroid, seen as increased weights and follicular cell hypertrophy; there is usually a dose response for these effects in the thyroid, which may mirror the liver changes. As a result, there may be the long-term consequence of thyroid tumors in carcinogenicity studies. The effects in the thyroid are elicited by increased metabolism of thyroid hormones by the enzymes induced in the liver. Reduced concentrations of thyroid hormones result in reduced negative feedback on the hypothalamic–pituitary–thyroid axis, leading to increased production of thyroid-stimulating hormone (TSH). The result is hypertrophy of the thyroid hormone–producing cells; there may also be a hypertrophic response in the anterior pituitary. Male rats are more commonly and severely affected than females because they have higher circulating concentrations of TSH.

This mechanism of toxicity is not thought to be relevant to man as the rat shows differences in protein binding of thyroid hormones, which tend to have a shorter half-life in this species than in man. In humans and monkeys, thyroxine and triiodothyronine are bound to a high-affinity thyroxine-binding globulin that is not present in other species; this lowers the percentage of free thyroid hormones. In rats, possibly, in part, as a consequence of the differences in protein binding and transport, the plasma half-life is between 12 and 24 hours compared with 5 to 9 days in humans. In addition, the follicular cells in the rat thyroid are very sensitive to increases in TSH; human thyroids are less sensitive to TSH changes than those in experimental animals (Gopinath 1999; Capen 2001).

Table 12.2 shows some of the problems encountered in general toxicology studies and suggested reasons.

TABLE 12.2
Troubleshooting in General Toxicology

Unexpected toxicity, compared with prior tests	Change in formulation or batch of test chemical. Poor predictivity of dose range finder studies due to factors such as differences in animal age, supplier, husbandry, or small group size.
Variation in individual response	Metabolic polymorphism or other genetic factor, social factors in group housing, e.g., nutrition status.
Low systemic concentration or area under the curve (AUC)	Poor absorption or poor formulation; isotonicity is important in parenteral formulations. Extensive first-pass effect. Short half-life.
Low toxicity	Low bioavailability; inappropriate route of administration or dose selection.
Interspecies differences	Different ADME; different mechanism of effect; species-specific mechanisms such as peroxisome proliferation; enterohepatic recirculation. Different expression of or affinity for pharmacological receptors.
Different response in males and females	Especially in rodents; due to different activities of metabolism enzymes in liver particularly but also physiological differences such as $\alpha 2u$-globulin excretion in males.

Source: Reprinted from Woolley A, *A Guide to Practical Toxicology*, 1st ed., London: Taylor & Francis, 2003.

One of the factors to be aware of in toxicity studies of all types is stress and the adventitious effects that stress may generate in any test system. In general toxicology studies, stress may be associated with decreased thymus and increased adrenal weight. This effect may be induced indirectly or directly. It is important to remember, however, that indications of immunotoxicity may be masked by this effect, and it is sensible to check other indicators of immune function before ascribing change in the thymus to stress.

A feature of general toxicity studies is the presence of specialist investigations, such as ophthalmoscopy, neurological examinations, and electrocardiography. In particular, electrocardiography poses a number of difficulties in interpretation because the method and conditions of collection are critical in defining the value of the data. In a restrained, anesthetized animal, the stress of examination will tend to increase heart rate and blood pressure, perhaps masking treatment-related effects that are present in the resting animal. To circumvent these problems, there is increasing emphasis on separate study of cardiovascular effects using animals with telemetry implants or telemetric jackets to monitor cardiac parameters such as heart rate, blood pressure, and electrocardiogram at regular intervals during the day. For the particular problem of QT interval prolongation, which was seen with drugs such as cisapride, there are *in vitro* studies that can give a reliable prediction of the presence or absence of this effect.

TOXICOKINETICS

Proof of exposure is one of the most important principles in toxicology and risk assessment; without exposure, there is no effect. Data presented can vary from a single sample taken post dose to a series of several samples taken over the dosing interval (usually 24 hours) to assess the course of systemic exposure in terms of absorption and clearance. Single samples cannot give any indication of the dynamics of exposure and simply give a snapshot of concentration at a particular time, usually in plasma. The chief question to ask in single-sample results is, What was the time of the sample relative to exposure? Toxicokinetic analysis or evaluation is the process of assessing the behavior of the analyte—parent and/or metabolites—following administration. There is some lack of clarity about the difference between pharmacokinetics and toxicokinetics. In essence, the latter generally refers to the assessment of exposure in toxicity studies. Definitions of toxicokinetics refer to the dynamics of toxic chemicals, and there may be some distinction from pharmacokinetics, which may be assessed at lower doses. Toxicokinetics looks at the time course of the first and last elements of absorption, distribution, metabolism, and elimination (ADME). Data from these two phases can give insight into the other two, if only in dynamic terms.

The basic parameters assessed include area under the concentration curve (AUC), half-life of elimination ($t_{1/2}$), and clearance and maximum concentration and time after dosing at which it occurs (C_{max} and T_{max} respectively). Usual practice is to take samples on day 1 of treatment and at the end of the study as it is common for differences in exposure to emerge with repeated treatment. Some of the factors involved in interpretation are considered in Focus Box 12.1.

FOCUS BOX 12.1   FACTORS TO CONSIDER
IN INTERPRETATION OF TOXICOKINETIC DATA

Dose proportionality	Exposure should increase with dose; if this is underproportional (the multiple between doses is less than the multiple between exposures), there may be reduced absorption at the higher doses. Care needs to be taken here, however, as lower exposure at high doses may be due to first-pass clearance.
	Exposure that is overproportional to dose may suggest some saturation of clearance at high doses.
Exposure after repeated dosing is higher than on day 1	Repeated dosing of a compound that is not completely eliminated within the dosing interval (typically 24 hours) is likely to lead to a steady increase in plasma concentration that eventually reaches an equilibrium between absorption and elimination; this gives higher concentrations than on day 1. There may also be differences between doses due to differential absorption and/or clearance. The toxicokinetics of metabolites may throw further light on these processes.
Exposure at the end of treatment period is lower than on day 1	Classically, this is the result of increased clearance or elimination due to enzyme induction in the liver. This may be associated with increases in liver weight, centrilobular hypertrophy, and, in rodents, effects in the thyroid glands. Again, the kinetics of metabolites should provide additional insight into these processes; greater exposure to the metabolite(s) after repeated dosing than on day 1 implies an increase in metabolism and may be correlated with the other endpoints described.
Half-life of elimination ($t_{1/2}$)	This is, classically, the time taken for the concentration of a drug to be reduced by 50%. It is a function of volume of distribution and clearance and, consequently, should be used with caution. However, it gives useful insight into toxicokinetics as short half-life should not be associated with accumulation.
Rapid absorption and clearance	Together, these may mean that the test system has not been exposed to the test material for long enough for toxicity (or any intended effect) to be expressed.
Systemic exposure increases with repeated dosing	Accumulating systemic concentrations may be associated with delayed onset of toxicity. Accumulation into specific body compartments, such as lipid tissue, may be associated with later toxicity when stored compounds are released.

CLINICAL PATHOLOGY

Two critical areas of investigation in general toxicity studies are clinical pathology and morphological (postmortem) pathology. Although both are capable of separate interpretation, their power is much greater when the data from both are combined. In this way, the presence of change in the blood or urine without associated change in morphological pathology can be put into perspective in terms of toxicological significance (Table 12.1). These investigations are useful during longer studies as they can indicate the target organ ahead of the terminal investigations and can lead to the use of specialist techniques for autopsy and microscopic examination. Generally, minor change in a single parameter, without any other correlative change, is unlikely to be of toxicological significance—especially if the values are within expectation (always assuming that there are enough historical data to give confidence).

Hematology

The critical groupings are indicated in Chapter 7. The normal life span of an erythrocyte in the blood is around 100 days (depending on species), and as it ages, it becomes less able to cope with oxidative stress. Aged or prematurely aged erythrocytes are removed by the spleen, and if this process is accelerated, anemia can result. If the bone marrow is healthy, there should be a compensatory increase in the immature forms of erythrocytes, particularly reticulocytes. If there is bone marrow toxicity, there may be a reduction in either erythrocyte or leukocyte counts or both; if erythropoiesis (generation of new erythrocytes) is affected, there should be no compensating increase in reticulocyte count. Anemia can also be induced by cell lysis in the peripheral circulation. Increased turnover of erythrocytes may be reflected in the presence of the pigment hemosiderin in the liver and increased plasma concentration of bilirubin. Changes in the leukocyte counts—total and differential—can indicate immunotoxicity or effects in the bone marrow. Changes in coagulation parameters are infrequent but can indicate liver change as the coagulation factors are synthesized there.

Clinical Chemistry and Urinalysis

These investigations cover a range of parameters that can be grouped loosely according to type (enzyme or nonenzyme) or target organ evaluated—typically the liver or kidney, although other organs and tissues can be evaluated by the use of targeted parameters, such as troponin to assess cardiac toxicity. The enzymes evaluated in the plasma typically give insight into liver effects, although urinary enzymes are increasingly used to assess renal toxicity. Nonenzymatic parameters that are examined routinely in toxicity studies include urea, creatinine, glucose, lipids, proteins, and electrolytes.

Changes in the plasma activity of enzymes can indicate the target organ for toxicity. It should be remembered, however, that early change in clinical pathology may not be reflected microscopically later in the study as the lesions may have resolved by that time. A further drawback of the theory of correlation of change in the plasma with morphological change is that it rarely seems to happen, except when the differences in the plasma are very large. Thus, increase in both alanine and aspartate aminotransferases (ALT and AST respectively) is a good indicator for hepatic toxicity. An increase in the plasma activity of AST alone suggests a different target tissue, for instance, muscle; this could be supported by increases in muscle enzymes such as creatine kinase or lactate dehydrogenase. Within the liver, alkaline phosphatase (ALP) is found on the biliary side of the hepatocytes, and increase suggests an effect on the biliary tree such as cholestasis. Increases in the plasma activity of aminotransferases may be due simply to increased permeability of the hepatocyte membrane, and this may not be associated with microscopic change. Where the enzymes are mitochondrial in origin, as with glutamate dehydrogenase or, in the rat, ALT, increased presence in the plasma may be indicative of necrosis, which should be evident microscopically. The presence of isoenzymes can complicate interpretation. ALP has isoenzymes that are specific for the liver (see earlier), bone, and gut. The bone isoenzyme decreases as growth slows, and the gut isoenzyme varies diurnally according to the feeding cycle. There is also an isoenzyme of ALP in the kidney in the brush border of the proximal tubule lumen, and the increased presence of this in the urine indicates renal toxicity. A change in the isoenzyme ratio can give an indication of target, though putting weight upon such an indication would depend on how accurately the change can be measured.

In the plasma, renal effects are indicated by increases in the concentration of urea and creatinine together. Increase in one of these alone is not usually indicative of renal change, especially urea, which may be increased due to inappetence and consequent nitrogen imbalance. Other indicators of renal toxicity are urinary volume and specific gravity or osmolality; total electrolyte output; and the presence of various other analytes including proteins, blood pigments, and cellular debris. Urinary enzymes such as ALP and N-acetyl glucosaminidase are good indicators of renal effects and are becoming more widely used in routine studies. The procedures used in urine collection should not be ignored in interpretation, including the duration of collection. Where the animals are placed in urine collection cages immediately after dosing, it is possible that the urine volumes and specific gravity will not reflect high water consumption recorded at other times of the study when free access to water is available. In this instance, in the absence of other evidence of renal change, the increased water consumption is probably a response to the dosing procedure rather than to any direct effect on the kidney.

A factor to consider in interpretation of enzyme activities is the variability of the values between animals and between examinations. For example, creatine kinase is very variable, and the levels can be affected by exercise and other factors, such as restraint. Lactate dehydrogenase, another enzyme with well-known isoenzymes, is also highly variable and is now less often examined. In studies in nonhuman primates, ALP may be considered to be too variable for meaningful interpretation and can be replaced by leucine aminopeptidase.

Changes in nonenzymatic parameters can lead to complexities in interpretation as they are often linked and codependent. At the simplest level, reduced food consumption and loss of body weight may be associated with increased plasma urea concentrations. Plasma concentrations of glucose and lipid may also be associated with differences in nutritional status, as may protein concentrations. However, protein concentrations may reflect changes in synthesis in the liver (which may result in increased coagulation times) or immunotoxicity (for which differential protein analysis may be useful). Concentrations of electrolytes in the plasma should be relatively stable as they are necessary for homeostatic mechanisms, particularly sodium and potassium. In the urine, however, interpretation becomes more complex as they become related to renal output and urine volume. Rather than look at individual concentrations, it is better to consider total output. Other urine parameters examined include proteins, glucose, and enzymes such as *N*-acetyl glucosaminidase and their ratios to creatinine.

During longer studies, where several examinations of blood and urine are undertaken, change with continuing treatment (or reversal of change following cessation of treatment) can indicate the progress of toxicity or the development of tolerance to the effects of the test compound. At the end of the study, these changes may be correlated with the presence of changes in the target organs when they are examined at necropsy, weighed, and then processed histologically for microscopic examination.

MORPHOLOGICAL PATHOLOGY

Interpretation in pathology is a specialist area, very much dependent on the experience of the pathologist performing the examination. Toxicological pathologists are famous for the divergence of their opinions, and it must be realized that an unwelcome conclusion will not be removed simply by getting another pathologist's viewpoint. What you then have is two opinions that are often slightly different and sometimes conflicting—such divergence can, if requested by regulators, lead to roundtable discussions of many pathologists (a potentially expensive proposition) to come to at least a majority opinion on the finding at hand. The key to pathological confidence is to ensure that the peer review, which is undertaken before finalization of the pathology report, is scrupulous, fully recorded, and agreed upon. The Organization for Economic Co-operation and Development (OECD) has provided guidance on this important process (OECD 2014), which has been reviewed by Fikes et al. (2015).

In examining differences in lesions, especially tumors, it is vital to ensure that the same number of sections have been examined from all the treated groups. Where macroscopic abnormalities are seen at necropsy in the highest dose group, more sections will be prepared and examined than in the controls. This can easily increase the recorded incidence of routine background findings, thus giving a false impression of treatment-related effect. Nonneoplastic findings are usually graded from minimal to severe or marked, and it is possible to see effects as increases in severity with increasing dose or as increases in incidence and severity. It is important that the interpretation takes this into account. In general, it is not practicable to provide historical control data for nonneoplastic pathology, because the data collection is to a degree subjective, for instance, in assigning grades. However, if there is an increase in incidence of a normal finding over the expected incidence in controls, it should be possible to offer an opinion on its significance. Considerations in this assessment include any increase in severity over expectation with increasing dose and the location and type of change.

Before attempting to interpret any data, it is important to have some understanding of its provenance and significance. For numerical data of simple parameters, this is relatively straightforward. Morphological pathology is more complex—more open to interpretation—and it must be acknowledged that the following discussion is not intended to equip you to interpret every pathology report that you read. This section is intended to be a stepping-stone, so that the changes detailed in a pathology report can be placed in an appropriate context; if it helps you to ask the right questions, it may have done its job successfully.

The intention of the following text is to review the types of change to look for in pathology, starting with the cell as the basic functional unit. With the exception of the nucleus, the normal cell is the smallest discrete unit that is visible to routine light microscopy. In terms of tissues and organs, many effects reflect changes in the cells that compose the tissue, either as direct effects on the cells themselves or as a result of changed cellular function in the extracellular tissues.

THE CELL

Death is always a crude endpoint whether an entire animal is involved or simply a focus of single-cell necrosis. Necrosis is the form of cell death that is associated with frank toxicity within the cell; in essence, it is unplanned and messy. This contrasts with apoptosis, which is programmed cell death and for which the causes may be much more subtle. In addition to cell death, effects at the cellular level may be seen due to changes in the following:

- Composition of the cytoplasm or cell contents (including extranuclear organelles such as the endoplasmic reticulum)
- Plasma membrane
- Hypertrophy (increase in cell size) and atrophy (decrease in cell size)
- Hyperplasia (increase in cell numbers)
- Death rate of cells—apoptosis or necrosis

SUBCELLULAR ORGANELLES

Subcellular change may be seen as effects on the cytoplasm or the nucleus. Effects on the cytoplasm may be seen at routine light microscopy as tinctorial or textural changes or as increases in the distance between the nuclei of adjacent cells. This is typically seen in hepatocytic hypertrophy. As a result of enzyme induction, the endoplasmic reticulum, which is the location of cytochrome P450 and associated xenobiotic-metabolizing enzymes, increases, and this is associated with increased cytoplasm in each cell. The result is that nuclei in the affected region—often the central zone of the liver lobule—become further apart, and this is evident in comparison with controls and with the adjacent periportal region of the liver.

The nucleus goes through a well-defined series of changes in response to two types of cell death, which may be programmed (apoptosis) or due to direct toxic insult (necrosis). Necrosis is generally considered to be a random event that may affect single cells (single-cell necrosis, often seen in the liver) or groups of cells. Nuclear change may be seen as karyolysis or pyknosis. In the former, the nucleus fades to a ghost outline, which may be due to nucleic acid degradation in response to a drop in cell pH. In pyknosis, the nucleus shrinks and stains more densely, followed by fragmentation; this is also known as karyorrhexis. Necrosis may be the result of gross changes in calcium distribution in the cell, between the endoplasmic reticulum or mitochondria and the cytoplasm.

In contrast, apoptosis is programmed, physiological cell death and is much tidier. While necrosis may be associated with macroscopic changes that are evident grossly, apoptosis is not evident at macroscopic examination. Apoptosis is an essential process in normal embryonic development, and disruption can lead to embryonic abnormalities. Later in life, disturbances in apoptosis may be associated with carcinogenesis. Apoptosis is more difficult to visualize, requiring specialist techniques, than necrosis because of its subtlety and relative rarity.

COMPOSITION OF THE CYTOPLASM OR CELL CONTENTS

Leaving aside the usually lethal changes in the physiological balance of the cell with respect to electrolytes such as calcium, sodium, or potassium, there are several changes possible in the cytoplasm, often as the result of changes in storage. Cells can store carbohydrate or lipid, and the levels of these may be affected by several toxicological mechanisms.

Carbohydrate storage is usually found in the liver, and increased amounts can lead to a foamy appearance of the cytoplasm due to its loss during the routine histological processing. Vacuolation due to the presence of fat—which is also lost during routine histological processing—is usually evident as microvacuolation or macrovacuolation. The accumulation of fat is a toxicological effect indicative of an effect on lipid utilization or export. This type of change is often seen in the liver or kidney.

Pigments may also accumulate in the cell due to toxicological processes:

- Lipofuscin is a "wear-and-tear" pigment that results from lipid peroxidation and reflects the polymerization of lipid peroxides.
- Hemosiderin is a breakdown product of hemoglobin and may accumulate in tissues such as the liver, spleen, and bone marrow in response to hemolytic anemia.

Although not strictly a pigment, calcium has affinity for dying or damaged cells. Usually the location is consistent with single cells, although more generalized calcification or mineralization may be seen. Hyaline deposits or droplets are indicative of protein deposition in cells and may be seen in renal proximal tubules as a result of reabsorption of protein not filtered out of the urine by the glomerulus.

PLASMA MEMBRANE CHANGES

Damage to the cell plasma membrane or to the membranes that enclose the organelles such as the endoplasmic reticulum is likely to lead to changes in permeability or function. While some agents such as anesthetics and detergents affect membranes in a general sense, there are also specific molecular events such as effects at receptors, protein channels, or enzymes embedded in the membrane. Binding to receptors and blockage of ion channels are frequent sources of toxicities. Membranes have high lipid contents that are affected by oxidative attack—lipid peroxidation—and are damaged accordingly.

Increased membrane permeability can result in entry or leakage of minerals or water and may be associated with leakage of enzymes into the plasma, which becomes evident in the results of clinical pathology measurements. Gradual leakage over a prolonged period should be distinguished from sudden release—and high-enzyme activities—that are associated with extensive cellular necrosis in organs such as the livers of rats treated with carbon tetrachloride.

HYPERTROPHY AND ATROPHY

At the cellular level, these terms relate to changes in size of cells but not in number. Hypertrophy is seen in hepatocytes in response to enzyme induction. Atrophy is seen in tissues such as the skeletal muscle but is more usually applied to complete tissues, such as the thymus. Hypertrophy in the zona reticulata and zona fasciculata of the adrenal is a frequent response to stress and is usually seen in conjunction with thymic atrophy.

Thymic atrophy is a normal process that is age related, otherwise known as involution. If it is seen without changes in the adrenal and earlier than in the controls, it may indicate immunotoxicity, either as a direct or indirect effect. Other data relevant to the immune system, such as white cell counts and function, should be assessed to confirm this.

HYPERPLASIA AND METAPLASIA

Hyperplasia—an increase in cell numbers—is seen as a response to a number of toxic agents. While hyperplasia may be a precursor to tumor formation, it is often reversible on cessation of treatment. Hyperplasia may be physiological, and this is often seen throughout a tissue or a region of a tissue, particularly in the endocrine system. Hyperplasia may also be seen in response to repeated toxicological insult, as in chronic irritation of skin or epithelia such as that in the bladder. Persistent hyperplasia may lead to tumor formation, which in these cases is usually a nongenotoxic response to treatment.

The converse of hyperplasia—when cell division ceases or is inhibited—has serious effects. At the most basic, it may result in lack of growth; at its most serious, it can result in rapid death. Agents used in the treatment of cancer are expected to reduce or halt cell division as the basis of their therapeutic effect on the tumor. However, the disadvantage of this is that tissues with high rates of cell division, the GI tract, bone marrow, and skin, are likely to be affected.

Metaplasia is a change in form of a normal tissue, such as an epithelium, to an abnormal form. This is seen classically in respiratory epithelia as a response, such as squamous metaplasia, to irritants such as tobacco smoke.

CHANGES TO WHOLE TISSUES OR ORGANS

These various cellular events are likely to be associated with change at the level of whole organs or tissues. However, the presence of so-called functional reserve in many organs is a factor in delaying the onset of overt change. This reserve varies from tissue to tissue and is an indication of the amount of damage that an organ can suffer before functional disruption and toxicity become evident; morphological change is probably one of the last processes to occur.

INFLAMMATION

Inflammation is a vital defense system in mammals and is divided into acute and chronic forms. Acute inflammation is a dynamic process that is characterized by erythema, heat, edema, and pain. It may have differing degrees of edema or structural tissue changes and may have differing dominant cell types associated with it. Inflammation may be initiated by allergens, infection, injury, or toxic insult. It is easiest to see in the skin, from which the classic descriptions of the signs—rubor, turgor, calor, and dolor—are taken. Initiation of inflammation can lead to a red, swollen area that is hot and painful.

Although inflammation is most visible on the skin, the process is the same in any affected tissue. The location of the inflammation may dictate which parts of the process are dominant. For example, inflammation in the lung is dominated by edema, which is not seen in bone; however, pain in bone inflammation is more noticeable.

Inflammation may end in the formation of an abscess or an ulcer, by resolution or fibrotic repair and scar formation. An abscess is an accumulation of dead polymorph leukocytes, surrounded by granulation tissue, while an ulcer represents the loss of epithelium and the formation of granulation tissue, which is a new connective tissue. Unless the tissue resolves completely to normal, which is unusual given that most injuries will include the local connective tissues, scar formation is inevitable.

The consequences (sequelae) of acute inflammation may include complete resolution or resolution with scarring or it may progress to chronic inflammation, which is seen if the stimulus persists or is repeatedly applied. Acute inflammation is of short duration (days or weeks), while chronic inflammation may persist for months or years. While acute inflammation is characterized by the presence of polymorphs, prominent granulation tissue and capillary formation, and an absence of granuloma, chronic inflammation has more macrophages and fibroblasts, low levels of granulation tissue and capillary formation, and prominent granulomas. Common causes of granulomas are foreign bodies, including dusts such as talc or silica.

CHANGES IN BLOOD SUPPLY

The blood supply to any tissue is, naturally, essential for maintaining the supply of oxygen and nutrients; disruption for anything other than transient periods may be critical.

There are clearly two possibilities in terms of blood supply, namely, an increase or a decrease. An increase is usually seen as congestion and may be a response to local irritation, inflammation, or similar factors. Congestion implies local dilation of the blood vessels, and this may be associated with reduced blood flow; this may then result in reduced oxygenation of the affected tissues. A decrease in blood supply in a critical location may be fatal.

Although this appears to be cut and dried, there are situations where blood supply may be interrupted temporarily or restricted so as to limit oxygen supply. This is classically seen in the papillary muscle of the beagle and minipig heart in response to agents that increase heart rate but reduce blood pressure. The papillary muscle is relatively poorly perfused when the dog is at rest; with increased heart rate, the muscle contracts more frequently but, with reduced blood pressure, is not perfused with enough blood to maintain itself. The result is necrosis, and this is an acknowledged effect of this type of agent in beagle dogs and in minipigs.

Blood supply is also an important factor in embryonic development and tumor growth, which have some analogies, as they are both situations where new tissue is forming rapidly and is dependent on blood flow. Reduced blood flow to the placenta would be expected to have potentially dire effects on the fetuses in the uterus and, at the least, could be expected to be associated with reduced fetal weights; effects beyond that might include early resorption, fetal death, or abortion. In tumor growth, angiogenesis—the process of formation of new blood vessels—is an essential factor; when blood supply to the center of large tumors starts to fail, the affected areas may become necrotic, giving a characteristic cut surface at necropsy.

REPAIR AND REVERSIBILITY

These two concepts are indivisibly linked, although it should be considered that repair does not necessarily mean reversibility. If a tissue is affected by a toxicant, the damage may be repaired but leave evidence of its repair in the shape of fibrotic lesions, which can be associated with functional change.

Repair may be affected by regeneration of lost cells or by fibrosis. Dead liver cells may be replaced by proliferation of adjacent healthy cells. Basophilic cells are characteristic of the early stages of regeneration and are often seen in the kidney, where damage repair may be seen as basophilic cells or tubules. When these cells become more differentiated to the normal function state, they lose their basophilic characteristics. In the kidney, death of the proximal tubular cells is followed by their shedding into the tubule lumen; the next stage is a gradual reconstitution of the tubular epithelium, firstly as flattened cells, which then become cuboidal before resuming the normal morphology of the proximal tubule. This type of repair is likely to be effectively invisible unless the tissues are harvested at the time that the damage or repair is in progress. It is possible to have early kidney damage that is not seen at subsequent tissue harvests weeks later. However, this process of repair in the kidney is heavily dependent on the integrity of the basement membrane of the proximal tubule epithelium, and if this is breached, this seamless process of repair is hindered, leaving the opportunity for persistent lesions that will effectively reduce renal function.

Fibrotic repair, however, is likely to leave persistent lesions that will be seen weeks or months after the event. It is characterized by the presence of new connective or granulation tissue and collagen. The consequence is scarring and a permanent lesion that may, if exposure is continued, accumulate to the extent that it inhibits normal function of the tissue, as in cirrhosis of the liver in response to chronic alcohol consumption.

The speed of repair may differ between lesions and tissues. Repair in the liver can be very rapid; following a toxic dose of carbon tetrachloride, hepatic repair may be complete by the time a normal 14-day observation period has been completed. On the other hand, changes due to accumulation of pigment or intracellular bodies generally take much longer to reverse, as the metabolic processes of removal tend to be slow. The consequence is that reversibility of pigment or protein inclusions may take much longer than the 4-week reversibility period that is routinely allowed in toxicity studies. In this situation, it is useful to have data from two dose levels available to get an estimation of any dose relationship in the extent or rate of repair.

Broadly speaking, the reversibility of morphological change is seen as a mitigating factor in the assessment of toxicity, and it may be cited as a reason for categorizing the change as not being of toxicological significance. However, before this can be done with confidence, the tissue affected, the nature of the original change, and its severity and incidence must be taken into account. Generally, hypertrophy of hepatocytes around the central vein, which is usually due to enzyme induction, is readily reversible and is seen as an adaptive change that is, depending on scale, unlikely to be of toxicological significance. The same may be said of minor change in the renal proximal tubule. Nevertheless, it may be necessary to view even reversible change in some tissues with suspicion; changes in the nervous system, either peripheral or central, may presage toxicity that will be limiting in longer studies or—more importantly—in exposed humans.

NEOPLASIA

Literally "new formation or development," neoplasia is a critical process in toxicology, and it may be classed benign or malignant, the latter being cancer. Neoplasia is distinct from hyperplasia, which is essentially an increase in cell numbers without differentiation from normal tissue.

The basic comparisons between benign and malignant tumors are given in Table 12.3. Diagnosis and classification of tumors is a contentious area for pathologists and can be a source of acrimonious dispute.

While the classification of tumors by the use of the suffixes, *-oma* for benign tumors and *-sarcoma* for malignant tumors, is apparently foolproof, there are exceptions, melanoma and mesothelioma both being aggressive malignant tumors. For most purposes in toxicity testing, neoplastic tissue is most likely to be seen in aged

---

TABLE 12.3
Benign versus Malignant Tumors

Benign	Malignant
Remains at the point of origin.	Invades adjacent tissues; metastasis to distant locations through blood and lymph.
Often encapsulated with defined margins.	Poorly defined margins.
May compress adjacent tissues.	May destroy adjacent tissue.
Necrosis is not common but may occur if blood vessels are compressed.	Necrosis is common.
Growth tends to be slow.	Growth may be very fast.
Tumors arising in epithelial and mesenchymal tissues are suffixed *–oma*.	Tumors of epithelial origin are suffixed *-carcinoma*; those from mesenchymal tissues are suffixed *–sarcoma*.

rodents, typically rats or mice. The appearance of tumors in other species is very rare, usually in single animals and not related to treatment.

Any tissue can be subject to a carcinogenic response; for example, Zymbal's gland in the rat ear is the site of sebaceous carcinomas induced by 2-acetylaminofluorene. There is no reason to suppose that any particular tissue is not going to be associated with a carcinogenic response. However, some tissues are more likely to be affected than others, the principal one being the liver, predicated on its central role in metabolism and the fact that it is the site of first systemic exposure following absorption from the gut.

Other rodent tissues that are frequently affected include the forestomach, lung, testis, kidney, and hemopoietic system. The actual background frequencies differ between strains and species and, in some cases, the sexes.

Overview of Interpretation in General Toxicology

In simplistic terms, the tissue, type, extent, and reversibility of the findings determine the significance of pathological change. A minimal, centrilobular hepatocystic hypertrophy, with a slight increase in liver weight but without associated clinical pathology change, is likely to be of low toxicological significance if it is shown to be readily reversible. This type of effect is seen typically with minor induction of hepatic metabolism. Where the mechanism for an effect is known, as for hemosiderin deposition in the liver in hemolytic anemia, the finding is also unlikely to be of toxicological significance, if the cause itself was not so great as to be of concern. Pigment deposits are unlikely to be quickly reversible, as pigments tend to be slowly metabolized and therefore more persistent than easily repaired changes in tissues such as the liver or kidney. Changes in tissues that do not repair readily, such as the nervous system, are of much greater concern. These have to be considered in terms of no-effect dose levels and the difference between toxic levels and those seen or expected in human populations. The levels of endocrine hormones are frequently not investigated in routine toxicity studies, in part because of the effects that stress of sampling can have on their plasma concentrations. However, examination of endocrine glands, or tissues under hormonal control, can indicate the presence of hormonal change and point to potential problems that may be seen in longer studies. This is due to the large influence that hormonal levels have on nongenotoxic carcinogenesis.

REPRODUCTIVE TOXICOLOGY

In general toxicology, the principal time-dependent change is growth, with increasing maturity and metabolic capability, which occurs over the lifetime of the animal. In contrast, reproductive toxicology adds extra layers of complication, because of its sensitivity to disturbance and the added dimension of transient, time-specific processes, which themselves have considerable complexity. The final outcome—offspring that can reproduce in turn—is influenced by effects on processes that start with male spermatogenesis and function, continue with mating behavior in both sexes and gestation, and culminate in parturition and postnatal care and development.

To this can be added, in rare cases, transplacental carcinogenesis, as expressed by diethylstilbestrol.

Reproductive toxicology is an area in which *in vitro* screening has been adopted in order to speed up the selection of compounds for further development or for chemicals that are already on the market but have not been examined previously. These tests, although reasonably predictive of effects *in vivo*, are not infallible, and it has to be asked whether the experiment has produced a result that is relevant to humans. Where the compound has been marketed for a significant time without problems and comes from a class of chemical that is not known for reproductive toxicity, it is probably reasonable to dismiss the *in vitro* data as not relevant—although it would be sensible to say why. Taking whole-embryo culture as an example, the major differences from the situation in life are the absence of a placental barrier and of maternal metabolism, both of which can have a protective effect for the fetus.

Although it may be relatively simple to conclude, from basic data, that a chemical reduces fertility or that it causes a reduction in postnatal survival, the root cause may not be obvious. In considering the results of fertility studies, it is useful to refer to the data from general toxicity studies for effects on reproductive organs or for data that might imply any hormonal effects. Depending on the data, and the existing knowledge of the compound's class and expected actions, it is important to confirm that the males were treated for long enough before mating to show any effects. Maternal toxicity is also a factor to consider. In general, guidelines require that the high dose be chosen so as to show toxicity, but excessive maternal toxicity can result in delayed development in the uterus, which may imply effects that are not immediately relevant to humans at low exposure levels. Typical of these are retarded ossification and reduced fetal weights.

TABLE 12.4

Troubleshooting in Reproductive Toxicology

Reduced fertility—male	Spermatogenesis or other testicular change (see histopathology), epididymal function changed, change in sperm quality (CASA results), behavioral change, and stress.
Reduced fertility—female	Lower implantation rate, increased postimplantation loss—possibly due to excessive maternal toxicity, behavioral change, and stress.
Wavy or extra ribs	Variant that is generally not thought to be significant.
Unexpected toxicity in rabbits	Inappropriate vehicle—oils or other vehicles that affect gastrointestinal function are not suitable in rabbits. Stress can also be a factor in this species.
Prolonged or abnormal parturition	Hormonal imbalance.
Poor survival of pups postpartum	Defective lactation or maternal care; excretion of test chemical or metabolites in milk.

Source: Reprinted from Woolley A, *A Guide to Practical Toxicology*, 1st ed., London: Taylor & Francis, 2003.

Note: The factors listed in Table 12.2 should also be considered. CASA, computer-assisted sperm analysis.

Much of the sensitivity of reproduction arises from the interdependence of factors such as hormonal balance, nutrition, behavior, physiology, maintenance of the placental barrier, and the complex balance in the embryo between growth, programmed cell death (apoptosis), and essential processes such as angiogenesis. Table 12.4 shows some of the problems that may be encountered in reproductive toxicity studies; it is by no means exhaustive. The conclusion is that interpretation of reproductive toxicity studies must be undertaken only when a full data set is available, including data for ADME. The data relating to fetal exposure and to excretion in the milk are particularly important in this, and these tests should be undertaken if there is any question that these factors may be relevant to the results in the routine testing program.

GENOTOXICITY

This is the one area of toxicological testing in which *in vitro* tests have been accepted by regulatory authorities, largely due to the relatively easy definition of the endpoint, which is essentially that of DNA damage at the level of either the gene or the chromosome (Tables 12.5 and 12.6). For secure conclusions to be drawn, all the genotoxicity studies need to be considered together. In contrast to other branches of toxicology, strength of response is not generally taken to be a prime factor in interpretation, as even a weak genotoxic response indicates mutagenic potential. Extrapolation from effective concentrations *in vitro* to those seen or expected *in vivo* is not sensible without caution, and as a consequence, no threshold is accepted for mutagenicity *in vitro*. Pragmatically, however, it is clear that there is a gradation of potency between mutagens, and a positive response at very high concentrations is less likely to be relevant to the situation in life.

Having said that, interpreting the results as positive requires some care. A positive result is indicated by a clear dose–response curve; if there is a sudden increase in effect at high concentrations, this may be due to physical effects or toxicity and hence irrelevant biologically. In addition, the difference from controls should be statistically significant. The weakness of the statistical approach, as for many of these tests, is that as n is usually only 3, the statistical method is inevitably not especially powerful; use of larger numbers of negative controls (e.g., 6) helps this situation. Finally, the results of the test for controls and positive controls should be compared with historical control data to confirm that they are within expectation. Positive results are sometimes found when testing early research batches of the chemical, due to the presence of impurities. These can also be introduced (or eliminated) by changes in production methods.

As with all toxicology, exposure must be demonstrated; this is generally not a problem *in vitro*, where even precipitates have been associated with genotoxicity. Physicochemical properties that prevent the substance crossing the cell membrane and solubility in aqueous media may become limiting factors. Proof of exposure of the target cells is a particular problem in the *in vivo* micronucleus test, especially with a negative result. In some cases, excessive toxicity or pharmacology limits the doses that can be achieved, leading to inadequate exposure of the bone marrow. In circumstances where it is not practicable to achieve high systemic exposure, a negative result *in vivo* cannot offset a positive *in vitro* result.

TABLE 12.5

Guide to Genotoxicity Interpretation

Negative result	No dose–response curve or statistically significant increase in effect compared with the negative controls, providing that the positive controls have performed as expected. *In vitro*, this is confirmed in a second experiment, sometimes with different harvest times. Evidence of exposure is essential in *in vivo* tests.
Positive results	There should be a statistically significant increase with dose response. A twofold or threefold increase over control values has been used.
Micronucleus test	Mean micronucleus count in controls and positive controls must be sufficient for the study to be acceptable; indicative values are 4 and 10 micronuclei per 2000 polychromatic erythrocytes per animal, respectively. There should be a dose–response curve with at least one point with a statistically significant increase in aberrations over the vehicle control.
Ames test (bacterial reversion assay)	There should be a dose-related statistically significant increase in numbers of revertant colonies in two separate experiments. The strains indicate the following: TA 1535, TA 100 → base substitution. TA 1538, TA 98 → frameshift. TA 1537, TA 97 → single frameshift.
Cytogenetics—Chinese hamster ovary (CHO) cells	Clastogenic effect is indicated by a dose–response curve with at least one point having a statistically significant increase in aberrations over the solvent control. Reduced damage scored at higher dose levels may result from complex interactions between cell cycle and induced damage, and the dose–response curve may not be a simple increase in damage with dose.
Mouse lymphoma assay	Small colony size may indicate slow growth due to DNA damage, while large size may indicate point mutation. The interpretation of large and small colonies is still debated.
Unscheduled DNA synthesis (UDS)	Increase in nuclear grain count indicates a positive result. Autoradiography allows correction of grain counts for cytoplasmic synthesis of nucleic acids.

Source: Reprinted from Woolley A, *A Guide to Practical Toxicology*, 1st ed., London: Taylor & Francis, 2003.

The initial response to a positive result should be to ask if it is biologically relevant and how it has arisen. Before a positive result can be dismissed, it is important to understand the underlying mechanism. Thresholds of response are a factor in assigning a negative result to a test. These may be due to interaction with non-DNA targets, for instance, through conjugation or lack of availability to DNA at low concentrations. This is seen with paracetamol, where the active metabolite is conjugated at low concentrations. There may also be metabolites that are not formed *in vivo* or in humans. Pharmacological activity, such as spindle inhibition, can also produce positive results.

As with other areas of toxicology, thresholds of effect are important in interpretation in genotoxicology as the presence of an effect beyond a particular, and preferably high, dose may indicate that the effect seen is not relevant to humans. Although the historical assumption has been that genotoxicity is a nonthreshold phenomenon—a single mutagenic event may trigger a carcinogenic response—it

TABLE 12.6
Troubleshooting in Genotoxicity

General Problems

Lack of toxicity or negative result	Possibly due to poor exposure. Mouse micronucleus test—limit dose, 2000 mg/kg orally → no effect → has it been absorbed? Try parenteral dosing. (Negative result may be due to excessive toxicity.)
No response in positive controls	Has the test system been correctly characterized?
Different results for different batches	Test substance purity. Manufacturing process changes.
Cytotoxicity	Excessive cytotoxicity may give a positive result in chromosome aberration studies. In the mouse lymphoma assay, positive responses at >90% cytotoxicity are not considered biologically relevant BUT need a close dose range to demonstrate reliable negative results.
Positive *in vitro*, not verifiable *in vivo*	Exposure *in vitro* cannot be replicated *in vivo* at target tissue. Can be due to poor absorption or excess pharmacology or different metabolism. If the positive result was with S9 mix, does this mimic metabolism *in vivo* or in test species or in humans? Perform new *in vivo* test (UDS) or *in vivo* mutation.
Positive result	Review of all data and assessment of cost/benefits. Choose additional assays that will help explain the result rather than simply add to the data set.

Troubleshooting in Specific Tests

Ames test	Lower colony counts at high concentrations may be due to toxicity, which can conceal a positive result.
In vitro mammalian cytogenetics or micronucleus	Chromosome damage at high concentrations in mammalian cells *in vitro* may indicate that the harvest times were inappropriate; different harvest times in the second experiment may help to clarify effects seen.
	Threshold effect or lack of dose response at high concentration may be without biological relevance, due to physical effects or toxicity or presence of metabolites at high concentration (in presence of S9 mix).
In vivo micronucleus	Excess pharmacology or toxicity or poor absorption. First-pass metabolism → poor systemic exposure to parent. Different active molecule at target tissue compared with *in vitro*. Excess stress may lead to a small increase in micronuclei.
Unscheduled DNA synthesis (UDS)	*Ex vivo* preferred over *in vitro*; autoradiography preferred over liquid scintillation counting.

Source: Reprinted from Woolley A, *A Guide to Practical Toxicology*, 1st ed., London: Taylor & Francis, 2003.

is increasingly recognized that there are thresholds of effect and that the concept of no-adverse-effect level (NOAEL) or no-observed-effect level (NOEL) can be applied to genotoxicity data as well as to other data.

Kirkland and Muller (2000) published a review on the importance of thresholds in the interpretation of the biological relevance of genotoxicity test results. They noted that there has been an increase in the numbers of positive results, especially in *in vitro* chromosomal aberration tests, but that few of these were associated with positive results *in vivo*. This lack of correlation calls into question the relevance of the *in vitro* result for either rodents or humans (the lack of concordance between rodent genotoxicity and/or carcinogenicity and the same endpoints in humans is another matter, which is dealt with elsewhere in this book). Although a threshold response at high concentration may not indicate any genotoxic risk at concentrations likely to be experienced by humans (or whatever target species), such effects should be explained and the mechanism of effect understood before they can be dismissed. Such mechanisms include extremes of pH, ionic strength or osmolality, indirect genotoxicity due to interaction with non-DNA targets, genotoxicity due to metabolites present at high doses that are conjugated or cleared effectively at low doses, and/or production of metabolites *in vitro* that are not present *in vivo*. Broadly, if the margin of difference between the threshold concentration and those expected in humans is very large, there is a good argument that the genotoxicity seen *in vitro* is not biologically relevant.

It is routine in genotoxicity testing *in vitro* to confirm the results of a first experiment in a second, preferably with slightly different conditions or harvest times. Because the assays are relatively inexpensive, it is easy to react to positive data by repeating assays or performing new tests. Among these, due to normal biological and statistical variation, there will be a proportion of results that are also positive. In these cases, as the data set grows, an overall interpretation becomes much more difficult. The moral of this is to be very careful about repeating tests or choosing supplementary ones.

Use of *in silico* technologies can be a useful aid to understanding the mechanisms behind any positive results and can, in some instances, predict potential toxicophores or areas of concern. This is further detailed in Chapter 5.

CARCINOGENICITY

Data relevant to potential carcinogenicity are contained in several different study types, including the classic rodent life span bioassay. As with genotoxicity studies, which are a critical part of carcinogenicity assessment, the data package should be viewed as a whole. Data relevant to carcinogenicity can also be derived from routine toxicity studies, and these can give valuable indicators for potential nongenotoxic carcinogenesis. Effects such as enzyme induction or the presence of hepatic foci with different staining characteristics may be associated with a later positive result in the carcinogenicity bioassays. As already said elsewhere, the genotoxic carcinogens are relatively easy to detect before getting to the stage of long-term studies; the chemicals that are carcinogenic indirectly are much more of a challenge. Classifications of carcinogens, for instance, by the International Agency for Research on Cancer (IARC), give useful background to interpretation in this often contentious area (Table 12.7).

The mainstay of carcinogenicity assessment is still the 2-year bioassay in rodents, although transgenic models are becoming more important, especially in the United States. In addition, the applicability of these assays to pharmaceutical development

TABLE 12.7

Carcinogen Classification

Good epidemiological evidence in humans; about 30 compounds	Known human carcinogens, e.g., arsenic, benzene, vinyl chloride, aflatoxin	IARC group 1
Limited epidemiological evidence, sufficient evidence in animals	Probable carcinogen, e.g., polychlorinated biphenyls, diethylnitrosamine, phenacetin	IARC group 2A
Insufficient human evidence, reasonable evidence in animals	Possible carcinogen, e.g., TCDD, DDT, diethyl-hexylphthalate	IARC group 2B
Not classifiable	Diazepam	IARC group 3
Not considered to be carcinogenic	Caprolactam	IARC group 4

Note: Compiled from sources in the References and reprinted from Woolley A, *A Guide to Practical Toxicology*, 1st ed., London: Taylor & Francis, 2003.

is under review, as contained in recent International Conference on Harmonisation (ICH) guidelines; although the process is dealing with the rat at the moment, it will progress to the mouse in due course.

The basic intention of a carcinogenicity assay is to demonstrate the presence or absence of an increase in tumor incidence or burden in treated groups compared with appropriate controls. This apparently simple objective becomes increasingly complex as the various supporting or influencing factors are considered. Differences in tumor burden or time of onset (latency), between control and treated groups, may be attributable to a range of factors other than the simple mechanism of action of the test substance. The effect of these factors may be additive or negative. Increased growth and body weight tend to increase tumor burden and to produce a different tumor distribution compared with animals that grow less or more slowly. This has been the subject of considerable debate in recent years and has resulted in the use of strains that do not eat and grow so much. In some models, the tumor incidence in the controls can approach 100%, which effectively reduces the information derivable from the study to an assessment of effect on latent period for tumors; this has been seen in some photocarcinogenicity protocols.

As with reproductive toxicology, carcinogenicity assessment is very dependent on the quality of the historical control database. If that is deficient, the assessment of the significance of rare tumors becomes much more difficult. Although use of a double control group will alleviate some of this, it cannot completely answer the problem. The use of mortality adjustment and statistics can only be of assistance with more common tumors and cannot address the single renal carcinoma that may be found in the high-dose group. In the lower-dose groups, a single rare tumor may not be a problem, providing that there is no evidence of a U-shaped dose–response curve or excess toxicity at the highest dose.

Other data that are available for assessment of carcinogenicity potential include the routine toxicity studies, genotoxicity, pharmacological actions (including those peripheral to that expected), and metabolism and pharmacokinetics (Table 12.8).

TABLE 12.8
Guide to Carcinogenicity Interpretation

Tumor increases in both sexes, both species	Clear carcinogen with probable relevance to humans. Review genotoxicity data.
Tumor increase in one sex in both species	Equivocal result: mechanistic studies may resolve this issue. Review genotoxicity and ADME data and all nonneoplastic pathology.
Negative genotoxicity data, with tumor increase in one tissue, possibly in one sex	Possible nongenotoxic, species-specific mechanism. Mechanistic studies should demonstrate (non)relevance to humans. Results of routine toxicity studies may show evidence of early change in the affected tissue. Tumor increase is often associated with nonneoplastic change predisposing to tumor formation. Possible class effect. Established classes of chemical and effect, e.g., peroxisome proliferation.
Low toxicity	Maximum tolerated dose (MTD) may not be achieved, leading to doubtful regulatory acceptance; a 10% decrease in body weight gain due to reduced food consumption is not evidence of MTD unless backed up by pharmacokinetics and/or metabolism. Presence of excessive pharmacology at higher doses may be a factor.
Lower food consumption in treated groups	Leads to lower tumor burden and increased life span through dietary restriction. Possibly due to poor palatability of diet if test substance offered with feed.
Higher food consumption	Increased tumor burden and reduced survival.
Increased survival in treated groups	Longer exposure may lead to different tumor burden of routine tumors in comparison with controls.
Decreased survival	Tumor rates must be adjusted for mortality to account for lower numbers of animals exposed for full test duration. If this is confined to high dose, this may be due to differential toxicity expressed only at high doses. Survival below 50% or 25 animals at completion of the study may invalidate the results for regulatory authorities.
Increased incidence of tumor and associated nonneoplastic changes not seen with other compounds of same class	Compare pharmacokinetics of other compounds from same class; achieving similar levels of exposure may show similar histopathological changes, indicating a class effect. Differential metabolism may be a factor.
Increase in rare tumor	Is the tumor in the high dose only? Is there a mechanistic explanation that is not applicable to humans? How recent and extensive are the background data at the test facility? Was this seen with other compounds of the same type? How frequent is this tumor in other historical control databases?
Increase in common tumor	Is the incidence within the historical control range? Is there a shorter or longer time to onset, or is there a difference in survival between the groups? Dose response?
Increase in tumors in mouse liver	May not be relevant to humans as the mouse liver is sensitive to nongenotoxic compounds.

Source: Reprinted from Woolley A, *A Guide to Practical Toxicology*, 1st ed., London: Taylor & Francis, 2003.

It should be borne in mind that pharmacokinetics and metabolism may differ at high doses from that seen in the lower-dose groups and that the pharmacokinetics for the test compound may well change as the animals get older. In particular, renal function declines with age, and if there is any subclinical nephrotoxicity, this decline may be accelerated by treatment. The assessment of toxicokinetics in long-term bioassays may not be available for the later stages of the study, as this is not always a regulatory requirement. This is a significant weakness of current guidelines.

Where a chemical has been shown to produce tumors in both sexes of both species or in several organs, it is a clear indication that this is a carcinogen of probable relevance to humans. Having said that, it has been difficult, with some known human carcinogens (such as arsenic) to produce tumors in animals. Where there is an increase in tumor incidence in one sex in a single tissue, it is possible that this may be due to a nongenotoxic mechanism that is unlikely to be of relevance to humans. This requires mechanistic studies for confirmation of lack of relevance. This category contains a large number of chemicals acting through well-established mechanisms, such as peroxisome proliferators and those that act on the α2u-globulin. These are associated with significant nonneoplastic pathology in the affected tissues and are, theoretically, easy to predict from the results of the routine toxicity studies. With this type of data, it may be possible to prepare for interpretation of the carcinogenicity studies in advance by performing appropriate mechanistic studies. Nongenotoxic mechanisms of carcinogenicity are often accompanied by a clear threshold dose below which no effect is seen. Within a class of chemicals, differential absorption of the class members may lead to unexpected differences in effects in the rodent studies. This may be investigated by comparative assessment of toxicokinetics, as was shown for the fibrate family of hypolipidemics; this is described in the ciprofibrate case study at the end of this chapter.

ENVIRONMENTAL TOXICOLOGY AND ECOTOXICOLOGY

Data for these linked disciplines relate to testing following deliberate (and known) exposure of individual test systems or specified areas (ecotoxicity tests), or investigation of unexpected effects in the environment as a whole (such as eggshell thinning or population changes) (Table 12.9). These data sets are distinct and set different challenges in interpretation. There is an implicit distinction between environmental events that affect people and those that affect the ecosystem as a whole. The former represents the interface between epidemiology and environmental toxicology and includes episodes that affect human populations. Ecotoxicology, by contrast, can be taken to include effects on other fauna and flora and their environment (the ecosystem), although the studies may well be similar to epidemiology, in that they concentrate on a single species. The presence of effects in humans due to pollution does not rule out effects on the local fauna and flora. In reports of smogs in Los Angeles or London, there are very few references to the effects on the urban wildlife. However, although there may be few reports of wildlife effects, it is counterintuitive to infer an absence of toxicity when such far-reaching human effects were seen.

Experimental ecotoxicology tests—of the type carried out to support registration of chemicals such as pesticides—produce data that follow similar rules of

TABLE 12.9
Factors to Consider in Environmental Toxicology

Controls	Are the controls correctly chosen and defined? If an area is selected as a control, is this area truly comparable with the study site?
Normality	How is normality or expectation defined, and how recently was this definition produced?
Population dynamics	Populations change naturally in the absence of effect from synthetic chemicals, and this may mask or enhance ecotoxicological differences. Population balance may be disturbed by factors outside the definition of the study limits.
Measured parameters	Was the correct parameter chosen for measurement—variability, normal levels, ease of measurement, and relevance?
Observed differences	Are these direct or indirect? Are they (a) real or (b) relevant?
Transient excess mortality	Was the period of record long enough to show effects that may have persisted beyond the study period? What were the effects of concurrent disease and increased susceptibility to subsequent disease? What is the differential mortality between polluted and nonpolluted areas?

Source: Reprinted from Woolley A, *A Guide to Practical Toxicology*, 1st ed., London: Taylor & Francis, 2003.

interpretation as for other toxicity studies. With single-species studies conducted in laboratory conditions, the data have to be extrapolated to the ecosystem or environment as a whole in much the same way that single-tissue studies *in vitro* have to be extrapolated to the whole organism and with the same uncertainties and weaknesses. Understanding the dynamics of the test environment and study apparatus is an essential for correct interpretation and subsequent efforts to predict ecotoxicity. Mesocosm tests based on replicas of ecosystems or parts of ecosystems may give a better insight into effects based on interrelationships between different fauna and flora. In all cases, it is important to define exposure and to chart distribution of the test chemical through the environment and then to follow its sequestration or elimination and to link this to the presence or absence of effect. One aspect of mesocosms is that the inherent variability increases with the complexity of the experiment. As a result, normality is more difficult to define. It is possible that clear-cut effects seen in the laboratory may be lost in the wider variation possible in a mesocosm experiment. At least with a mesocosm study, the source of all the components should have been characterized before they were added to the system. With a field study, this is less feasible, but characterization of the components must be as scrupulous, to avoid masking of treatment-related differences, which might confound interpretation.

When an unexplained environmental effect is observed, there are similarities with the problems encountered in epidemiological studies. These include poor definition of exposure, difficulty of choice of controls and a multiplicity of interrelationships, and dependencies that complicate interpretation. One of the first questions to be answered relates to exposure to synthetic chemicals, which may be previously identified markers of exposure or effect. Analysis of relevant tissues, corpses, soil samples, or whatever sample is appropriate or available for chemicals is a specific indicator of exposure. One problem with such specificity is the greatly increased sensitivity of analytical technique. Presence at low levels is not necessarily causative, and it must be decided if a chemical is present in sufficient quantity to be responsible for the observed effect. This is not always cut and dried. At one time, Americans contained quantities of DDT, but this was not linked to significant toxicity in the population as a whole, probably because most was sequestered into adipose tissue. One reason for the absence of human effect is specificity of toxic mechanism; another is the usual epidemiological difficulty of assigning cause and effect in conditions that are widely present in the general population. In predatory birds that showed eggshell thinning in response to accumulation of DDT up the food chain, it was found that a metabolite, DDE, reduced calcium deposition in the eggshell; this has so far not been a significant problem in Americans. Although markers of effect may also be indicative of exposure to a xenobiotic or environmental factor, they are not necessarily specific, and the question has to be asked: "Were the markers relevant?" As with epidemiological studies, it is important that a proposed cause and effect be linked by a credible toxicological mechanism; otherwise, other causes should be considered.

Population decline, which can be a first indication of an ecotoxicological effect, may be due to a variety of causes such as reproductive failure or incapacity, habitat destruction, and direct toxicity, among others. Accumulation of toxins in the study population in comparison with suitable controls from a similar location is a good indication of cause, but it must be considered that such accumulation could be responsible for a different effect than the one being studied. Populations are dynamic and can respond to changes in pollutant levels quite quickly; this can make interpretation of studies carried out in successive breeding years much more complex. Any ecosystem is subject to a range of pressures, and changes in species distribution and population are likely to be influenced by more than one simple factor. In the Great Lakes, populations of fish-eating birds have been affected by DDT-induced eggshell thinning, changes in fish populations in terms of numbers and species due to fishing practices, and habitat; these various factors have worked together or independently to make year-to-year comparisons more difficult.

Data sets that accumulate following known pollution events such as oil spills or high-concentration chemical releases into rivers pose their own particular challenges. The first of these is that a trustworthy preincident characterization of the local ecosystem is not always available. Spillage into a bay with low tidal exchange of water may mean that the effects will be localized and comparison with nearby or similar sites may be possible. Release into a river can be associated with long-distance transport of pollutants, as has been illustrated in the Rhine on at least two occasions. Interpretation of all such data is dependent on prior knowledge of the affected areas; the discovery of a difference from expectation does not necessarily imply relationship to recent high-profile pollution. Chronic low-level release— leakage from old mining activities or water reservoirs for holding washings from mines—is likely to have as significant (but possibly more insidious) effects as sudden release in large amounts. Furthermore, such low-level release may not become evident until long after the pollution started. One challenge is that a pollution incident seldom provokes only one study and that the data from different studies may not be collected in a manner that allows easy comparison of results or pooling of data to enable more powerful analysis. Uncoordinated study can simply lead to a larger database that is not conclusive.

EPIDEMIOLOGY AND OCCUPATIONAL TOXICOLOGY

The differences between epidemiological and occupational toxicology, which are in some ways different parts of the same field, are subtle and relate in large part to the size and definition of the population under study. Both involve the study of chemically induced effects in populations, the epidemiological population usually being larger and more diverse than a worker group. This may be summed up by the difference between the workers in a chemical facility and the effects of discharge from the same facility on the surrounding community. The difference is also exemplified by smoking and vinyl chloride. The former investigation involved the population at large, and in the latter, a small, defined group of chemical workers was examined. The overlap between the two disciplines occurs at the point where a small group becomes a population—a community versus small focused groups of workers in an occupational setting. Epidemiology tends to highlight an effect after it has happened, as was the case with asbestos, rather than indicate its probability beforehand. It is often a retrospective tool, although prospective studies are undertaken.

In both areas, it is important to ensure that the correct comparators are used, with avoidance, as far as possible, of confounding factors such as the healthy worker effect, where workers tend to be healthier than the general population (Table 12.10). Investigation in a wider context may be complicated by small differences from controls and differences in protocol. In the examination of the influence of the Mediterranean diet on longevity and health, there have been suggestions that the genetics of the local population may be a significant factor and that the influence of olive oil and red wine may be less than hoped for by interested parties. Clearly, a lot more research is needed here, conducted locally in appropriately smoke-free bistros.

EPIDEMIOLOGY

In contrast to laboratory toxicology, epidemiological study is conducted in the field with a diverse population in which the exposure is often poorly defined and outcome is often compromised by other factors such as smoking or alcohol. Good epidemiology is dependent on rigorous control of variables and of confounding factors that may invalidate the conclusions if not fully appreciated and allowed for. Thus, in a study of respiratory disease, due to an occupational hazard, smoking could be expected to influence the results and is therefore a confounding factor. Because the natural variability among humans is so large, it is difficult to detect minor deviations from controls without using vast numbers of subjects. Epidemiology is therefore good for detecting clear effects that may be associated with exposure to a specific substance. Clarity of effect may be due to rarity of the observed disease or to numbers affected. Thus, cause-and-effect linkage of vinyl chloride with hepatic hemangiosarcoma was made easier by the rarity of the tumor and the distinct population in which it was seen.

TABLE 12.10
Factors in Epidemiology and Occupational Toxicity

Controls	Healthy worker effect. Population chosen—influence on study outcome.
Confounding factors	Alcohol, smoking, and occupational exposures. Effects that are synergistic or additive to that of the investigated substance.
Faulty or inconsistent differential diagnosis	Poor distinction between conditions having similar symptoms but different etiology, e.g., bronchitis may be bacterial or viral or associated with smoking, atmospheric pollution, or occupational exposure.
Questionnaire	If questionnaires were used for data collection, was the wording structured so as to avoid bias?
Definition of exposure	In epidemiology by history of persons; in occupational toxicology by personal monitoring equipment and by urine and blood collection for analysis.
Biological markers	Is the chosen marker specific for the chemical of concern or for the same group, e.g., cholinesterase inhibition? Is it a measure of exposure or effect or susceptibility?
Statistical significance	Spurious significances due to numbers of relationships being examined in some studies—leads inevitably to a number of false positives.
Data accessibility	In some data sets, there is a temptation to collect and analyze only the more easily accessible data; this can lead to bias.
Significant contributing factors not considered	Genetics, lifestyle, intercurrent disease. Recent papers have suggested that the greater survival seen with the Mediterranean diet may be due to genetic factors.

Source: Reprinted from Woolley A, *A Guide to Practical Toxicology*, 1st ed., London: Taylor & Francis, 2003.

The importance of numbers is illustrated by the association of smoking and lung cancer, which has been taken further to show that smoking exacerbates respiratory disease in occupational exposure to asbestos or in uranium mining. For "normal" diseases such as leukemia, variation from normality is more difficult to define, especially as the human population tends to be naturally inhomogeneous. The result of this inhomogeneity is the presence of clusters of diseases in particular areas, for example, a high incidence of meningitis in some villages or the presence of leukemia clusters around a nuclear plant. The normal presence of a disease in a population makes it extremely difficult to separate low-level effect, which might be due to exposure to a chemical, from normal variation. One product of this uncertainty is often a succession of studies, each seeming more authoritative than the last and each with a different conclusion. For many years, it has been considered that all alcohol should be avoided in pregnancy because alcohol in large amounts has been clearly associated with fetal alcohol syndrome; recently, it was suggested that small amounts might be beneficial. Currently, the pendulum has swung the other way. Equally, it has long been suggested that saturated fats should be avoided in favor of polyunsaturates; then it has been countersuggested that too much polyunsaturated fat is not good. Although it would seem sensible to take all the data from all the studies and pool them to establish the cause of the effect or correctness of attribution, this is usually made difficult or impossible by variations in experimental technique, population differences, differing criteria for differential diagnosis, etc.

Establishment of a relationship between cause and effect by epidemiology is dependent on comparison of an exposed population or one showing the effect with a control that is unaffected or unexposed to the chemical or agent of interest. Correct choice of controls is crucial as they are the "normal" population against which the test group will be compared. If you are looking at minor differences between individuals or study groups, the definition of normality is very important. Move the normality goalposts, and the conclusions will change.

Where the condition investigated in epidemiological studies is associated with a naturally existing background incidence or with a wide range of values, it is quite possible to have different studies indicating different and opposite effects. The influence of interest groups is also a factor to consider. The information that red wine is good prophylaxis against cardiovascular disease was good news for the red wine producers of Bordeaux but less wonderful for white wine sales. Within a relatively short time, a study emerged showing the benefits of white wine.

Another example of contradictory epidemiological study results is provided by examination of the relationship between electromagnetic radiation from power lines and leukemia in children. In this work, the definition of the exposed population varied between studies, some including houses up to 100 m either side of the power lines while others used a smaller distance. Lack of comparability between study protocols made an overall assessment of the data impossible, leaving little scientific proof but a lingering public perception that they had been misinformed or led astray. The absence of any mechanism by which the leukemia could be induced was also a crucial weakness. In the final analysis, no amount of epidemiological research into a fuzzy problem will overcome public perception.

The most contentious aspect of epidemiology is the interpretation of small differences from expectation or controls. This is seen with conditions or events that have a significant natural background incidence in populations or environments that are inherently variable and often poorly controlled. This leads to poor reproducibility of results from one study to the next, interpretation of which is made more difficult by differences between protocols and chosen populations. These methodological differences reduce the extent to which data can be pooled for extended analyses of the whole database. It is only when associations are very strong that the results of epidemiological study are accepted with anything resembling speed. The postulated decline in sperm counts in response to environmental estrogens is a case in point. Interpretation has been hampered by analysis and reanalysis of data with conflicting results, and variability in sperm counts due to seasonal factors, donors, health, occupation, counting techniques, and sample quality. Although there may be, intuitively, a toxicological mechanism that can be held responsible, the differences from controls have not been large enough to satisfy epidemiologists of cause and effect. The overriding problem here is the variation that is inherent in the population and the consequent inability to produce a sufficiently robust definition of normality.

A crucial task for epidemiological toxicologists is the definition of the extent and duration of exposure to the chemical of concern. Human life is seldom challenged by a single chemical at high doses but is subject to exposure by a mixture of many chemicals at individually low doses, although the total exposure may be huge when expressed in milligrams per day. In general, mixtures, especially undefined mixtures, are much more difficult to assess due to antagonistic and synergistic interactions of the various components. The effects of oxidative attack by chemicals, present naturally or as contaminants in the environment, can be offset by high levels of dietary antioxidants, absorption of which may itself be compromised by the presence or absence of other dietary components.

Added to this is the problem that disease is seldom an immediate response to exposure, although asthma and other allergies are notable exceptions to this. The time lapse between exposure and response, especially to long-term (chronic) low-level exposure, makes attribution nearly impossible unless the response is unusual or rare. Therefore, it is generally not possible to ascribe a common tumor such as breast cancer to a specific cause in a patient for whom there is no predisposing exposure defined. In contrast, despite the long latency period, it is relatively straightforward to ascribe mesothelioma to occupational exposure to asbestos because it is rare tumor found in an easily characterized population. One approach is to use the concept of excess mortality. That is, in specific conditions, it may be possible to attribute deaths above the normal rate to those conditions. A challenge for epidemiologists, therefore, is knowing where to draw a line between clear toxic effect and normal background and showing when change from normality is due to toxicity.

OCCUPATIONAL TOXICOLOGY

While epidemiological data may relate to hundreds of people, the occupational toxicologist may be presented with data from a single individual, for instance, DNA adduct analysis in the urine of an employee handling a potentially reactive chemical. Unless the difference from expectation is large, interpretation of a single data point is difficult. Ideally, there should be baseline data from the same individual before exposure took place or from unexposed workers in the same plant or area. It should be possible to chart the exposure from baseline and start of work, to abnormal DNA adduct levels during work, and then a return to normal when the shift stops. Before any trustworthy assessment is attempted, it must be confirmed that the increase in DNA adducts is a marker not only for exposure but also for the adverse effect attributed to the chemical being handled; without this vital linkage being made, valid conclusions cannot be drawn. One of the principal advantages that an occupational toxicologist has over an epidemiologist is the relative strength of definition of exposure that is available in the workplace, in terms of identity and, often, of dose. In either field, the criteria for attribution include definition of exposure, exclusion of confounding factors, a clear significant connection between exposure and condition, and a supportable mechanistic explanation. With such data to hand, it should become possible to attribute an individual case or group of cases to a particular cause. The confidence with which this can be done becomes critical when there is a legal case to answer, especially in occupational health cases.

When attempting to interpret the relationship between exposure to a particular chemical and effects in the workplace, it is important to consider alternative sources of exposure to the substance of interest. For example, formaldehyde, a commonly used chemical, has a wide presence in the home, in carpets, furniture, clothing, and home insulation products. Other factors must be considered; for example, radon, smoking, and other agents such as diesel exhausts may make a synergistic or additive contribution to the incidence of lung cancer in underground miners. When these additional external factors are taken into consideration, it should be possible to draw a conclusion as to whether the cancer is connected with occupational exposure to the chemical. For individual cancers, which are not part of a group or specific to a particular agent, it is highly unlikely that a connection between it and a specific exposure can be drawn.

In contrast to epidemiology, interpretation of effects in the workplace may be facilitated by the relative certainty of what the exposure was, the presence (usually) of a condition in a specific group of workers, and the timing of effect in the affected individuals. Thus, response to a chemical in the workplace may be seen at a higher level or incidence toward the end of the working day, at particular times, or during particular processes and may resolve on weekends. Ventilation, especially the recirculation of air in new buildings, is an important consideration and can be associated with sick building syndrome.

Biological markers are an essential tool of the occupational toxicologist, but, as with all tools, their limitations have to be accounted for in interpretation of their data. The basic contention is that exposure to a chemical is associated with change that may be seen as variations in the concentration, expression, or activity of a biological marker. This type of change, which in all probability is subclinical in the early stage of exposure, may lead to organ dysfunction later in life, with associated clinical consequences. If levels of the biological marker are monitored, preemptive action may be taken to prevent further exposure. From the data, it should be feasible to extrapolate backward to the exposure and forward to the prospective clinical outcome. In order to be of use in the investigation of potential effects, biological markers should be as specific as possible to the chemical of concern. In essence, they are markers of exposure, effect, or susceptibility to effect. The most specific marker available is analysis of blood or urine for the presence of the chemical or its metabolites. Certain markers of effect are specific for particular groups of chemicals, such as inhibition of acetyl cholinesterase due to exposure to organophosphate or carbamate insecticides. Other markers, such as DNA adducts, are produced by a wider range of chemicals and are not peculiar to any single chemical unless specifically identified as such.

Once a biological marker has been validated as being specific or indicative for exposure to a particular chemical or type of exposure, it should be possible to interpret data from individuals. A change in the value for a marker from baseline for that individual (beyond normal variation) or from general expectation indicates excessive exposure and the possibility of subsequent clinical effects if exposure is not minimized or halted. In validating biological markers in animal experiments, it should be remembered that the thresholds of toxicity may differ between the experimental animal and humans; this is especially significant if humans are more susceptible than the test animals.

There should be interactive interpretation between toxicology and epidemiology. Increasing use of biological markers should encourage the interaction between the two disciplines through consideration of toxicological sequelae of exposure and the mechanism of effects attributable to chemicals.

The following case study on ciprofibrate (Case Study 12.1) may be a little elderly but nevertheless illustrates a good way to interpret data. The chapter on errors in toxicology contains a number of poor ways to go about interpretation (Chapter 20). In addition, this case illustrates an excellent investigative strategy and is a powerful answer to the accusation that can be made (especially in pharmaceutical development) that toxicology kills good compounds.

CASE STUDY 12.1   CIPROFIBRATE

Ciprofibrate, a derivative of phenoxyisobutyrate, is one of a series of hypolipidemic compounds (fibrates), which includes clofibrate, bezafibrate, and fenofibrate. All have marked hypolipidemic activity in humans and animals, reducing both plasma cholesterol and triglycerides through effects on low-density or very low-density lipoproteins. The animal toxicity of fibrates has been reviewed by Bonner et al. (1991). They are well known as peroxisome proliferators, an effect that is known to be associated with hepatocarcinogenicity in rodents; the class as a whole is nonmutagenic. The safety evaluation program for ciprofibrate, which was conducted in the light of previous work on earlier members of the series, showed up a number of toxicities in rodents, which required explanation. These consisted primarily of liver changes, with associated effects in the thyroids and the presence of a low incidence of carcinoid tumors in the glandular mucosa of the rat stomach.

Peroxisome proliferation was noted in both rats and mice (Rao et al. 1988) together with increased liver size. This was associated with hepatic adenomas and hepatocellular carcinomas, which were seen in the long-term carcinogenicity studies. There were also functional and morphological changes in the thyroid of rats (Bonner et al. 1990, 1991; Spencer et al. 1988), which were associated with decreased plasma concentrations of thyroxine (T4) and with minimal-to-mild thyroid follicular hyperplasia. The morphological changes were considered to be consistent with increased thyroid activity. Increased metabolism of thyroid hormones as a result of hepatic enzyme induction is often associated with increased plasma concentrations of TSH as a result of the absence of the negative feedback provided by normal T4 or T3 levels (see earlier). Although TSH was shown to be increased over short administration periods, this was not demonstrated in longer studies, a situation that is not unusual, due to compensatory mechanisms.

To demonstrate the rodent specificity of the hepatic effects, a long-term study in marmosets showed a lack of peroxisome proliferation; although liver changes were seen, these were an order of magnitude lower than in the rat (Graham et al. 1994). In view of the long-standing association of peroxisome proliferation with hepatocarcinogenesis and of hepatic enzyme induction with thyroid change, these various effects were not unexpected. They are known to be specific to rodents and to have no relevance to humans, a point underlined by the absence of effect in the marmoset.

Although the liver and thyroid effects were expected and explainable, the presence of carcinoid tumors in the glandular fundus of the stomach of Fischer rats posed a problem that was potentially more serious. The incidence of gastric carcinoid tumors was 5/59 males and 1/60 females seen in animals that survived for the whole study at 10 mg/kg/day in the diet. Marked hyperplasia of fundic neuroendocrine cells was seen in non-tumor-bearing animals of this group with other changes in microscopic gastric morphology. These changes were not seen in the mouse and were not reported with other fibrates. However, this type of carcinoid tumor had also been seen with long-acting gastric antisecretory compounds such as omeprazole. An investigation was mounted to discover whether there was a secondary pharmacological action of ciprofibrate on gastric secretion and to determine the sequence of events in tumor formation. Two other objectives were to look for this effect in other species and to ask if other fibrates had the same effects. The duration of antisecretory activity is proportional to the likelihood of tumor formation. Thus, the long-acting H2-antagonist loxtidine and proton pump inhibitor omeprazole have both been associated with gastric carcinoid formation. Shorter-acting compounds such as cimetidine, in once-daily regimens, are not associated with this change.

Following treatment of rats with ciprofibrate, changes were seen in the acid-secreting oxyntic cells—hypertrophy, with eosinophilia and reduced vacuolation of the cytoplasm, associated with reduced secretory cell organelles. In separate studies, ciprofibrate was shown to decrease acid secretion and the volume of gastric juice, an effect that was also shown with other fibrates. However, in a 26-week comparative study with once-daily dosing of bezafibrate in rats, no similar changes were seen. Investigation of the pharmacokinetics of the two compounds (Eason et al. 1989) showed that the elimination half-life of ciprofibrate was significantly longer than for bezafibrate—3 to 4 days and 5 hours, respectively. To reproduce the systemic exposure pattern for ciprofibrate, bezafibrate was administered twice daily at 12-hour intervals; this was successful in producing similar changes to those seen with ciprofibrate given once daily. Furthermore, ciprofibrate given at 10 mg/kg every 48 hours gave similar sustained plasma concentrations to those produced by bezafibrate given at 125 mg/kg every 12 hours.

Gastrin stimulates acid secretion, with low gastric pH acting as a negative feedback mechanism. It is also involved in regulation of mucosal growth and exerting a trophic action on neuroendocrine cells. Reduced acid secretion can therefore lead to hypergastrinemia, which, if sustained, may produce neuroendocrine cell hyperplasia in the gastric mucosa. Ciprofibrate given at 20 mg/kg/day to rats gave a modest but statistically significant hypergastrinemia over a period of 56 days. Investigation of other fibrates showed that twice-daily administration of bezafibrate at 150 mg/kg also produced increased plasma levels of gastrin after 12 weeks. Clofibrate twice daily at 75 mg/kg was shown to have similar effects. That these changes in gastrin concentration were dependent on pharmacokinetics was illustrated by the finding that ciprofibrate given once every 48 hours at 20 mg/kg produced less hypergastrinemia than with daily dosing. Morphologically, 6 to 9 months of treatment two or three times daily with bezafibrate or clofibrate at up to 150 mg/kg produced similar changes in neuroendocrine cells in the stomach to those seen with ciprofibrate, consistent with prolonged mild hypergastrinemia.

The species specificity of the changes in gastrin concentrations was examined in mice and marmosets. While mice showed a transient increase in plasma gastrin concentrations, this was not sustained, and there was no evidence of change in the morphology of the gastric mucosa. These findings are consistent with the absence of gastric carcinoids from the 2-year carcinogenicity study in mice. Similarly, there was no change in plasma gastrin level in the marmoset over a 26-week treatment period. Although there were some minor changes in the oxyntic cells in marmosets after 26 weeks of treatment at 100 mg/kg, there was no evidence of change in the neuroendocrine cells. In addition, hypergastrinemia was not seen in humans.

A hypothesis was constructed that prolonged antisecretory activity induced by ciprofibrate (in contrast to the transient effect seen with bezafibrate and clofibrate in once-daily dosing regimens) led to hypergastrinemia and a persistent trophic stimulus with hyperplasia of the neuroendocrine cells. This latter effect is responsible for the gastric carcinoid tumors seen in rats. Prolonging the antisecretory activity of bezafibrate and clofibrate by twice-daily administration leads to similar effects as seen with ciprofibrate.

From this case study, it may be seen that rapid clearance of drugs can mean that toxicity is not manifested and that toxic potential is not predicted adequately. Determination of the duration and consistency of exposure is critical in the interpretation of the data, and this should be correlated with the pharmacodynamics of the compound. Persistence of an otherwise easily reversible change, as seen with the fibrates and hypergastrinemia, may lead to unexpected effects that have significance for the development of the compound. Equally, it is important to look at the species specificity of these changes to assess their relevance to the ultimate target species, which is usually humans. In terms of interpretation, it is clear from this investigation that, at each stage, consideration of the accumulating data and of relevant literature allowed a logical progression of studies and a solidly based interpretation of the findings when the process was complete.

SUMMARY

The following are suggested as a basic set of rules for successful interpretation of toxicological data:

- The whole picture is needed for secure conclusions to be drawn.
- A definition of normality, provided by adequate controls or historical control data, is essential.
- The experimental protocol must be sufficiently robust to achieve the stated experimental objectives.
- Demonstration of exposure—duration and extent—is essential.
- Confounding factors due to husbandry or experimental technique or procedures must be excluded or accounted for.
- Do not overinterpret or extrapolate from small or poorly controlled data sets.
- In the immediate aftermath of a crisis or incident, it is difficult to achieve correct interpretation, as data will continue to emerge as the situation progresses.
- Above all, you cannot interpret the data unless you understand what they mean.

When you know the meaning of the results, it should be possible to interpret them and then to perform an extrapolation to humans. Interpretation is an evolutionary process and is supported by appropriate additional experiments to test developing hypotheses for mechanism of effect and species specificity. From this basis, it should be possible to perform a prediction of the compounds toxic potential in humans, and this is addressed in the next chapter; needless to say, the bedrock of an accurate prediction is secure interpretation of the full set of toxicity data and all other supporting information.

REFERENCES

Abi-Gerges N, NHolkham H, Jones EMC, Pollard CE, Valentin J-P, Robertson GA. hERG subunit composition determines differential drug sensitivity. *Br J Pharmacol* 2011; 164(2b): 419–432.

Bonner FW, Eason CT, Spencer AJ. Phenoxyisobutyrate derivatives: A review of animal toxicity. *Res Clin Forums*. 1990; 12(1): 23–45.

Bonner FW, Deavy L, Astley N et al. Investigation of the mechanism of ciprofibrate induced thyroid hyperplasia in the rat. Proceedings of the 30th Annual Meeting of the Society of Toxicology. *Toxicologist* 1991; 11: 152.

Capen CC. Toxic responses of the endocrine system. In: Klassen CD, Ed. *Casrett & Doull's Toxicology—The Basic Science of Poisons*, 6th ed., New York: McGraw-Hill, 2001.

Eason CT, Powles P, Henry G, Spencer AJ, Pattison A, Bonner FW. The comparative pharmacokinetics and gastric toxicity of bezafibrate and ciprofibrate in the rat. *Xenobiotica* 1989; 19(8): 913–925.

Fikes JD, Patrick DJ, Francke S, Frazier KS, Reindel JF, Romeike A, Spaet RH, Tomlinson L, Schafer KA. Scientific and Regulatory Policy Committee Review: Review of the Organisation for Economic Co-operation and Development (OECD) Guidance on the GLP requirements for peer review of histopathology. *Toxicol Pathol* 2015; 43(7): 907–914.

Gopinath C. Comparative endocrine carcinogenesis. In: Harvey PW, Rush KC, Cockburn A, Eds. *Endocrine and Hormonal Toxicology*. Chichester: Wiley, 1999.

Graham MJ, Wilson SA, Winham MA, Spencer AJ, Rees JA, Old SL, Bonner FW. Lack of peroxisome proliferation in marmoset liver following treatment with ciprofibrate for 3 years. *Fundam Appl Toxicol* 1994; 22: 58–64.

ICH. Harmonised Tripartite Guideline: Safety Pharmacology Studies for Human Pharmaceuticals S7a. November 2000. Available at http://www.ich.org.

Kirkland DJ, Muller L. Interpretation of the biological relevance of genotoxicity test results: The importance of thresholds. *Mutat Res* 2000; 464(1): 137–47.

OECD: OECD Series on Principles of Good Laboratory Practice and Compliance Monitoring Number 16. Advisory Document of the Working Group on Good Laboratory Practice Guidance on the GLP Requirements for Peer Review of Histopathology, 2014.

Pollard CE, Abi Gerges N, Bridgland-Taylor MH, Easter A, Hammond TG, Valentin J-P. An introduction to QT interval prolongation and non-clinical approaches to assessing and reducing risk. *Br J Pharmacol* 2010; 159(1): 12–21.

Rao MS, Dwivedi RS, Subbarao V, Reddy JK. Induction of peroxisome proliferation and hepatic tumours in C57BL/6N mice by ciprofibrate, a hypolipidaemic compound. *Br J Cancer* 1988; 58(1): 46–51.

Redfern WS, Carlsson L, Davis AS, Lynch WG, MacKenzie I, Palethorpe S, Siegl PK et al. Relationships between preclinical cardiac electrophysiology, clinical QT interval prolongation and torsade de pointes for a broad range of drugs: Evidence for a provisional safety margin in drug development. *Cardiovasc Res* 2003; 58(1): 32–45.

Spencer A, Eason CT, Pattison A et al. Functional and morphological changes in the thyroid gland of Fischer 344 and Sprague-Dawley rats given bezafibrate and ciprofibrate. *Hum Toxicol* 1988; 8: 400.

Woolley A. *A Guide to Practical Toxicology*, 1st ed. London: Taylor & Francis, 2003.

13 Prediction of Hazard

INTRODUCTION

Defensive toxicology, which is showing lack of relevance of a result to humans or other target species, is a major driver for continuing the development of a chemical, particularly new pharmaceuticals. The previous chapters have covered the background and process of testing that seeks to show differences from normality, which may represent toxicity and the interpretation of the resulting ocean of data. This chapter sets out to explore the process of hazard prediction that takes place once the results of toxicological investigations have been collated and reported. In the context of this book, it is seen as high-level interpretation of the whole data set that takes place after interpretation of individual studies, but before risk assessment and management. However, it has to be said that defensive toxicology does not mean saving a compound at any cost; clearly, there is an ethical and moral line that should not be crossed. Data should be explained, not ignored or concealed.

The key word in this is *prediction*, which has often been used to mean the use of *in silico* methods. However, while toxicity studies can be said to be broadly successful in identifying toxicological hazard, it is the failures—TGN1412, Vioxx—that get the publicity because they are associated with clear ill-health or deaths. When viewed dispassionately, it becomes evident that all of toxicity is predictive, whether it is an *in vitro* study or a 26-week study in rats. Human clinical trials of pharmaceuticals are frequently found to be less than perfect in their prediction of low-incidence effects, and cynically, it should be recognized that relatively small groups of humans are often a poor model for the wider population. There is no single system that can be said to be infallible when used in isolation for hazard prediction; the only recourse, as there is no choice but to test chemicals for safety, is to use a range of tests and to take all the evidence into account in assessing likely toxicological hazard and thence risk.

The safety evaluation studies, which should be seen as including all three major test systems—*in silico*, *in vitro*, and *in vivo*—identify hazards and are used for hazard characterization. After this, it should be possible to predict which hazards are relevant for risk assessment. Inevitably, there is substantial overlap between these processes. Risk assessment is carried out on one hazard at a time, and so it is important to identify and prioritize human-relevant hazards before embarking on the next stage.

There are a number of settings in which this process of prediction is required, including preparation of applications for field trials with pesticides, for first dose in humans with novel pharmaceuticals, and when establishing best work practices with chemicals. Assessment of toxicities as relevant hazards for humans has taken on much greater significance as the cost of chemical development has soared, encouraging investigation of toxicity to show lack of human relevance to recoup the huge costs of development. As *in vitro* methods of investigation have improved, these investigations have become ever more focused. Frequently, these take the form of *in vitro* comparative experiments in which human tissues are used with tissues from the test species. If the effects seen can be shown to be irrelevant to humans, the chemical may still be worth developing, other factors being favorable.

HAZARD, RISK, AND HUMAN-RELEVANT HAZARD

It is important to distinguish between hazard and risk. Hazard is the description of the adverse effects of a chemical; it is not quantitative and does not take exposure, dose, or form into consideration. Risk is the probability that this hazard will occur; clearly, if there is no hazard, there is no risk. Hazards of concern to humans are, broadly, cancer, reproductive effects, debilitating illness or disease, and workplace-related effects that might prevent them from working. In addition, any potential progressive, degenerative change that may be due to acceleration of normal age-related decline in function should be considered. Needless to say, this type of insidiously progressive change is extremely difficult to predict from toxicological data because the circumstances for each individual at risk are so different, in terms of genetics, exposure, diet, and other factors that influence individual responses.

A hazard identified for animals is not necessarily a hazard for humans—for example, peroxisome proliferation, which is associated with hepatic carcinogenesis in rodents, is not considered to be a hazard for humans, due to differences in hepatic metabolism. The object is to predict which toxicological effects detected in the safety evaluation have significant potential to be expressed in humans, and so, to identify them as being relevant for the risk assessment process. This indicates the need for accuracy in prediction, as it is important that false positives and false negatives are avoided; the former may divert attention from real hazard, and the latter may expose people to unacceptable toxicity. Hazard characterization is reliant on correct overall interpretation of the various studies, and this, in turn, is dependent on the interdependent interpretation of studies within each toxicological discipline, taking account of any overlap with other areas. For instance, the interpretation of data from reproductive studies may be influenced by findings in general toxicity studies, such as testicular atrophy. It is vital, therefore, that all evidence be considered and that the interpretation of all studies be used to extrapolate a prediction of effect in the target species, which is usually humans.

The difficulty lies in extrapolating effects seen at high doses in animals to those expected in humans at lower doses. The number of animals used in testing will be small relative to the numbers of humans exposed to the chemical, and it is important to be able to extrapolate the findings to the much larger human population to assess their relevance to humans. The fundamental flaw in this paradigm of extrapolation from effects at high dose to those predicted at low dose is that pharmacokinetics, exposure, and mechanisms of effect are very likely different at low dose, and accordingly, high-dose or high-concentration experiments may not be predictive of human-relevant effect.

Such extrapolation is dependent on a thorough understanding of the chemical's effects on the individual, whether animal or human (dynamic and toxic changes), and of the effects of the individual on the chemical (kinetics and biotransformation). Understanding these processes, dependencies, and influences is key to successful prediction of effect in humans and then to the process of risk assessment that follows. The conduct of a carefully designed program of toxicity studies, in a range of test systems, assists this understanding and makes risk assessments for human use more secure.

The case of ciprofibrate, given in the previous chapter, is an example of the type of stepwise approach that may be taken in assessing the human relevance of hazards identified in nonclinical studies. Another example is lamotrigine, an antiepileptic drug, which was found to accumulate in kidneys in the male rat, causing progressive nephrosis and mineralization. These effects were attributed to action on α2u-globulin, a mechanism that is specific only to male rats and not relevant to humans. In dogs, lamotrigine is extensively metabolized to the 2-*N*-methyl metabolite, which is associated with dose-related effects on cardiac conduction, leading (at high doses) to complete atrioventricular conduction block. In humans, production of this metabolite (found in urine) equates to less than 0.6% of a dose (*Physicians Desk Reference*, 2007), and the relevance of this hazard to humans is considered to be minimal. However, it was suggested that in patients with liver disease and/or reduced glucuronidation capacity, the concentrations of this metabolite may be increased.

CIRCUMSTANCES OF HAZARD PREDICTION

In evaluating chemicals, especially new synthetic chemicals, it is important to distinguish among hazards that are specific to the various test systems used during safety evaluation and those that might affect humans or the environment. The objective is to protect humanity, the environment, or any specific target from potential adverse effects that might arise from the use of novel chemicals. This can also apply to natural chemicals that are proposed for use in unnaturally high concentrations or circumstances.

Predictions of hazard are made to support the use of new drugs in clinical trials (especially for first administration to healthy volunteers), for pesticide field trials, workplace exposures during production, environmental effects, and the use of food additives. Differences in target populations may modify hazard assessment or significance; teratogenicity is not a hazard for an exclusively male population, although many teratogens, such as thalidomide and diethylstilbestrol, also have effects on the male reproductive system. The presence of disease may modify the response to a chemical, particularly a drug. Thus, patients may benefit from taking a drug, but healthy workers may show an adverse response, usually an unwanted pharmacological effect. Although there are populations for which particular hazards are not relevant, factors that affect the response of the individual do not affect the relevance of the hazard, but they do modify the risk. Uranium miners are all subject to the hazard of lung cancer as a result of their workplace exposures, but for smokers, the risk is much greater than for nonsmoking miners. Likewise, the toxicities expressed by slow and fast metabolizers of isoniazid may be different, but it is sensible to consider that the hazard of both is relevant to both populations of patient; it is simply the risk that is different.

Prediction of hazard to the environment or the ecosystem is more complex than that for human populations as the scope for interactions is much greater. There may be some sense in drawing a distinction between an environmental hazard that affects humans directly, for example, release of estrogenic substances or discharge into drinking water, and those that affect the ecosystem as a whole. The latter will affect humanity indirectly but is likely to be much less emotive to the general population than a perceived direct effect such as cancer.

PRINCIPLES OF PREDICTION

There are two "simple" stages of hazard prediction, firstly, identification of the hazard—usually from the animal and *in vitro* studies that are available—and then assessment of the relevance of the hazard to humans. There have been two approaches to these linked questions. The first (and least discriminating) was that any hazard identified in animals was relevant to humans and that the second question was therefore irrelevant. It was this type of assumption that spawned the Delaney amendment, by which any substance shown to cause cancer in animals should not be allowed as a food additive in the United States. The problem with this is that practically anything can be shown to cause cancer in animals, if you are sufficiently dedicated and the dose levels are high enough. Such dedication has shown that a natural constituent of mushrooms, 4-hydrazinobenzoic acid, can cause tumors in mice when administered in the drinking water at high dose levels (see Focus Box 1.2, Chapter 1). This might become relevant to humans if people start drinking mushroom ketchup in gargantuan quantities.

The second approach weighs all the evidence and subjects it to a process of expert judgment to arrive at a conclusion as to the relevance of the changes seen. In particular, the inadequacy or appropriateness of experiments should be taken into account when assessing the data and the credibility of conclusions reached in individual studies. Data from inadequate or poorly conducted studies should carry significantly less weight than those that are clearly robust scientifically.

The terms *strength* and *weight* of evidence have been used to describe assessment approaches to data, but it is extremely difficult to find a satisfactory definition of either. In view of the ambiguity possible with the use of such similar words as strength and weight in this context, it is probably best to ignore attempts to name the process by which the data are assessed. The clear essential is that all data should be assessed for adequacy as well as for scientific content and that there should be an expert judgment of their relevance to humans.

Identification of hazard is essentially independent of dose and formulation, but this must be considered within reasonable limits. Thus, for the case of 4-hydrazinobenzoic acid in mushrooms and its carcinogenicity in mice, the relevance of the hazard needs to be assessed. Given that there was a question of formulation relevance in the various studies (the material was given in the drinking water), the inadequate design of the study, and the large daily intake of whole mushrooms that would be necessary to produce tumors, it is likely that this hazard is not relevant to humans. This conclusion is supported by the absence of any epidemiological evidence of carcinogenic effect of mushrooms in humans. Although folklore cannot be considered to be scientific evidence, much is based on historic experience, and it may be an indicator of effect; this may become politically embarrassing when there is public belief but no demonstrable mechanism of action.

Toxicities or hazards that are revealed in safety evaluation studies are usually placed in areas of effect, such as reproductive or genotoxicity. This is simply a reflection of the fields into which toxicology, particularly regulatory toxicology, has been divided for evaluation of the functions that are considered to be significant to the consuming public. These comfortable divisions tend to ignore the fact that some substances have undesirable activities across the whole toxicological spectrum and, conversely, that some very toxic substances do not have toxicity predicted for them in certain areas. Conventional toxicology teaching tends to address one aspect of a compound's toxicity at a time, for instance, emphasizing the hepatotoxicity of paracetamol (acetaminophen) while not mentioning its renal effects, which may be seen independently of overdose and particularly in combination therapy with other nonsteroidal anti-inflammatory drugs. TCDD (tetrachlorodibenzo-p-dioxin) is reported to be carcinogenic, immuno-toxic, and acutely toxic in animals, and to affect male reproductive capacity. In geno-toxicity, however, results have been largely negative, and evidence suggests that it is not genotoxic. For many compounds, there is overlap between findings in general toxi-cology and those in other areas such as reproductive toxicology or carcinogenicity; effects in one area may indicate potential effect in another. The corollary of this is that although it may appear neater to pigeonhole the various effects into simplistic categories, this may not be the best option from the point of view of hazard prediction.

Having identified a hazard from the safety evaluation data, relevance to humans may be assessed by knowledge of the mechanism by which the effect was achieved. For this to be successful, there has to be thorough understanding of the comparative physiology of the test systems and of humans. For instance, the action of hepatic peroxisome proliferators in rodents has been shown to be a rodent-specific effect, through comparative studies, including long-term studies in marmosets (see the ciprofibrate case study in Chapter 12) and *in vitro* studies in human hepatocytes. Similarly, the renal toxicity seen in male rats with compounds that complex with $\alpha2u$-globulin has no human relevance due to the absence of these proteins in humans. Although an effect seen in animals may be expected to be absent in humans, due to differences in pharmacokinetics or quantitative differences in metabolic pathways, this does not necessarily remove it as a potential human hazard, but it may reduce the risk to vanishingly small levels. Due to the wide variation in the human population, it is possible that metabolic polymorphisms and other individual differences may be able to reproduce the effect in susceptible individuals. The likelihood of this happen-ing is assessed through risk assessment.

In summary, the overriding principle of hazard prediction is that all the data should be assessed and a mechanistic explanation sought for any effects seen. If there is a scientifically acceptable explanation for an effect, an assessment may then be made of the relevance for humans. If the specific mechanism of toxicity is absent in humans, it is probably reasonable to conclude that the hazard is not relevant to humans. Where there is no explanation of effect, other aspects of the data must be considered, including dose response and comparative absorption, distribution, metabolism, and elimination (ADME).

STEPS IN THE PREDICTION PROCESS

Prediction of hazard is an evolving process. Initial predictions made from the data of early studies are tested in further studies and then "finalized" when the program is completed. These predictions may be revised as more data are gathered. Epidemiological study in target populations is often the source of such data and is used to test the earlier predictions made from toxicity studies and any trial data in humans. Because epidemiological studies or marketing surveillance is initiated after the release of the compound onto the market, they are not predictive unless used to support changes in use of the test compound. However, epidemiological studies for similar chemicals may be used to support predictions of safety (or hazard) made for the test chemical.

BASIC PRELIMINARY QUESTIONS

Before embarking on the process of hazard prediction (or characterization), the objective has to be clearly defined by asking the question, "What are you attempting to predict?" The reason for the prediction has to be considered in the light of the intended use of the chemical and, as a result, what the expected target population is. The objective should indicate the type of data that are necessary (or optimal, as there are often gaps or deficiencies in the available data) for successful prediction. A further consideration is the level of prediction required. The process is influenced by the specificity required, whether the whole population is concerned, a selected part of that population (e.g., farm workers), a patient group, an individual, or the environment and ecosystem. Prediction in the early phases of chemical development may simply relate to test system choice, for instance, using data from *in vitro* comparative metabolism studies for species selection.

Prediction may also use computer models or expert systems to predict hazard, and a choice has to be made of which should be used. Using one system in isolation may give a skewed perception of the real hazards involved, while using every system in existence will cloud the issue irretrievably. System selection should be carried out in the knowledge of the weaknesses of the available options and the desired endpoints for prediction (Table 13.1).

The next question to be asked relates to the available database from which the prediction is to be made. How extensive and how reliable are the data? Are there animal data (pharmacology, toxicology, or ADME), human clinical data, or results from *in vitro* experiments? Furthermore, were the data derived from studies conducted as part of the basic package required for registration with regulatory authorities, or were they performed to explain the results of such studies? This database review should also indicate if any further work is needed to clarify the results of any of the existing studies, for instance, through an *in vitro* study of toxic mechanism. From these questions, the uncertainties involved may be assessed; for instance, is an extrapolation from an *in vitro* experiment to humans being requested? The ease of prediction increases with increasing biological proximity to the target species.

TABLE 13.1

Endpoints Predicted by Expert Systems

Acute inhalation toxicity LC_{50}	log P
Acute oral toxicity LD_{50}	Maximum tolerated dose (MTD)
Acute toxicity LC_{50}	Methemoglobinemia
Acute toxicity EC_{50}	α2u-globulin nephropathy
Ames mutagenicity	Mutagenicity
Anticholinesterase activity	Neurotoxicity
Carcinogenicity	Phototoxicity
Chronic lowest-observable-adverse-effect level (LOAEL)	Respiratory sensitization
Corrosivity	Rodent carcinogenicity
Developmental toxicity	Skin and eye irritation
Hepatotoxicity	Skin sensitization
hERG channel inhibition	Testicular toxicity
Irritancy	Teratogenicity
Lachrymation	

In this instance, "biological proximity" includes experimental design as well as taxonomic considerations, although precise targeting of mechanistic studies *in vitro* may mean that this latter concern is less important in some cases.

DATABASES FOR PREDICTION—QUALITY AND COMPOSITION

Unsurprisingly, the accuracy of hazard prediction is critically dependent on the quality and extent of the database that is used, and on the interpretation and conclusions that have been drawn. The available data set may be large or small and, in some cases, may not relate directly to the chemical of interest, but to a member of the same chemical class; this is often the case with workplace-related assessments when few data are available. Although it has been said that the whole data package needs to be taken into account in hazard prediction, it is important that the data be relevant to the question asked. Any safety evaluation study can be said to be predictive, and generally, the security of prediction increases with the increasing database. However, large amounts of inadequate or inappropriate data will not help the process and will simply add unwanted complications. Klimisch et al. (1997) proposed a systematic approach to the evaluation of data quality, reliability, and adequacy, which is discussed in more detail in Chapter 15.

While it is reasonable to assume that a contemporary safety evaluation program conducted to modern standards is likely to be reliable, this should not be taken as a certainty. In contrast, older studies, especially those performed before the inception of Good Laboratory Practice in the late 1970s, should be viewed with some caution. This is not because they are likely to be scientifically inept, but because standards of conduct and examination have improved to such an extent that changes dismissed then as irrelevant may be viewed differently today. Older studies need to be assessed in terms of the group size, data records and reporting, and husbandry and treatment procedures. The presence of audit reports by a Quality Assurance Unit working under Good Laboratory Practice will add a degree of reassurance to the exercise. In comparing older studies with more recent ones, possible variation in the quality of the test material should be considered. Changes in quality can occur over a period of years due to evolution of production methods; sudden changes in production can lead to unexpected impurity, sometimes associated with unwelcome toxicity, as seen with tryptophan. Equally, advances in analytical techniques, generally in the direction of vastly increased sensitivity, can reveal impurities in modern batches that intuition indicates must have been present from the outset, but undetected. The composition of an optimal database is reviewed in Focus Box 13.1.

**FOCUS BOX 13.1   DESIRABLE DATABASE
FOR PREDICTION OF HUMAN-RELEVANT HAZARD**

For a recently developed chemical, the normal and desirable database would contain information on the following:

- *General toxicology:* Target-organ effects resulting from repeated administration should be highlighted in these studies. They may identify progressive or chronic changes, which can indicate significant hazard. These studies provide data relevant to many areas of effect.
- *Genotoxicity:* Shows potential for genotoxic effects; any positive results are indicative of hazard, as the experimental conditions often do not reproduce *in vivo* conditions.
- *Carcinogenicity:* With indication of mechanism if appropriate.
- *Reproductive toxicology:* One of the major hazards to look for; endpoints examined that are of concern include effects on fertility, embryotoxicity, and postnatal development.
- *Skin sensitization and hypersensitivity:* Should show potential for dermal effects that could be of importance in production personnel and that are, of course, critically important for dermal preparations.
- *ADME data and information on pharmacokinetics and toxicokinetics:* There may be information on the particular P450s that are involved in metabolism, and this part of the package will act as an anchor for the *in vivo* data, particularly in interpretation and mechanistic work.
- *Safety and efficacy pharmacology:* These studies should identify transient, reversible hazards, e.g., cardiovascular or respiratory changes.
- *Human data:* These may relate to clinical experience (with drugs) or (very occasionally) to volunteer studies with pesticides.

In addition, there should be data on the physical and chemical characteristics of the parent molecule and on the physical form used in the evaluation. Although these are important, they have a greater significance in assessing risk. For instance, lead poses a number of hazards that do not change with physical form; however, lead on church roofs carries much less risk to the public than lead in paint or drinking water. There may also be predictions derived from computer-based models and systems, which can cover a number of endpoints. These are useful when there is only a small database to work from, but as with all other test systems, they should be considered to be tools, which can be misused all too easily.

There is a stark contrast between what might be seen as a desirable database, as outlined above, and the type and extent of data that are often available. This is especially the case for workplace assessments of chemicals used as intermediate steps in the synthesis of the final product. In these cases, the hazard prediction process has to be conservative and is often based on proximity of the molecule to the final product in the synthesis pathway. For a chemical produced late in the synthesis, it may be possible to relate structure to expected pharmacological effects or toxicity; in these cases, the use of computer models becomes more important. Early in synthetic pathways, it is likely that the compounds used or produced will be commercially known or sufficiently similar to known chemicals to be assessed for hazard by literature searches or similar means. Database deficiencies are also frequently encountered when chemicals that have been in use for years, often decades, are considered. Concerns expressed by new producers or people looking for new uses are sometimes greeted with indifference—"We've used it for years without any [recorded] problems"—and it is very difficult in these cases to come up with a rational approach that is based on science rather than comfort factors. The production in tonnes of a chemical may also affect the size of the available database because the amount of testing increases with intended annual production.

DATA HANDLING

At first sight, this might seem to be an oversimplistic item to be included. However, for a full-scale review of a complete data package, an ordered approach is essential as the amounts of data that are available can be enormous and not all of them are necessarily relevant or useful. Although it has been said earlier that putting studies into areas of investigation may be counterproductive, it is an essential first step when there are large numbers of reports or papers, as it allows you to see what there is and gives an initial indication of any deficiencies. It is useful to decide early which studies are pivotal to the assessment and which provide supporting evidence. The quality of study design and reporting comes into consideration at this point; if there are studies that are not as good as others, these may be useful as support rather than being seen as definitive or pivotal. A definitive study may be defined as one that completes a series, confirms a set of findings, or offers a mechanistic explanation and that, crucially, has been conducted to high standards of design and interpretation. The term *pivotal* usually refers to the study that is used as the basis for a risk assessment; the term also implies quality of conduct and interpretation. As the report is usually the only evidence of this available, it has to be complete and has to have all the data and details of personnel responsible for the study conduct and reporting. Further layers of comfort for the reviewer may be provided if the testing facility is well known and independent of the developer, although with current controls exerted through enforcement of Good Laboratory Practices, the latter point is less significant than it used to be, even allowing for regulatory cynicism.

The next step is to identify toxicities in the various study areas covered by the reports and check for potential overlap and interdependencies between the various areas. It is also important to check for consistency. For example, where studies have been repeated, were the effects reproducible and consistent among studies or laboratories, and if not, why not? There is a degree of interpretational variation among toxicologists in both contemporary and historic terms. Historically, interpretation may have been different due to lack of knowledge of the significance of changes seen; effects dismissed at one time may acquire new meaning, as research continues. In some cases, interpretation may have been weakened by standards of study design and conduct that were acceptable at the time the study was commissioned but are now outdated. It may be possible at this point in the review process to indicate what extra studies are needed to facilitate the hazard prediction. This may save a fair amount of effort in reviewing essentially useless studies.

FACTORS FOR CONSIDERATION IN PREDICTION

In considering toxicity seen in test systems used in safety evaluations, the primary questions relate to the effects seen and the mechanisms by which they occurred. Although it is possible to point out toxicities in humans that are not easily reproduced in animals, it is not safe to say that the reverse is true. Much research grant money has been spent in investigation of the toxicity of TCDD, after experience in the Vietnam war and at Seveso, in Italy. Although it is clear that it is highly toxic in animals and that guinea pigs are extremely sensitive to it, with lethal doses measured in micrograms, it has been said that the only proven effect in humans is chloracne. Having said that, however, no one is queuing up to say that TCDD is safe.

However, there are a number of toxicities that are seen in animals that are acknowledged to be specific to the species, for example, peroxisome proliferation seen in rodents treated with hypolipidemic compounds such as ciprofibrate or plasticizers like diethylhexyl phthalate. Although these are "standard" toxicities, there still needs to be proof that they are responsible for the changes seen. Once the mechanism of an effect has been established, the relevance or otherwise to humans may be assessed. However, it should be borne in mind that these assumptions may—and should—be challenged as the research base expands.

In some cases, the test system used in the evaluation may be said to be irrelevant to humans. Although it is clear that bacteria are phylogenetically remote from humans, a positive effect in the Ames test should not be ignored, as it shows a potential for genotoxicity that may be reflected in other systems. If there is a particular mechanism by which this was achieved, then the relevance of the effect may be assessed. Historically, it has been usual to indicate an order of increasing human relevance with increasing evolutionary complexity. Thus, a progression from bacteria to *Drosophila*, to mouse, to rat, to dog, to nonhuman primate, might be set up to suggest that data from rats are more relevant to humans than those from mice. This may be so as a general rule of thumb, but it is no more than that. Although it may be intuitive to assume that nonhuman primates will give a better indication of human effect than other species, this is not necessarily the case. The increasing use of transgenic animals will further challenge these traditional (and falsely) comforting beliefs. As an additional complication, a general prediction of the effects in humans is unlikely to be completely applicable to the whole population due to genetic variation between individuals and their circumstances (lifestyle, disease, etc.).

As has been pointed out above, dose is not a primary factor in hazard prediction, as it is considered during risk assessment. In the same way, the form of the chemical does not alter the hazard, merely the risk of expressing that hazard. However, if the margin of safety is very large—expressed as a multiple of the expected human exposure needed to reach the no-effect level in the most sensitive species tested—it may be possible to say that the hazard is not predicted to be relevant to humans and that further risk assessment is not needed. Furthermore, if there is a clear threshold below which the toxicity is not expressed, this may be used to determine relevance to humans. A large multiple between the toxicity threshold and the expected exposure in humans is a significant driver in this assessment.

Another factor that might appear to reduce the significance of an effect from the point of view of human relevance is reversibility. In toxicological terms, an easily reversible effect, such as a mild increase in liver size due to enzyme induction, is often flagged as being of minor toxicological significance. In any assessment of the relevance to humans of such change, the type of change and the speed and extent of reversibility have to be considered. In hazard assessment terms, a transient change in the liver (which has considerable recuperative powers) will be rated as less significant than a transient change in the central nervous system, which has poor repair capabilities.

The mechanism by which a systemic toxic effect is produced is, in broad terms, a function of physiology or biochemistry and the disposition and elimination of the chemical (ADME), and interspecies differences in toxicity are often attributable to these factors. The nephrotoxicity of $\alpha 2u$-globulin complexes is attributable to the large amount of this protein that is produced in in the livers of male rats, in comparison with females. There are various hormonal differences between laboratory animals and humans that can be invoked to explain toxicities in test animals. Overproduction of growth hormone in dogs following progestogen administration resulted in an increased incidence of mammary tumors. In rats, increased prolactin concentrations are also associated with mammary tumors. Neither of these hormonal pathways and mechanisms is present in humans, and both are therefore not human relevant. There are also differences in the hormonal control of reproductive processes, including parturition, between laboratory animals and humans, and such differences may mean that some effects seen in reproductive toxicity studies are not relevant to humans.

The processes of ADME in the test system should be considered when attempting to relate effects seen to those expected in humans. Differences in toxicities seen among species may be due to inherent differences in metabolism; the task for the toxicologist then is to assess which of the species is more relevant to humans. Acetylaminofluorene is a potent animal carcinogen, which causes tumors in the liver, bladder, and kidney through *N*-hydroxylation followed by production of a sulfate conjugate. However, guinea pigs are resistant to the effects of acetylaminofluorene because they have low activities for *N*-hydroxylation and sulfation; this resistance is overcome by giving *N*-hydroxyacetylaminofluorene. Acetylaminofluorene has been variously designated as a suspected, potential, or probable human carcinogen.

If the pharmacokinetics in animals are grossly different from those in humans, the effects seen may not be human relevant, but this does not entirely remove the hazard as a risk. Much metabolism of xenobiotics is carried out through the cytochrome P450 family, and there are differences in activities between the various laboratory species and humans. It needs to be pointed out that these are usually the differences in activity rather than presence or absence, and that the toxic metabolites may still be present in humans, albeit at much reduced concentrations compared with those in the test species. This may then be considered in a more formal risk assessment, if this is considered appropriate.

If toxicity seen in animals is due to a metabolite that can be shown to be absent in humans, it is unlikely to be human relevant. Equally, the absence of toxicity in a test system that metabolizes the chemical differently from humans, either by prediction or observation, does not indicate that the chemical will be safe in humans. Studies with the major human metabolites should be considered, if they are not present in normally available laboratory animals. (Although it has been said that toxicity studies should be conducted in a metabolically and pharmacokinetically relevant species, this is usually no more than a Holy Grail, due to expense and practicality.)

In assessing the significance of the effects of one chemical, knowledge of the properties and toxicity of chemicals from the same class or with the same mode of action is also an invaluable aid. Although such knowledge is useful, it has to be treated with some circumspection as toxicity can vary widely across a group. This is illustrated by the organophosphates, which have a very wide range of active dose levels as shown by the three examples given in Table 1.1, in Chapter 1. When comparing chemicals across groups, knowledge of the structure–activity relationships is also important. In organophosphates, the bond types around the central carbon atom of the phosphate group affect whether the compound will be associated with "aging" of the bound enzyme and possible delayed-onset neuropathy. Aging involves *in situ* metabolism of the bound organophosphate molecule with consequently increased binding affinity. The presence of a P–O–C bond between the phosphorus and one of the side groups of the molecule, as in triorthocresyl phosphate, is associated with rapid aging, while a P–C bond makes this impossible.

Some of the more complex hazard prediction situations are provided by *in vitro* data, from which an extrapolation to humans is necessary. This is often seen with genotoxicity data, where a single positive in an *in vitro* test, usually (but not always) a chromosome aberration study in Chinese hamster ovary or mouse lymphoma cells, can cause a variety of problems. If this is offset by a negative *in vivo* study in the mouse micronucleus test and a negative Ames test, it used to be that the single positive result would be dismissed. However, a more questioning approach has evolved where the circumstances of the various results are considered very carefully before reaching a conclusion on relevance. Once it has been accepted that the positive result is not associated with a threshold of effect or is due to excessive toxicity, a number of questions can be asked in order to clarify the meaning of the data. These are not only directed particularly at exposure of the test cells but also at mechanism. Partly, this is driven by the difficult question of whether negative data in genotoxicity tests constitute adequate proof of nongenotoxicity. For chemicals that have low oral bioavailability but that are reasonably soluble in routine parenteral vehicles, it should be possible to achieve adequate exposure of bone marrow by intravenous administration. If the results of this test are negative and it can be shown that the bone marrow was exposed to a greater degree than that achieved *in vitro*, this will add to evidence that the positive result *in vitro* is not relevant *in vivo*.

A poorly soluble compound may still be associated with genotoxicity *in vitro*, as it has been demonstrated that precipitates may still give positive results. However, such a compound is often associated with poor absorption from the gastrointestinal tract and is usually very difficult to be given intravenously at high-enough doses to duplicate the exposures seen *in vitro*. Both these factors mean that exposure of the bone marrow cells in the mice is likely to be less than of the cells *in vitro*. Low bone marrow exposure to an active mutagen may also be seen where there is extensive first-pass metabolism following oral administration, especially if this results in a conjugated metabolite that is not dissociated in the target tissue. Although *in vitro* tests use S9 mix as a metabolic activation system, in standard protocols, this is prepared from the livers of rats treated with enzyme-inducing agents, which may not be the most appropriate tissue or system for the test chemical. At this point, mechanistic studies looking at the activities of specific enzymes in target tissues may be invoked to determine relevance of the results; if there is significant reversion from conjugate to parent in human tissues, the absence of mutagenicity in the various tests may be deceptive.

Another approach is to examine tissues that may be expected to have had maximal exposure to the test substance, whether as a precipitate or as a saturated solution; these would normally be the stomach (in oral administration) and the liver. A positive result in the comet assay indicates DNA damage in the target tissue and is a clearer indicator of mutagenic potential. As with any toxicity, an understanding of the mechanism by which the result was produced is essential to overall interpretation of the various studies. Interpretation of genotoxicity data is facilitated by the presence of other data relating to the carcinogenic potential of the test or similar chemicals.

In summary, for secure prediction that a toxic effect seen in safety testing is not human relevant, there has to be knowledge in the following areas:

- Mechanism of effect and whether this is species specific.
- Comparative physiology.
- Whether there is a clear threshold of effect, below which toxicity is absent.
- Comparative ADME; these studies are useful but may not imply absence of risk.
- Relevant data from other chemicals of the same class and action.

In final analysis, it has to be recognized that it may not be possible to predict that an observed toxicity is not a hazard for humans from the available data, in which case, appropriate mechanistic studies should be conducted to demonstrate specificity of effect. In the event that these studies are not conclusive, a conservative approach must be taken with a formal risk assessment.

PREDICTION FROM MINIMAL DATABASES

All too frequently, as implied by the discussion of what a desirable database should include, the amount of data available is less than completely ideal. This is usually not only the case for an untested intermediate used during the synthesis of a final product but is also often found with well-established chemicals that have been used without (reported) problem for many decades. With the early parts of a synthetic pathway, the chemicals used are sometimes well known and characterized by existing research. The problem becomes more acute in the later stages when the end products of each reaction are themselves novel chemicals. The distinction between these intermediates and the final product is that formal regulatory testing is not required, so the database for the former is small. One exception to this would be that when the synthetic process is carried out at more than one location, the toxicity of the transported chemicals would have to be assessed according to the annual amounts produced. Another setting in which the database may be expected to be small is found at the very early stages of evaluating a new chemical. Factors to predict the toxicity of an unknown chemical and then extrapolating that prediction to a situation of human exposure are summarized in Focus Box 13.2.

If the intermediate is synthetically remote from the final product, other factors have to be considered. These include an assessment for the presence of chemical

---

FOCUS BOX 13.2   PREDICTION FROM MINIMAL DATABASES

The following are among the factors that should be considered in maximizing the database:

- Physicochemical properties of the molecule: partition coefficient and solubility, molecular weight, pKa (the pH at which it is 50% ionized), volatility. These properties may be compared with the final molecule.
- Structure–activity relationships known for the final molecule; if structure associated with pharmacology or toxicity in the final product is present in the intermediate, it is sensible to assume similar activities for the intermediate.
- Expected dose levels compared with no-observed-effect levels for the final product.
- Metabolism of the final product; if there are metabolites that are similar to the intermediate, this may point to toxicity.
- Properties of chemicals related by structure or intended action.

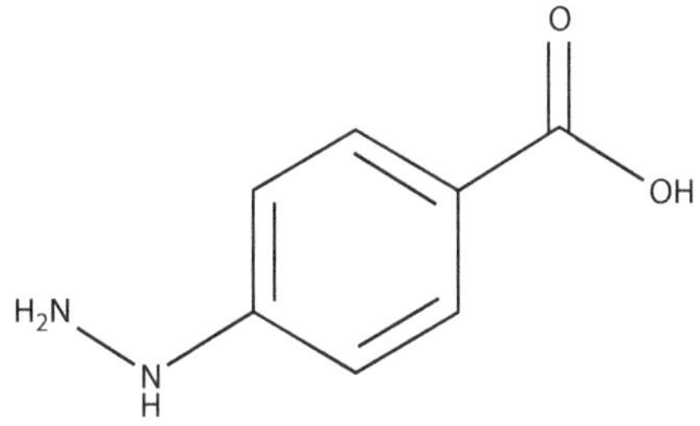

FIGURE 13.1 4-Hydrazinobenzoic acid (CAS no. 619–67-0).

groups or structures that are associated with known toxicity in other chemicals. Thus, 4-hydrazinobenzoic acid (Figure 13.1), the contentious component of mushrooms, is an aryl hydrazine, a structural configuration that has been associated with mutagenicity, carcinogenicity and skin sensitization. There are a number of tools available that assist in this process of identifying these structure–activity alerts. These are seen in the various toxicity prediction software systems that are available and in systems that can conduct literature searches based on structure as opposed to key words.

If the structure is known, the *in silico* predictions using structure–activity relationship (SAR) programs can add excellent weight to read-across or give predictions in their own right (such as bacterial mutagenicity). The prediction of toxicity using *in silico* is discussed in Chapter 5.

Ultimately, the decision process is similar to that used for large databases. The evidence is reviewed and an assessment made as to the probable toxicity and consequent hazard. The inevitable difference is that the conclusions of such an assessment must be conservative until supported by more trustworthy evidence. The use of computer-based expert systems has a great part to play in these processes, and as they evolve and become more interlinked, their reliability will increase.

PREDICTION FOR INDIVIDUALS

The prediction of effect in individuals—as opposed to populations—is becoming more important, although it may be characterized as individual risk assessment. Such assessments can cover the likely outcome of overdose to the probable response to treatment with specific drugs and the likelihood of interactions among prescribed drugs. The results of many clinical trials, which are the ultimate basis for regulatory acceptance for new drugs, are adversely skewed by lack of response of some of the patients entered into the trial. If it can be predicted in advance that a patient will not respond due to the presence of a metabolic polymorphism or some other phenotypic aspect, there does not seem to be much sense in exposing him/her needlessly to a drug that may actually harm him/her. The implication of this is that patient populations for clinical trials can be selected on the basis of their likely responses and that the trial data may be much more favorable as a result. The corollary of this is that the suitability of patients must be assessed by physicians before drugs are prescribed;

the time- and money-saving implications of this are considerable, as appropriate treatment can be selected immediately and without lengthy experiment. This is the nascent science of pharmacogenomics. The difficulty arises when it is realized that biological situations are rarely black or white, and that a patient's phenotype is probably an expression of varying rates of genetic expression and not simply a matter of presence or absence.

As well as phenotype, the reaction of an individual to chemical exposure— intentional or otherwise—is affected by personal circumstances. Thus, pregnancy or malnutrition will affect predictions such as the relevance of defined toxic hazards. In overdose, additional risk factors to be considered, apart from the dose taken, include the presence of alcoholism, smoking, or other drug abuse. The factors relevant in paracetamol overdose are reviewed in Case Study 13.1.

Diclofenac, a nonsteroidal anti-inflammatory drug, has also been associated with idiosyncratic liver toxicity, although in contrast to paracetamol, this may have a delayed onset of up to 3 months in chronic use (Boelsterli 2003). Prediction of

CASE STUDY 13.1 PREDICTION OF TOXICITY IN PARACETAMOL (ACETAMINOPHEN) OVERDOSE

Paracetamol is a widely used and available pharmaceutical that is available in the majority of countries as an over-the-counter drug, i.e., without prescription. The *ad libitum* availability of paracetamol coupled with its low cost means that accidental or intentional overdoses are relatively common. Paracetamol toxicity arises when the usual metabolic phase II pathways are overwhelmed and the phase I pathway—which produces a reactive quinone metabolite—comes to the fore. If the exposure is sufficient, the liver cells glutathione stores are depleted, and toxicity can occur unchecked.

Clinically, symptoms in the first 2 days do not reflect the seriousness of the situation. Nausea, vomiting, anorexia, and abdominal pain are possible in the first 24 hours and may persist for a week or more. Hepatic damage becomes clinically manifested in 2 to 4 days; plasma transaminases, bilirubin concentration, and prothrombin time are increased. Renal failure may be noted in the final stages; paracetamol is associated with nephrotoxicity as well as pancreatic toxicity. The following points detail the risk factors and predictive outcomes of overdose.

- The lethal dose is approximately 16 g in a normal 70 kg human; outcome is influenced by hepatic status (e.g., coadministration of enzyme inducers, such as phenobarbital). Dose is always difficult to establish accurately, in part due to poor reporting from the patient.
  - Patients who consume excessive quantities of paracetamol in multiple doses usually present with toxic blood concentrations. It is not easy to predict paracetamol toxicity when there has been repeated use.

- Starvation, which depletes glutathione stores in the liver, exacerbates paracetamol toxicity.
- Chronic alcohol ingestion is additive in effect; chronic alcohol ingestion upregulates the CYP2E1 isozyme; it is this isozyme that produces paracetamol's reactive metabolite.
  - However, acute single alcohol ingestion, concomitantly, is protective.
- Hepatic damage is seen as centrilobular necrosis.
  - Hepatic recovery has been noted in biopsies of people who recovered.
- Use of the Rumack–Matthew nomogram gives the probability of hepatotoxicity from estimation of time of ingestion and plasma concentration of paracetamol; a plasma half-life or more than 4 hours is associated with a high probability of hepatotoxicity, which is often fatal.

Although there is an antidote (*N*-acetylcysteine), which is given within 16 hours of ingestion of large single doses, this is of questionable value in cases where repeated doses have been taken.

Source: Adapted from Dart RC et al., Medical Toxicology, *3rd ed., Philadelphia: Lippincott Williams & Wilkins, 2004*

adverse effect is complicated by the lack of any simple relationship to dose, and it is necessary to examine individual patient factors, such as metabolism, to reactive metabolites including 4-hydroxylation secondary to glucuronidation. In some cases, the effects may be immune mediated. Some concurrent diseases, such as osteoarthritis, may also increase the susceptibility to diclofenac hepatotoxicity, but the reasons for this are not clear. Boelsterli suggests that cumulative damage to mitochondria may explain the delay in onset of symptoms. Clearly, there is a case for identification of patients at risk before treatment begins, or at least as soon as possible after it starts.

SUMMARY

The following points are given as an overall summary for successful prediction of human-relevant hazard:

- Be sure of the parameters for which you are predicting: population group, environmental area, etc.
- What is the purpose for prediction? Before you ask the question, you should know what you would do with the answer(s).
- Be sure of your database from which you are predicting hazard.

- Use or consider all data but dismiss those that are not interpretable with any security.
- Predictions are not static; today's prediction may be questionable tomorrow and can change completely if new data become available.

REFERENCES

Boelsterli UA. Diclofenac-induced liver injury: A paradigm of idiosyncratic drug toxicity. *Toxicol Appl Pharmacol* 2003; 192(3): 307–322.

Dart RC et al. *Medical Toxicology*, 3rd ed. Philadelphia: Lippincott Williams & Wilkins, 2004.

Klimisch H, Andreae M, Tillmann U. A systematic approach for evaluating the quality of experimental, toxicological, and ecotoxicological data. *Regul Toxicol Pharmacol* 1997; 25: 1–5.

14 Background to Risk due to Toxicity

INTRODUCTION

The use of any chemical is associated with risk, whether it is a novel drug, an established pesticide, an intermediate in a synthetic pathway, an industrial by-product, or sugar in cola. The next four chapters outline the basic tenets of risk analysis (risk assessment and risk management) with respect to toxicity. How risk is perceived and described is critical in successful communication and subsequent management of risks due to toxicity, either in the workplace or in a wider context. Risk is a product of hazard and exposure; without exposure to the hazard, there can be no risk.

OVERVIEW OF RISK ANALYSIS

Risk analysis is the study of the overall process that includes risk assessment and risk management. Before starting to discuss this process, it is important to define the stages, as they can be easily confused. Various definitions of risk analysis and its components have been given. The most internationally recognized are those of the Inter-Organization Programme for the Sound Management of Chemicals [IOMC, a cooperation of the United Nations Environment Programme (UNEP), Food and Agriculture Organization of the United Nations (FAO), World Health Organization (WHO), United Nations Industrial Development Organization (UNIDO), United Nations Institute for Training and Research (UNITAR), and Organization for Economic Co-operation and Development (OECD)], with the more accessible versions usually being from the USA. It is also important to note that the framing of the risk analysis process (the framework of legal instruments under which the risk analysis is conducted) and the undertaking of the risk evaluation (explained later) are at the intersection between the science of toxicology and policy—and this is not always a comfortable or smooth relationship.

Firstly, it is important to separate risk assessment from risk management. The US federal government definition of risk assessment is "the characterization of the potential adverse effects of human exposures to environmental hazards." Risk assessment is, classically, the process of characterizing and evaluating the potential adverse effects (e.g., carcinogenicity) of a particular or generic type of human exposure to a chemical and assessing the probability that these hazards will be expressed in a target population (or their offspring). It therefore includes a consideration of the type of exposure (e.g., use as a therapeutic agent) and the existing (or likely) exposure (the exposure being that with the current or proposed management controls in place—e.g., requiring a prescription from a medical practitioner). Classically, this is a high-level process that extrapolates from the results of some *in vitro* assays and animal testing to the target species. It should include an assessment of likely exposure—the dose to be administered and bioavailability. It may include the use of sophisticated mathematical models. The relationships of the various stages of risk analysis are illustrated in Figure 14.1.

It is generally agreed between the various sources that there are four steps in risk assessment, namely

1. Hazard identification
2. Hazard characterization (or determination of type of toxicity and dose–effect/response relationships)

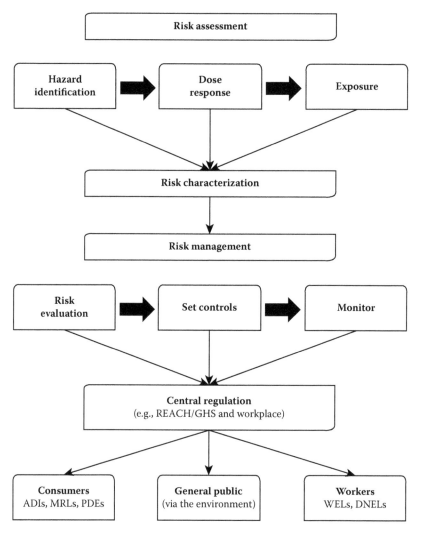

FIGURE 14.1   Risk assessment and management. GHS, Globally Harmonized System of Classification and Labelling of Chemicals; MRL, minimal risk level; REACH, Registration, Evaluation, Authorisation and restriction of Chemicals.

3. Assessment of expected exposure
4. Characterization of the risk

The process of risk characterization was defined by the WHO/FAO in 1995 as the integration of hazard identification, hazard characterization, and exposure assessment into an estimation of the adverse effects likely to occur in a given population, including attendant uncertainties. This includes qualitative and quantitative aspects and is the preliminary to risk management, which is seen as a separate entity.

Evaluation of the risks defined in the risk characterization is the initial step in risk management. There are three steps in risk management.

- Evaluating the risks
- Setting appropriate control measures
- Monitoring whether the risks are adequately controlled (e.g., for drugs, pharmacovigilance and postmarketing surveillance for consumer products)

The first of these risk management stages, risk evaluation, seeks to establish that the risks are "broadly acceptable" or that a satisfactory relationship exists between the risks and benefits (i.e., the risks are "tolerable"). To do this, it is necessary to determine the relevance and significance of any hazard identified in the risk analysis and of the risks to the target species or biological system that is or may be exposed. Naturally, where possible, the potential benefits of any such exposure should also be considered, whether they be a reduction in fungal damage to stored crops or therapeutic benefit for an anticancer agent. Benefits may be socioeconomic as well as pathophysiological: the former are often more open to political and social debate, and the latter are usually more easily quantifiable. Risk evaluation may be conducted simultaneously with the earlier stage of risk characterization, although the two processes are conceptually separate.

Following evaluation of the risks, risk management processes should ensure that there is "adequate" control of the exposure and emission and that the consequences of any remaining risk can be managed, if necessary, through emergency planning. The final stage, risk monitoring, is essential to ensure that the risks are adequately controlled and that adverse effects are not occurring in consumers, workers, or patients.

Although high-level risk assessment and evaluation is briefly considered in this chapter, the prime intention is to look at what might be termed its everyday use, for example, in the workplace and when choosing the first doses to be used in human volunteer studies. This type of assessment considers the possible expression of any of the hazards identified as human relevant following the overall interpretation of the safety evaluation data, whether, practically, these are phenomena that are either present or absent (stochastic or all-or-nothing) phenomena like cancer or effects such as respiratory allergy (asthma), reproductive abnormality, or other toxicities. The risk characterization should lead to a quantitative estimate of the *margin of safety* (or equivalent) between the exposure level likely to be encountered and the exposure level at which a certain low frequency of effects may be seen. That estimate is then evaluated in terms of its acceptability in the risk evaluation.

If certain assumptions are made concerning the acceptability of a risk (usually by defining a standard as a level of exposure that is the maximum acceptable or tolerable risk for the risk being examined, and therefore omitting the need for actual exposure data), these can lead to generalized risk evaluations in the absence of exposure data. Such risk evaluations lead to the setting of parameters such as acceptable daily intakes (ADIs), tolerable daily intakes (TDI), permitted daily exposures (PDE), derived no-effect levels (DNEL), workplace exposure limits (WELs), and maximum atmospheric concentrations in indoor or outdoor air. This is a clear case where risk characterization and risk evaluation are conducted simultaneously. The European guidelines on setting health-based exposure limits for risk identification in the manufacture of different medicinal products in shared facilities use the approach taken by the International Conference on Harmonisation (ICH) Q3C guideline. The guidelines use default values and assumptions to accommodate for variations in the quality, duration, and species used to assess a product coupled with a point of departure and known exposure hazards to calculate a value below which no hazard is expected to be observed in patients unwittingly exposed to the substance.

Risk characterization and risk evaluation may need to be performed several times, either for different groups or populations or as more data accumulate. It may be necessary to set controls on exposure in order to obtain an acceptable (or tolerable) exposure level. Those controls may lead to different exposure levels and hence different risk characterizations and evaluations. If risk monitoring (or other new information, such as new interpretations of hazard data or new data) indicates that there is a problem with the risk evaluation or the controls, then, as with the setting of controls, the evaluation has to be repeated with the new information incorporated into the risk characterization. Periodic reviews of the data and of the risks (e.g., by review of the license for a medicine, a plant protection product, or a biocide) are therefore to be expected for the chemicals and uses that pose the highest risks.

There has been a long-standing assumption in toxicological risk analysis that the results of nonhuman experiments, whether in animals or in a petri dish, are relevant to humans. A second assumption often, but not universally, made is that the effects of high dose or exposure will be seen at lower levels, that is, that dose response is always linear. Risk analysts also have to pay attention to public opinion concerning "safe levels of exposure" and "acceptable risk" when setting limits. The assumptions made have weakened risk analyses in the past and resulted in the imposition of unrealistic limits on chemical exposures. Risk analysts have considerable responsibility to get both their technical (toxicological) analyses and their assessments of societal opinion right, so that the limits of exposure that they propose are reasonable and derived from all the available data without the use of unrealistic assumptions. Setting a limit too high may expose people to toxic concentrations of a chemical with unacceptable effects. Setting a limit too low may mean loss of benefit from use of the chemical or may impose cleanup processes that are excessively costly in relation to the marginal increase in benefit. In risk analysis, it is necessary to apply Occam's razor—to make no more assumptions than necessary—to accomplish the declared purpose.

LEVELS OF RISK AND FACTORS THAT AFFECT RISK

Risk is present at various human levels ranging from personal; through specific populations, such as farm workers; to the general population of an area; to national and international; and thence to global. Personal levels of risk are determined by factors such as occupation, lifestyle, and home environment. The lifestyle of those about you may also be significant. Smoking is clearly associated with a range of diseases, the most emotive of which is lung cancer. Although there have been epidemiological attempts to correlate the effects of passive smoking with disease, these have generally been unsuccessful. However, to say that there is no quantifiable risk or demonstrated correlation is counterintuitive. Passive smokers are exposed to sidestream smoke and to exhaled smoke. Although processing smoke through the lungs of the smoker might be expected to remove a large proportion of the chemicals and particulates, this will not be totally efficient, and consequently, other people in the vicinity will be exposed to the same toxic mixture, albeit at lower concentrations. In a similar manner, the vapor exhaled from e-cigarettes cannot be said to be risk free; but it is sensible to expect that the hazards and risks associated with this are very different from those associated with tobacco smoke.

Genetic profile is also important and may determine the individual responses to chemicals both short term and long term. Skin color and associated sensitivity to ultraviolet (UV) light is a well-known determinant of susceptibility to skin cancers. Biochemical or physiological factors such as metabolic polymorphisms influence individual responses to drugs and pesticides such as organophosphates (OPs). Such individual characteristics may offer an explanation for differences between farmers' long-term responses to occupational exposures, sheep dips being a point of heated discussion in recent years. Focus Box 14.1 looks at cholinesterase inhibition as a toxicological target with specific reference to the use of OPs. It is worth noting that the referred paper by Stephens et al. provoked two critical commentaries that were published in the same issue of the *Lancet*. The authors' response (also in the same issue) addressed the concerns by reference to the full report, neatly illustrating the pitfalls in attempting to draw conclusions from a partly complete account or data set (Stephens et al. 1995).

Occupation has always been a factor to consider in personal risk as well as for particular working populations such as miners. Cancers associated with painting radium onto watch faces and working with 3-naphthylamine or vinyl chloride are well known, and it is fair to say that the incidence of occupational cancers has been much reduced by appropriate application of risk analysis and risk management. In addition, there are synergistic factors that affect individual risk, such as the increased cancer risk associated with mining and smoking.

Diet (individual and national) is also a factor of great significance, both in terms of cause and in terms of prevention; as a result of this counterbalancing act, it can be difficult to sort out what the significant factors are in a population. High salt intake in Japan was linked to a high incidence of stomach cancer; high intake of rye bread in Finland has been associated with a reduced incidence of gastrointestinal cancers. Increased intake of antioxidants has been associated with reduced cancer risk. Such factors have clear impact on individuals, but, because diet is usually influenced by national traditions and circumstances, the risks associated with it are relevant nationally. While dietary imbalance confers risks in various, sometimes contrary directions—in terms of sucrose, fiber, fatty acid composition, antioxidants, and deficiency or excess of factors such as trace elements—the actual chemical composition

FOCUS BOX 14.1 CHOLINESTERASE INHIBITION: FACTORS IN RISK DETERMINATION

Cholinesterase inhibition came to prominence with development of OP nerve gases and insecticides and of the carbamate insecticides, which have been widely used in agricultural and domestic pest control. Cholinesterase inhibition has been used as a therapeutic objective in Alzheimer's disease and is a property of natural chemicals, such as solanine (see Focus Box 14.4).

- OPs show a wide range of acute toxicities; the nerve gas sarin has a parenteral acute median toxicity (Lethal Dose 50; LD_{50}) less than 0.05 mg/kg, while malathion is lethal at around 1000 mg/kg orally (Table 1.1, Chapter 1). Carbaryl, a carbamate insecticide, has broadly similar acute toxicity to malathion; carbamates tend to be less toxic than OPs due to differences in reactivation rates for the enzyme.
- OPs have been widely used in dipping sheep; this, which was compulsory in the United Kingdom between 1976 and 1992, poses significant practical problems in controlling occupational exposure. It is strenuous, dirty work; sheep are inherently uncooperative, and as in all such circumstances, personal protection equipment is difficult to wear with any comfort or, as a result, hope of real benefit. Although the acute effects of OPs are well known, long-term neurological effects have come to prominence as the history of use has increased.
- A study in 146 sheep farmers indicated deficits in attention and speed of information processing and susceptibility to psychiatric disorder (Stephens et al. 1996). Another study of OP-related change in sheep dippers has indicated evidence of neurological effects, especially in those with high exposures over a long period (Pilkington et al. 2001).
- Various papers have examined polymorphisms in paraoxonase, which metabolizes OPs in humans; this is present on a genetic level and as different affinities/activities between substrates (Mallinckrodt and Diepgen 1988; Mutch et al. 1992).
- These papers indicate potential differences in susceptibility to OPs and may explain differences in response seen between farmers. Low paraoxonase activity has also been reported in Gulf War veterans complaining of Gulf War Syndrome, in which OPs were one of the implicated factors (Mackness et al. 2000).
- Although OPs are clearly acutely toxic and pose significant long-term risk, alternatives are not necessarily better. One possibility is to use synthetic pyrethroids, but they have potential environmental problems, and resistance by sheep scab mite has been shown in some areas.

- Other factors to consider include runoff from dips into surface water, which has been shown as a significant environmental problem (Virtue and Clayton 1997). OP use in salmon farming has also been criticized.

These various papers refer to OP use in the so-called developed world. Standards of use and acceptance of risk are almost certainly different in other environments.

of the food is also worth considering. Leaving aside chemicals that are clearly of nutritional benefit, there is a host of other chemicals present, some of which are artificial (pesticides or chemicals introduced during cooking) and some of which are endogenous to the food consumed (see Focus Box 14.2). Most plants, such as cabbage, contain their own chemical defenses against pest attack—so-called natural pesticides—which are present at far higher concentrations than the carefully regulated amounts of synthetic pesticides that may have been sprayed on them. As a further contrast, the synthetic chemicals have generally been far better characterized in toxicological terms than many of the natural (so they "must" be safe) chemicals in our diet.

While individual and national diets are potential sources of adverse effects, they are at least reasonably focused. In contrast, air quality is an international and global concern that has the potential to affect everyone. Poor air quality standards have clear impact in individual cities—Mexico City and Beijing being good examples. As airborne pollution can cross international borders, risk becomes a matter of international concern, which requires agreement on standards of emissions from factories and so on. The unintentional export of atmospheric pollution from the United Kingdom to countries of northern Europe and the subsequent environmental damage through acid rain caused a lot of acrid debate.

RISK PERCEPTION

Perception of risk is an essentially unquantifiable concept that is influential at three levels.

- The framing of the regulatory system is such that it asks the right questions concerning the risk.
- The evaluation of risks is such that societal concerns are properly taken into account.
- The management of the risk is such that the perception by the risk taker is congruent with that of the risk manager.

Understanding how risks are perceived is vital to their management—especially in the workplace, where disregard of risks by employees can rebound on company management in a welter of litigation and financial regret. Lack of risk appreciation, coupled with factors such as uncomfortable or ill-maintained personal protection equipment, can have serious consequences for a company and its employees. Before a risk can be managed successfully, it must be understood by the people who are at risk and who need to put the desired measures into practice. If their perception and appreciation of the risk is different from that of the managers, then the management process is unlikely to be successful.

Defining risk in terms of the probability of the occurrence of an adverse event is often easier than getting the significance of that possibly remote event accepted by those affected by it, and hence an acceptable course of preventative action agreed upon. The perception of risk, particularly by the public, affects responses to risk reduction initiatives or to new introductions of processes or chemicals and cannot be ignored. This is due to the ways in which risk is perceived by the people involved. Risk that is remote in time or distance is less threatening than immediate risk; a familiar risk is more likely to be accepted than one from an unfamiliar source or one that has been poorly communicated and explained. For an observer or someone else who is not directly affected, even severe risks may not carry personal significance, whereas for the person about to fall off the cliff, the risk is real and immediate. However, if the person is told that there is a risk that he/she will fall off the cliff after 5 years, there is much less immediacy and so less likelihood of corrective action. In understanding risks of diseases normally present at low incidences, it could be said that any disease is rare—until you get it yourself. If a disease is brought into the personal sphere—through a friend or relative or yourself—it immediately assumes much greater significance. Furthermore, if there is no obvious cause, responsibility may be assigned to any local factor or circumstance, especially if more than one case is present in a small group. Such local incidences may not climb above the levels of normality in a wider context, but they may well be perceived as critical.

Risk perception is also influenced by the position or identity of people who are given the task of explaining it. Explanation and assessment from clearly interested parties, generic "scientists," or (especially) government will tend to be more closely questioned and ignored than that from so-called independent sources, such as campaigning organizations, that depend on creating a climate of suspicion for their funding. Increasingly fervent dismissal by government of concerns about pesticide residues in food simply leads to greater levels of disbelief in the public perception, ably fueled by the relevant campaigning organization.

There are psychological and sociological views about risk and attitudes to risk that have to be allowed for. Slovic (1999) emphasizes the subjective and value-laden nature of risk assessment and indicates that it is not a purely scientific process. Risk may not be equated with danger. Risk assessment blends science and judgment and is inherently subjective; it has to take into account factors that are psychological, social, cultural, and political. In addition, any individual or institution tends not to be trusted in risk assessment by at least some of the subjects.

The major health concerns of the public, such as cancer, are driven by personal understanding of risk and knowledge of the mechanisms; however, the smaller the understanding, the greater the concern. While cancer as an aspect of tobacco smoking is well understood, it is chiefly of concern to those who do not smoke, as the smokers have accepted the risk either by acknowledging or ignoring it. This acceptance might not prevent a smoker buying organically grown produce in the belief that it has fewer pesticides and carries less risk as a result; there is even a market for organic tobacco. The disparity between the risks associated with smoking and the risks associated with minimal amounts of safety-tested pesticides is not a factor to be considered. One risk (the lethal one) is acceptable, while the other is not.

Another factor is that risks are often perceived and considered in isolation, without considering that risk in one sense may be offset by benefit (risk/benefit) or vice versa; in other words, risk is often two sided, but only one is considered. Organically grown or stored peanuts may have higher levels of aflatoxins due to the uncontrolled presence of mold, which is usually absent in peanuts treated with pesticides. Aflatoxins are highly potent carcinogens and carry a higher risk than the pesticides. Removing one risk often means promoting another to the same or a higher level. One risk may be enhanced by another; the risk of occupational cancer is often increased in smokers or by excessive alcohol consumption.

Data that appear to indicate differences in risk are another source of false perception. Deaths from cancer may be higher in a poor industrial town than in one that is academic and rich. However, the incidence of cancer could well be the same in the two areas, and the apparent difference is due simply to a disparity in health-care standards.

Risk perception is one of the factors in the widespread rejection of genetically modified crops and foods. In this case, poor communication has exacerbated the problem to the extent that any benefit from this technology is likely to be wasted on a wave of public rejection that is—in part at least—ill-informed. Focus Box 14.2 gives a simplistic overview of this contentious area. Normal agricultural development of new strains of animal or crop is genetic modification through selection of desirable characteristics via breeding programs; the critical difference is that normal agricultural development does not involve the insertion of genes from different species. Although genetic modification can produce plants with excessive concentrations of endogenous chemicals or with allergenic proteins (although this has not been proven in any marketed food), these risks could be managed. For foods that are unacceptably hazardous to certain individuals—for instance, nuts—genetic modification holds out the possibility for the removal of unwanted genes from foods; nonallergenic peanuts might have considerable attraction to some people.

The above should not be taken as evidence for banning genetic modification. Hazards have been identified in terms of food content and environmental effect, but risk is determined by local circumstances, such as individual susceptibility and the concentrations of proteins or chemicals expressed in the crop or food, together with local conditions. With appropriate management and some lateral thought, the risks can be reduced and managed; with looming global food shortages, this is becoming imperative.

ACCEPTABILITY AND TOLERABILITY OF RISK

Clearly, risk perception has a considerable influence on the acceptability of risk, which may be seen from two angles, that of the public and that of regulators; these are frequently different. The type of hazard and its characteristics define the

FOCUS BOX 14.2   GENETICALLY MODIFIED FOODS

- Genetic modification introduces new genetic material, from bacteria or other species, into a plant's genotype to express a characteristic such as longer shelf life, better flavor, or resistance to pesticides or pests. The action of these inserted genes and their proteins can be very specific, as in the Bt protein expressed by *Bacillus thuringensis* DNA inserted into genetically modified maize to give resistance to corn borer.

- Toxicological risk may be expected in two areas—environment and human health. The former may be seen through effects on wild populations or increased use of pesticides; the latter might be associated with food intolerance, toxicity, or factors such as induced resistance to antibiotics.

- Differences from "traditional" crops are loosely classifiable according to type of chemical expressed—similar or dissimilar to those in normal human metabolism. *Similar* would include nucleic acids and proteins. *Dissimilar* would include small molecules such as endogenous alkaloids, which can be present at toxic concentrations in new strains; a new strain of potato (produced without genetic modification) was found to have acutely toxic concentrations of solanine and chaconine and was withdrawn (Ames et al. 1990). Similar chemicals are likely to be lost safely through natural biochemical pathways; dissimilar chemicals are subject to the normal processes of ADME for small molecules and associated with adverse effects such as the anticholinesterase activity shown by solanine (Ballantyne et al. 2000).

- Adverse effects on human health are dependent on the composition of the foods and the crops from which they are derived and are most likely to arise from ingestion. However, dermal exposure can also be significant. A new insect-resistant celery was associated with rashes and burns when handling was followed by exposure to sunlight; subsequently, it was found that the new variety contained sevenfold more psoralens than normal celery.

- Genetic modification of food crops might be expected to be associated with allergic reactions. Allergens are mostly proteins, and only a small percentage of dietary protein is allergenic. If the protein expressed is similar to a known allergen (e.g., a nut-derived protein), then there is a risk that the new food will also be allergenic. As a basic rule, genes should not be transferred from known allergen sources.

- Food has always been associated with certain risks. Genetic modification may also pose risks to the environment, where effects may include gene transfer or toxicity to nontarget insects such as butterflies, as a result of excessive expression of proteins such as Bt.

- Pollen from a variety of Bt corn expressing high levels of Bt protein was reported to be toxic to monarch butterfly caterpillars when it fell on the leaves of milkweed plants growing among the crop (*New Scientist* 2001).
- Transfer of genes for herbicide resistance has been reported in sugar beet and in some weeds so that weeds and crops have become resistant to several herbicides (*New Scientist* 2000).
- Narrow spectrum of action of some genetic inserts may mean that use of pesticides is still needed when the crop is attacked by untargeted pests (*New Scientist* 1999).

acceptability of its expression. Increased risk of irreversible change such as birth defects is unacceptable to a greater degree than a minor effect that is seen to be transient. When a transient effect turns out to have long-term consequences, the acceptability of the risk is likely to change. In official terms, acceptability of risk is expressed in incidences; 1 additional death in 1 million due to cancer appears to be acceptable, although, in practice, as the incidence of cancer is more than 30%, this would not be detectable unless the cancer was unusual.

The relationship between risk and benefit (or risk/benefit ratio) is a critical factor in determining the acceptability of risk due to toxicity. An acceptable risk is one acceptable in all circumstances. A tolerable risk is one that can be accepted because some benefit is obtained by tolerating the risk. As a broad rule of thumb, tolerability of risk increases with increasing benefit, including for instance, in medical circumstances, reduction in debility due to severe disease. Tolerability is very difficult to quantify and is influenced by perception of the risk. Often, it is a matter of judgment, which tends to be conservative and hence to disproportionately favor minimizing risk. At either end of the risk/benefit spectrum, this judgment is relatively easy; significant toxicity is acceptable in cancer drugs but is not tolerated in analgesics sold without a prescription. In the middle of the spectrum, however, the choices can be much harder. For more serious inflammatory conditions such as rheumatoid arthritis, where long-term treatment of a wide range of patients could be expected, a greater degree of risk would probably be allowable, but, for example, reproductive effects or human-relevant carcinogenicity would not. Although significant toxicity may be acceptable in a cytotoxic anticancer agent, as cancer drugs become more receptor specific and closer to being a cure, it is likely that they too will have to conform to expectations of low toxicity. According to indication, some toxicities may be more acceptable than others. Diabetes is associated with a number of clinical effects, and any toxicity that might act to enhance these effects or accelerate the progress of the disease would not be acceptable.

COMPARATIVE RISK

Risk usually has a comparative element that needs to be considered in any assessment, most simply as a risk/benefit analysis. When considering a chemical that has been developed for a particular purpose, the advantages of its use should be considered as well as the disadvantages. This is also true when looking at chemicals that have been used for many years but then are shown to be associated with toxicity. Replacement of a toxic chemical with an essentially unknown substitute is not necessarily better, as long-term experience with the substitute can show different hazards that may also be undesirable. Thus, benzene was replaced by toluene in the 1970s, when it was found that benzene is a carcinogen; long-term experience with toluene has demonstrated a number of chronic effects including neurological deficits that include central nervous system (CNS) depression, peripheral neuropathy, encephalopathy, and optic neuropathy together with a variety of other toxicities (Dart 2004).

Different forms and concentrations of the same chemical will probably be associated with different levels of risk. Hazards posed by different chemicals may be similar but have different risks attached to them; also, hazard does not necessarily imply significant risk. Comparative risk is valid for single uses of chemicals but should also be considered across different origins, chemical groups, or boundaries of use, such as when a pesticide is reassessed for medicinal use. Risk, therefore, should not be considered in isolation; Focus Box 14.3 looks at elements of comparative risk and seeks to put some perspective on this.

Comparison of risk is also valid across location; a risk that is unacceptable in the United States may be tolerable in a less developed country due to local circumstances. Use of a carcinogenic pesticide is unlikely to be acceptable in a developed country because life span is long enough to allow expression of the cancer; where life span is shorter, this risk could possibly be acceptable (Illing 2001). Other circumstances may be more important, such as the cost of alternatives, for example, in the control of malarial mosquitoes and consequent reduction in disease. The environmental fate of the pesticide and the associated environmental risks may well be similar in both locations, but the consequences for human longevity may be completely different.

It may be concluded that concentrating effort on reducing one risk is pointless if other factors pose similar or different risks that are greater. There comes a point beyond which effort to reduce risk becomes an expensive waste of time and money.

Differing national standards of water quality have also been a factor in bringing to light unexpected hazard and risk. In the United States, it was indicated that chlorine in the drinking water could be associated with a small increase in the risk of bladder cancer. Chlorination of drinking water was found to be associated with the formation of chlorinated organic compounds, some of which were mutagenic in the Ames test. The US Environmental Protection Agency (EPA) concluded that chlorination of the public supply was no longer necessary. This decision was noted in Peru, and it was decided that what was good in the United States should also be reflected in local policy and that water should not be chlorinated. This did not take into account the differing microbiological properties of the local public water supplies, and as a result, large numbers of people died in the cholera epidemic that followed (Kellow 1999).

Risks associated with the differing uses of warfarin have already been mentioned in Box 14.3. Another topical example is the use of cannabis- and cannabinol-related compounds in therapeutic applications, as against use as a drug of leisure. As always

FOCUS BOX 14.3 ELEMENTS OF COMPARATIVE RISK

The following list is not exhaustive:

- *Similar hazard, different risk.* This may be seen for single chemicals and between different chemicals or chemical groups and may be a function of relative potency, formulation, place of use, physicochemical characteristics, bioavailability, etc. Comparative risks for lead are seen in location of exposure (paint, car fuel, drinking water or on church roofs) and form (organic or ionic).
- *Origin: endogenous/nonendogenous or natural/synthetic.* Natural chemicals are not less toxic than synthetics; they may act as precursors to safer synthetic chemicals. Synthetic chemicals include pyrethroid insecticides developed from pyrethrum in chrysanthemums and antibiotics developed from penicillin (Ames et al. 1990).
- *Location.* Risk that is unacceptable in one environment may be offset by benefits in other places. Ceasing chlorination of the public water supply in Peru, due to perceived risk of cancer, was offset by the major cholera epidemic that resulted (Kellow 1999). The cost/benefit of pesticide use changes with different circumstances of the environment or country in which it is used.
- *Intended use.* Warfarin is an effective rodenticide but is also used as an antithrombotic drug. The form (bait preparations) in which the use as a pesticide is permitted may therefore be controlled; its use as a drug is also controlled but is subject to other factors, such as comedications, that alter the risk factors for the anticoagulant hazard of its use, for example, by competing for protein-binding sites and increasing the amount of free warfarin in the plasma.
- *Production.* Organic produce is widely marketed as "better" than produce grown or treated with pesticides or preservatives. Untreated peanuts may contain unacceptably high levels of aflatoxin due to contamination with the mold *Aspergillus flavus.*
- *Substitutes.* An apparently nontoxic substitute may have unsuspected long-term effects of a different kind to the original. In proposing a substitute, there is an onus to ensure that it is less toxic than the original chemical. (See Chapter 18 and associated text on methylene chloride.)
- *Type of exposure.* Risk due to low-level radiation from nuclear plants may be contrasted with the risk of melanoma due to sunbathing.

in risk assessment and management, it is useful to invoke Paracelsus, that the dose makes the poison. Given the pharmacological activity of cannabinoids and evidence of their benefit in a range of disorders, it seems less than sensible not to investigate their use more extensively. In considering the risks associated with use of cannabis, the comparative risks of alcohol and nicotine should also be considered. Once again, it seems that science and politics are, to all intents and purposes, incompatible.

SYNTHETIC VERSUS NATURAL

One of the greatest public debates in comparative risk has been on the merits (or otherwise) of natural versus synthetic chemicals. Many synthetic chemicals have similar counterparts in nature to which humans have been unwittingly exposed—sometimes endogenously—for centuries or millennia, without epidemiologically perceptible effect. Two good examples of this phenomenon are the classifications by International Agency for Research on Cancer (IARC) of processed meats and estrogens as known human carcinogens. Although there is a great suspicion—fear even—about synthetic pesticide residues in food, it is seldom remembered that these synthetic chemicals are far outweighed by those that are present naturally. This emphasis is disproportionate; the synthetics are present in low, regulated concentrations and have all been thoroughly tested for safety, unlike the vast majority of natural chemicals. Concentrations of synthetic pesticides are further reduced by setting the interval between spraying and harvest. Natural chemicals in foods are essentially unregulated; for the most part, there are no epidemiological data that indicate necessity for regulation, and in any case, the database is not present to allow any sensible limits to be put in place. In fact, when natural chemicals are tested, a potentially alarming number of them are associated with unwelcome toxicity, often with a low margin of safety (Focus Box 14.4).

Given the structural similarities between synthetic and natural chemicals, there is no scientific future in trying to draw a toxicological distinction between them: synthetic does not mean toxic, and natural is not always beneficial. The structures of some natural chemicals are so complex as to make them very difficult to synthesize in a laboratory, yet these same comfortingly natural chemicals include some of the most toxic substances known. Batrachotoxin is a structurally complex alkaloid found in the skin of the South American frog, *Phyllobates aurotaenia*, which has an LD_{50} in mice that is in single-figure micrograms; few chemicals that are exclusively synthetic approach this level of lethal toxicity. Another consideration is that there are many times more natural chemicals than there are synthetic ones, and also that many toxic chemicals considered to be artificial, like dioxins, are present in the natural environment through processes such as burning wood.

Against this background, the Delaney amendment for food additives—that no additive found to cause cancer in animals after oral ingestion shall be deemed safe—seems a little redundant.

Use of pesticides or preservatives has two sides in terms of relative risk, that is, risks associated with their use and those associated with not using them. A decision not to use pesticides because they are toxic ignores the natural presence in our diet of vast amounts of naturally present chemicals that have evolved as endogenous defenses against insect attack (see Focus Box 14.4). Poor preservation of food—in an effort to maintain organic standards or production or for lack of facilities—can be associated with the growth of molds.

When considering risks associated with pesticide use, it is also worth looking at the increasing consumer enthusiasm for organic produce, driven by the perception

FOCUS BOX 14.4   RISK AND CHEMICALS
NATURALLY PRESENT IN FOOD

In 1999, the American Council on Science and Health published a holiday dinner menu to coincide with the Christmas festivities, to demonstrate just how much of our regular diet is made up of potentially toxic chemicals, which are there naturally and in larger quantities than artificial chemicals such as pesticides.

- Endogenous plant chemicals include an assortment of alcohols, aldehydes, isothiocyanates, heterocyclic amines, carbamates, psoralens, caffeic acid, hallucinogens, and large numbers of known rodent carcinogens such as benzo(a)pyrene and ethyl alcohol (which is also a human carcinogen).
- It has been shown repeatedly that high intake of fruit and vegetables protects against cancer, despite the fact that they contain chemicals that have been shown to be rodent carcinogens.
- White bread contains furfural, which is a rodent carcinogen. The carcinogenic dose in rodents was 197 mg/kg/day; the equivalent human dose would be 13.79 g/day *for life*; given that a slice of white bread contains about 167 µg of furfural, you would have to eat 82,600 slices of bread a day to achieve an equivalent carcinogenic dose.
- Although the emphasis is often placed on carcinogens, "ordinary poisons" are also present. Potatoes contain the glycoalkaloid solanine, which is a cholinesterase inhibitor and teratogen (Friedman et al. 1991). Concentrations are much higher in green potatoes—about 2 mg/g compared with 0.1 mg/g in normal potatoes; the human lethal dose is about 500 mg (Lappin 2002). Concentrations of such alkaloids can increase after harvesting through exposure to light or damage (Friedman et al. 1999). There has even been a suggestion that the potato may be the "environmental culprit" in schizophrenia (Christie 1999). The safety margin between normally present concentrations and those that are toxic in humans is not large.
- Severe toxicity is associated with improperly prepared food plants such as cassava root, in which cyanogenic glycosides react with stomach acid to release cyanide. Red kidney beans produce toxicity unless boiled before eating.
- Edible mushrooms contain various hydrazines that have been associated with cancer in mice (Focus Box 1.2, Chapter 1).

that "pesticides are bad for you." This has led to production of increasingly insect-resistant varieties so as to avoid the use of pesticides. This can result in insect-resistant crops that contain higher-than-normal amounts of endogenous chemical, leading to adverse effects. Thus, organic produce has at least the same amounts of natural chemicals as nonorganic food and may have more; the absence of pesticides makes a small difference to the overall chemical burden (Ames et al. 1990). In the final analysis, there is no dispute that pesticides are toxic, but as with every other chemical (natural or not), this is very much a question of dose, and the margins of safety between toxic concentrations and those acceptable in foods are regulated. Frequently, these regulated margins are larger than for endogenous chemicals, such as solanine.

Consideration of comparative risk is essential in risk assessment of new or existing chemicals; there is no scientific point in setting stringent limits of exposure on a new chemical, if similar levels of hazard and risk are posed by an endogenous chemical present at greater concentrations. Too frequently, the public and politicians are blind to such comparison. It is one of the responsibilities of toxicologists in general to communicate this aspect of risk assessment in a more effective manner.

RISK EXPRESSION AND QUANTIFICATION

Risk is the probability of harm and could, therefore, be expressed as a number between 1 and 0. In practice, risk is expressed in terms of incidence per unit of population or as a percentage. Expression as an incidence—y cases per 100,000—is useful in terms of the general population, within which the wide range of risk determinants that affect specific groups or individuals can be accommodated without too much problem. A refinement of this is to compare risk in an exposed population with that in an unexposed population. Expression of relative risk implies some knowledge of normality, that is, the unexposed population. Normal mortality in Scotland is about 11 per 1000 population (Scottish Executive 2014), possibly a little higher than in England. The same sources give access to a plethora of normal data for the incidences of disease and resulting mortality.

There may be circumstances in which it is possible to say that a percentage of people exposed to a chemical will probably show a particular adverse effect—as in patients receiving monotherapy with a specified drug—but this tends to be an exception. Although the dose makes the poison, it is the individual who makes the response, and doses that leave many people unaffected can leave others severely disabled. It is this breadth of characteristic and potential response that complicates numeric expression of risk in terms of particular groups or individuals. In assessing the risks associated with production of a drug, it is simple to look at the incidence of reported adverse reactions in patients receiving known doses and to extrapolate these data to new patients. However, extrapolating these effects to a group of healthy production workers, who do not have the disease target for the drug and who will be subject (theoretically) to lower-than-therapeutic systemic exposures, becomes so imprecise that numeric expression is not possible, even if it was legally sensible.

Another, sometimes useful, but easily devalued, method of risk expression is to use the doses associated with effect in animal studies and to extrapolate from these to the anticipated human dose. This can lead to seemingly ludicrous similes, such as drinking 400 bottles of cola a day while standing on top of a mountain (see Focus Box 14.4). However, this method is particularly useful when looking at the risks associated with pesticides or natural chemicals. It is sometimes also useful to consider the factors that might increase the probability of death by one chance in a million. These include living with a cigarette smoker for 2 months, eating 40 tablespoons of peanut butter, or living for 150 years within 20 miles of a nuclear power plant (Kellow 1999). In any case, an increase in risk of one in a million is so small as to be undetectable, given that cancer, for example, will occur in about 30% of the normal population. Only if a cancer is particularly rare and occurs in a particular identified group can it be attributed to a particular chemical, as with hepatic hemangiosarcoma and occupational exposure to vinyl chloride.

Although it is desirable to quantify risk when looking at high-level assessments, for instance, in terms of carcinogenicity, for the most part, this is not particularly easy or necessary. For a workplace assessment, it is enough to know that there is a hazard to be controlled and that a reasonable estimate of risk can be made in general terms. This is arrived at by consideration of the various factors that contribute to the risk of the hazard being realized.

SUMMARY

Risk is a product of hazard and exposure—without exposure, there can be no hazard or risk.

- Risk is seen at different levels, from individuals to worker groups to national and global populations.
- Risk is increased or decreased by factors that include occupation, diet and lifestyle (smoking or alcohol consumption), genetics, and local circumstances.
- Risk of use may be offset by risks associated with nonuse.
- Perception of risk, which is critical to successful risk management, is not necessarily subject to logical analysis. Perception is influenced by clarity and perceived honesty of communication and the acceptability of the risk expected.
- Risk may be quantified, but this is usually only done for endpoints such as carcinogenicity after extensive mathematical modeling. For general purposes, a qualitative assessment is enough.

REFERENCES

Ames BN, Profet M, Swirsky Gold L. Nature's chemicals and synthetic chemicals: Comparative toxicology. In: Proceedings of the National Academy of Sciences of the USA. Revised 1990, and Dietary Pesticides (99.99 per cent All Natural).

Ballantyne B, Marrs T, Syversen T. *General and Applied Toxicology*, 2nd ed. London: Macmillan Reference, 2000.

Christie AC. Schizophrenia: Is the potato the environmental culprit? *Med Hypotheses*. 1999; 53(1): 80–86.

Dart RC, editor. *Medical Toxicology*. Lippincott Williams & Wilkins, 2004.

Friedman M, Rayburn JR, Bantle JA. Developmental toxicity of potato alkaloids in the frog embryo teratogenesis assay—Xenopus (FETAX). *Food Chem Toxicol* 1991; 29(8): 537–547.

Friedman M, McDonald GM. Postharvest changes in glycoalkaloid content of potatoes. *Adv Exp Med Biol* 1999; 459: 121–143.

Illing P. *Toxicity and Risk—Context, Principles and Practice*. London: Taylor & Francis, 2001.

Kellow A. *International Toxic Risk Management*. Cambridge: Cambridge University Press, 1999.

Lappin G. Chemical toxins and body defences. *Biologist* 2002; 49: 33–37.

Mackness B, Durrington PN, Mackness MI. Low paraoxonase in Persian Gulf War veterans self-reporting Gulf War Syndrome. *Biochem Biophys Res Commun* 2000; 276(2): 729–733.

Mallinckrodt MG, Diepgen TL. The human serum paraoxonase polymorphism and specificity. *Toxicol Environ Chem* 1988; 18: 79–196.

Mutch E, Blain PG, Williams FM. Interindividual variations in enzymes controlling OP toxicity in man. *Hum Exp Toxicol* 1992; 11(2): 109–116.

New Scientist. 11 September 2001.

New Scientist. 19 February and 21 October 2000.

New Scientist. 18 December 1999.

Pilkington A, Buchanan D, Jamal GA, Gillham R, Hansen S, Kidd M, Hurley JF, Soutar CA. An epidemiological study of the relations between exposure to organophosphate pesticides and indices of chronic peripheral neuropathy and neuropsychological abnormalities in sheep farmers and dippers. *Occup Environ Med* 2001; 58(11): 702–710.

Scottish Executive. Health in Scotland 2014. Available at http://www.scotland.gov.uk.

Slovic P. Trust, emotion, sex, politics and science: Surveying the risk assessment battlefield. *Risk Anal* 1999; 19: 689–702.

Stephens R, Spurgeon A, Calvert IA, Beach J, Levy LS, Berry H, Harrington JM. Neuropsychological effect of long-term exposure to OPs in sheep dip. *Lancet* 1995; 345(8958): 1135–1139. Commentaries and authors' reply.

Virtue WA, Clayton JW. Sheep dip chemicals and water pollution. *Sci Total Environ* 1997; 194/195: 207–217.

15 Risk Assessment in Practice and Setting Exposure Limits

INTRODUCTION

We are now at the point where the hazards have been predicted to be human relevant (see Chapter 13). The information from these earlier stages is assessed to indicate the probability that the toxicities seen will be expressed in the target population under the anticipated conditions of exposure; these are usually assessed and evaluated on a worst-case basis. This probability is governed by factors such as safety margins, working practices, and form of chemical (see Focus Box 15.1). It should also be considered, in many instances, that the experience, judgment, and prejudices of the assessor also play a large part in such probability assessments. In essence, in a risk assessment, the toxicity of the chemical, related to dose levels in safety tests, is considered in conjunction with anticipated exposure levels for the target population. This should lead to an assessment of the likelihood that toxicity will be expressed in the target population (a margin of safety), facilitating decisions on risk (risk evaluation) and hence exposure limits (which are set as the maximum exposure considered to represent a "broadly acceptable" risk/safe for that particular type of exposure). These limits and controls must be reevaluated in the light of the intended measures to be taken to control exposure in the target population and when new information/new interpretations come to light.

Firstly, it is necessary to define hazard and risk. Hazard is the property expressed by a chemical or mixture—for example, peripheral neuropathy due to solvents in paints. Risk is essentially a probability that the hazard will be expressed (in a given population/set of circumstances). With the evolution of understanding of the effects of solvents such as n-hexane, which is a peripheral neurotoxin, the composition of paints has changed so that they now contain little or no n-hexane. Consequently, the risk of acquiring peripheral neuropathy due to n-hexane exposure from paints is much lower than it used to be. In contrast, the understanding of the effects of toluene, currently a common constituent of nonaqueous paints, is growing with an appreciation that chronic exposure to this solvent may also be associated with long-term neurological effect. Unlike the situation with n-hexane, however, there is currently no neat molecular mechanism for these effects.

RISK ASSESSMENT AS A PROCESS

Risk assessment is a well-defined process, consisting of four interrelated stages, the first three of which are hazard identification, definition of dose response, and assessment of exposure. The information gained from these three processes is then fed into the fourth risk stage: characterization. These processes are distinct from risk management, which is the process of evaluating the risks; deciding on whether those risks are acceptable, tolerable, or intolerable; and deciding on risk management procedures. Risk management puts the risk assessment into effect, determining acceptable daily intakes (ADIs) for consumers or workplace exposure limits for production workers. It also considers precautions, such as personal protection equipment (PPE), to be taken when these chemicals are used; considers any restrictions on who can purchase and use them; and monitors to ensure that the risk management proposals are effective and are being enforced. Registration, Evaluation, Authorisation and restriction of Chemicals (REACH) is the umbrella process that will drive much chemical risk assessment and management in the European Union in the coming years, and it uses the hazard characterization descriptors contained in the Globally Harmonized System for classification and labeling of chemicals.

DATA QUALITY

Data quality is one key to successful risk assessment. Another is knowing what you are seeking to achieve: a margin of exposure (MOE) or a maximum acceptable/tolerable level of exposure—the former is a risk characterization, and the latter is a risk characterization and evaluation rolled into one. The former requires exposure data, but the latter does not. Risk assessment is the end stage of toxicological evaluation, and it uses data from a wide variety of study types, ages, and provenances. It may be complete or exiguous, of uniformly good quality, or so poor as to be unusable. It goes without saying, therefore, that the quality of data available is fundamental to the success of any risk assessment; this is particularly an issue for older chemicals. Poor data will inevitably result in a poor, or at the least conservative, risk assessment.

For every review, but especially where the database is old and of dubious provenance, an assessment of data quality is an integral part of the process. A loose classification of data may run from high to medium to acceptable or unacceptable, based on aspects such as compliance with Good Laboratory Practice (GLP), experimental design and reporting, and age of publication. Klimisch et al. (1997) proposed a systematic approach for evaluating the quality and reliability of toxicological and ecotoxicological data. The main points they raised were reliability, relevance, and adequacy, as follows:

- Reliability: data may be described as reliable without restriction, reliable with restriction, not reliable, and not assignable.
- Relevance: based on factors such as *in vivo* versus *in vitro*, test material, endpoint studied, test system, etc.
- Adequacy defines the utility of the data for risk assessment. If there is more than one study for an endpoint, use the more reliable one. Are the data fit for purpose?

Relevance may be influenced by factors such as test system or age of the data. For instance, historical data may not reflect current reality, as with lead exposure. If an inappropriate test system or human population is used in the study, it is possible that its relevance may be reduced. Likewise, exposure by parenteral injection is unlikely to be relevant to a situation where exposure is exclusively dermal. If it becomes necessary to support your risk assessment with data from another compound, it is clearly important to ensure that the compounds are comparable in structure and use.

Variable quality of data may be attributed to a number of causes. Klimisch et al. suggested the following factors that should be considered:

- The use of test guidelines not compatible with modern standards
- Poor characterization of the test substance (e.g., for purity or other physico-chemical parameters)
- Use of techniques superseded by more modern methods
- Absence of measurement for endpoints that should normally be expected
- Completeness of reporting

Incomplete reporting of data is a perennial problem either because the paper was targeted at particular endpoints, excluding the ones of modern interest, for lack of space, or a host of other reasons.

Data quality drives the conservatism of the assumptions and the security of the whole risk assessment. Poor quality data may influence the selection of the endpoint on which the risk assessment is based and may result in choice of an inappropriately high or low no-observed-effect level (NOEL) or no-observed-adverse-effect level (NOAEL); this in turn is likely to affect the margins of safety used and any acceptable exposure limits that may be decided. At worst, it can result in extra testing and may affect the commercial viability of continued use and production of the chemical.

In looking at a data set, it is important, therefore, to ask a number of questions. These include checking if the studies were conducted to Good Laboratory (or scientific) Practices or the clinical equivalent; if the description of the methods and study design is adequate; if the data are reported completely; or if selected group means with or without standard deviations or standard errors of the mean are reported. For human reports, does the paper relate to a case report or to a full study?

DATA SELECTION FOR RISK ASSESSMENT

For new notifications of chemicals or modern data sets, selection is not—or should not be—a problem. Frequently, however, the chemical being assessed is old or well established, and the amount of data available can range from close to nothing to vast numbers of dubious academic papers from the 1960s and 1970s.

When there is a large data set available, choice of study or report becomes critical in determining the success of the risk assessment. It is important to identify a critical endpoint and a pivotal study that demonstrates the most sensitive species or most relevant endpoint and gives a clear NOEL or NOAEL. For small data sets, there may be little choice as to which studies to select, and it may become necessary to look outside the data set at comparable compounds.

In looking at the types of data available, these are broadly classifiable as experimental (usually in nonhuman species), epidemiological (including case studies of individual exposures, usually accidental), and, of course, physicochemical data. Experimental data are often the strongest part of the database and may result from a focused program of studies. Specialist regulatory studies should be reliable if conducted in laboratories of good reputation, and newer academic work is often well designed and reported. In modern work, test systems are usually well chosen and understood.

However, such robust data are usually present only for new substances. Older substances are often associated with poorly designed studies, which have been poorly conducted and incompletely reported. Old substances are often associated with a mass of unacceptable reports of low quality and little relevance. In these circumstances, choice of publication becomes critical.

Given the stark reality of many risk assessment situations, it is sensible only to give a listing of the types of data that are desirable rather than lay down a list from which deviation is not acceptable. These include

- Prediction using computer-based and *in vitro* models
- Physicochemical properties
- Relevant pharmacology and safety pharmacology
- Pharmacokinetics
  - Bioavailability
  - Absorption, distribution, metabolism, and elimination (ADME)
- Single- and repeat-dose toxicology
- Reproductive and developmental toxicity
- Genotoxicity and carcinogenicity
- Sensitization, irritation, or corrosion
- Human experience, clinical data, or epidemiological studies
- Exposure estimates
- NOELs and/or NOAELs

Comparison of one structure with another—known as read-across—may become necessary but should not be used to support an entire risk assessment. If both data sets are weak, however, it should be realized that the risk assessment itself will be weakened. Read-across is appropriate with similar chemical classes or structures and if the endpoints assessed are the same or if they have similar uses or targets.

Read-across has been extensively covered by guidance from the European Chemicals Agency (ECHA 2008). This makes the point that read-across is based on identifying similar compounds to the one to be assessed. A first step is to assess if the target chemical is a member of a chemical category; this is a groups of chemicals with similar physicochemical and toxicological profiles or that follow a regular pattern. Similarity of pattern may be identified across a chemical series, such as straight-chain alkanes (being careful to avoid the trap set by *n*-hexane—see below). The Organization for Economic Co-operation and Development (OECD) QSAR Toolbox has been specifically developed for aiding in read-across in such situations;

however, the complexity of its operations requires specialist users to perform evaluations. Broadly, the toxicological profile of members of a series may be similar but is
likely to change with increasing molecular weight or chain length, usually in terms
of potency or dose. If this is accepted, it should be possible to take account of known
toxicity at either end of the series and to predict toxicity for a relatively uncharac-
terized chemical in the middle of the series. Furthermore, considerations of steric
hindrance (i.e., the occlusion of a functional group with another) should be made. To
be eligible for read-across, the comparator chemical should be similar to the target
in the following respects:

- Physicochemical properties (pKa, log P, molecular weight)
- Structural groups
- Stereochemistry
- Mechanism of toxicity

The mechanism of toxicity or molecular initiating events (MIEs) is a critical
consideration. The MIE is the initiating molecular interaction that is the origin of
an adverse outcome pathway, and this can be very different for apparently similar
chemicals. For example, n-hexane is a known neurotoxin due to the formation of
a toxic metabolite 2,5-hexanedione (which is also a metabolite of methyl-n-butyl
ketone). This cross-links axonal neurofilaments, leading to peripheral neuropathy.
n-Hexane is a six-carbon straight-chain alkane; however, this mechanism of toxicity
is not common to others in this series. Another example of differing mechanisms
of toxicity, shown by similar molecules, is cinnamaldehyde in comparison with
3-phenylpropanal. These have the structures shown in Figure 15.1.

While both bind covalently to protein, 3-phenylpropanal does so via Schiff's base
formation via the carbonyl group, while cinnamaldehyde does so through a Michael
addition to a protein thiol group via the alkene group. Both are skin sensitizers but
by different mechanisms.

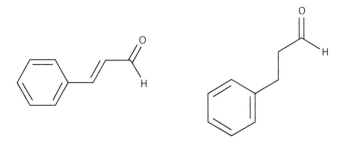

Cinnamaldehyde (CAS 104-55-2)   3-Phenypropanal (CAS 104-53-0)

FIGURE 15.1   Structures of cinnamaldehyde and 3-phenylpropanal. (Images from ChemIDplus
from the US National Library of Medicine.)

Read-across is employed in (quantitative) structure–activity relationship (Q)SAR software, which gives percentage similarity scores based on Tanimoto scores. While it may be reassuring to know that the comparator and target chemicals are 85% similar, it is still necessary to perform a visual assessment of the structures being compared.

SINGLE EFFECT VERSUS GENERAL RISK ASSESSMENT

The end goal of risk assessment is an expression of risk in a quantitative or qualitative form. Although all hazards are assessed, risk assessments tend to fall into two overlapping types—broad assessments, which look at all the relevant endpoints, and those focused on a single effect. The latter may be described as high-level assessments, typically conducted for carcinogenicity and using data from rodent bioassays in conjunction with mathematical modeling to give a numeric estimate of risk for a general population. This focused type of assessment, which is an extension of the broad overview, is discussed briefly here but is not the main focus for this book; it is covered in detail in other texts (see References). The general multi-endpoint approach forms the basis of all assessments and is more likely to produce a qualitative gradation of risk estimate from likely to unlikely. The output is dependent on expert interpretation and discussion of all the data, and this essentially opaque and indefinable process concludes with an overall interpretation that results in proposals for maximum exposure levels. Due to the lack of easy definition of process and decision pathways, it is important that records are kept on how and why the decisions were reached. These records may well become significant in a court of law or in discussion with regulatory authorities. They also help in the communication of the decision to those at risk, particularly if the decision is unwelcome in any way.

Although assessments may focus on particular hazards or effects, it is quite usual for a compound to show several different toxicities, and these may be differently expressed according to dose, concentration, route of exposure, and where appropriate, design and type of the experiment. All may be more or less relevant to humans, although only one may be relevant to the target population for whom the risk assessment is intended.

The above gives an indication of the broad approach to risk assessment, which should give a basic foundation on which to build a set of assessment practices. There is no sense in laying down dogmatic rules for such assessments because the circumstances of each type (whether for agrochemicals, industrial chemicals, drugs, or food additives) and for each compound differ to such an extent that they cannot be covered in detail here.

As stated earlier, the single-endpoint assessment is an extension of the general assessment on which it is firmly based. These assessments are conducted typically for carcinogenicity and are often more relevant to the population in general rather than specific groups or individuals. They rely on numeric data from carcinogenicity bioassays and extrapolation from the low dose to the exposures that may be allowable in humans. For situations where the general population are already in contact with the chemical—for instance, in the diet—these assessments can give an estimate of the existing risks associated with continued use. The weakness of this single-endpoint approach is that the other risks associated with the chemical may be forgotten, and the data on which they are based and the way in which they are manipulated may be of questionable relevance to humans. They are, however, based on defined mathematical models and produce numbers via a traceable process. While a numeric output may give a degree of comfort to the assessors, regulators, and the general public, it does not necessarily mean that the conclusion drawn from these numbers is any more secure or sensible than one arrived at by a more flexible or broader method. Judgment, which is rarely traceable and sometimes difficult to explain, should always play a part in the final conclusion. Various aspects of this approach are covered in a brief discussion of models used (below) and later in a section on risk assessment in carcinogenicity (Chapter 18). For those who would like greater detail, there are specific and general texts cited in the References.

TOOLS AND MODELS IN RISK ASSESSMENT

There is a range of data-handling tools and models that can be used to assist with risk assessment. This short section looks briefly at physiologically based pharmacokinetic (PBPK) modeling, allometric pharmacokinetic scaling, and models for carcinogenicity data treatments.

PBPK MODELS AND SCALING

The principle of PBPK models is relatively simple, although their design, validation, and mathematical complexity are not. If the factors known to affect the ADME of a chemical (liver blood flow, partition coefficient, distribution into tissue compartments, and the kinetics of metabolism and excretion) are known for one or more species at several doses, it is possible to extrapolate the dynamics from known (tested) doses to higher or lower doses and from one species to another. One of the objectives is to predict the behavior of small doses in humans from the behavior of higher doses used in animals. PBPK models can also be used to study the relationship between predicted tissue concentrations and toxic effect. Initially, a model is constructed, which consists of a series of compartments that are linked by blood flow, the whole being represented by a diagram (Figure 15.2). Basic rules are applied to the design so that tissues that play a prominent role in the pharmacokinetics or toxicodynamics of the compound are individually specified in the model. Other tissues can be grouped together as single compartments, distinguishing tissues with high blood flow from those that have low perfusion rates. These are then linked by a series of kinetic expressions describing the movement of the chemical or its metabolites between compartments. Values for parameters such as blood flow through the liver or kidneys or pulmonary characteristics can be obtained from the literature. These are linked to parameters for the chemical or its metabolites such as partition coefficients and binding affinities for proteins or receptors and to biochemical values for ADME. It should be borne in mind that much toxicity is associated with thresholds and that such thresholds are in turn associated with saturation of a process such as elimination or with exhaustion of a protective agent like glutathione. The presence of thresholds and the consequent drift from linear kinetics complicates the

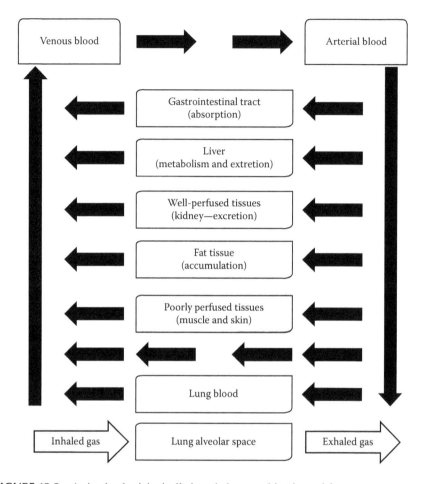

FIGURE 15.2 A simple physiologically based pharmacokinetic model.

mathematics considerably. The model can then be validated against experimental data (or such data could be used in the design stages).

By adjusting parameters, such as dose, in the model, it is possible to predict concentrations in various tissues either of the parent compound or its toxic metabolites and so to predict effect. In this way, the effects at doses used in toxicity tests may be used to predict the dynamics at low doses in the same species or by extrapolation to another species.

Extrapolation of pharmacokinetic parameters from animals to humans is a critical part of risk assessment. An adjunct to PBPK modeling is the use of allometric scaling, which takes account of the differences in pharmacokinetics that are seen with increasing body weight in animals. Broadly, small animals such as mice tend to have shorter half-lives for chemicals than larger animals such as dogs or sheep. Allometric scaling offers a method of relating body weight or surface area to parameters such as blood flow through organs such as the liver and to pharmacokinetic parameters, particularly clearance, volume of distribution, and half-life. This allows prediction of these parameters by extrapolation from one species to another. This deceptively tidy and simple concept has encountered a number of problems. Clearance has not been well predicted (Mahmood and Balian 1999), and the body weight model works best for renally excreted compounds (Ritschel et al. 1992). The concept of neoteny—originally meaning the retention of juvenile characteristics in an adult animal—has been invoked to make the process more reliable by taking into account the larger brain weight and longer life span of humans.

Scaling is used in the US Food and Drug Administration (FDA) approach to calculation of first-in-man doses (FDA 2005). This is based on the calculation of a human equivalent dose (HED) from a NOAEL. The assumption is made that toxic endpoints for drugs administered systemically to animals scale acceptably between species when doses are normalized to body surface area (mg/m²) (Table 15.1). The point is made that this tends to give a more conservative starting dose, thus enhancing clinical safety. On the other hand, scaling based on body weight may be more appropriate; in this case, there should be evidence that the NOAEL in mg/kg is similar across species.

When the HED has been calculated, safety factors are applied to provide a margin of safety for clinical trial participants. These uncertainty factors allow for differing pharmacological sensitivity between humans and animals, the presence of toxicities hard to detect in animals (headache, mental disturbance), receptor affinities, unexpected toxicity, and interspecies differences in ADME.

Traditionally, two factors of 10-fold have been used, to account for variation between species and to account for human variability. In addition, other factors may be added to take account of particular toxicities or endpoints (e.g., developmental

---

TABLE 15.1

Conversion of Animal Doses to Human Equivalent Doses Based on Body Surface Area

Species	To Convert Animal Dose in mg/kg to Dose in mg/m², Multiply by km	To Convert Animal Dose in mg/kg to HED in mg/kg, Either	
		Divide Animal Dose by	Multiply Animal Dose by
Human (60 kg)	37	–	–
Mouse	3	12.3	0.08
Rat	6	6.2	0.16
Dog	20	1.8	0.54
Monkey (e.g., cynomolgus)	12	3.1	0.32
Minipig	35	1.1	0.95

Source: Adapted from FDA, Guidance for Industry: Estimating the Maximum Safe Starting Dose in Initial Clinical Trials for Therapeutics in Adult Healthy Volunteers, 2005.

effects or carcinogenicity). These factors are explored in more detail when calculation of permitted daily exposure levels is covered (see below).

There is a continuing search for more mathematically based coefficients than body-derived measurements, which tends to demonstrate the ruthless quest of the numerate for greater "accuracy" of prediction. It must be asked how the resultant accuracy is to be judged, given that the model can only be set up on one set of characteristics and the target population is composed of thousands of highly variable individuals. In view of the degree of variation within a normal human population, 100% accuracy—or even 95% plus or minus 15%—is a chimera that cannot be achieved. Furthermore, a drive toward greater mathematical complexity could mean that the benefits of scaling are lost to a significant section of the toxicological risk assessment community, and the disadvantages of this outweigh the advantages of any illusory increase in numerical accuracy. As with all such substitutes for real experimental data, the processes should be used rationally, with understanding of their weaknesses and with full explanation of how the conclusions were reached. Used appropriately, allometric scaling is a valuable tool that, as with all other methods in toxicology, provides useful data that can be used to support the whole database used for assessment.

MODELS FOR RODENT BIOASSAY DATA

Mathematical models for the treatment of data from rodent bioassays for carcinogenicity risk assessments have proliferated over the years (Table 15.2). The intention of these models is to extrapolate a line or curve from high dose to low dose, so as to estimate dose levels that are safe. There is a number of such models, each one adding another layer of mathematical complexity to assessment of a single endpoint from data that are, in many cases, of questionable relevance to humans. To the nonnumerate, the credibility, in terms of relevance and general application, of many of these models is hidden behind abstruse equations and figures and assumptions that

TABLE 15.2
Models for Risk Assessment

Model	Comments
One hit	Based on the theory of single-cell origin for cancer. Simple but producing conservative results.
Probit	Assumes a normal distribution of log tolerances. Gives an S-shaped dose–response curve.
Multistage	Assumes that carcinogenesis has various stages necessary for the development of cancer and that effects are additive.
Multihit	A generalization of the one-hit model.
Weibull	Another generalization of the one-hit model.
Logit	Leads to an S-shaped dose–response curve.
Log-probit	Assumes a lognormal distribution of individual responses.

are not relevant to all data sets. The result of this is a set of numbers that can prove to be less than perfect, when viewed in the light of subsequent experience. Because of the breadth of the target at which these models are aimed, the assumptions that go into them are necessarily imprecise and tend toward the conservative. They seem to offer a one-size-fits-all approach that, intuitively, is flawed because of the data they are based on (typically single endpoints such as tumors in mice) and the inherent absence of assessment of all the data that is necessary in "real" toxicology. The end result can be an overestimate of risk, leading to an exposure limit that is conservatively low but is not scientifically justifiable. The cost to society of such conservatism is probably considerable, through increased cost of reduction and cleanup programs. One possible reason for the existence of such models, and their continued use with irrelevant data, is the fact that judgment is essentially unquantifiable and can be disputed—and frequently is. The provision of a conservative number, by a defined route and model with documented assumptions and record of the data used, gives a bulwark behind which some shelter can be found; this may be a poor shield, however, to use in response to litigation.

These models clearly have a valid role to play in the process of risk assessment, but that validity is driven not by the elegance of the mathematical gymnastics but by the assumptions and the data that are fed into them. There seems to have been a tacit recognition that the data used were not always especially relevant or uniformly suited to the models available, and this has been approached by attempts to make the models more able to cope with them. However, this has not been uniformly successful, because the data are the problem, not the basic concept of the models themselves. The objective should be to use data that are relevant to the problem under consideration and to have a model that does not rely on unjustifiable (or overconservative) assumptions or fudge factors. The various models used in risk assessment are discussed in detail by Gad (2000). One aspect of the utility of these models should be remembered: although it may be concluded that a particular dose or level of exposure would be associated with 1 additional cancer in 1 million, it is highly unlikely that this increase would be detectable.

TARGET POPULATION, DOSE, AND EXPOSURE

In toxicology, risk is largely a function of exposure to the chemical under examination; if there is no exposure, there is no risk. For a normal dose–response curve, risk increases with increasing dose, although dose is not necessarily the same as exposure for purposes of expression of toxicity, as a high dose may not be reflected in high exposure systemically or at the site of toxicity. Assessing the expected or actual dose and exposure levels is therefore fundamental to the whole process of risk assessment.

There are, theoretically, two basic types of exposure, controlled and uncontrolled. Control may be exerted through the means of maximum allowable concentrations, recommended doses of drugs, or preset workplace concentrations. In reality, there is a sliding scale of control, the highest level being that of a patient who is given a drug and watched while he/she swallows it and the least during accidents. Control is usually a remote phenomenon, and assumptions of regulating authorities may be made meaningless by the practices of the end user or by unpredicted spillage. Exposure can be assessed prospectively or after the event and may be to known or unknown chemicals; in addition, it may be voluntary or involuntary. Each case poses its own challenges in assessing exposure—involuntary exposure to unknown chemicals being the most complicated to evaluate.

The expression of toxicity is determined by availability of the parent compound or metabolite at the target (or site of expression); therefore, while the ambient concentrations of the chemical are important, the concentrations achieved internally or at the site of toxicity are also critical. The extent of systemic exposure is dependent on the bioavailability of the chemical for the principal route and circumstances of exposure. To distinguish these, it is probably convenient to think in terms of external exposure (e.g., an atmospheric concentration) and internal exposure, as shown by maximum systemic concentrations or area under the concentration curve.

For controlled exposure, which is the exception, the analytical techniques and matrices—blood or urine usually—are very similar, the main difference being that the dose is known and can be related to the concentration data (exposure) that emerge from the analyses. Methods of assessing exposure are explored briefly in Focus Box 15.1. For purposes of risk assessment, the exposure levels are assessed to answer two questions: "What hazards are associated with a measured concentration?" and "What is the safe concentration of the chemical?" The outcome should be a proposed concentration or exposure level that should not be exceeded; these are usually expressed as workplace or occupational exposure limits (WEL or OEL) or ADI. The first question may be asked with respect to an existing environmental contaminant or natural chemical. The second question is typically asked about concentrations in production facilities, so as to regulate exposure of the workforce or for setting acceptable residual concentrations of pesticides or veterinary pharmaceuticals in foods.

PREDICTION OF EXPOSURE

Prediction of dose or exposure is necessary when there is no prior knowledge of the chemical, as with a new food additive or on new or excessive release of a known chemical. The process of prediction may need to chart a chemical's progress from release into the environment or a situation from which exposure can be experienced to the time it comes into first contact with people, and what amounts are available for dermal or systemic exposure. The more steps between release and actual availability to the target population, the more difficult it is to predict exposure accurately. For example, for a chemical carried on or in food, uncertainties such as inconsistent daily intake, absorption effects due to other foods eaten at the same time, and preparation losses make the prediction more complex and imprecise.

It is worth briefly revisiting the factors that affect the systemic levels found following external exposure to a chemical, bearing in mind that the duration of exposure may be over several hours each day or irregular and may not perfectly reflect data in safety studies. Following ingestion, the situation is relatively simple compared with dermal or inhalation exposure; systemic exposure (internal dose) is likely to be similar to that achieved in toxicity studies by oral administration. However, it is necessary to make allowance for differences in formulation and duration of exposure

FOCUS BOX 15.1   ASSESSMENT OF EXPOSURE— BASIC PRINCIPLES

Uncontrolled exposure may be assessed by prediction or measurement, either direct or indirect. For most purposes of dose calculation, a 70 kg person is assumed, although 50 kg is more conservative, especially for female target populations.

Prediction:

- By ingestion: Knowing the concentration of the chemical in the diet or water, together with a reasonable daily intake, allows simple calculation of dose level. For example, a pesticide present as residues on apples at 1 ppm might result in a daily dose of approximately 2 µg/kg, based on an average apple weight of 150 g, with a pesticide burden of 0.15 mg/apple, eaten at the rate of one a day by a 70 kg person.
- For inhalation exposure, the normal pulmonary tidal volume at rest is approximately 0.5 L, and the respiration rate is between 12 and 20 breaths a minute (Witschi and Last 2001). For a normal working day of 8 hours at a gentle work rate, a convenient nominal volume of air breathed is 10 m³, based on 20 breaths a minute each of 1 L. On this basis, a concentration of 3 mg/m³ translates into a daily dose of approximately 0.4 mg/kg in a 70 kg person. The total daily volume of respired air may need adjustment according to the type of work.
- Bioavailability data can then be used to predict systemic exposure from the expected dose.

Measurement:

- Direct measurement may be made of the parent compound or its metabolites, usually in blood or urine but occasionally in expired breath, as with ethanol. These data should be specific for the chemical being assessed.
- Indirect measurements use biological markers or biomarkers of effect or response, which include changes in enzyme activity or biochemical parameters such as inhibition of cholinesterase or changes in coagulation times.
- DNA or protein adducts are a useful indication of exposure to reactive chemicals but are produced by numerous chemicals and, unless identified specifically, do not necessarily imply exposure to the study chemical.
- Choice of matrix for analysis should be made according to the chemical under study or the type of marker to be used.

(once-daily bolus versus probably constant low-level ingestion), but it should be possible to make a reasonable estimate of the amount of chemical that will be absorbed and reach the target site.

For dermal exposure, the local concentration, duration of contact, local humidity, extent of any local vasodilation, and formulation or form of the chemical should be considered; in addition, the local effects may be more significant than the systemic toxicity. Local concentration may be more important than dose expressed either as a total or in milligrams per kilogram body weight, and the effects, which may well be local, are likely to be enhanced when high workloads act to increase body heat, peripheral vasodilation, and local humidity. The presence of abrasions or skin disease, such as psoriasis, is also likely to enhance absorption across the skin. Furthermore, the location of exposure on the body coupled with the number of hair follicles may influence absorption. In some cases, it may be necessary to assume a worst-case scenario and assume total bioavailability, even if this is unlikely.

When exposure is likely to be by inhalation, the atmospheric concentration may be assumed to be constant, but dose level will be varied by changes in the breathing rate and volume of air respired, which are normal responses to changes in work rate or load. This is further complicated in that atmospheric concentrations may well vary, depending on local ventilation and distance from the source. For solids or aerosols, the particle size determines the region of the respiratory tract in which the particles or droplets are deposited; the physicochemical or pharmacological properties of the material may also affect dose levels by causing avoidance behavior in people exposed to it. Clearance of insoluble solids is via the mucociliary escalator, whereas liquids or soluble chemicals are likely to remain in the tract and be absorbed. For purposes of estimating the volume of air breathed in a typical 8-hour working day, 10 m^3 is often used as an arbitrary figure, being about twice the volume anticipated for a person at rest (Focus Box 15.1).

Ultimately, the assessment is likely to produce an exposure level that may not be achieved systemically, due to factors such as genetic polymorphisms or differences in ADME. Once an applied dose has been predicted, use of the bioavailability data for the compound should result in a reasonably accurate estimate of exposure for a given dose.

MEASUREMENT OF EXPOSURE

Without doubt, the best method of assessing exposure is directly by measurement of the parent compound or a metabolite in the blood or urine. This should also be used as a check of predicted exposure levels. Where this is not possible, the use of indirect methods may give good data from which exposure levels may be extrapolated. These measure biological markers or biomarkers of response or effect as indicators of exposure to a chemical. These include inhibition of cholinesterase as an indicator of exposure to organophosphates and prolonged coagulation times due to coumarins. They are usually much less specific than the direct methods; for instance, cholinesterase inhibition is produced by all organophosphates and carbamates and, as a result, is an indicator only of exposure to a chemical with this activity, not to a specific agent. This lack of specificity increases as the remoteness of the measurement from the original chemical increases. Normally, however, there is a history of exposure to a particular chemical, and it is relatively easy to make the connection between effect and cause. The extent to which the marker is changed from baseline data or from normal gives a broad indication of the extent of exposure. However, it cannot be used to calculate the dose or plasma concentration of the chemical responsible (if it has been identified), because the severity of response is so much influenced by the individual in terms of genetics and circumstances under which exposure took place. Another factor, working against calculation of dose for unattributed exposures, is the differing potencies seen between members of the same chemical class.

Reactive compounds or their metabolites interact with DNA or proteins, giving another method of assessing exposure. The total adduct concentration may be measured, which gives an indication of total exposure to reactive chemicals without assigning responsibility to any individual agent. However, where a specific adduct—a nucleotide complexed with an identified additional chemical group—can be demonstrated, this is a clear indication of exposure to the parent compound.

A further group, biomarkers of susceptibility, indicates differing susceptibility to effect, as shown by genetic polymorphisms. A concentration of chemical that is relatively risk-free for the majority of the target population may carry an unacceptable degree of risk for susceptible individuals. This is exemplified by allergic responses to very low concentrations, which leave the majority of people unaffected. They do not indicate exposure but are a significant factor in determining individual risk.

PROCESS AND FACTORS IN RISK ASSESSMENT

Risk is influenced by innumerable factors that act together or against each other in the process of delivering toxic chemicals or metabolites to their site of action. Although it is possible to list these factors, it is probably unwise to do so as this may be unduly restrictive. Risk assessment and risk evaluation should not be seen as box-checking exercises, and all the circumstances surrounding the expected use of the chemical must be considered as these may well be unique to the proposed use in the intended location. Some of the factors relevant to risk assessment are reviewed in Focus Box 15.2; these points are expanded in the text that follows.

PHYSICAL FORM AND FORMULATION

The physical form of a chemical is significant in determining risk associated with it (cf. nanotoxicology in Focus Box 15.3). A chemical generally poses the same hazard in whatever form it is present, but different forms carry different risks, as with lead in paint, in drinking water, or on church roofs. As stated earlier, if the toxic moiety cannot get to its site of action, it cannot cause toxicity at that site. Dermal exposure to a chemical that is not absorbed through the skin will not cause systemic toxicity, although there may be local effects such as irritation. A low probability of exposure is usually associated with a low level of risk. For example, in production facilities, a granulated, dust-free product poses less risk than the same chemical as a low-density, easily dispersible powder. A liquid form may be less risky to handle, providing that aerosols are not generated. Particle or droplet size affects availability

FOCUS BOX 15.2   FACTORS TO BE
CONSIDERED IN RISK ASSESSMENT

The following is not in any particular order of importance and should not be seen as complete, as other factors may be relevant to particular chemicals or circumstances of use.

- Physical form: The risks associated with dusty powders, liquids, or granulated products are different as the likelihood and type of exposure are different with each. A low-density, easily blown powder offers significantly greater risk than a solid or a viscous, aerosol-free liquid.
- Formulation: In general, diluted formulations pose less risk than concentrated forms; formulation also can have profound effects on absorption and bioavailability.
- Expected exposure, in terms of route, dose, and duration.
- General consideration of safety evaluation data for the shape of any dose–response curves and if there are thresholds for toxicity.
- Any human data that exist as a result of controlled experiment or accidental exposure; data from similar compounds may also be useful.
- Type of toxicity or hazard expected.
- Target organ and mechanism for toxicity, reversibility of any effects.
- Species differences in ADME and pharmacological potency, and likely impact of any such difference in humans; human or animal polymorphisms in metabolism.
- Safety margins: The relationship between NOEL, NOAEL, or lowest-observed-adverse-effect level (LOAEL), and expected levels of exposure in humans. A wider safety margin will be desirable for toxicities such as allergy or carcinogenicity than for other effects such as transient, reversible change.
- Target population for whom the risk assessment is being prepared, in terms of sex, age, disease status, etc.
- Purpose of the proposed risk assessment.

through inhalation, micron-sized particles being respirable, while larger ones are trapped in the upper respiratory tract. Physical form is also a consideration in dealing with accidental spillage: a free-flowing volatile chemical goes further when spilled and is more hazardous to clean up than a nondispersible solid.

Formulation of a chemical with a carrier or excipient also changes risk levels, as this can have significant effects on bioavailability. A solution in a solvent that enhances transdermal absorption is inherently more hazardous than a mixture of the same chemical with a solid excipient such as lactose. A solution combined with a process that allows formation of aerosols is potentially as hazardous as working with a dust. Dilution has the potential to reduce concentrations at the target site and so has benefits in reducing risk.

FOCUS BOX 15.3　NANOTOXICOLOGY

The study of toxicity resulting from exposure to very small particles is not a new phenomenon, but it is only relatively recently that the potential significance of very small particles—nanoparticles or nanomaterials—has been realized.

Nanoparticles are broadly defined as those with a median mass aerodynamic diameter of less than 100 nm (0.1 μm); to put this into perspective, it is generally considered that particles of 5 μm or less are respirable and can reach the deep lung. In contrast, a nanomaterial requires at only one dimension to be below 100 nm. People have always been exposed to atmospheres that contain particulate material, be it smoke or dust from farming or in the home; much of this would not enter the respiratory tract, due to size and filtering mechanisms in the respiratory tract, such as in the nose. Nanoparticles are a subset that can enter the lungs and that can exert unwanted effects. The other two significant routes of exposure are skin and ingestion; the former should act as a barrier preventing absorption, and the latter should minimize absorption other than by phagocytosis into the gut epithelium.

Although exposure to nanoparticles a long-term phenomenon, the nature of the particles has changed. This change from the essentially "natural" type—organic based particles resulting from livestock, earth, or burning natural materials—to synthetic particles in from multiple sources such as cosmetics (e.g., sunscreens), agrochemicals, medicines, and medical devices. The latter two examples add a "nontraditional" route of exposure to nanoparticles—parenteral. This picture is complicated by the fact that in addition to exogenous sources of nanoparticles, the body makes its own, for instance, lipid chylomicrons, which are present in the blood at all times.

In practical terms, each nanoparticle and/or nanomaterial must be considered on its own merits. It is not sufficient to indicate that a previously inert macro substance is of low risk when in the form of a nanoparticle or nanomaterial—shape and size are key. Broadly, the toxicity of a particulate material is dependent on its surface area and the chemical properties of its surface. This is exemplified by the difference in toxicity of carbon and carbon nanotubes and the formation of asbestos-like growth in experimental animals (Seaton et al. 2010; Kendall and Holgate 2012).

The expected form, and consequent bioavailability, of the product to which people will be exposed has to be compared with the formulations used in the safety evaluation studies because these are frequently different. A judgment has to be made as to how the form or formulation of the chemical compares with that in the safety data and whether it will be associated with greater systemic exposure levels than those seen in toxicokinetic data.

ROUTE OF EXPOSURE

The most likely routes of exposure must be considered together with the likely dose levels achieved. Also, the extent of control over the exposure is a significant factor;

thus, while it is difficult to control exposure to chemicals present in the environment, much greater control is possible with prescribed drugs (in theory) and in the work-place. This can highlight two populations of exposed people for whom the risks are different. For an orally active drug, patients receive a tablet or capsule of known size at a predetermined frequency; exposure by other routes is expected to be minimal. Workers producing the same chemical, who do not have the disease target for the drug, can expect to be exposed to it dermally and by inhalation. Following inhala-tion, a degree of gastrointestinal tract exposure can be expected via the mucociliary escalator in the bronchi and, as humans are mouth breathers under exercise, by direct swallowing of deposits left in the mouth. The doses achieved clinically and in the workplace are also likely to be different, as is the duration of exposure. It is possible to have much higher local concentrations of undiluted chemical in a production facil-ity than those encountered in other situations.

BIOAVAILABILITY

When the likely routes of exposure have been assessed, the bioavailability by the most significant route should be considered. Although risk assessment is often based on dose levels used in the safety data, where possible, the plasma concentrations and systemic exposure should also be considered. Bioavailability is usually considered in connection with systemic exposure, but availability for local effects is also impor-tant, given that the skin forms a reasonably effective barrier to many chemicals. It should be remembered that the concentration on the skin is likely to be higher than anywhere else and that local concentration on the skin is usually more important than dose expressed in milligrams per kilogram of body weight. This principle has relevance wherever the compound comes into contact with the body at high concen-trations, including the respiratory tract and eyes, where locally high concentrations can be associated with significant irritation.

DOSE RESPONSE

The relationship, defined by the dose–response curve, between the anticipated expo-sure levels and the dose levels at which effects were present or absent is absolutely critical to the risk assessment process. The difference between toxic concentrations and the expected dose in the target population gives the safety margin; the steep-ness of the dose–response curve gives an indication of how quickly the spectrum of effect is likely to change for small increases in dose level. Risk assessment is simplified where there is a large margin of safety between effect and target popula-tion exposure levels. This is seen in situations where limited toxicity is seen at very high exposure levels but exposure in the target population is, for example, lower by a factor of 1000 or more. Simplistically, large margins of safety are associated with lower risk.

The presence of a concentration or dose above which toxicity becomes evident should not necessarily be taken to mean that the hazard is not relevant, as individual circumstances can act synergistically with the chemical to reduce the threshold. While it may be statistically sensible to take in the 95% of the population who can be expected to fall within 2 standard deviations of the mean, this does not mean that the other 5% can be ignored.

For drugs, low-dose effects may result from the compound's intended activity, but these are still likely to be undesirable when expressed in a healthy group of production workers or in the population at large. Where toxicity is expressed only after a threshold level of systemic exposure has been exceeded, this is a clear indication that the effect will probably be absent at the (theoretically) lower levels expected in the target group. This is reinforced if the exposure levels above which the effect is seen are similar in all the test systems in which it is present.

SAFETY EVALUATION AND HUMAN DATA

Having considered the data that relate to the physicochemical characteristics and form of the chemical and the effect that these and any excipients and concentration factors have on the likely risk, it is time to consider data relating to safety evaluation and human exposure (if any). First consideration should be of any dose–response curve (see earlier) that has been generated for the effect under consideration, particularly if there is a threshold for toxicity below which effects have been shown to be absent.

Although species specificity of response has been cited as a major reason for saying that a hazard is not relevant to humans, a cynical view is that this means simply that the risk of the effect being seen (or detected) in humans is very small. Although the normal approach is to look for differences between species, in some ways, a simpler option is to look for uniformity of response. While differences in response can be useful in assessing risks, there is an onus to define these differences mechanistically, and this is not always immediately practicable. Lack of definition of mechanism can lead to uncertainty, and this must be taken into account in the risk assessment. However, if every test species shows the same effect, it is very likely that humans will show a similar response, and this gives a degree of certainty—although this is unlikely to be welcomed.

Expression of the same effect in different species does not mean necessarily that it will be present at the same dose levels or concentrations of exposure. ADME will probably be different across the species and possibly between individuals—especially in a genetically diverse population. Much toxicity is due to metabolism by the family of cytochrome P450 enzymes, which show a substantial degree of diversity between species and individuals. While a P450 that produces a toxic metabolite in rats may be absent or nearly absent in humans, this does not necessarily mean that the metabolite cannot be produced in humans, as the activity needed for the reaction can be expressed by other P450s. Such activity is usually lower than that in the affected species or target tissue.

Where toxicity is expressed through interaction with a receptor, the interspecies differences in affinity for the target need to be taken into account. These differences affect the time of onset of the effect and the speed with which it can be reversed. High-affinity binding can lead to a prolonged effect, whereas transient binding to a low proportion of the receptor population is usually associated with transient effect, especially if it is combined with rapid clearance from the target tissue.

In addition to the factors considered earlier, the safety evaluation studies should also define the hazard in terms of extent, mechanism, and reversibility, as well as the factors covered above such as ADME and any differences between species or individuals. One of the most important data points to come from these studies is the level at which no effect has been detected, the NOEL. If treatment-related effects were seen at every dose level, it might be possible to assign one as a level at which no adverse effects were seen (NOAEL). For this purpose, an effect that is not adverse is generally one that is reversible, is slight in extent, does not affect the well-being of the organism, and is not associated with any permanent consequences. Such slight effects include transient increases in plasma enzymes, reversible increase in liver size due to hepatocyte hypertrophy, or transient pharmacological action. Failing this, the LOAEL may be used, but this is not usual.

Each case needs to be considered individually, as an overprecise definition of what constitutes "adverse" may not be helpful. If the hazard for which the risk assessment is being conducted is always adverse, the NOAEL will not be acceptable, and a NOEL becomes essential. Circumstances are also important: for agrochemicals, for which no effect in humans is desirable, a NOEL is more appropriate than a NOAEL, which is the usual starting point for risk assessment for pharmaceuticals.

Although the acceptability of the risk and the specific hazard under consideration do not affect the numeric process of risk assessment, it is relevant in setting safety margins and acceptable exposure limits. This is influenced by the target organ for toxicity, the extent and type of effect, and its reversibility. The last point is a critical aspect of the safety evaluation and has significant impact on the acceptability or otherwise of the risks. Some effects, such as birth defects, cancer, or lesions in the central nervous system, are not reversible and demand a wider margin of safety than those that do not have any long-term consequences. Some effects, such as allergy, demand wide safety margins, as the reaction to even very low concentrations can be life threatening. Allergy is a particularly difficult hazard to deal with because, once established, the concentrations required to provoke a reaction are much lower than those needed for induction. Another factor here is that it is not necessarily easy to predict and can take a long time to develop, as shown by countless animal workers who have become sensitized to animals after years of problem-free work.

Any human data, which may have resulted from clinical exposure, accidental spillage, overdose, or even experience gained through working with the chemical, are also an important aspect of any risk assessment. For uncontrolled exposure, these data have to be assessed carefully for any estimate of dose achieved, and this is not always possible. On top of this, the circumstances and adequacy or extent of the data associated with each report have to be considered, as not every exposure is reported with future risk assessments in mind. Where several epidemiological studies are available, care must be taken that the protocols and diagnostic criteria used were consistent (if the data are to be pooled) and that other epidemiological pitfalls have been avoided.

Another source that should not be ignored is information from any similar compounds, either of the same chemical class and toxicological profile or having the same mode of action. It should be borne in mind that pesticides can have pharmacological effects (not always welcome) and this does not simply relate to drugs.

In looking at such data, the relative potencies of the chemicals should be taken into account, together with any other toxicities expressed, as similarity of toxicological profile can be used to back up conclusions from the risk assessment of the hazard under consideration.

PURPOSE AND TARGET POPULATION

Finally, the purpose of the risk evaluation, and the population at which it is aimed, should not be forgotten. The composition of the population that is liable to exposure is relevant, as risk factors differ according to age, sex, disease status, occupation, and expected circumstances of exposure. Some risks are not relevant to some populations; teratogenicity is not usually a risk factor for all-male working groups, although it may be indicative of other reproductive hazards relevant to males. Dermal exposure and irritation may be a problem in production workers but should not be significant in patients taking a capsule by mouth. Inequality of risk between populations is demonstrated by consideration of diethylhexylphthalate (DEHP). Exposure to DEHP was significantly higher in US patients receiving dialysis (50,000 patients receiving 4500 mg/year intravenously) than in the general adult US population (220 million), who were each exposed to 1.1 mg/year through dietary contamination. Patients who received irregular blood transfusions, hemophiliacs, and young children who were exposed orally are additional subgroups that were exposed to DEHP to different extents (Gad 2000).

The composition of the target population for whom the risk evaluation is being generated will define the responses that may be expected. The reaction to any chemical is likely to be different between healthy production workers, a group of patients, and the general population (and between the general populations in countries with long life expectancies and those with short life expectancies). A judgment has to be reached to cover the circumstances of each situation. In a region with short life expectancy and food shortages, it may appear to be pragmatic to say that use of toxic pesticides is not of significance, or that teratogenic effects are not important if there are no women in the target population. However, such approaches should be used with caution.

Pharmaceuticals offer a number of examples of the way risk evaluation works and illustrate the dynamics of the whole process, showing that an evaluation for one group is unlikely to suit another. The first contrast can be drawn between patients and production workers; the former have a disease and would be expected to benefit from exposure. Furthermore, this exposure is more controlled than for the production workers who do not have the disease and for whom the pharmacological effects may be unwelcome if not actually adverse. Another group, which can be expected to be exposed to a new pharmaceutical, are the healthy volunteers who, for non-life-threatening diseases, are the first humans to be purposely exposed to the drug. The assessment of risk for this group has to be very conservative because of the significant step that is being taken—from the laboratory animals of the safety evaluation to a first dose in humans. For this reason, first doses in humans are usually conservatively low, with a large margin of safety from the NOEL identified in the most sensitive species. However, for drugs intended for life-threatening conditions such as cancer, it is normal to start human studies in patients rather than volunteers, especially if the drug is toxic, which is frequently the case for cancer therapies. For these patients, the risk/benefit ratio is clearly different from other drugs, and as a result, the starting dose is often close to those associated with toxicity in the safety studies. The risk evaluation is complicated in these patients by the general expectation that they are likely to be seriously ill and taking comedications that are likely to affect responses to the test drug. As the clinical evaluation continues, the risk evaluation of the drug must be continuously revised as the human data accumulate. Thus, a risk evaluation conducted in the final stages of clinical trials is likely to be very different from that carried out for the human volunteers in the first trial.

SETTING SAFETY FACTORS AND MARGINS

Safety factors are set with the intention that the hazard being assessed will not—with reasonable confidence—be seen (at unacceptable levels in some cases) in the target population. The use of safety factors in risk management has been routine for many years to establish a margin of safety between the doses used in safety evaluation tests (normally in animals) and the levels to which humans will be exposed. Therefore, agreement on safety factors is a basic requirement before exposure limits can be set. The traditional (i.e., questionable) approach has been to use a factor of 100, being a factor of 10 lower to take into account possible differences between species and a further factor of 10 lower to take into account variation among the human population. Although this is a simple approach requiring limited numeracy to put into effect, it was based on the fact that there was little knowledge to justify any other method. In reality, therefore, the two factors of 10 are uncertainty factors that have to allow for differences in toxicokinetics and toxicodynamics within the test species and between human individuals. In either case, the traditional approach may not give a large-enough safety factor or, equally, may give one that is larger than is warranted by the actual data. With greater understanding of mechanisms of toxicity, differences in ADME, and toxicodynamics, a more refined approach is becoming possible.

Focus Box 15.4 summarizes the factors that need to be taken into account when deciding on safety factors with respect to a particular hazard and target population.

HAZARD WEIGHTING AND SAFETY FACTORS

The type of toxicity expressed—the hazard—affects the weighting of the subsequent decisions on how the risks are managed. Reproductive effects and frank carcinogenicity are seen as more undesirable than transient effects and so carry more weight in any subsequent risk assessment—and demand larger safety factors in deciding permissible exposure limits. With increasing dose and toxicity, it is normal for the number of changes to increase also, producing a range of effect from low- to high-dose level. The various dose levels associated with change in degree of effect—from NOEL to NOAEL to LOAEL—may be relevant for different populations, and as a

FOCUS BOX 15.4   CHOOSING SAFETY FACTORS

A safety factor is a desired margin of safety over a dose that has been shown to be without effect or shows change that is not adverse. When deciding this margin, factors considered should include the following, although each case is different and other factors may well be relevant.

- Database available: Increasing study length and statistical power (sample numbers or individuals) and quality increase confidence in the data; small-scale, short studies are a poor basis for extrapolation and so an extra factor should be added.
- Type of effect under consideration—knowledge of mechanism, duration, and reversibility: An additional safety factor may be added, but apply this to the toxicity of concern, not to another lesser effect seen at a lower dose level or in another species (Renwick 1995).
- Type and use of chemical: The class of compound and its mode of action—toxic or pharmacological—should influence the size of safety factor chosen. Replacement therapies such as hormones or potent pharmacological disruptors of normal physiology (e.g., bisphosphonates), which could have significant adverse effects in a healthy population, need particular care.
- Precedent with chemicals of similar structure, class, or mechanism of action: This information, taken with data on relative toxicity in terms of NOEL or NOAEL and knowledge of their effects, should allow more precise choice of desired safety margin.
- Acceptability and perception by the target population of the risk being assessed.
- The traditional method—10 for species plus 10 for individual variation and a further multiplier to take into account the factors above; e.g., reproductive effects in safety studies might add a further factor of 5.
- Additional factors may be added for different absorption between test route and exposure route and between different ADME and any other clear differences.

result, any targeted risk assessments should take this into account. The result may be to set exposure limits that differ for particular groups and exposure situations, for example, production facility workers and consumers.

The use of safety factors is illustrated by the International Conference on Harmonisation (ICH) Q3C guideline on residual solvents and the recent European Medicines Agency (EMEA) guideline for setting exposure limits for contamination of drug substances by other drugs that may have been produced in the same facility (EMA 2012) and is discussed further in Case Study 15.1.

Safety Factor Rationale

Looking at the traditional approach of two multiples of 10, the first is intended to account for potential (probable) differences in response between species. In setting exposure limits, it is normal to use safety evaluation data from the most sensitive species, which may be a nonrodent. Due to the greater variability between individuals of nonrodent species and the smaller data sets in comparison with rodents, a factor of 10 to allow for extrapolation between species may not be enough. If there is not much difference in response or sensitivity between the species that have been investigated, a lower factor than 10 may be reasonable. For the second factor of 10, to account for variability within the human population, there may be huge potential differences in response, a situation that is seen with allergens. It is possible to be exposed over a number of years to an allergen without any evidence of adverse reaction, a situation that is seen repeatedly with animal allergy. The problem for this type of hazard is that the majority of the population will be able to tolerate quite high concentrations, whereas a relatively few sensitive individuals—whom it may be difficult or impossible to exclude from the target population—could be sensitive to concentrations 100-fold lower. In this case, the only approach is to select a safety factor that should protect everyone, even though this may place costly restraints on the production or containment processes.

To remove some of the uncertainty from the use of the traditional 100-fold safety margin, Renwick (1993, 1995) argued that the two factors of 10 should be split to take account of differences in toxicokinetics and toxicodynamics in the test species and differences in pharmacokinetics and pharmacodynamics in the human population. The first (for interspecies variations) is split into 4.0 for toxicokinetic and 2.5 for toxicodynamic differences; the second 10 (for extrapolation to humans) may be split in the same way or as two factors of 3.2. This approach is intended to remove some of the uncertainty in the use of the traditional factors but is dependent on detailed knowledge of the kinetics and toxicodynamics of the material, not only in animals but also in humans, and on the correct choice of starting dose (see later).

Whatever approach is used, a safety factor must not be derived from an exposure limit chosen simply because it can be achieved. It is not acceptable to say that exposure limits cannot be reduced because, for example, the equipment in use is not capable of greater containment levels or the PPE is inadequate or wrongly used. The safety factor should be decided first, based on the best data available, and the exposure limit is then derived from the safety evaluation data, as described later.

EXPOSURE LIMITS

One of the first hurdles to overcome in this discussion is the plethora of abbreviations that are used for the various types of exposure limit. These can become confusing when all are discussed together; Table 15.3 provides definitions of the more usual limits. These are broadly divisible into those that are relevant in the workplace (WELs, OELs, and occupational exposure bands) and those that are relevant to the human population in general, although there is some overlap [ADIs and threshold of toxicological concern (TTC)]. The WEL is a UK term that has similarities to the OEL, a term used in the United States. The WEL/OEL comes in two forms—either the exposure

TABLE 15.3

Definitions of Exposure Limits

WEL/OEL	Workplace exposure limit/occupational exposure limit	Average airborne concentrations of a chemical to which workers may be exposed over a defined period (see text).
PEL	Permissible exposure limit or permissible dose	The dose that has no adverse effect on a worker. Mostly US term for limits for industrial chemicals that are enforceable by a central authority such as the Occupational Safety and Health Administration.
OEB	Occupational exposure band	An absolute upper limit of exposure based on categories to which compounds with few data are assigned based on hazard.
TLV	Threshold limit value	The upper permissible airborne concentration.
TLVC	Threshold limit value—ceiling	An airborne concentration that should not be exceeded at any time.
STEL	Short-term exposure limit	The upper airborne concentration that is acceptable for short-term exposure (e.g., not longer than 15 minutes experienced no more than 4 times in a day at intervals of not less than 1 hour) without prolonged or unacceptable adverse effect.
TWA	Time-weighted average	The average concentration to which nearly all workers may be exposed repeatedly without adverse effect, during a working day of 8 hours or a 40-hour week.
MRL	Maximum residue limit	The maximum acceptable concentration in foods for pesticides or veterinary drugs.
ADI	Acceptable daily intake	The daily intake of a chemical that is expected to be without adverse effect when ingested over a lifetime.
TDI	Tolerable daily intake	Used in similar contexts to ADI, for residues and food contaminants.
PDE	Permitted daily exposure	Used in the context of exposure to pharmaceutical contaminants in drug substances produced in shared facilities; may also be call ADE or acceptable daily exposure (EMA 2012).
TTC	Threshold of toxicological concern	A threshold for human exposure for all chemicals below which there would be no significant risk to health.

Source: Lewis' *Dictionary of Toxicology* and other texts cited in the References. Woolley A, *A Guide to Practical Toxicology*, 1st ed., London: Taylor & Francis, 2003.

level at which no adverse effect would be expected to occur, based on the known and/or predicted effects of the substance, and which is reasonably practicable or, if this is not possible, the exposure level that is achievable with good control, taking into account the nature and severity of effect and the costs/benefits of the control solutions.

Those that cover the workplace generally assume intermittent exposure, usually for no longer than a typical working day of 8 hours in a 40-hour week, and can set average exposure limits [time-weighted averages (TWAs)] or maximum concentrations that are tolerable transiently within the working day, usually for no longer than 15 minutes [threshold limit value—ceiling (TLVC), short-term exposure limit (STEL)]. The occupational exposure band, as distinct from the OEL, places compounds into four or five bands of acceptable concentrations based on their known or expected toxicity. They are used when there is little information about the chemical, particularly for human effects. A further concept increasingly used in risk assessment is the TTC, which is used in assessment of chemicals in food and, increasingly, in areas such as genotoxic impurities in pharmaceutical drug substances.

Exposure limits that are applied to the general population cover chemicals such as food additives or pesticides (ADI), although a similar limit, the tolerable daily intake (TDI), is used for residues of veterinary drugs and food contaminants (which may be pesticides but may also be other unintended contaminants such as aflatoxins). The TTC is increasingly used to establish a threshold for chemicals in food and other substances such as pharmaceuticals; these are discussed in more detail at the end of this chapter. The maximum residue limit (MRL) is an offshoot of the ADI/TDI and refers to the upper limits of residual drug or pesticide that is allowable in food that reaches the supermarket; in turn, this limit influences the interval that is allowed between application or treatment and harvest. These various limits refer to the chemical of concern or, in some cases, to metabolites or other degradation products. A further measurement is becoming more common, the biological threshold limit value (TLV), in which a biological marker (such as cholinesterase inhibition) is used to define the limit of effect beyond which exposure should cease or be reduced.

Many chemicals are already officially regulated and have set OELs/WELs, under European regulations such as Control of Substances Hazardous to Health (COSHH) or the US Occupational Safety and Health Administration (OSHA) and the American Conference of Governmental Industrial Hygienists (ACGIH). If the chemical of concern is not novel, the regulations from these or similar bodies should be checked before setting your own limits that may fall outside (higher than) official ones. Standard texts such as Casarett and Doull have listings of such exposure limits (see References).

A recent development, encapsulated in a recent European Medicines Agency (EMA) guideline (EMA 2012), has been the realization that if production equipment and, perhaps, facilities are used to produce more than one active drug substance, it is possible that there may be cross-contamination of the active substances. For instance, acetaminophen (paracetamol) may be produced in the same equipment used for production of other substances and may be present at low levels in the substances produced next in the same equipment. There is no easy way to avoid such contamination, and accordingly, it makes sense to estimate a daily dose that would not be associated with adverse effects; this has some similarity to a TWA and is known as the permitted daily exposure (PDE). The starting point for this calculation is a NOAEL from an appropriate study, to which safety factors (see above) are applied to arrive at a PDE (see Case Study 15.1). Care has to be taken in assessing contamination levels. While acetaminophen is relatively benign and used clinically at high doses, other drugs, such as bisphosphonates, have very different potencies and toxicities; low level contamination of a bisphosphonate with acetaminophen may well be acceptable, but similar contamination of acetaminophen with bisphosphonate would not be. This process behind the generation of a PDE is examined in Case Study 15.1.

CASE STUDY 15.1 CROSS-CONTAMINATION
AND PERMITTED DAILY EXPOSURE (PDE) CALCULATION

Cross-contamination in shared manufacturing facilities has always been a concern for medicinal product manufacturers and regulators. Unintentional patient exposure to low levels of unprescribed pharmaceuticals is unlikely to give benefits but may still confer risk. To address this issue, the European Medicines Agency (EMA) adopted the guideline on setting health-based exposure limits for use in risk identification in the manufacture of different medicinal products in shared facilities—which came into effect in June 2015. This document details the risk assessment procedure that should be undertaken for all medicinal products (human and animal) produced in shared facilities; however, the end user should always be considered to be human. The PDE report itself should follow a format similar to a nonclinical overview and may include (if relevant) clinical details.

Determination of a PDE follows the similar principles discussed for risk assessment, namely,

1. Hazard identification
2. Identification of "critical effects"
3. Determination of NOAEL
4. Use of several adjustment factors to account for uncertainties

The guideline requires that, on the cover sheet accompanying the report, the author identifies (checking a box for yes, no, or unknown) whether the substance is a genotoxicant, a reproductive developmental toxicant, or a carcinogen, or has highly sensitizing potential. The latter term is subjective and is further defined as a substance that "shows a high frequency of sensitizing occurrence in humans; or a probability of occurrence of a high sensitization rate in humans based on animal data or other validated tests." In the absence of data for reproductive and developmental toxicity data, the guideline allows for the use of a subchronic or chronic study accompanied with an additional (scientifically justified) adjustment factor (e.g., 10). In the total absence of data, the guideline indicates that the TTC of 1.5 µg/person per day should be used—in line with the EMA guideline on the limits of genotoxic impurities.

The key step in deriving a PDE, following identification of the hazards and critical effects, is the choice of the NO(A)EL as a point of departure. The point of departure should be scientifically robust and be taken from a relevant species. Generally, the value chosen is from the most sensitive species (if relevant) in a study conducted over the longest period. Where possible, clinical data may be used in place of nonclinical NO(A)EL, though such data should be used with caution and it must be ensured that the data are of good quality.

The calculation given in the PDE guideline is adapted from the ICH Q3C guideline on residual solvents. The standard body weight used in human medicinal products is 50 kg. The PDE is calculated using the following formula:

$$PDE = \frac{NOEL \times weight\ adjustment}{F1 \times F2 \times F3 \times F4 \times F5}$$

where
> F1: A factor (values between 2 and 12) to account for extrapolation between species. (If clinical data are used, then this factor becomes 1.)
> F2: A factor of 10 to account for variability between individuals.
> F3: A factor 10 to account for repeat-dose toxicity studies of short duration, i.e., less than 4 weeks. (ICH Q3C specifies separate factors depending on study length.)
> F4: A factor (1 to 10) that may be applied in cases of severe toxicity, e.g., nongenotoxic carcinogenicity, neurotoxicity, or teratogenicity. (ICH Q3C specifies which value should be used for developmental toxicity studies dependent on the toxicity observed and the presence or absence of maternal toxicity.)
> F5: A variable factor that may be applied if the no-effect level was not established. When only an LOEL is available, a factor of up to 10 could be used depending on the severity of the toxicity.

The choice of scaling factors is highly dependent on the judgment of the person undertaking the assessment; as such, all choices should be justified in the text and be scientifically valid.

Route-to-route extrapolation should be based on the best available bioavailability data, either human or animal, and should try to take into account the known routes of exposure of the shared products. If there are clear differences between the route of the chosen point of departure and the additional route of administration, appropriate scaling factors should be applied to take account of this.

Note: Substances that are known to degrade and/or denature (such as therapeutic macromolecules and peptides) under cleaning conditions and hence become pharmacologically inactive may not require a PDE—however, each product should be considered on a case-by-case basis.

DOSE LEVEL SELECTION FOR STARTING RISK ASSESSMENT

NOEL AND NOAEL

Dose selection for exposure limit setting is usually carried out by assessing the data from a single species and often one study. If the effect is seen in more than one species, the most sensitive species is taken for dose selection, using the longest toxicity study performed (or based on 90-day study values for EU risk assessment of repeated dose toxicity). The first step is to ask at which dose level the effect of interest becomes apparent or likely in the test species used in the study being assessed. Traditionally, there have been two values to look for, namely, the NOEL and the NOAEL. These are both doses that have been used experimentally and are, inevitably, subject to the imprecision of such an approach. Ideally, in a typical study of three treatment groups in which progressive liver toxicity is seen with increasing dose, there should be no effect at the low dose (the NOEL), clear toxicity at the highest dose, and minimal change at the mid-dose levels. If the effects at the mid dose are clearly minimal (essentially functional) and reversible, it may be designated as a NOAEL. In this instance, use of the NOAEL is sensible because it is a dose that you know to be associated with some effect. In cases where effects at the mid dose are too severe to allow its use, it may be necessary to use the NOEL in the calculation.

There are problems with both these approaches. If the NOEL is used, it may be much lower than the next dose up. Although scientifically supportable, dose choice in toxicity studies is often also apparently consistent with numerical convenience or the borders designated for classification criteria, figures such as 3, 10, and 30 mg/kg/day or a similar variation being common. For relatively nontoxic chemicals, however, the intervals between doses can be very much larger. The result is that if there is an effect at the mid dose that renders it unsuitable for this use, the low dose may be so much lower as to give little confidence in estimation of the dose at which minimal effects would become apparent. Thus, the NOEL may give a lower dose for calculation of exposure limits than is actually justified by the true toxicity of the chemical. Exposure limits on this basis may be more conservative than is desirable, leading to containment or cleanup measures that impose higher-than-necessary costs on those responsible for managing the risk.

The use of the NOAEL is also flawed in other ways. The NOAEL is dependent on the ability (statistical power) of the chosen study to detect differences between a control and a test group; the use of larger group sizes is likely to result in lower NOAELs. It does not take sample size (numbers per group) into account or variability around the mean that may be due to inhomogeneity in the sample population. Larger group size will tend to be associated with lower NOAELs. Also, because animals tend to tolerate higher doses over short studies than over longer treatment periods, the NOAELs in subchronic studies are likely to be higher than those in chronic studies. This is particularly true for toxicities that accumulate with continued treatment; nephrotoxicity may not be seen in 4-week studies, may be only minimal after 13 weeks, but may be life threatening after 26 weeks—even if dose levels for the successive studies are reduced according to the data from the previous study. In addition, the NOAEL takes no account of the shape of the dose–response curve and does not take into account the degree of toxicity at higher doses or the variability in the data. Similar arguments may be leveled at the NOEL. In these cases, the benchmark-dose (BMD) approach may be better, as this calculates a starting dose (effectively a derived NOAEL) from all the available data. This useful concept is gaining ground in the United States and is being introduced into Europe.

The Benchmark Dose

In some studies, where there is a progressive dose–response curve for an adverse effect, there may be no NOEL. In this case, it may be necessary to use the LOAEL or the BMD. The LOAEL is the lowest dose at which adverse effects were seen and does not refer to a dose at which a defined percent response is seen—either in terms of numbers of animals affected or in extent of response relative to controls.

The BMD, which has been extensively reviewed by Filipsson et al. (2003), is a calculated value that seeks to define a dose at which a low response is seen, for example, 10%. This may be estimated by extrapolation or calculation from the upper confidence limits of the dose–response curve. It has been seen as a point of departure for the onset of toxicity and replaces the NOEL or LOEL/LOAEL. The drawback of the BMD concept is the need to define a response rate and the statistical weakness of shorter toxicity studies for this purpose. As a result, this is probably best suited to relatively large data sets such as those that result from chronic toxicity, reproductive toxicity, or carcinogenicity studies in rodents. If using smaller data sets, as found in shorter rodent studies and those using nonrodents, the uncertainties of this become more significant. The BMD approach used for threshold versus nonthreshold effects is also slightly different, the method in the latter being to draw a line from the BMD to the origin of the curve; this is used for genotoxic carcinogens. With the BMD dose, it is also necessary to assume an acceptable risk level. It is important to bear in mind that the BMD still requires expert judgment in choosing the endpoint, pivotal studies, and levels of response.

The principle of the BMD is illustrated in Figure 15.3. The dose of interest is the lower confidence limit on the BMD, the BMDL, or benchmark dose low; this is a more conservative approach than using the BMD value itself and serves to account for any uncertainty and variability in the data. Figure 15.4 illustrates a dose–response

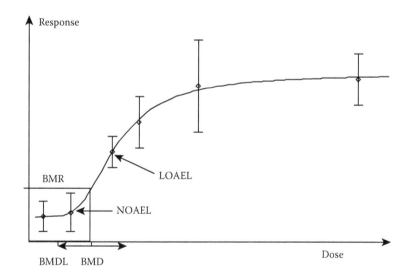

FIGURE 15.3 Comparison of NOAEL/LOAEL and BMD approaches. (From Filipsson AF et al., *Crit Rev Toxicol* 33(5): 505–542, 2003.)

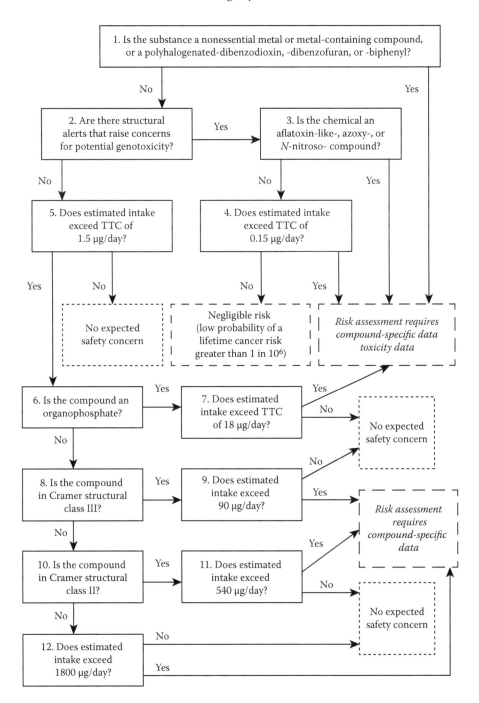

FIGURE 15.4 Decision tree for low-molecular-weight compounds for which limited toxicity data are available, which incorporates different TTCs related to different structural characteristics. (Modified from Kroes R et al., *Food and Chemical Toxicology* 42: 65–83, 2004.)

relationship onto which the LOAEL and NOAEL derived from a study have been plotted.

The lowest figure plotted is the ADI, and the gap between the NOAEL and the ADI is a function of the safety factor used. The BMD response (BMR, a term used by Filipsson et al. in an attempt to end terminological confusion in this respect) is a predetermined increase in response (e.g., 10%). The point on the dose–response curve indicated by the BMR corresponds to the BMD, and the lower confidence limit of the BMD (indicated by the double-headed arrow) is the BMDL.

SETTING EXPOSURE LIMITS

There are two basic steps in setting an exposure limit: first take the chosen safety factor relevant to the lead effect that you are guarding against and then apply it to the dose level that you have selected as the basis for your calculations (Focus Box 15.5). Both these steps imply choice, and both therefore require a degree of expert judgment. Choice of safety factor has been reviewed earlier, and we have assumed that this far into the process, the effect of concern has also been chosen. However, as indicated by Renwick (1995), there is little scientific sense in choosing a safety factor for microscopically evident liver toxicity if it is then applied to a lesser effect such as functional, transient increase in urine volume without pathological correlate.

Occupational exposure bands may be set for chemicals for which there is no official set limit or in-house OEL (Chemical Industries Association 1997). These set bands of acceptable airborne concentration; for dusts, these are <0.1 mg/m^3, 0.1 to 1 mg/m^3, and 1 to 10 mg/m^3. Gases and vapors are assigned to four bands. There is also a lower, unspecified band for very active substances that cannot be assigned to these bands and for which special arrangements have to be made.

USE OF THE THERAPEUTIC DOSE FOR SETTING WELs/OELs FOR PHARMACEUTICALS

For pharmaceuticals, the minimum therapeutic dose may be used in conjunction with data from clinical trials or adverse reaction reports. However, it is necessary to remember the type of indication the drug is used for, given that toxic treatments are more acceptable for cancer than for non-life-threatening diseases. For non-life-threatening diseases, the minimum therapeutic dose may be expected to be relatively nontoxic; for cancer treatments, this cannot be assumed. Examples of these various methods are given in Tables 15.4 and 15.5. Note that because the doses of thalidomide and ethinyl estradiol are given as the human dose, there is no need to multiply by body weight.

For chemicals that have a long history of use or production, it is essential to find out what the existing knowledge base is to indicate currently experienced dose levels. This may not be easy when production is being set up in competition with a rival company. If there is a reliable estimate of current exposure levels—for industrial chemicals, food contaminants, or endogenous dietary compounds—a MOE or safety can be calculated as a ratio by dividing the NOEL, NOAEL, or LOEL by the highest dose experienced in the target population. This may indicate that the MOE gives a sufficient safety margin and that no further action is needed; alternatively,

FOCUS BOX 15.5   SETTING EXPOSURE LIMITS

The first step is the choice of dose level with which to start the process. Taking the toxicity or effect of concern, as seen in the longest toxicity study available in the most sensitive species, assess the data to decide which type of dose level is most appropriate, usually selected from the following:

- NOEL, NOAEL, or LOAEL—These do not take account of the shape of the dose–response curve or the risks and effects present at higher dose levels.
- The BMD is the dose at which there is a defined low response (e.g., 10%), observed or estimated from the upper confidence limits of the dose–response curve. It is probably best suited to larger data sets and requires some estimate of what constitutes an acceptable risk to provide a safety factor.
- For known pharmaceuticals, the minimum therapeutic dose may be taken and a safety factor applied to produce an exposure level that can be expected to be without adverse effect or undesired pharmacological consequences.

Then, taking the chosen dose, apply the safety factors selected in Focus Box 15.4 and put into the basic formulae as follows:

Occupational exposure limit =

$$\frac{NOAEL\,(mg/kg/day) \times 70\,kg\,(human\ body\ weight)}{a \times b \times c \times breathing\ rate}$$

Accpetable daily intake =

$$\frac{NOAEL\,(mg/kg/day) \times 70\,kg\,(human\ body\ weight)}{a \times b \times c}$$

where NOAEL represents the dose chosen from the safety data, 70 kg is a standard human (50 kg gives a more conservative figure and may be more appropriate for females) and a, b, and c are safety factors (a for extrapolation from the test species to humans, b for human variability, c for other considerations such as nature of toxicity, etc.). The breathing rate is usually assumed to be 10 m^3/8-hour working day for a medium work rate. The OEL here is for airborne particulates, and the ADI would be suitable for foods or drinking water. Occupational exposure bands are set for chemicals without official or in-house OELs and for which there are not much data (see text). Some basic rules of thumb may be added to this:

- Although it may be politically correct to choose a dose level at which there is absolutely no effect, this may not be relevant to the toxicity of concern if the NOEL/NOAEL for that is higher.
- All activities of the compound should be considered, not simply the toxicities. An undesired effect other than toxicity may indicate a lower ADI.
- Overconservative selection of exposure limits can lead to unnecessary expenditure in cleanup or containment. Setting them too high may mean unacceptable adverse effects.

it may indicate that additional containment or cleanup measures are needed to reduce exposure.

THRESHOLDS OF TOXICOLOGICAL CONCERN

The concept behind the TTC is that it should be possible, for all chemicals, to establish a threshold for human exposure below which there is no appreciable risk to human health (Barlow 2005). This assessment does not necessarily have to be based on experimental data for the compound as data from compounds of similar structure may be used. TTCs have a long history of use by the US FDA Joint FAO/WHO Expert Committee on Food Additives (JECFA) and for screening food packaging

---

TABLE 15.4
Calculating an OEL for Thalidomide Using Different Methods

Database	Therapeutic Dose	BMD
Toxicity seen as birth defects	Assume effect level is 25 mg/day oral	25 mg/day assumed as BMD
NOAEL unavailable in animals or humans	100× factor for extrapolation from effect level to NOAEL	Response at BMD assumed to be 50%; linear extrapolation to origin
50 mg/day oral: 10% to 50% affected	10× for human variability	Acceptable risk assumed 1:10,000
25 mg/day oral: % response unknown	2× to adjust for route difference (oral instead of inhalation)	2× to adjust for route difference (oral instead of inhalation)
Breathing 10 m³ in an 8-hour workday	OEL 1.25 µg/m³	OEL 0.25 µg/m³

Source: Presentation by Ku, Robert H (SafeBridge Consultants Inc.) at the American Chemistry Society annual meeting in San Diego, 2001.

TABLE 15.5
OEL Calculations for Ethinyl Estradiol, a Synthetic Estrogen

Method	Calculation
NOAEL and safety factors	Human NOAEL is 3.5 µg/day divided by (safety factor of 10 multiplied by breathing rate at 10 m³/8-hour day) = OEL 0.035 µg/m³. The factor of 10 is for human variability.
Therapeutic dose and safety factor (100×)	Lowest therapeutic dose = 20 µg/day. Divided by (100× factor multiplied by breathing rate at 10 m³/8-hour day) = OEL 0.02 µg/m³.
1% increase over endogenous production	Endogenous production of 17-beta estradiol in humans is about 70 µg/day. Ethinyl estradiol is about 2× more potent than 17-beta estradiol. A 1% increase in activity would be equivalent to a daily exposure to 0.035 µ/m³ if breathing at 10 m³/8-hour day.

Source: Presentation by Ku, Robert H (SafeBridge Consultants Inc.) at the American Chemistry Society annual meeting in San Diego, 2001.

migrants and flavoring substances (Renwick 2005), and Europe International Life Sciences Institute (ILSI) has proposed a systematic approach to assessing low levels of chemicals in food, using a decision tree (Kroes et al. 2004). The TTC is a risk assessment tool for which conservative, generally oral, thresholds are derived based on the chemical structure of the substance. The TTC was developed from extensive analysis of chronic toxicity data of multiple chemicals by regulatory bodies (such as the FDA) and research groups—initially, these were substances contained within food. Substances are divided into three classes, termed Cramer classes in ascending order of toxicity, class I being the lowest toxicity, while with Cramer class III compounds, there is no initial presumption of safety. The values for Cramer classes I, II, and III are as follows: 1800 µg/day, 450 µg/day, and 90 µg/day, respectively. It is considered that in the majority of substances, exposure at or below these thresholds should not constitute undue risk to the individual, provided that the substance has not been excluded on the basis of high toxicity (Kroes et al. 2004). The TTC concept has also been expanded to other routes of exposure and has been developed by the Product Quality Research Institute (PQRI) into guidelines for the presence of extractables and leachables in products for oral inhalation, ophthalmic products, and parenteral products. Rather than TTC, they refer to a safety concern threshold, and pitch this at 0.15 µg/day, which is more conservative than the TTC.

The approach used is to examine potential genotoxic carcinogens first, using the presence of known structural alerts as a guide. Potent compounds such as those similar to aflatoxin and nitroso compounds are not considered for TTC, on the basis that setting a practical limit would not be possible. Compounds with other structural alerts for genotoxicity may be assigned a TTC of 0.15 µg/person per day. The TTCs for compounds that do not have structural alerts are assessed using data from the NOAELs from chronic toxicity studies of compounds of similar structures, together with an uncertainty factor of 100. Other properties of the chemical may be taken into account. If a compound's intake is below the threshold and it is predicted to be metabolized to innocuous metabolites, there should be no safety concern; however, if intake were above the threshold, more data on the compound or structural analogues would be desirable (Renwick 2004).

The TTC approach is essentially pragmatic and has a lot of merit based on the potential to use data from similar compounds, while excluding compounds that are known or expected to be potent carcinogens. Although it may be pragmatic, setting TTCs should not be seen as a shortcut for risk assessment. Using this approach to risk assessment still requires careful judgment of data sets and realistic assessment of exposure. Low-level exposure to toxic chemicals cannot be dismissed and accepted without due care.

Despite the reliance on judgment, the process still depends on the generation and use of numbers. Numbers are widely used in toxicology to describe threshold values such as ADIs, which translate into maximum permissible concentrations in food and similar exposure limits. As Wennig (2000) has pointed out, the rationale for such numbers has to be completely understood, and they should be applied only by people with sufficient toxicological knowledge and expertise. They should be used with care as misuse can have serious consequences. The increasing sensitivity of analytical methods and technology means that lower and lower concentrations of chemicals can be detected, and this can cause concern; what is the significance of the 4000 molecules of dioxin that have been found in the local swimming pool?

As with any tool—and the TTC is just a tool—appropriate use has potential benefits and wide application. Although the TTC concept has been used mainly in the field of foods and diet up to this point, other applications are becoming apparent. For example, the EMEA has adopted this approach in assessing genotoxic impurities in pharmaceuticals and suggests that a TTC of 1.5 μg/day, corresponding to a 10^{-5} lifetime risk of cancer, can be justified where there is a pharmaceutical benefit. The 1.5 μg/day value was also the limit chosen for DNA-reactive pharmaceutical impurities in ICH M7, when exposure is expected to exceed 10 years. In ICH M7, however, the TTC value can be altered depending on the duration of exposure (Chapter 5, Case Study 5.1).

SUMMARY

The following may be seen as the basic steps in risk assessment and the subsequent setting of exposure levels or acceptable intakes:

- The expected exposure of the target population in terms of environmental concentration and dose must be predicted or measured; predictions should be confirmed by measurement once the target population comes into contact with the chemical.
- The level of risk posed by a chemical is affected by physical form and formulation, the route(s) of exposure and the bioavailability by that, or the most significant route (either in terms of gross absorption or in terms of maximal effect).
- The dose response shown by the safety evaluation data should be considered together with the effects expressed. Where there are human data available, these should be examined carefully for utility, given the sometimes

imprecise reporting of dose and timing of effect relative to the time of exposure. Reliable human data are a luxury.
- The purpose and target population of the risk assessment may well influence the final outcome of the assessment. The qualitative factors of perception and acceptability of the risk due to the hazards should also be taken into account; successful management of risk is more likely if the target population is kept informed and in agreement.
- The setting of safety factors and margins is dependent on the weight given to the different effects or hazards, carcinogenicity requiring greater weighting than minor transient effects. Safety factors should be chosen scientifically wherever possible, although the traditional approach may be necessary where appropriate pharmacokinetic and mechanistic data are absent.

Exposure limits are calculated by using the selected safety factors and the dose level from the selected safety study or studies that indicates a NOEL, NOAEL, or LOAEL. Limits should be chosen on the basis of what is necessary rather than what can be achieved. This is taken as the basis of workplace risk management, which is considered in Chapter 17.

REFERENCES

Barlow S. Threshold of toxicological concern (TTC); a tool for assessing substances of unknown toxicity present at low levels in the diet. ILSI Europe. 2005.

Chemical Industries Association. Guidance on Allocating Occupational Exposure Bands, Regulation 7. London: Chemical Industries Association, 1997. Available at http://www.cia.org.uk.

ECHA: Guidance on Information Requirements and Chemical Safety Assessment Chapter R.6: QSARs and Grouping of Chemicals, 2008.

EMA: Guideline on Setting Health Based Exposure Limits for Use in Risk Identification in the Manufacture of Different Medicinal Products in Shared Facilities. EMA/CHMP/CVMP/SWP/169430/2012.

EMEA (European Medicines Agency). Guideline on the Limits of Genotoxic Impurities. EMEA/CHMP/QWP/251344/2006.

FDA: Guidance for Industry: Estimating the Maximum Safe Starting Dose in Initial Clinical Trials for Therapeutics in Adult Healthy Volunteers, 2005.

Filipsson AF, Sand S, Nilsson J, Victorin K. The benchmark dose method—Review of available models, and recommendations for application in health risk assessment. *Crit Rev Toxicol* 2003; 33(5): 505–542.

Gad SC. Trends in toxicology modelling for risk assessment. In: Salem H and Olajos EJ, *Toxicology in Risk Assessment*. Taylor & Francis, 2000.

Kendall M, Holgate S. Health impact and toxicological effects of nanomaterials in the lung. *Respirology* 2012; 17: 743–758.

Klassen CD, ed. *Casarett and Doull's Toxicology*, 6th ed. New York: McGraw-Hill, 2001.

Klimisch H, Andreae M, Tillmann U. A systematic approach for evaluating the quality of experimental toxicological & ecotoxicological data. *Regul Toxicol Pharmacol* 1997; 25: 1–5.

Kroes R, Renwick AG, Cheeseman M, Kleiner J, Mangelsdorf I, Piersma A, Schilter B et al. Structure-based thresholds of toxicological concern (TTC): Guidance for application to substances present at low levels in the diet. *Food and Chemical Toxicology* 2004; 42: 65–83.

Ku RH. An overview of setting occupational exposure limits (OELs) for pharmaceuticals. Chem Health Saf. January/February 2000.

Mahmood I, Balian JD. The pharmacokinetic principles behind scaling from preclinical results to phase 1 protocols. *Clin Pharmacokinet* 1999; 36(1): 1–11.

Renwick AG. Data derived safety factors for the evaluation of food additives and environmental contaminants. *Food Addit Contam* 1993; 10: 275–305.

Renwick AG. The use of an additional safety factor or uncertainty factor for nature of toxicity in the estimation of ADI and TDI values. *Regul Toxicol Pharmacol* 1995; 22(3): 250–261.

Renwick AG. Toxicology databases and the concept of thresholds of toxicological concern as used by the JECFA for the safety evaluation of flavouring agents. *Toxicol Lett* 2004; 149(1–3): 223–34.

Renwick AG. Structure-based thresholds of toxicological concern-guidance for application to substances present at low levels in the diet. *Toxicol Appl Pharmacol* 2005; 207(2 Suppl): 585–91.

Ritschel WA, Vachharajani NN, Johnson RD, Hussain AS. The allometrical approach for interspecies scaling of pharmacokinetic parameters. *Comp Biochem Physiol* 1992; 103(2): 249–253.

Seaton A, Tran L, Aiken R, Donaldson K. Nanoparticles, human health hazard and regulation. *J R Soc Interface* 2010; 7: S119–S129.

Wennig R. Threshold values in toxicology—Useful or not? *Forensic Sci Int* Sep 11, 2000; 113(1–3): 323–30.

Witschi HR, Last JA. Toxic responses of the respiratory system. In: Klassen CD, ed., *Casarett and Doull's Toxicology*, 6th ed. New York: McGraw-Hill, 2001.

Woolley A. *A Guide to Practical Toxicology*, 1st ed. London: Taylor & Francis, 2003.

16 Safety Assessment of Extractables, Leachables, and Impurities

INTRODUCTION

The issue of transfer of chemicals from packaging to pharmaceuticals in solution or into solutions passed through medical devices came under regulatory scrutiny relatively recently, with the realization that it was quite possible for chemicals in, for example, an ink on a label to be transferred gradually to a solution in a polymer vial. Although the transfer of chemicals from packaging to food products is not considered here, the toxicological solutions are broadly similar in terms of assessment of hazard and risk. For example, a saline solution for inhalation via a nebulizer may be stored in a polyethylene polymer vial, with an adhesive label with printed instructions. There is scope for transfer of components of the ink, adhesive, and any free processing chemicals into the solution, and it makes sense to identify and quantify these chemicals so that any toxicological hazards and risks can be assessed. In a similar way, there is the potential for transfer of chemicals from medical devices to solutions passing through them, for example, from dialysis tubing. These substances have been classified as extractables and leachables and are somewhat different from other sources of impurities in pharmaceuticals in that they are external to the product.

The presence of extractables and leachables is assessed in a range of studies using extraction or elution techniques of varying intensity or aggression. Impurities are assessed by prediction of degradation pathways, from the synthetic pathway and by analysis of the final active substance and drug product.

Another type of impurity found in drug substances and products are active pharmaceutical substances that have been carried over in the production equipment from one synthetic run to the next. These are the subjects of predicted daily exposure levels and assessment and are discussed separately in Chapter 15, Case Study 15.1.

DEFINITIONS

Extractables are assessed in extraction studies in which the packaging or device is immersed or extracted into a solvent system that is intended to maximize elution of potential contaminants and to illustrate a possible worst-case situation. Solvents used can include nonaqueous ones such as hexane, and conditions can be high temperature for long periods.

Leachables are a subset of extractables, which are extracted from packaging or devices under conditions that are more realistic of conditions in use, typically into aqueous fluids related to the use of the device.

Examples of either category would include process residues, degradants, solvents, plasticizers, antioxidants, colorants or pigments, and residual monomers or oligomers.

Impurities in pharmaceutical substances and products are differently derived in that generally, they arise from the synthetic process (residual solvents or unreacted intermediates and nonintended reaction products) or equipment (particularly elemental impurities), or, postsynthesis, from normal processes of degradation during storage.

REGULATORY BACKGROUND

Guidance on the control and assessment of impurities have been issued by International Conference on Harmonisation (ICH) under its Quality banner and by organizations such as the British Standards Institute within the guidelines on medical devices, specifically 10993:17, and the European Medicines Agency (EMA) in a guideline on plastic packaging materials (EMA 2005).

ICH guidelines that are relevant to this are ICH Q3A to D; these cover aspects such as reporting and identification thresholds. A specific guideline (ICH M7) was developed to cover genotoxic impurities in pharmaceuticals, and this includes limits for these substances, which differ according to the stage of development. In general, higher levels of such impurities are acceptable earlier in development than after marketing and in drugs for which administration is short compared with products for chronic, lifetime use.

Another source of recommendation in this area is the Product Quality Research Institute (PQRI), an industry initiative produced in collaboration with the Food and Drug Administration (FDA), although they make the point that the opinions expressed are those of the PQRI—it is a catch all phrase the FDA uses on all non-official documents. Their documents have produced recommendations for safety thresholds for extractables and leachables in orally inhaled and nasal drug products (PQRI 2006 et seq.), and in ophthalmic and parenteral products.

The issue of thresholds—the concentrations or exposures above which an impurity should be identified and qualified (in toxicological terms)—has been differently addressed in the various legislative regions. The EMA and others have accepted a threshold of toxicological concern (TTC) for genotoxic impurities of 1.5 µg/day, based on the expectation that the drug is intended to be of benefit to the patient (for dietary exposure, this level is 0.15 µg/day). This is also the threshold used in ICH M7. For inhaled, ophthalmic and parenteral products, the PQRI has recommended a safety concern threshold (SCT) of 0.15 µg/day. In either case, these thresholds are expected to present negligible safety concerns from carcinogenic and noncarcinogenic toxic effects.

TOXICOLOGICAL ASSESSMENT

It is rare for specific empirical studies to be undertaken for an individual impurity, extractable or leachable. Instead, impurities in pharmaceutical drug substances are, ideally, qualified—that is, tested and assessed toxicologically—during the nonclinical testing of the substance, providing that they are present at concentrations at least equal to and preferably above those present in the marketed form of the drug. Frequently, however, this is not the case, and the impurity profile for a substance is likely to change as synthesis is scaled up and as storage and stability studies are conducted. As a result, it is often the case that other toxicity studies have to be relied on for their toxicological assessment.

Qualification of extractables and leachables as contaminants in early batches is not possible, as they are essentially external to the product. For these, as with many impurities in drug substances and products, other resources have to be used, such as those provided by the US National Library of Medicine [TOXNET, Hazardous Substances Data Bank (HSDB), Chemical Carcinogenesis Research Information System (CCRIS), etc.].

There are three situations of knowledge when these contaminants or impurities are considered:

- Impurities that have been identified and for which there are enough data available for toxicological hazards to be identified with confidence
- Identified impurities for which there is no toxicological information
- Unidentified impurities with no structural information, unless categorized into a chemical class, for example straight-chain alkanes or siloxanes

For impurities with an identified structure the process of assessment starts with a search for a CAS Registry Number from the Chemical Abstracts Service of the American Chemical Society. This is an essential step, as although a structure may have been identified, it is possible that more than one CAS number may be associated with it, and having a number anchors the subsequent toxicity research to a firm base. The CAS number can then be used to interrogate databases such as the HSDB and International Uniform Chemical Information Database [IUCLID, from European Chemical Agency (ECHA)] or other sources such as the International Programme on Chemical Safety (IPCS), which gives access to a range of documents, including Screening Information Data Sets, International Agency for Research on Cancer (IARC) publications, and more.

Well-known substances do not pose a problem in terms of toxicological assessment, beyond the need to filter the data set to ensure that only the most relevant and reliable studies and publications are considered.

Unfortunately, although some substances, such as antioxidants or components of label ink that may be found as extractables or leachables, may be widely used with a long history of use, this does not imply that they are well known in any toxicological sense. For these substances, there are two main options that need to be explored:

- Read-across
- *In silico* analysis

As discussed in Chapter 15, read-across is an attractive option that has some potentially serious drawbacks. However, with judicious choice of comparators, it should be possible to use toxicological data from the query substance and the comparator to bring together a tenable toxicological profile of the impurity. Thus, the data sets for the impurity and comparator may both be incomplete but between them have elements from each of the main areas of concern: toxicity, genotoxicity, reproductive and developmental toxicity, carcinogenicity, and sensitization. When these are compared and intermeshed, a picture may emerge from which an overall assessment or prediction can be made. This type of approach may be combined with *in silico* methods to fill the gaps.

In silico analysis is a very useful tool to use in these situations and, in the case of impurities in pharmaceuticals, has regulatory blessing for prediction of genotoxicity. Like all tools, though, it needs to be used appropriately and the results interpreted with caution. The training sets are critical to the predictive success of these softwares, and if the test data that are used in them are faulty, the predictions will be faulty too. Although written regulatory acceptance of the predictions from these softwares has been confined to genotoxicity, they have wider utility in situations where the output is supported by streams of evidence from other sources, such as toxicity studies, literature review, and read-across. In some situations, it may be useful to use *in silico* methods to run a full screen for toxicity assessing all the available endpoints, using both the impurity and the comparator structure(s); in this way, a picture of prediction can be built in which successful prediction of the known toxicity of a chemical can be extrapolated to the unknown substance. *In silico* methods are also useful for predicting physicochemical properties, such log *P* and Cramer class. Furthermore, for most pharmaceutical impurities, examination for DNA reactivity using *in silico* methods is now mandatory as per ICH M7 (see Chapter 5, Case Study 5.1). These methods are discussed in more detail in Chapter 5.

For unidentified substances, the assessment is pragmatically confined to comparison of predicted daily dose to the TTCs or SCTs of the region at which the substance is targeted. For extractables and leachables from medical devices, it may be possible to relate a lack of effect in toxicity studies with extracts, as for cytotoxicity or genotoxicity, and imply lack of effect for the substance, but this is tenuous and not easy to recommend as the concentrations in the eluates are likely to be well below those at which toxicity might be seen, and in any case, the scientific utility of a study of extracts from a device is itself of dubious scientific utility in the sense that it offers little more than a fig leaf of reassurance for safety assessment.

Unless the data set is very complete and has reliable studies to refer to, it is likely that a combination of these approaches will be necessary. As with any toxicity profile, the threads of evidence are drawn together to come to an overall conclusion from which risk assessment can be started.

RISK ASSESSMENT

Risk assessment for these substances has to take into account factors such as intended use of the device or drug, the toxicity of the chemical, expected daily dose, and the risk/benefit ratio. For example, low levels of an identified chemical with low pre-dicted toxicity are likely to be more acceptable in some products than in others.

Slightly different questions need to be answered for extractables and leachables. Extractables are found following aggressive extraction procedures that are done to establish a worst-case situation; in assessing the relevance of these to the final use of the device or product, the extraction conditions have to be considered (nonaqueous solvent, conditions for the extraction process, etc.) and compared with the intended-use condi-tions. The results of extractable studies should always be considered with those from leachate studies. When assessing dialysis tubing, for example, the presence of low con-centrations of a toxic material extracted only with the use of hexane in an extractables study would probably be irrelevant if it was absent in a leachables study using aqueous media that were more closely relevant to blood. If necessary and applicable, consider the physiological relevance of the methods used to perform the extraction and determine their applicability to the use of the actual product. A compound may be found if the packaging is boiled for 3 hours in isopropyl alcohol at 80°C, but the presence of this substance may not be relevant to the overall risk assessment if this does not happen in real life.

In each case, the following questions should be asked:

- What is the expected daily exposure?
- Has the substance been identified, with a CAS number and a structure found from which to conduct assessments?
- Is the extractable or leachate likely to cause undue harm to patients?
- Is the impurity genotoxic or a substance of significant concern?
- If the substance is not genotoxic, does it fall into a Cramer class associated with a higher TTC?
- Is exposure expected to be acute, subacute, subchronic, chronic, or over the lifetime of the user?
- What is the intended use of the product, and how is it used?

Medical devices such as dialysis tubing may only be used on 3 days/week, so assuming daily exposure from one session may be unrealistic as the chemical will, in fact, be given in three divided doses. Accordingly, it is good practice to calculate daily exposure by taking the predicted exposure from one session of dialysis, multi-plying that by 3, and dividing by 7 for a daily total.

This value should then be compared with predicted Cramer class and the associated TTC. Cramer classification is a long-standing method of assessing likely toxicity of chemicals and assigning TTCs to them (Cramer et al. 1978). For nongenotoxic end-points, a TTC may be derived using a decision tree; the structural class of the chemical is derived and a safe level of exposure assigned based on the makeup of the structure. The TTC is a risk assessment tool for which conservative, generally oral, thresholds are derived based on the chemical structure of the substance. The TTC is derived from extensive analysis of chronic toxicity data of substances by regulatory bodies (such as the FDA) and research groups. Substances are divided into three classes, termed Cramer classes, in ascending order of toxicity [class I being the lowest toxicity (sim-ple chemicals, with structures associated with known metabolism and end products), while with Cramer class III compounds, there is no initial presumption of safety]. The values for Cramer classes I, II, and III are as follows: 1800 µg/day, 450 µg/day, and 90 µg/day, respectively, in a 70 kg human. It is considered that in the majority of substances, exposure at or below these thresholds should not constitute undue risk to the patient (Kroes et al. 2004; Leeman et al. 2014). So for example, a Cramer class III substance may be present in dialysis tubing at concentrations that could result in a per-session dose of 175 µg, which is higher than the daily TTC. However, the total weekly exposure over three sessions would be 525 µg, and dividing this by 7 gives a prospective daily total of 75 µg/day, which should be acceptable in this particular set of patients.

IMPURITIES IN PRACTICE

The presence of impurities in active pharmaceutical ingredients (APIs) arises from several sources, which include the equipment used; prior use of the equipment for other APIs; unreacted or residual intermediates in the synthesis pathway (e.g., residual solvents); unwanted reaction products from the synthesis; and postsynthesis degradation products that accumulate during storage. For the most part, these impurities can be predicted or are apparent at an early stage in development. Occasionally, however, there are incidents during manufacture that lead to unexpected contamination with unrelated substances; the contamination of Viracept with ethylmethane sulfonate (EMS) is an example of this. The background to this and the subsequent study and risk assessment is covered in Case Study 16.1 below.

CASE STUDY 16.1   CONTAMINATION OF VIRACEPT WITH ETHYLMETHANE SULFONATE (EMS)

Background: In June 2007, Roche, the market authorization holder for Viracept (nelfinavir, a protease inhibitor for HIV treatment), notified the (then) European Medicines Agency (EMEA) of contamination of the product with EMS, a genotoxic substance. Patients had complained of strange smell of the product, and one reported nausea and vomiting. EMS is a known mutagenic DNA-ethylating agent and carcinogen and is a developmental toxicant in animals. The contamination originated in a tank used to store methanesulfonic acid (MSA), which was used in the last step of the synthesis to convert nelfinavir base to nelfinavir mesilate. The tank was cleaned with ethanol, but no drying took place, leaving residues of ethanol, which reacted with MSA to form an ester, EMS, at concentrations of up to 2300 ppm in the tank and up to 120 ppm in the product. The dose to patients of EMS was estimated to be 2.75 to 50 µg/kg/day.

Studies: Extensive studies were undertaken to establish threshold for mutagenic and clastogenic activity *in vivo*, using bone marrow micronucleus test (MNT) and MutaMouse test systems. The objective was to reassure patients that their accidental exposure to EMS at up to 0.055 mg/kg was not associated with toxicological risk.

- A no-observed-adverse-effect level (NOAEL) of 20 mg/kg was found in a 28-day study with EMS in rats.
- Dosages were between 1.25 and 260 mg/kg/day given orally.
- Groups treated with N-ethyl-N-nitrosourea (ENU) at between 1.1 and 22 mg/kg/day were included.
- There was no mutagenicity in the lacZ gene in bone marrow or gastrointestinal (GI) tract at up to 25 mg/kg/day or in the liver at up to 50 mg/kg/day. There were no micronuclei in the bone marrow at up to 80 mg/kg/day, given over 7 days.
- Genotoxicity of EMS was only evident at higher dose levels.
- Further evidence of a threshold for effect was found by comparing the effects of a single dose of 350 mg/kg, with the same dose divided over 28 days (12.5 mg/kg/day); there was no accumulation of mutation below the threshold.
- No threshold was found for ENU; dividing the dose showed that the effects of individual doses were additive (Gocke et al. 2009).
- Ethylation of hemoglobin at the N-terminal increased linearly with dose (Gocke and Wall 2009).
- Gocke and Wall (2009) concluded that cells can repair large amounts of DNA ethylation induced by EMS without increased mutation frequencies.
- A difference noted between adduct formation with DNA and protein was attributed to repair of DNA adducts that became saturated above a threshold concentration of EMS.

Risk assessment:

- There were clear no-effect levels in animals, allowing confident calculation of safety margins.
- The no-observed-effect level (NOEL) (25 mg/kg/day) divided by the calculated maximum daily dose for patients (1068 ppm EMS in 2.92 g Viracept, equivalent to 2.75 mg EMS or 0.055 mg/kg for a 50 kg person) gives a safety factor of 454-fold for oral intake.
- Absorption, distribution, metabolism, and elimination (ADME) studies in mice, rats, and monkeys and with human surrogates *in vitro* allowed estimation of safety factors for the calculated highest exposure [area under the concentration curve (AUC) and C_{max}] of patients to EMS.
- Modeling of patient EMS exposure using AUC indicated a safety factor of at least 28-fold, based conservatively on a predicted half-life of EMS in man compared with animals. However, based on the

estimated human C_{max} the safety factor was calculated as 370-fold; this was based on C_{max} being mainly dependent on volume of distribution, which does not vary much for EMS between species.

• The potential adverse effects of EMS (including cancer, birth defects, and heritable effects) were considered to be consequent to its genotoxicity, indicating that the threshold dose relationships could be applied to these endpoints (Müller et al. 2009).

• The EMEA considered that there was no increased cancer risk for patients who took the contaminated product.

This was covered comprehensively in a special supplement of *Toxicology Letters*: Assessment of human toxicological risk of Viracept patients accidentally exposed to EMS based on preclinical investigations with EMS and ethylnitrosourea, *Toxicology Letters*, vol. 190, issue 3, 2009 (Lutz 2009).

Some contaminants are universal in their occurrence and, in some cases, attract venomous comments that can change as time passes and the knowledge base evolves. An example of one of these demon chemicals is bisphenol A, one of the highest-production-volume chemicals in the world, which is covered in Case Study 16.2. The case of bisphenol A illustrates a general trend of evolution in risk assessment and the generally conservative outcome of reassessment.

CASE STUDY 16.2   BISPHENOL A

Background: Bisphenol A (BPA) is used in making plastics for food contact, including water bottles and coatings for cans and in the thermal paper used in till or cash receipts, as well as in products such as CDs and DVDs and in epoxy resin lining for water pipes. It is one of the highest-production chemicals in the world. It has caused controversy as being associated with hormone-like properties that raise potential questions about its use in products for children, particularly baby bottles, and in addition, has been shown to have effects on the liver and kidney in laboratory animals.

The public are exposed to BPA in the diet, drinking water, by inhalation, and by dermal contact in cosmetics and thermal paper. An European Food Safety Authority (EFSA) opinion (EFSA 2015) stated that infants and toddlers had the highest BPA intake in their diet at up to 0.875 µg/kg/day. Women of childbearing age were exposed to similar amounts as men of the same age group at up to 0.388 µg/kg/day. Adolescents had the highest aggregated (from all sources) exposure of 1.449 µg/kg/day. Data from biomonitoring data were in line with estimated internal exposure to total BPA.

Toxicology and Registration, Evaluation, Authorisation and restriction of Chemicals (REACH): EFSA (2015) indicates that BPA has the potential to affect the liver and kidney and, to an uncertain extent (being explored at the time of writing), may affect reproduction and related tissues and processes. The ECHA website gives the following Classification Labelling and Packaging (CLP) classification phrases—"H317: May cause an allergic skin reaction. H318: Causes serious eye damage. H335: May cause respiratory irritation. H361f: Suspected of damaging fertility." In addition, ECHA states that BPA is toxic to aquatic life with long-lasting effects and is suspected of damaging fertility or the unborn child.

Risk assessment: EFSA (2015) carried out a new risk assessment based on new data and using the benchmark dose (BMD) approach. A weight-of-evidence approach was used to evaluate BPA toxicity. It was not possible to assess a BMD for the mammary effects, and a Benchmark dose low 10% ($BMDL_{10}$) was calculated to estimate a dose at which there was a 10% change in the mean relative kidney weight in mice in a two-generation toxicity study; the result was a $BMDL_{10}$ of 8960 µg/kg/day. Toxicokinetic data were used to convert this to a human equivalent dose of 609 µg/kg/day. Then a total uncertainty factor of 150 (for interspecies and intraspecies differences and uncertainty in mammary gland, reproductive, neurobehavioral, immune, and metabolic system effects) was used to give a temporary tolerable daily intake (TDI) of 4 µg/kg/day. Comparison of this TDI with the exposure estimates led to the conclusion that there was no health concern for any age group from dietary exposure and, in addition, a low health concern from aggregated exposure. It was noted that there was uncertainty in the exposure estimates to BPA from nondietary sources.

EFSA noted that the TDI had been lowered from 50 µg/kg/day in the light of new data, refinement of the risk assessment process, and continuing uncertainties with respect to effects on the immune system and mammary gland and on reproduction, metabolism, and neurobehavioral effects.

SUMMARY

Evaluation of extractables, leachables, and impurities follows the standard risk assessment paradigm of hazard identification coupled with exposure considerations. However, unlike the main product, it unusual for these compounds to have empirical assays conducted on them by the sponsor. As such, an *in cerebro* approach using existing literature and *in silico* approach using computational models should be used.

REFERENCES

British Standards Institute: Biological evaluation of medical devices Part 17: Establishment of allowable limits for leachable substances (ISO 10993-17:2009).

Cramer GM, Ford RA, Hall RL. Estimation of toxic hazard—A decision tree approach. *Food Cosmet Toxicol* 1978; 16: 255–276.

EFSA. Scientific Opinion on the risks to public health related to the presence of bisphenol A (BPA) in foodstuffs. *EFSA J* 2015; 13(1): 3978.

EMA: Guideline on Plastic Immediate Packaging Materials CPMP/QWP/4359/03, EMEA/ CVMP/205/04 2005.

EMEA: CHMP Assessment Report for Viracept, EMEA/CHMP/492059/2007.

Gocke E, Ballantyne M, Whitwell J, Müller L. MNT and Muta™Mouse studies to define the in vivo dose response relations of the genotoxicity of EMS and ENU. *Toxicol Lett* 2009; 190(3): 286–297.

Gocke E, Wall M. In vivo genotoxicity of EMS: Statistical assessment of the dose response curves. *Toxicol Lett* 2009; 190(3): 298–302.

Kroes R, Renwick AG, Cheeseman M, Kleiner J, Mangelsdorf I, Piersma A, Schilter B et al. Structure-based thresholds of toxicological concern (TTC): Guidance for application to substances present at low levels in the diet. *Food Chem Toxicol* 2004; 42(1):65–83.

Leeman WR, Krul L, Houben GF. Reevaluation of the Munro dataset to derive more specific TTC thresholds. *Regul Toxicol Pharmacol* 2014; 69(2): 273–278.

Lutz WK. The Viracept (nelfinavir)—Ethyl methanesulfonate case: A threshold risk assessment for human exposure to a genotoxic drug contamination? *Toxicol Lett* 2009; 190(3): 239–242.

Müller L, Gocke E, Lavé T and Pfister T. Ethyl methanesulfonate toxicity in Viracept—A comprehensive human risk assessment based on threshold data for genotoxicity. *Toxicol Lett* 2009; 190(3): 317–329.

PQRI. Safety Thresholds and Best Practices for Extractables and Leachables in Orally Inhaled and Nasal Drug Products, 2006. Available at http://pqri.org/.

17 Risk Assessment and Management in the Workplace

INTRODUCTION

Exposure to chemicals occurs in all workplaces, whether they are in industry or agriculture (including chemical production plants), offices, shops, builders, or railway premises. Although the home is not classified as a workplace, exposure to chemicals occurs there as well, in the form of air fresheners, cosmetics, disinfectants, cleaners, and do-it-yourself materials.

The workplace is distinct from other arenas where chemicals may be encountered, as in theory, there should be a degree of exposure control in a limited and defined space. However, control can vary substantially from nothing special to purposefully built high-containment, positive-pressure facilities. Control includes substitution, engineering controls, and personal protective equipment as well as management systems. Fundamentally, the difference between a worker and a consumer or end user is that the worker accepts a higher degree of risk as part and parcel of his/her job but is assumed to be trained in minimizing risk.

Some very toxic chemicals are required by society in the preparation of end products that are benign [e.g., vinyl chloride monomer in the preparation of polyvinyl chloride (PVC)]. To be effective, control should also include consideration of the knowledge and skills of those exposed. This is exemplified by the differences between industrial, professional, and general public users of biocides and plant protection products. As a rule of thumb, exposure in the production facility is often better controlled, from the producer's viewpoint, than at the point of use.

The conditions of exposure, especially in production facilities, are often very different from those to which an end user is subject. Potentially, in comparison with the end user, the production worker is exposed to high concentrations of the undiluted chemical for long periods in conditions of heat and humidity. This is particularly so for pharmaceuticals where the occupational dose for potent substances may be higher than those needed to achieve the desired therapeutic effect. Furthermore, the ultimate user normally uses the chemical diluted by excipients. However, this does not apply in the case of chemicals used as intermediates in the pharmaceutical or agrochemical industries, which are handled undiluted.

The following discussion is largely aimed at workplaces in the so-called developed world. The situation in the developing world may be very different, and the workplaces there may be much less regulated than in, for example, the United States or Europe—Focus Box 17.1 looks at the history of such regulations. Workers in developing countries

FOCUS BOX 17.1   HISTORICAL WORKPLACE REGULATIONS

The causative role of occupation in progressive and debilitating diseases has long been recognized, although scientific documentation of the hazards due to particular occupations started comparatively recently. One of the earliest and best-known examples was Percival Potts, whose observation in the eighteenth century of the connection between chimney sweeping and cancer of the scrotum is seen as a milestone in this process. Occupational disease is a roll call of suffering and—in some cases—corporate irresponsibility based on ignorance and lack of understanding. The illnesses with occupational causes include pneumoconiosis in coal workers, asbestosis, silicosis, farmer's lung, solvent-induced neuropathies, cancers due to agents such as β-naphthylamine in dye-stuff manufacture and processes such as the Mond process for nickel refining, sensitization, and allergy including asthma. Skin disease (irritant and allergic contact dermatitis), although not lethal, is a particular problem. Although there may some truth in lamenting that the rising tide of sensitization and asthma is partly due to the ruthless pursuit of hygiene in childhood and consequent immune incompetence, the unpalatable fact is that today's managers have to cope with this and reduce exposures. They also have to cope with psychologically mediated illness attributed by the patient to chemical exposure at work.

One characteristic of some occupational diseases is that they develop over many years and may not become apparent until after exposure has ceased, a situation seen with mesothelioma due to asbestos, which has a particularly long period between exposure and onset of the cancer. In many cases, it has required careful epidemiological study as well as experimental work in order to demonstrate that exposure to particular chemicals or processes is responsible for a disease. The corollary of this is that it has often taken a long time for diseases of this type to be associated with their cause, especially where there is an existing normal background incidence. Often it also leads to problems of tidying up once the ill-health has been identified. Asbestos delagging for railway coaches in the 1980s, for example, required specially designed enclosed housing, air-fed positive-pressure suits, and a maximum limitation on hours of work. Since the mid-nineteenth century, there has been a vast increase in range of chemicals produced and used by industry. The increase in toxicological understanding of the long-term consequences of exposure to these chemicals has awaited epidemiological studies and therefore lagged behind the initiation of use of a chemical. The predictive use of toxicology for industrial chemicals started in the 1920s and 1930s. Greater appreciation by legislators of the problems of industrial chemicals has led to increasingly stringent regulation in developed countries.

Occupational exposure is controlled by a government department or agency in every developed country. In the United Kingdom, the Health and Safety at Work Act of 1974 introduced a new comprehensive approach that covered all aspects of workplace safety including exposure to chemicals. This is now administered by the Health and Safety Executive (HSE, http://www.hse.gov.uk/), by agencies such as US federal agency National Institute for Occupational Safety and Health (NIOSH, http://www.cdc.gov/niosh/) and the Occupational Safety and Health Administration (https://www.osha.gov/), Health Canada (http://www.hc-sc.gc.ca/), and organizations such as ACGIH (formerly the American Conference of Governmental Industrial Hygienists, http://www.acgih.org/).

may be exposed occupationally to a variety of toxic chemicals, substances, and elements through work such as dismantling electrical goods (for example, computers) and major items such as ships in conditions that would not be tolerated in the developed world.

CLASSIFICATION AND EXPOSURE LIMITS

One of the cornerstones of chemical risk management is the classification of chemicals, which until recently has been a process conducted under slightly different rules in Europe and America. This produced inconsistencies, whereby a chemical could be labeled toxic in one country but not in another. This dichotomy is now being rectified by implementation of the Globally Harmonized System of Classification and Labelling of Chemicals (GHS), an initiative of the UN starting in 1992 and now enshrined in EU laws. The GHS classifies chemicals according to types of hazard and proposes common aspects of hazard communication through the use of consistent labels and safety data sheets. One of the objectives is to ensure that information about the hazards and toxicity of chemicals is available to ensure that the safety of humans and the environment are protected as far as possible.

As only existing data are used for the classification of existing substances, the data set available for workplace assessments can be less than the ideal situation outlined in Chapter 14. Frequently, the route of administration in safety studies is not the same as that expected in the workplace, and an adjustment has to be made for differences in absorption and pharmacokinetics between the routes. Initiatives such as those by the International Council of Chemical Associations and the Screening Information Data Set (SIDS) program operated under the auspices of the Organization for Economic Co-operation and Development (OECD) have filled many data gaps for high-production-volume chemicals. Registration, Evaluation, Authorisation and restriction of Chemicals (REACH) is aimed at formally requiring this information from manufacturers and importers into the European Union.

Occupational exposure levels have been set by a number of bodies. The Control of Substances Hazardous to Health (COSHH) regulations were introduced in the United Kingdom in 1988. The 2002 reenactment of these regulations and the regulations concerned with lead and asbestos reflected the Europeanization of legislation on workplace safety and health, and was based on EU Directives. Limit setting has a much longer history, including the earlier (and continuing) attempts at limit setting by the American Conference of Government Industrial Hygienists and the German

Research Society (known by its initials as the DFG). In Europe, these functions are now covered both at the EU level and by national organizations.

FACTORS IN WORKPLACE RISK ASSESSMENT

Risk assessment in the workplace differs in a number of respects from risk assessment in the wider environment.

- The agents to which the workforce is exposed are known, and the level of exposure can be controlled by containment of processes or by use of personal protection equipment (PPE).
- The exposure limits can be set centrally by legislation or locally by management but are potentially higher than would be acceptable in other situations. Accidental spillage of undiluted chemical can pose significant hazard.
- The workforce is a selected population, whose long-term or day-to-day composition can be controlled by management; susceptible (or potentially susceptible) individuals can be reassigned or not employed in the first place. Women of childbearing potential or who are pregnant may be excluded from certain production processes, notably those involving lead.
- The use of nonsensitive people does not necessarily mean that higher exposure levels will be acceptable as susceptibility can develop with time in some cases.
- The level of risk is affected not only by the substance but by the process it is subject to. A fully contained milling operation may pose very little risk— no exposure, no risk; however, at the end of this process, if the substance is transferred into polythene bags topped up by someone using a shovel, the risk can be significantly greater.
- There is strict control legislation that can result in heavy fines or closure of the plant if appropriate measures are not put in place.

These various factors, with the managerial responsibility of protecting the workforce (and so avoiding prosecution), mean that there is little margin for error. An overconservative assessment may mean that it is not possible to work on a particular chemical due to cost of unnecessary containment or cleanup. An assessment that is too generous may have serious consequences for worker health.

WORKPLACE RISK ASSESSMENT

There is a large amount of support available on the Internet offering information and assistance with risk assessment in the workplace. A document produced by the International Programme on Chemical Safety, Assessing Human Health Risks of Chemicals: Derivation of Guidance Values For Health-Based Exposure Limits (Environmental Health Criteria 170) (IPCS) has much useful information that can be used to supplement this and the following discussion. In addition, COSHH Essentials, produced by the UK HSE, offers guidance and worked examples (http://www.hse.gov.uk/coshh/). COSHH requires that employers assess health risks due to chemicals and decide on controls. These controls must then be used; management has to ensure that workers use them and that they are working properly. As corollaries to this, the workers must be informed about risks to health, and they must be trained appropriately.

The process of risk assessment in the workplace is similar to that in other contexts. The human-relevant hazards are predicted from the safety data, and information for other compounds is considered as appropriate. These data are then assessed for dose response and mechanism, and the likelihood of human toxicity is predicted. In the workplace, the physical properties of the compound—particle size, powder density, aerosol formation, etc.—and the process involved are important. For example, milling a compound to produce micronized powder is associated with significant risk as there is greater potential for exposure by inhalation to a respirable form of the chemical, which may also be more easily absorbed by any of the other possible routes of exposure.

Although classic, numerical risk assessment may focus on the most serious hazard, in the workplace, all the hazards due to a compound should be considered, according to the context of use or exposure. If a compound is carcinogenic at high doses in animals but is also associated with significant acute toxicity at lower doses, the risk assessment could be based on both aspects of its toxicity. The acute toxicity may determine the takeoff point for setting the workplace exposure limit (WEL) (it would have to be below the level at which acute toxicity was seen); the carcinogenic response means that there would be use of rather greater uncertainty factors in determining the level of the WEL (which would only be a "virtually safe dose"), and the WEL would have to be achievable with good control. There would also be a continuing duty to reduce exposure levels "as low as is reasonably practicable" below the WEL. Each compound assessed poses a different set of problems, but broadly, the most difficult problems from a workplace limit-setting viewpoint are irreversible effects associated with carcinogenicity, reproductive toxicity, and asthma.

WORKPLACE EXPOSURE ASSESSMENT

There is a whole discipline whose aim in life is workplace exposure assessment and control—occupational hygiene. In addition, most workplaces nowadays have access to health and safety officers/advisors, who should be able to seek appropriate occupational hygiene advice when they are not capable of doing these assessments themselves.

In considering worker exposure to the chemicals being used or produced, a range of factors have to be taken into account, including the following:

- The physical form of the chemical, whether a fine particle, aerosol, gas, or liquid.
- The method of use or handling. Exposure to chemicals in production plants may be different from exposure to workers who use the finished chemical, for example, pesticides (the exposure of any bystanders is another issue here).

- The point at which exposure occurs—e.g., during transfer of finished chemical from a reaction vessel to a storage container; mixing of a pesticide for application or during use.
- The use of any PPE.
- Dermal absorption of the chemical.
- Frequency of exposure.

There are computer models available for these assessments such as that provided by the European Union System for the Evaluation of Substances (EUSES, available from the EU Joint Research Centre) or the COSHH e-tool produced by the HSE.

A further factor is the circumstances of exposure in the workplace. Conversion of exposure between routes is often a requirement when dealing with workplace exposure. Standard conversion factors are available, but it should be noted that these factors are usually worst case, and uptake information will permit modification of them.

A higher work rate is associated with increased breathing rate and larger inhaled volumes. Thus, breathing rates used in calculations have to take into account the nature of the work, and breathing rates higher than the standard figure of 10 m^3 for an 8-hour day (which is itself higher than the 22 m^3/day used for 24 hours' exposure) may be needed. Because humans are mouth breathers at high work rates, the opportunity for ingestion via the gastrointestinal tract is also increased, either directly or indirectly by clearance from the bronchi via the mucociliary escalator and swallowing. In addition, these increased work rates will probably be associated with increased temperature, moisture, and peripheral vasodilation, which work together to increase local dermal effects and absorption into the systemic circulation. A side effect of these latter factors is that personal protective equipment is likely to be discarded or not worn correctly. Working in tropical regions can have the same effect; hence, the PPE used in Western Europe for plant protection product application may not be practicable in these climates.

For pharmaceuticals developed for oral use, there is unlikely to be extensive toxicology by inhalation or dermal routes; in addition, any dermal toxicity studies may not have been conducted with the compound as it appears in the workplace. However, good human data by other routes are likely to be available. While the finished product will be well characterized in terms of all the data necessary for registration and marketing, any intermediates are often relatively unknown, particularly toward the end of the synthetic pathway, when they are unlikely to be standard off-the-shelf chemicals. In these cases, their similarity or otherwise to the final product has to be considered, and it will usually be necessary to conduct a small set of safety studies. Of course, if the processes in which the intermediate is formed and consumed are totally enclosed and exposure avoided (even for maintenance workers through the use of PPE), such studies may not be necessary.

Where a facility has been producing a chemical for years with minimal precautions or reported effects, instituting a formal risk assessment may encounter some difficulties, particularly if the safety data indicate an assortment of hazards at exposure levels lower than those actually encountered in the workplace. Although there may have been many years of human exposure, it is quite probable that this has not been quantified or monitored and that any effects cannot be separated from the background by the usual epidemiological methods. In these cases, instituting a system of personal exposure and health monitoring over a period of months may well prove useful in defining effects and the precautions to be taken, as indicated by the safety data.

RISK MANAGEMENT IN THE WORKPLACE

The goal of workplace risk assessment is to ensure that work practices do not result in ill-health. The setting of occupational exposure limits (OELs) or workplace exposure limits (WELs, see Chapter 15) is one part of the management of the risks that have been identified as significant for the particular worker population and the processes involved.

There is a hierarchy of risk management procedures for workplace health risk management, set out in COSHH.

The first preference is substitution of a less safe chemical with a safer chemical. If this is not possible, then control will be needed. Control should be undertaken preferably using engineering techniques (containment in totally enclosed systems, use of local exhaust ventilation, and use of good general ventilation). If this is not possible, then personal protective equipment (and training in how to use it) is the option. In all cases, to be successful, this must be undertaken in a supportive management philosophy and using adequate monitoring and enforcement.

The following sections look at risk management in the workplace, starting with compound hazard categories.

Compound Categories for Containment

Where a chemical falls outside the normal mechanisms of classification and labeling, which would otherwise help in deciding what risk management steps were necessary to minimize risk associated with using it, it may be necessary to set up an ad hoc system of classification; this is a useful approach to take with pharmaceuticals, pesticides, and their intermediates but does not follow the GHS classification system. This system relies on the creation of a series of categories that can be used to define the extent of containment for the chemical and any PPE. This simplifies risk management as it provides a set of basic controls that should work in the majority of cases with minimal modification. There is, however, the possibility that there will be pressure to keep chemicals in as low a category as possible because the higher categories are usually associated with significant containment and cleanup costs.

The system of categorization in Table 17.1 is based loosely on schemes from several sources; although there are differences between schemes, these are essentially only of detail, and the broad outline is the same. This system should be workable without significant modification, although minor adjustment may be necessary to take account of local circumstances and preferences. In this schema, a category 5 substance represents a low health risk, while category 1 is a severe toxicant with serious health implications on exposure.

The criteria for classification given in Table 17.1 are what we consider to be the most relevant, but others could be added, such as no-observed-adverse-effect level

TABLE 17.1
A Basic System of Compound Categorization

Category	Criteria for Classification
5	Very low acute oral toxicity: Acute median toxicity (Lethal Dose 50; LD_{50}) > 2000 mg/kg. Minimal, transient toxicity due to functional change seen at high doses without pathological findings. No effect on reproductive function. Not mutagenic. No evidence of carcinogenicity by any mechanism. Nonsensitizing, nonirritant. Poorly absorbed through skin or by inhalation. Fast elimination without accumulation into any body compartment. No evidence of human effect. No-observed-effect level (NOEL) for relevant effect > 100 mg/kg/day in animal studies (equivalent to 5000 to 7000 mg/person per day).
4	Low acute oral toxicity: LD_{50} > 300 to 2000 mg/kg. Toxicity in single organ system or species seen only at high doses on repeated dosing, without progression of effect with longer dosing; pathological findings no greater than slight or minimal; all effects fully reversible within 4 weeks without treatment. Dose–response curve shows presence of threshold of effect. No effect on reproductive function. Not mutagenic; no evidence of carcinogenicity in animals. Nonsensitizing, nonirritant. Transient pharmacological effects present at >5 mg/kg that do not affect ability to work machinery. Pharmacokinetics show no accumulation; short half-life. Poor dermal absorption. NOEL shows high margin of safety. NOEL in animal studies is 10 to 100 mg/kg/day.
3	Moderate acute oral toxicity: LD_{50} > 50 to 200 mg/kg. Reversible, slight toxicity seen at mid to high doses in more than one species or more than one organ system. Not mutagenic. Evidence of carcinogenicity by a clearly nongenotoxic mechanism without human relevance. Minor effects on reproductive function associated with toxicity and not human relevant. Low potential for irritancy or sensitization. Transient pharmacological effects present between 1 and 5 mg/kg. Some potential for dermal absorption. Pharmacokinetics show longer half-life with incomplete elimination within 24 hours but without significant accumulation. NOEL is 1 to 10 mg/kg/day in animals.
2	High acute oral toxicity: LD_{50} > 5 to 50 mg/kg. Potentially moderate to severe human-relevant toxicity, with pathological change that is only slowly reversible. Mutagenicity *in vitro* but not *in vivo*. Carcinogenicity in more than one animal species. Reproductive effects that may be human relevant, including transient fertility reductions or changes in postnatal care, fetal toxicity without malformation. Pharmacological effects that may be irreversible or debilitating present below 1 mg/kg. Good dermal absorption with or without irritancy; potential for delayed sensitization; corrosive. Pharmacokinetics indicative of slow elimination and possible accumulation. NOEL is 0.1 to 1 mg/kg/day. Dose–response curve with little margin between NOEL and toxicity.
1	Very high acute oral toxicity: LD_{50} ≤5 mg/kg. Potential for severe irreversible toxicity at low doses. Mutagenicity *in vitro* and *in vivo*. Evidence for human carcinogenicity. Embryotoxicity seen as malformations in the absence of maternal toxicity; clear effects on fertility. Pharmacological effects present at microgram doses. Severely irritant or sensitizing, with potential for anaphylaxis or other severe allergenic reaction. Pharmacokinetics show strong binding in a particular body compartment. NOEL < 0.1 mg/kg/day.

(NOAEL), half-life, structural alerts, percentage protein binding, tissue-specific accumulation, the presence or absence of specific toxicities, or thresholds for specific pharmacological effect; atmospheric concentration may also be relevant. There may also be some utility in including bands of effect on specified markers, e.g., cholinesterase inhibition, which could indicate differing potencies for groups such as organophosphates. Another practice is to link the categories to the European Union risk phrases that are used to indicate specific hazards, particularly for transport of chemicals.

Although such schemes look very good and undoubtedly have great utility, actually using them is not necessarily straightforward. The central problem is that chemicals do not fall easily into individual categories; they can be category 4 according to one criterion but category 4 for another. The trick is to judge the circumstances and risks of the individual hazards and arrive at a consensus opinion. It may be necessary to categorize one form of a chemical at one level but a more hazardous form at a higher level. In broad terms, the greater the number of classification criteria, the greater will be the complexity of applying the scheme to individual chemicals. On balance, where there is uncertainty of relevance or effect, the bias should be toward a higher category rather than a lower one.

For chemicals, COSHH Essentials is based on this process. The Material Safety Data Sheet should contain classification and labeling information. That information leads to a categorization of the chemical and hence to an indication of what are appropriate management procedures. Obviously, certain categories lead to the evaluation "seek expert advice," which, if available, may lead to alternative approaches to managing the chemical safely.

THE PROCESS OF RISK MANAGEMENT IN THE WORKPLACE

The intention is to manage the risks associated with the chemical so as to avoid adverse effects in the workforce or in the wider environment. The responsibility for this lies in the first instance with facility management; however, there is usually government oversight of the process, which typically becomes active when things go wrong. In the United Kingdom, the facility management is usually assisted by health and safety advisors and may have access to occupational hygienists; the enforcement function is undertaken by the HSE, which has published extensive guidelines on various aspects of occupational safety. In theory, the risks that have been identified are assessed and acceptable levels of exposure agreed against appropriate levels of containment and PPE, remembering that PPE is not the primary means of avoiding exposure. These risks are then managed to prevent expression of the hazards identified, while maintaining awareness of costs and risk/benefit ratios. Sadly, practice is often very different.

Assuming that hazards have been identified, the first step is to identify whether substitution is possible. If not, then it may be necessary to consider exposure control, using the OELs/WELs, either locally or from central legislation or regulation, as the guideline as to whether current exposure is acceptable or whether further control is needed. If necessary, an exposure limit may have to be derived. Having set an exposure limit, it is important to identify how it is intended to meet it. This may include knowing what the processing equipment is capable of in terms of containment and knowing what additional equipment may be required ("bolt-on technologies") to achieve adequate control. This can be derived either from experience or manufacturer's technical data or by experiment. It should be noted that using advanced equipment inappropriately to achieve a much lower limit than is actually required may have unacceptable cost implications.

Look at the processes involved and the characteristics of the compound or the formulation to be used; if the main risk is by inhalation and the material is a liquid without chance of aerosol formation, the level of risk will be lower than for a low-density powder that can be easily blown about. It is normal to work to a worst-case scenario and then to manage risks within those limits. Although appropriate measures may be put in place, it may be necessary to institute a regimen of measurement to monitor exposure of the workforce. The first step would be assessment of actual concentrations in the workplace by the use of static and personal air samplers, and assessment of the residue on surfaces, during working and after cleaning. If necessary, it may then be useful to analyze urine or blood samples from workers before and after shifts. These analyses can be for the compound itself or a metabolite or for a biological marker of effect.

When all the above factors have been adequately controlled, the diversity of the workforce should be considered. There will probably be a range of people available for the work, ranging from young to old, from the healthy to those on medication of various kinds. They are likely to have varying susceptibilities to drugs or chemicals, defined by their genetic polymorphisms or lifestyles; alcohol can interact with some chemicals encountered in the workplace, and smoking can increase the likelihood of occupational disease. The workforce is likely to contain women of childbearing potential, and the reproductive effects of any chemical should be taken into account before allowing such people to work with it (Focus Box 17.2).

Monitoring for Exposure or Effect

It should be normal to monitor atmospheric concentrations in the workplace to ensure that the equipment is operating according to expectation and that OELs for substances of concern are not being exceeded. It should be borne in mind that concentrations from one location to another—even in the same room—are likely to be different and that some processes are associated with higher exposure levels than others. This serves also as a check on the original predictions made during the various assessments.

If adverse effects are suspected or expected, or if managing risks in certain industries, it may be appropriate to institute a scheme of health surveillance of the workforce. This has been carried out in occupations involving exposure to carcinogens, as with the workers involved in nickel refining (Sunderman et al. 1986). Health surveillance includes biological monitoring; biological effects monitoring; medical surveillance; inquiries about symptoms; inspection, e.g., for chrome ulceration; and reviews of medical records and occupational records.

Biological monitoring and biological effects monitoring have a place in monitoring exposure. Evidence of exposure may be direct (plasma or urinary concentrations

FOCUS BOX 17.2 RISK MANAGEMENT IN THE WORKPLACE

The following sets out a basic listing of risk management factors to be considered after hazard identification, exposure assessment, and agreement on the extent of control required but before starting work with the substance:

- Examine whether there is a less toxic substitute for the chemical (but beware the hazards of substitution, as shown by the story of methylene chloride in Chapter 18).
- Examine if the available equipment is capable of meeting the safety requirements, or does it require modification or replacement? If performance is known, the correct equipment can be chosen to achieve the necessary level of control, including equipment to monitor personal exposure levels, before and during processing.
- Decide on appropriate containment measures such as the use of isolators or cabinets and/or local exhaust ventilation. Engineering controls and PPE tend to increase in complexity (and expense) with increasing hazard category.
- Decide on appropriate levels of PPE, bearing in mind the type of work necessary, the efficiency of containment of the equipment, and the risks due to substance and process. PPE does not replace containment of the risk and is always a second-line approach.
- Ensure that the PPE is regularly tested, where appropriate. It is often faulty, and due to complexity or lack of comfort in the working conditions, it may not be properly used. Use a what-can-be-achieved approach rather than blind faith that compliance with impractical standards can be achieved.
  - Bear in mind that incautious removal of contaminated PPE can lead to greater exposure than the process itself. Cleaning equipment after use can also be hazardous. Contaminated PPE can contaminate other things—for instance, personal clothing—and may need to be disposed of rather than cleaned or reused. There may be a need for clean/dirty systems and changing on entry/exit from the workplace.
- Handling—if the process is completely enclosed, there is not so much cause for concern. However, it is useless to enclose everything and then to handle powder manually out of a polythene sack using a hand shovel—even if the operator is wearing vast amounts of PPE.
- Consider any possible individual susceptibilities to the effects of the substance amongst the workforce.
- Consider instituting a scheme of monitoring the workforce for systemic levels of the substance in the blood or urine or for a biological marker of exposure. This will be important to demonstrate compliance with COSHH.
- Have an appropriate set of COSHH assessments, and hence standard operating procedures that are relevant to actual practice.

of parent compound or a specific metabolite) or indirect as in markers of effect (cholinesterase inhibition or DNA adducts). The further removed the marker is from the actual cause, the less reliable it becomes (see discussion of biological markers in Chapter 12, under "Occupational Toxicology"). Bear in mind, however, that health surveillance is a potential minefield, if there is any doubt about the effect being investigated. This is the case where the effect may be other than that intended for the finished dosage form (for medicines) or is otherwise unexpected. Precursor intermediates may cause different effects than the finished product.

There is little sense in monitoring for effect if the resulting data are not understood or their consequences fully appreciated. Monitoring of workers can indicate the effects of elements or compounds to which they are exposed—but may not shed light on the significance of such exposure. Chen et al. (2006) monitored 25 workers in a storage battery factory for three endpoints using the micronucleus assay, comet assay, and a T-cell receptor gene (TCR) gene mutation test; 25 controls were age-matched for gender, age, and smoking. The level of lead in the air in the workplace was 1.26 mg/m³, which is somewhat higher than the levels indicated by the American Conference on Governmental Industrial Hygienists.

This source indicates a time-weighted average for the various compounds of lead in the region of 0.5 mg/m³. The UK WEL for lead in air under The Control of Lead at Work Regulations 2002 (CLAW) is 0.15 mg/m³ (lead alkyls, 0.10 mg/m³) 8 hours time weight average (TWA). The workers' blood concentration of lead was significantly higher than in the controls (0.32 mg/L compared with 0.02 mg/L). This study showed statistically significant increases in workers in micronuclei and micronucleated cells, and in tail length and moment in the comet assay, relative to the controls. However, the TCR gene mutation test did not show any differences. The results of the micronucleus test and the comet assay indicate a degree of damage to DNA, which may be attributed to the relatively high levels of lead in these workers' blood, although other occupational factors cannot be excluded. Dart (2004) indicate that there is evidence for adverse effects of lead at less than 0.2 mg/L and that moderate lead poisoning, indicated by blood concentrations between 0.25 and 0.55 mg/L, may be associated with neurological effect. Above these levels, in mild lead toxicity, myalgia or paresthesia, mild fatigue, irritability, lethargy, and occasional abdominal discomfort may be experienced. Rearrangement of the TCR gene has been noted in various leukemias. The results of two of these tests indicate a degree of DNA damage, the significance of which is somewhat diminished by the absence of change in the TCRs. While it is possible to say that such DNA damage in the peripheral blood may be due to exposure to particular substances or mixtures, it does not necessarily indicate any quantifiable additional risk of cancer or other ill-health in the affected workers. The UK suspension-from-work level for blood lead is 30 µg/dL (0.3 mg/L; women of childbearing age), 50 µg/dL (young people), or 60 µg/dL (general population) (HSE 2015).

SUMMARY

- The workplace is a distinct area in terms of risk assessment and management, which is apparently simple (ease of containment, knowledge of chemicals involved, a defined population, and known processes) but, on closer acquaintance, has a complexity that can be daunting.
- The process of risk assessment in the workplace is similar to that in other areas but has to comply with specific legislation.
- Many of the compounds used are precursor intermediates, which have relatively unknown toxicity in comparison with the final product.
- Data sets directly related to the chemical are often limited; this is a routine situation and requires some ingenuity in obtaining relevant information for a viable risk assessment.
- Correct choice of exposure limits is critical to successful risk management in a commercial setting. Too high, and adverse effects may be seen in the workforce; too low, and the costs of containment and engineering measures may make the process too expensive to be financially viable.
- Where a facility is routinely using many different chemicals, it is useful to categorize chemicals according to the hazards they pose.
- The complex legislative background must not be forgotten and is central in setting exposure limits. In putting management controls in place, it should also be remembered that the use of PPE is seen as a last resort in reducing worker exposure, not as the primary method.

REFERENCES

Chen Z, Lou J, Chen S, Zheng W, Wu W, Jin L, Deng H, He J. Evaluating the genotoxic effects of workers exposed to lead using micronucleus assay, comet assay and TCR gene mutation test. *Toxicology* 2006; 223(3): 219–226.

Dart RC, Ed. *Medical Toxicology*, 3rd ed. Philadelphia: Lippincott: Williams & Wilkins, 2004.

HSE: COSHH. Essentials. Available at http://www.coshh-essentials.org.uk/.

HSE Exposure to Lead in Great Britain, 2015. Available at http://www.hse.gov.uk/statistics/.

IPCS (International Programme on Chemical Safety). Assessing Human Health Risks of Chemicals: Derivation of Guidance Values For Health-Based Exposure Limits (Environmental Health Criteria 170). Available at http://www.inchem.org/documents/ehc/ehc/ehc170.html.

Sunderman FW Jr, Aitio A, Morgan LG, Norseth T. Biological monitoring of nickel. *Toxicol Indus Health* 1986; 2(1): 17–78.

18 Risk Assessment

Carcinogenicity, the Environment, Evolution, and Overview of Risk Assessment

INTRODUCTION

In simple terms, assessment of risk due to toxicity is the process of extrapolation from a limited data set to a wider situation, such as the environment or the general population or a specific target group such as a workforce. The data set may contain human data, derived from accidental exposure, clinical or epidemiological studies, or safety studies in animals and/or *in vitro*. The application of the assessment may be local, national, or global. It should be remembered that relevance and utility are likely to decrease as the brief becomes wider because conditions differ from one place or population to another, either in exposure, collateral conditions, or, significantly, sociological factors. While it may be possible to arrive at an objective risk assessment for any given (local or regional) situation; application of the conclusions through risk management cannot be separated from local factors such as living standards including income, risk perception, and acceptance. The ultimate use of the risk assessment—which should be based on an objective appraisal of the data and the indicated risks—should take into account any benefits of using the chemical and all the collateral risks and factors. As with chlorination of drinking water in Peru (see Focus Box 14.3 in Chapter 14), the risks of nonuse may be greater than those due to use. Ultimately, environmental risk must be determined according to local conditions because global assessment is not always appropriate. Environmental risk assessment should take into account the risks that follow any cleanup process; what is the intended fate of the concentrated toxic chemical residues that result? It is better to optimize the production process and prevent the problem in the first place.

Above all, risk assessment has to be communicated to people at risk and to risk managers in an understandable format; it must be user friendly not user hostile, which is not easily demonstrated for some of the more complex models. Depending on the audience, analogies and comparisons can be useful; for example, something can be said to be equivalent to drinking a small beer once a year for life, or causing as much (or more) of the same toxic effect as a lower concentration of another chemical known and widely acknowledged as seriously toxic.

It is also important that any risk assessment should be honest and not distorted by undue emphasis on one aspect of the problem or by self-interest. An example of this is the "assessment" of genetically modified foods and organisms. For a genetically modified food, there are two main areas of risk—environmental release and ingestion. If the composition of the new food is similar to the existing variety without significant change in concentration, it seems likely that the risk from ingestion will be little different from the risks due to eating the normal strain. The environmental risks are, however, potentially different. The role of environmental pressure groups is very important as they tend to concentrate on one aspect of a problem to the exclusion of all else, thereby devaluing their own arguments. All too frequently, risk/benefit and collateral factors are not considered.

It is quite possible to arrive at an objective risk assessment, given a valid set of data and appropriate knowledge of the local situation. The real challenge comes when this assessment is brought into the political and sociological context of its use; science and politics are uneasy companions. The following is suggested as a list of desirables for a successful risk assessment:

- Look at all the data dispassionately.
- Take into account collateral risks and local conditions, including any background presence or incidence of effect.
- Include a risk/benefit analysis where appropriate.
- Make it usable, user friendly, and easy to communicate to those at risk.
- Keep it honest, without concentration on a single aspect.
- Combine this with local sociological factors that may increase or decrease the risk of use or how it is perceived and its consequent acceptability to those at risk.
- Produce management proposals that are achievable and are themselves not associated with significant risk.

RISK ASSESSMENT AND CARCINOGENICITY

When looking at carcinogenicity as a toxicological endpoint, it is worth remembering that a very small proportion of cancer in humans is actually attributable to a specific chemical to which we may be exposed in our diets or at work. Given that cancer is a high-incidence disease of old age with significant links to the normal human environment (in all its forms), it is very difficult to partition the risks attributable to individual aspects of that environment and then pinpoint a cause for a particular cancer in a single individual without clear evidence of exposure. Risk assessment for carcinogenicity is further complicated by the natural presence in the environment of well-known human carcinogens such as arsenic or the products of combustion. In addition, individual chemicals cannot, for the most part, be classified black or white as carcinogens or noncarcinogens. The classification of carcinogens drawn up by International Agency for Research on Cancer (IARC) (Table 12.7 in Chapter 12) indicates the gradation of certainty from the clear human carcinogens, through those that are carcinogenic in one sex of one rodent species, to those few that are not considered to be carcinogenic.

The risk of cancer is increased by a number of factors, including

- Diet: e.g., high processed meat consumption, high fat, high sugar
- Genotype/phenotype: e.g., xeroderma pigmentosum, BRCA1 and BRCA2 genes
- Lifestyle choices: e.g., drinking, illegal drugs, smoking, lack of exercise, sunbathing
- Occupation: e.g., industrial chemicals, mining, exposure to asbestos or radiation
- Reduced immunological competence
- Age
- Ignored exposure or novelty of mechanism

These are crudely divisible into those that can be avoided—lifestyle choices such as diet, exercise, and smoking—and those that are unavoidable, due to phenotype or to ignorance of their significance. Occupational exposure to carcinogenic chemicals or processes is not necessarily avoidable, due to personal circumstances and other sociological factors. Most occupational cancers have been due to lack of knowledge or understanding or simply ignoring risk. In some cases, workers learnt to take actions that effectively reduced the risks, such as washing out nasal passages each day after nickel refining to reduce the risk of nasopharyngeal cancer (Doll 1970; Kaldor 1986; Morgan 1994; JR Pincott, personal recollections of his grandfather in South Wales).

Inevitably, ignorance has been a significant factor in occupational carcinogenesis, from the scrotal cancer of unwashed chimney sweeps to the more recent exposures to chemicals such as benzene or cyclosporine. These later effects and slow attribution have been due to lack of understanding of carcinogenic mechanisms, poor prediction of effect, and the inevitable slowness and imprecision of epidemiological study, when there is a low incidence of an effect that is present as part of the normal background. There is also, apparently, a role for serendipitous observation by professionals; there is a clear tendency to be suspicious of any initiative that comes from the untrained public or that is in any way associated with old wives' tales. It follows that these last perceptions are the most difficult to deal with because of the eternal triangle of conflicting interests: industry, seen as wanting to avoid costly cleanup or compensation; government, keen to avoid expenditure on research or diversion of resources; public, interested in finding out the cause of effect, apportioning blame, and receiving compensation, while maintaining suspicion of the other two sides. In this kind of atmosphere, the necessary growth of knowledge and understanding of all aspects of the case is unlikely to be smooth or progressive.

DNA Vulnerability, Genotype, and Phenotype

DNA, and the control of its expression, is central to carcinogenesis, either through direct attack or by changed regulation. Although much is made of genotoxicity due to low levels of synthetic chemicals, the level of naturally occurring damage should be considered in any overview of risk due to low levels of synthetic chemicals; this is explored in Focus Box 18.1.

Although change to DNA may be theoretically avoidable, there is no avoidance of genotype and its phenotypic expression—at least until the advent of designer babies (although that is beginning to happen). Some aspects of this are readily characterized, such as the DNA repair deficiency that is associated with xeroderma pigmentosum. Many others, however, cannot be defined because of the diversity of influence and effect that is possible in an individual. For any individual, there is a balance between the processes of absorption, distribution, metabolism, and elimination (ADME) of the chemicals that are naturally present in the diet, and those taken as medicines or habit (alcohol or nicotine), coupled with any synergistic, additive, or inhibitory effects that any of them have on the others. The balance between these factors may result in different exposures to active metabolites or systemic levels between apparently similar people.

EPIDEMIOLOGY AND BACKGROUND INCIDENCE

The net result of the various risk factors is a background incidence of unattributable cancers, above which any new cause has to rise before it can be unequivocally identified by normal epidemiological techniques. Despite the inherent weakness of the epidemiological process, it is still human data that are the most easily accepted basis for risk assessment of human carcinogenicity. Identification is reliant on initial observation and study of incidence in the target population and in an appropriate control group, coupled with evidence of exposure. There may be evidence of a dose–response curve, where dose is indicated by degree or duration of exposure (years worked), bearing in mind that some workers are more heavily exposed than others due to differences in job, for instance, between production line and packers. The great value of human data is just that—it is human. However, the likelihood of getting all the foregoing factors in place, so as to facilitate a risk assessment based on human data alone, is small and decreases as the potency of the carcinogenic effect decreases. The type of effect being modeled is critical in terms of incidence in the target population compared with naturally occurring background. Contrast the relative certainty of vinyl chloride attribution (an unusual cancer in a defined population) versus a chemical causing a range of cancers in the general population with undefined exposures.

The obvious problem here is that humans have to be exposed to the chemical before the assessment can be made. Problems arise when there is a significant background incidence of the cancer, if differential diagnosis is poor, and if there are unaccounted confounding factors, such as smoking, intercurrent disease, or prior exposure to other agents. For marginal carcinogens, the quality of the data is significant, and the vast amount of data required to achieve statistical significance becomes limiting. Many small studies, conducted to different protocols with different assessment criteria, do not form a secure database from which to make an assessment of any precision, as it is usually not possible to combine all the data together to make a statistically sound basis for bulk analysis.

FOCUS BOX 18.1   DNA VULNERABILITY—
ENDOGENOUS DAMAGE AND REPAIR

DNA is a deceptively simple molecule, composed of only four nucleotides arranged in a regular primary structure, but having great complexity in controls on secondary and tertiary structure, replication, repair, transcription, and hence gene expression. It is highly vulnerable to oxidative or other attack or to changes in repair efficiency or gene expression.

- Alkylating agents, such as dimethylnitrosamine and cyclophosphamide, introduce methyl or ethyl groups into bases, leading to base pair changes or dysfunctional DNA.
- Oxidative attack on DNA bases can lead to base pair changes; ultraviolet (UV) radiation produces thymidine dimers.
- There are numerous cellular sources of oxygen radicals and hydrogen peroxide, including mitochondria, peroxisomes, and some enzymes.
- Fe^{2+} associated with DNA reacts with hydrogen peroxide as follows:

$$Fe^{2+} + H^+ + H_2O_2 \rightarrow Fe^{3+} + H_2O + HO$$

- The hydroxyl radicals damage DNA, and Fe^{2+} can then be regenerated through NADH, making a self-perpetuating cycle of damage.
- Asbestos carcinogenicity has been attributed to generation of hydroxyl radicals in the presence of hydrogen peroxide and Fe^{2+}.
- Daily oxidative damage to DNA has been estimated at 100,000 oxidative hits per cell per day in rats and 10,000 hits per cell per day in humans, assessed by analysis of urine samples for oxidized bases.
- Increased levels of 8-oxo-guanine have been noted in the lymphocyte DNA of smokers.
- Oxidative damage to DNA accumulates with age, associated with a decline in DNA repair.
- Planar molecules (such as estrogen metabolites) interact with DNA by intercalation into the structure, producing disruption to processes such as repair and transcription.
- Infidelity of DNA synthesis and repair leads to abnormalities of gene control or expression.
- Prevention of DNA damage is enhanced by antioxidants such as glutathione, ascorbic acid, and tocopherols, and enzymes such as superoxide dismutase.
- DNA repair is provided by a range of enzyme systems. Defective DNA repair is seen in the skin cancer, xeroderma pigmentosum, where repair of UV damage is deficient.
- Mutation of DNA may be passed on to daughter cells, producing heritable defects in cellular control. For example, the p53 protein arrests cell growth and protects against neoplastic responses; mutation in this gene is relevant to 50% of human tumors.

Data Used in Carcinogenicity Risk Assessments for Novel Chemicals

The use of long-term human exposure data is not an option for novel chemicals, although it may be possible to draw analogies with closely related chemicals already present in the marketplace or environment. Here, the backbone of carcinogenicity risk assessment is still (currently) the long-term bioassay in rodents, supported by other data derived from general toxicity, genotoxicity, and ADME studies. As indicated previously, the credibility of this fragile prop is being steadily eroded; in time, this system of assessment is gradually being replaced by a more human-relevant set of tests.

However, for the moment, data from rodent bioassays play a significant role in carcinogenicity risk assessment, particularly for the mathematical models where the statistical power of the large group sizes and any dose–response curve can be taken into account. While these models and statistical power may give a fig leaf of numeric security, this is reduced by the need for judgment to assess the influence of mechanism and other factors such as differences in ADME or pharmacokinetics between humans and rodents. This is quite apart from the fact that the doses in the two species will be radically different, and the high doses used in rodent bioassays may unduly influence the carcinogenic response. For proven human carcinogens, such as aflatoxins or diethylstilbestrol, there is good agreement between the affected tissues in animals and those that show cancer in humans. The problem is that this is true for proven human carcinogens, but this cannot be assessed a priori (from a point of epidemiological ignorance) for the vast majority of novel chemicals that are subjected to routine carcinogenicity bioassays followed by risk assessment.

The use of bioassays in two species has been debated for some years, suggestions being made that the use of the mouse could be abandoned or that testing could be reduced by using one sex each from the rat and the mouse. Neither approach has achieved regulatory acceptance. In fact, among toxicological pathologists, the use of two species has been seen as an advantage, as the results in one could be used to offset the results from the other. Thus, the presence of increased tumor incidence in the livers of male mice could be discounted from human relevance by citing the absence of similar findings in female mice and both sexes of rats. Equally, if a chemical is carcinogenic in two species in a similar manner, it is very likely to have carcinogenic potential in humans at similar dose levels. This clarity decreases as the potency of effect decreases and as the mechanism moves from direct genotoxicity to indirect effect on the control of DNA expression, and hence, apoptosis and cellular controls. The presence of thresholds in the dose–response curves of many nongenotoxic rodent carcinogens is accepted as evidence that any tumorigenic effect expressed only at high dose levels is unlikely to be relevant to expected human exposure levels. This can be backed up by data from other safety evaluation studies and investigation of mechanism.

Thresholds in Carcinogenicity and Genotoxicity

In nongenotoxic mechanisms, where cellular control is deranged, a threshold indicates a point beyond which the cells can no longer cope with the mechanistic strains imposed upon them, as in the accumulation of protein in α2u-globulin nephropathy in male rats. Such mechanisms are usually tissue specific and often seen in one sex only at high doses. Tumors thus produced in rodents may be dismissed as irrelevant to humans due to lack of an equivalent mechanism, as with tumors in the lungs of mice exposed to methylene chloride (Focus Box 18.2) and hepatic tumors due to peroxisome proliferation.

In rodents, genotoxic carcinogenicity is roughly proportional to general toxicity, is usually associated with a clear dose–response curve, and can produce tumors in several tissues in males and females. As, theoretically, a genotoxic event in a single cell can lead to cancer, it has been generally considered that there is no dose threshold for this type of effect. However, biology is very rarely black and white, and such assumptions are increasingly challenged, as indicated in Focus Box 18.3. Although damage to DNA may become fixed, no cancer will arise if the affected cell does not divide and if that process is not continued by further proliferation: mitosis is as important as mutation. To this may be added influences such as apoptosis and immune surveillance. Some carcinogens induce a reduction in cell division at low doses, showing J-shaped dose–response curves. Studies with 2-acetylaminofluorence (2-AAF), in which treatment with 2-AAF was followed by treatment with the tumor promoter phenobarbital, showed evidence of threshold effects (Lutz 1998; Lutz and Kopp-Schneider 1999).

The role of genotoxicity in carcinogenicity assessment is changing rapidly. Firstly, there is evidence of thresholds in carcinogenicity, as with 2-AAF. While it is entirely possible that a single genotoxic event could trigger a cancer, in view of the host of other factors—fixation of the mutation and DNA repair, division of the cell affected and subsequent progression—it is highly unlikely. Intuitively, therefore, single-event genotoxicity has always been an extrapolation based on risk-averse fear rather than rationality. The interpretation of genotoxicity tests has not helped in this. Until recently, a positive result in any test was considered to be evidence that the test chemical was genotoxic. This blanket embargo was replaced by an overview of the whole data set and thought about the mechanisms involved and the conclusion that a chemical might be positive in some assays but, overall, was probably not a human-relevant genotoxicant.

The gradual reassessment of the assumption of linear response to genotoxicity—that there was no threshold and, hence, no safe dose—is exemplified by various authors (Lutz 1998; Lutz and Kopp-Schneider 1999; Henderson et al. 2000). Henderson et al. pointed out that some chemicals may be genotoxic at high doses by mechanisms that do not occur at low doses or concentrations. Thus, they may be active in the constrained systems of *in vitro* tests, at high concentrations, but inactive at low concentrations, which may well be more relevant to human exposure. While the authors admit some uncertainty in this general hypothesis, they indicated the start of regulatory acceptance of the concept of genotoxicity thresholds.

This was explored in more detail by Jenkins et al. (2005), who examined the presence of thresholds for genotoxic alkylating agents. Alkylation is one of the archetypal mechanisms of DNA damage and genotoxicity, and the presence of an alkylating structure in a molecule has been considered to be an indication of geno-toxicity and resulting regulatory control, if not proscription. However, it has become

FOCUS BOX 18.2   METHYLENE CHLORIDE AND CANCER

Methylene chloride has been in use since the 1940s in various industrial applications and as a domestic paint stripper (see Case Study 18.1); as a result, there is a large amount of human and safety data.

- A US National Toxicology Program study completed in 1986 showed an increase of benign and malignant tumors in mice and benign tumors in rats, following inhalation exposure; studies using drinking water exposure or intraperitoneal administration were negative (Riley and Fishbeck 2000).
- Various epidemiological studies have suggested increased incidence of pancreatic, biliary, and liver cancer, while others have refuted these findings (Riley and Fishbeck 2000).
- Increased mortality from prostate and cervical cancer was reported among cellulose fiber production workers with more than 20 years' exposure. The same study did not confirm earlier findings of increased biliary tract and liver cancer (Gibbs et al. 1996).
- A meta-analysis of data published between 1969 and 1998 indicated weak increases in risk for methylene chloride workers with respect to pancreatic cancer but judged that a strong causal link could not be drawn (Ojajarvi et al. 2001).
- Two cohorts of photographic film workers were studied, having received exposures averaging 39 ppm (8-hour TWA) for 17 years (1311 men) or 26 ppm for 24 years (1013 men). There was no increase of death from any cause including cancer and no evidence for effects on target organs identified in animal studies. Combining these results with other studies showed that exposure to methylene chloride does not increase the risk of death from any cause (Hearne and Lifer 1999).
- A review of 10 years of work on the mechanism of methylene chloride carcinogenicity in mouse liver and lungs indicated that this is specific to the mouse. In the lung, this is probably due to DNA damage in the Clara cells, through interaction with a high-activity glutathione S-transferase unique to the mouse, which is present in the nucleus. DNA damage was not detected in other species, including human hepatocytes. Therefore, the mouse is not a good model for humans for methylene chloride (Green 1997).
- A critical review of the epidemiology literature concluded that cancer risks associated with methylene chloride exposure are small and limited to rare cancers (Dell et al. 1999).

FOCUS BOX 18.3:   THRESHOLDS IN CARCINOGENICITY AND GENOTOXICITY

There is widespread agreement that nongenotoxic carcinogenicity is associated with thresholds of exposure, below which there is no increase in tumor incidence. The long-standing belief that genotoxic carcinogens are not associated with such thresholds (i.e., that the dose response is linear at low dose levels) and that even low levels are associated with cancer risk is being reassessed.

Using a promotion protocol with 2-AAF (12 weeks treatment with AAF followed by 24 weeks phenobarbital to promote liver tumors), no tumors were seen in the low-dose group and one at the mid dose; at the high dose, all animals had hepatocellular neoplasia. Nonlinearity was also seen for cell proliferation and hepatocellular altered foci (Williams et al. 1998).

A more recent example is provided by the incident in which the contamination in 2007 of Viracept (nelfinavir, a protease inhibitor for HIV treatment) by ethyl methanesulfonate (EMS), a known mutagenic DNA-ethylating agent, resulted in an EMS dose of 2.75 mg or approximately 50 µg/kg/day. Extensive studies indicated a NOAEL of 20 mg/kg in a 4-week study with EMS in rats and a threshold-like response for chromosome damage in a bone marrow micronucleus test and for gene mutation in a lacZ transgenic MutaMouse test in several tissues of mice treated for 4 weeks (Gocke and Wall 2009). The outcome was threshold risk assessment based on an estimated $C_{\max}$ for EMS, which provided a safety factor of 370-fold for the patient exposures. This is discussed in more detail as a case study in Chapter 17.

Mechanisms for thresholds in genotoxic carcinogenicity could be as follows:

- Inhibition of DNA repair; effects on cell cycle; interference with apoptosis; meiotic and mitotic recombination; direct interaction with the spindle apparatus; DNA methylation. Low-level DNA damage may delay the cell cycle leading to lower cell turnover.
- Response in some tissues for the same chemical may be linear but nonlinear in others. Indirect mechanisms of genotoxicity may result in thresholds.
- A carcinogen may show a J-shaped curve if it increases cell division or oxidative stress at high dose but inhibits them at low doses; this can result in a decrease in tumor incidence at the low dose (Lutz 1998).
- Modeling on the cell cycle shows the possibility of thresholds for genotoxic carcinogenesis (Lutz and Kopp-Schneider 1999).
- Linearity may be hidden within the background variability.
- The relative potency of genotoxicants should be considered, as defined by comparison of NOELs or concentrations in *in vitro* tests.
- Practical or pragmatic thresholds probably exist at background levels below which effects cannot be estimated practically. Saccharin

epidemiology is said, by epidemiologists, to be compatible with a small but *undetectable* risk of bladder cancer (despite work showing that rodent metabolism is not relevant to humans) (Tomenson 2000).

Thresholds may not exist for potent genotoxic carcinogens, but they do (probably) exist for some and may exist for individual tumor responses.

apparent that while an alkylating agent may induce different endpoints, there may be differences in dose response for each of these endpoints.

DNA repair was identified as a primary contributor to thresholds for genotoxicity, along with deactivation of active metabolites by processes such as conjugation and the impact of processes such as exclusion of the active moieties from the cell or nucleus, the location and functional impact of any mutation, and the impact of apoptosis induced by chromosome damage. Furthermore, saturation of DNA repair mechanisms at high concentrations may result in expression of genotoxicity. Another factor to consider is the potential that cells used in *in vitro* tests may be deficient in DNA repair compared to wild-type cells and so show different dose responses.

In other words, although there may be considerable chemical reasons to suggest the potential for DNA alkylation in insolation, the presence of a range of other factors means that the expression of such damage is very much reduced. Finally, the authors point out that thresholds for genotoxicity are described for acute or single exposures to individual agents, often *in vitro*, and imply that dose response data for chronic exposure to complex mixtures have not been assessed.

The corollary of this line of argument is that, if there are thresholds for genotoxicity, then it may become relevant to consider the presence of no-observed-effect levels (NOELs) and/or no-observed-adverse-effect levels (NOAELs) in genotoxicity tests. This line of reasoning and its impact on risk assessment is still in an embryonic stage at the time of writing. However, it does raise the more general question of potency in interpreting genotoxicity data, as, clearly, some chemicals are more potent genotoxicants than others, for example, nitrosamines compared with saccharin.

Where a carcinogen adds progressively to a mechanism that is associated with a background incidence of tumors, a true threshold will probably not exist. When there is no association between background mechanism and tumor incidence, it is likely that there will be a threshold. However, it is likely that there is a practical threshold below which the increase in incidence is indistinguishable from background. In effect, this is a no-detectable-effect level (NDEL) below which cancers caused by the chemical will remain unattributed to that chemical. Arsenic is widely present in the environment at low concentrations, and everyone is exposed to it at low dose levels; however, it is clearly associated with carcinogenicity at occupational exposure levels. Although a chemical may be associated with an NDEL, does this mean that its use at low concentrations is acceptable, and should we add to the carcinogenic burden that is already present in the environment, even if that is at low levels? Paradoxically, the

NDEL would have to be defined by epidemiological study—a science that is inherently not sensitive enough to make such distinctions.

Low-Dose Extrapolation

The tenuous basis of carcinogenicity risk assessment, using rodent bioassay data, is the extrapolation of effect from the high dose levels used in the short-lived animal to the much lower doses expected in long-life-expectancy humans. There are two central problems to this. First, the effects seen at the highest dose level may be a result of pharmacokinetics and metabolism or mechanistic overload that are not present at lower dose levels. The significance of this is that the effects at the high dose cannot necessarily be extrapolated to lower doses and that the dose–response curve is not linear in the section defined by the data. The second challenge, given lack of linearity in the upper levels, is that linearity at dose levels lower than those tested cannot be assumed.

As indicated in Focus Box 18.3, there may be nonlinearity in responses to genotoxic carcinogens at low doses, and as a result, the shape of the response curve at low doses cannot be predicted without extensive experiment. The result of this unpredictability is a strong trend to conservatism in risk assessment, as models tend to use the upper confidence limits of the dose–response curve, and as additional cover, a safety factor is added to that. Various models have been used, such as the Mantel–Bryan and Weibull models, and derivatives of these have been developed into mathematical monsters that try to take everything into account, including time to tumor and spontaneous tumors. The reliability of the final result is inversely proportional to the number of assumptions that are made in producing it. Such mathematical complexity renders these models unsuitable for day-to-day use, returning risk assessment to a point where there is no false security offered by overconservative numbers produced by opaque processes and where expert judgment is essential. The gradual realization that the results of carcinogenicity bioassays conducted at high doses in a model that is not human relevant are a poor basis from which to extrapolate human effects will eventually make this type of model redundant. The problem is that judgment is open to challenge—scientific and legal. Having said that, a peer-reviewed assessment of a full data package that includes comparative ADME and mechanistic data is likely to produce a more realistic assessment of low-dose effects and risks.

Overview of Carcinogenicity Risk Assessment

For chemicals already in the market or in the environment, carcinogenic risk assessment is the subject of academic research and debate, conclusions constantly evolving or changing as the database grows and understanding of specific and/or general mechanism deepens. For new chemicals, the situation is more difficult, especially where there are no human exposure data to assess ADME or actual exposure levels. For these chemicals, the main database is the safety evaluation conducted *in vivo* or *in vitro*. There is some official guidance; for example, an addendum on dose selection for carcinogenicity studies for the International Conference on Harmonisation (ICH) suggests that a positive result for tumorigenesis in rodents at 25 times the human exposure is probably not relevant as a risk for humans. This type of statement does not mean that such an argument will be accepted by every regulatory authority; therefore, it will still be necessary to have evidence of mechanism to back up any marketing application.

In assessing the carcinogenic risk due to a chemical, especially where there is significant environmental exposure or contamination expected, it may be useful to consider how much extra risk (additional cases of cancer) would result if it was

CASE STUDY 18.1   METHYLENE CHLORIDE
AND EVOLUTION IN RISK ASSESSMENT

Methylene chloride ($MeCl_2$) was first discovered in the nineteenth century. Large-scale production started in the 1940s (Riley and Fishbeck 2000).

- *1940s:* Used as a paint stripper, replacing lye—a caustic alkali—with advantages of speed, nonreactivity, and, it was thought, safety. In 1946, the American Conference of Governmental Industrial Hygienists (ACGIH) set an 8-hour TWA of 500 ppm. Safety advice said that employees should keep their hands out of the solvent, because of skin irritation and skin absorption potential.
- *1960s:* Approved for preparation of hop extract; residue maximum set at 2.2%. Use in decaffeination of coffee approved with a maximum residue of 10 ppm.
- *1970s:* $MeCl_2$ was linked to formation of carboxyhemoglobin (COHb); the 500 ppm limit was associated with greater levels of COHb than the limit for carbon monoxide. There was evidence that 1000 ppm (allowed as a short-term exposure limit) led to central nervous system (CNS) depression. It was also linked to cardiac arrhythmias, which was proved in 1976. In 1974, a limit of 75 to 100 ppm was proposed, and in 1975, ACGIH indicated a change to 100 ppm as an 8-hour TWA, although 500 ppm was maintained as a limit by the US Occupational Safety and Health Administration (OSHA). The US National Institute for Occupational Safety and Health set a limit of 75 ppm in line with limits set for carbon monoxide.
- *1980s:* The ACGIH reduced its limits to 50 ppm. Links to cancer were suggested.
- *1990s:* OSHA proposed a change to a 25 ppm limit. This was the first OSHA assessment to use physiologically based pharmacokinetic modeling. Industrial pressure was mounted for a 50 ppm limit.

Use of $MeCl_2$ as a paint stripper has declined as knowledge of its toxicity has grown (with a return toward alkali-based strippers and alternatives), illustrating neatly the hazards of replacing a supposedly hazardous compound or process with one that is supposedly safer, but unknown.

introduced into the environment. An acceptable figure appears to be one in a million, although people may ask if they are the individuals likely to be affected. (With the advent of proteomics and genomics, individual risk assessment of this type is becoming more possible.) Where the risk of cancer is greater than one in a million there, is a moral question to answer: should it be accepted that the additional risk—in comparison with that already present naturally in the diet or in the wider environment—is tiny and so can be ignored or that imposing any additional risk, however small, is unacceptable? In any case, given the idea of one in a million cases, how can this be quantified and assessed in the face of the background incidence?

In such cases, comparative risk and risk/benefit analysis become important and subject to judgment that cannot easily be supported scientifically. In Scotland, it is normal for schoolchildren receiving free lunches to be given fruit as an alternative dessert, which almost certainly contains trivial residues of pesticides. There might be argument in favor of organically produced fruit to avoid such residues, but the cost could make supply of fruit financially impossible. Quite apart from the endogenous chemicals present at far higher concentrations than the pesticides, the counterrisk of reduced cancer prevention through not getting a daily shot of vitamins, trace elements, and dietary fiber would hugely outweigh the risk due to synthetic chemicals.

Finally, life is about mixtures, and the carcinogenic impact of a single chemical has to be viewed in the context in which it will be used and consumed. No chemical is taken in isolation; even a medicine taken on an empty stomach is subject to the gastric environment and the excipients in the formulation. For chemicals in the diet, the biological matrix in which they are found has far-reaching effects on bioavailability, and there may well be synergistic or inhibitory effects due to other chemicals. These interactions cannot be incorporated into routine assessments, as they are too complex to model or predict; they represent a final layer of uncertainty, which may always be present. The evolution of risk assessment for individual chemicals is illustrated by the history of methylene chloride (Case Study 18.1), which shows the changing emphasis in risk assessment and regulation that occurs over several decades as a database for risk assessment is expanded. It also highlights differences between different regulatory bodies in the same country.

RISK ASSESSMENT AND THE ENVIRONMENT

Environmental risk assessment is a somewhat fraught area, in part not only due to its high public profile but also due to the complexity of the data and wide range of interactions that must be predicted and taken into account. The usual response to environmental disasters—especially from pressure groups—is invariably pessimistic, although the outcome is often less horrendous than initially expected. Recovery is possible and can be quite quick, as with some recent oil spills. For prospective assessments, there is the possibility that the risks are understated and that a larger problem may arise as a result. It is extremely difficult to maintain a balance between angry prediction of irreparable harm and a reasoned assessment of data that may indicate safe concentrations of a chemical.

Environmental risk assessment may be divided into retrospective examination of chemicals already present in the environment and prospective prediction of risk for new chemicals, such as agrochemicals. The difference between intentional and unintentional release into the environment should also be considered. Unintentional release may be from a single point such as an industrial facility or of widespread origin such as traffic pollution, the release of chlorofluorocarbons (CFCs) from consumer goods, or the use of contraceptive pharmaceuticals, and subsequent environmental release of metabolites or unchanged drug. In theory, intentional release is more controlled or predictable, as with the use of pesticides, but this is not always the case.

There is also a need to differentiate between chemicals present naturally and those that are introduced by human activity. While it is easy to dismiss any artificial chemical as pollution, it is less easy to do so with a compound present in the normal environment. In general, the natural chemicals only become a toxicological problem when they are present at concentrations significantly greater than normal. Combustion is a case in point here; naturally induced forest fires are a source of transiently high local concentrations of combustion products, from wood and other organic matter, which include dioxins. It can be argued that similar products produced from burning fossil fuels are not natural and so constitute pollution, especially as these are present at higher-than-normal concentrations and usually for longer periods. There is, therefore, a concept of excess discharge; using general toxicology as an analogy, there is an exposure level beyond which adverse effects may be expected and below which there will be no significant (or detectable) adverse effects. The environment has the capability to cope with limited release of chemicals in much the same way that an animal deals with a low dose of chemical.

There is the added complication that environmental risk assessment cannot be separated from sociological factors, and so it is very much more difficult to bring forward a purely scientific solution that will prove acceptable to the people who feel themselves to be at risk. There is also an element of lack of control in environmental discharges, especially due to unintentional release from industrial facilities. The original concept of pollution avoidance according to the principles of best available technology not entailing excessive cost (BATNEEC) was too easily replaced by the unofficial and unstated concept of cheapest alternative technology narrowly avoiding prosecution (CATNAP). BATNEEC has been replaced by best available technology, although there is still consideration of costs. An unexpected aspect of local pollution is that contaminated sites may become wildlife refuges due to restricted human access and that, paradoxically, major cleanup operations may produce more harm ecologically than leaving them alone.

Successful environmental risk assessment is dependent on appreciation of the interrelationship of many factors and the consequent prediction of the outcome. While risk assessment for other purposes may focus on one aspect of a chemical's toxicity, this is not so easy in environmental terms, due to the complexity of the ecosystem and the dependency of the whole on its individual components. Although it may be predicted that a pesticide or a genetically introduced chemical resistance may have little effect in a general sense, prediction of effects on single species and, through that, on the whole ecosystem may well be less easy.

Factors in Environmental Risk Assessment

Much of the difficulty with environmental hazard prediction lies with the simplicity of the test design and data compared with the complexity of the ecosystem and the difficulties encountered in assessing or predicting exposure. Single species tested in a laboratory environment do not necessarily give a sound basis for hazard characterization and risk assessment. Mesocosm studies may make this process easier but are likely to be undertaken toward the end of the development process due to cost. Certain substance properties make prediction easier, such as estrogenic activity, and these can be relatively easily tested for and related to the likely persistence of the chemical in the environment. Persistence is a significant factor in environmental risk, as shown by the relative persistence of TCDD (Tetrachlorodibenzo-p-dioxin) and atrazine in the soil at 10 and 2 years, respectively. Where a process of degradation is identifiable, associated with a short half-life, this is an indicator of lower risk than for nondegradable chemicals. This presupposes that the degradation products have been identified, remembering that DDE, a metabolite of DDT, is also very persistent. In a manner analogous to that in protein binding in mammals, sequestration of chemicals into compartments, such as clay soils, implies potential for long environmental half-life and possible toxicity if there is a sudden release to produce high concentrations. However, high-affinity sequestration may reduce immediate risk levels slightly.

In terms of legislation in Europe, the regulatory framework for environmental risk assessment is based on the risk quotient, which is the ratio of the predicted environmental concentration (PEC) to the predicted environmental no-effect concentration (PNEC). Typically, the PEC is modeled using data on expected market volume and usage data, together with estimations of diffuse or point-source introduction, degradation, distribution, and fate. In some cases, these predictions are supported by analytical measurement. The PNEC is then estimated by using empirically derived effect or no-effect data from laboratory experiments, applying safety factors of up to 1000 depending on the uncertainties inherent in the test data. A risk characterization ratio (the PEC divided by the PNEC) of less than 1 indicates low risk, while a ratio greater than 1 may indicate a relevant risk. The margins of safety (MOSs) are also considered; the risk decreases with increasing MOS. This process and the reasoning involved are nicely outlined in an environmental risk assessment of methyl tertiary butyl ether (MTBE) carried out by a team from the European Fuel Oxygenates Association (EFOA), WRc-NSF National Centre for Environmental Toxicology (NCET), and European Centre for Eco-toxicology and Toxicology of Chemicals (ECETOC). This assessment is summarized in Case Study 18.2.

Environmental risk assessment, in common with other areas of toxicological investigation, should be a dynamic process and is unlikely ever to be static, in view of the continually increasing database. This is particularly true of high-profile chemicals such as MTBE and TCDD. The latter has acquired a dire reputation that has made it into a toxicological icon of all that is chemically evil and (supposedly) man-made. However, even this is being reassessed in the light of new data and perspectives. Bruce Ames and colleagues have reviewed the effects of TCDD in comparison with those of other natural chemicals, particularly indole carbinole and ethanol, and the indication is that, although TCDD is very toxic, its effects should be seen in

CASE STUDY 18.2   ENVIRONMENTAL RISK
ASSESSMENT OF MTBE USE IN EUROPE

An environmental risk assessment (Ahlberg et al. 2001) was carried out for
the use of methyl tertiary butyl ether (MTBE) in Europe, using the European
Union System for the Evaluation of Substances (EUSES). MTBE is a highly
water-soluble octane enhancer used in petrol, at concentrations up to 14%.
Leakage of MTBE into groundwater has caused concern in the United States
due to potential contamination of drinking water; it has a pronounced taste
and odor. This environmental risk assessment was performed for three uses of
MTBE: as a fuel additive, in production of isobutylene, and as a pharmaceutical solvent.

- *Environmental distribution and fate:* Most of the MTBE was expected
  to end up in the air, with a significant percentage in water but virtually
  zero in biota with no bioaccumulation. MTBE appeared to be degrad-
  able in some circumstances but not in others; expected half-life in air
  was less than 6 days.
- *Predicted environmental concentrations (PECs):* PECs vary with site
  and type of use. Background concentrations are <5 µg/L. The annual
  average local PEC for production facilities of 172 µg/m^3 was similar
  to values reported for worker exposure, which may be up to 1 mg/m^3.
  The highest local PEC of 37.7 mg/L was estimated for processing use.
  The highest reported concentrations in urban air were about 60 µg/m^3
  and were generally 10 µg/m^3 or less. There were few data for MTBE
  in soils in Europe.
- *Effects assessment and predicted no-effect concentration (PNEC):*
  Acute toxicity tests indicated low toxicity to aquatic organisms.
  The amount of acute data justified the use of a safety factor of 100
  (rather than 1000) applied to the lowest EC$_{50}$ value to generate a
  PNEC. Using the lowest acute EC$_{50}$ value for a freshwater organism (184 mg/L for *Selenastrum capricornutum*) gave a PNEC for
  the aquatic compartment of 1.84 mg/L. Chronic aquatic toxicity test
  data showed a 5-day no-observed-effect concentration (NOEC) for
  Ceriodaphnia dubia of 202 mg/L and a 21-day IC$_{20}$ value of 42 mg/L
  for *Daphnia magna*; a chronic NOEC value of 26 mg/L was reported
  for the marine shrimp *Mysidopsis bahia* and a NOEC for reproduction in *Daphnia* of 51 mg/L. There were also further chronic toxicity
  data for *Daphnia*, fathead minnow, and algae, although not all were
  completely compliant with Organization for Economic Co-operation
  and Development (OECD) test guidelines. In view of the amount of
  data available, use of a factor of 10 to derive a PNEC from the lowest
  chronic (NOEC) value was justified for continuous (chronic) release.
  This gave a PNEC$_{aquatic}$ of 2.6 mg/L, in line with the PNEC from

acute data, which was used in the EUSES risk assessment modeling. The EC_{50} value for *M. bahia* of 136 mg/L was used for intermittent releases, with a safety factor of 10, giving an intermittent $PNEC_{aquatic}$ of 13.6 mg MTBE/L.

- *Risk characterization ratios (RCRs PEC/PNEC) and margins of safety (MOSs):* Except for the sediment and water environmental compartments, all of the RCRs were less than 1, and all of the MOS values were greater than 1. The RCRs that were greater than one were for the use of MTBE as a feedstock for high-purity isobutylene manufacture. From monitoring data for production, it was known that the PECs for the aquatic compartments were overestimates and that the true RCRs were probably lower than those calculated by the model.

It was concluded that the environmental risk of using MTBE as a fuel additive, process intermediate, or solvent was low. Where MTBE is released into the environment from production and processing, it was considered that more data and testing were required, including sediment toxicity testing and a sampling and analysis program to measure concentrations of MTBE in wastewater from sites producing isobutylene.

perspective with those of other chemicals (Ames et al. 1990). TCDD is an example of an environmental contaminant that is present both naturally and as a result of human activities. It is characterized by extensive animal toxicity but by few proven effects in humans. The doses humans ingest are, however, far lower than the lowest doses that have been shown to cause cancer and reproductive damage in rodents. The environmental concerns about TCDD have produced stockpiles of this potentially lethal chemical that, if spilled, could have devastating local effects. Any incineration of biological material can produce dioxins, as was complained about in a recent foot-and-mouth outbreak in the United Kingdom. This holds true for human crematoria as well: Should they be closed down?

INTERNATIONAL MANAGEMENT OF ENVIRONMENTAL RISKS

The environment is global, and there is ready potential for transfer of toxic chemicals among countries either intentionally by transport of toxic waste or by natural processes such as river flow or atmospheric pollution and precipitation as acid rain. There have been various attempts to manage toxic risks internationally, with varying success. In general, agreements made with an objective of stopping environmentally bad practice (usually for the benefit of developed countries) are often significantly weakened by the economic or humanitarian need to continue the same bad practices in less developed countries. Thus, it may be acceptable to ban the use of DDT in the developed world because acceptable substitutes are available (albeit at higher cost). In contrast, in the third world, expensive alternatives are not economically available, and the environmental risks are seen as less important than the benefits. Likewise, it may appear sensible from a Western point of view to ban the transport of toxic material across borders, but if the result of this ban is large, ill-managed dumps of toxic waste, the environmental costs may well be greater than the risk of accidental spillage in transit. As a result, this type of agreement is often ineffective due to the influences of interested parties and countries; the wider the proposal, the more difficult it is to reach an agreement that is effective. At this point, the mixture of politics with toxicology becomes unstable, to the extent that common sense and science lose out. Aspects of international toxic risk management are considered in Focus Box 18.4.

FOCUS BOX 18.4   NOTES ON INTERNATIONAL MANAGEMENT OF TOXIC RISKS

Management of toxic risks in an international context is fraught with difficulty and frustration; a risk that looks terrible in the West is likely to be acceptable in less developed countries due to local conditions. Differences in risk perception among countries lead to different approaches and priorities. Toxicology has an initial role in this but then becomes subsumed in politics (Kellow 1999; Illing 2001).

- Toxicology can identify hazard, which will probably vary little for a single form, but risk should be assessed according to local conditions and the use to which the material will be put.
- There is often significant difference between international policy agreement and local implementation.
- International policy can be constructed on inappropriate or incomplete data or flawed premises with little consideration for the side effects of such policy; this can lead to an unbalanced agreement that has undesirable side effects in other related areas.
- The assumption that all waste is immoral and hazardous is not a sensible starting point for effective policy construction.
- The most successful international agreements are regional, involving few countries; tackle an acknowledged definable pollution problem; have little cost impact on industry; and affect rich countries, which can administer them.
- Risk reduction cannot be equated with risk abolition.
- Agreements that are unfocused and have differing standards between developed and developing nations will probably be unsuccessful, as the good effects in developed countries will be balanced out by the continuing abuses in the developing nations. This is becoming increasingly apparent with the continuing arguments about atmospheric concentrations of carbon dioxide.

THE EVOLUTION OF RISK ASSESSMENT

The science of risk assessment is dynamic and evolving constantly. For toxicological risks, this has been due to increasing knowledge and understanding of interdependencies in toxicity and mechanisms of action, and to increased appreciation that risks should not be viewed and assessed in isolation. Risk is positive or negative, balanced on a host of supporting or dependent factors, to which must be added the perceptions of the people at risk and those who are attempting to manage that risk. Risk assessment results in regulation and management; greater knowledge and understanding should result in better regulation, although this is offset by the inherent (and understandable) conservatism of regulators working in the shadow of the principle that it is almost impossible to prove a negative—that chemical A is safe.

An inescapable factor in risk assessment is the increasing refinement and sensitivity of analytical techniques. When a chemical has been branded toxic and harmful to health, the presence of tiny amounts, revealed by new methods, can result in huge efforts to produce a cleaner environment, even when the chemical is present naturally in greater amounts. Doull (2000) points out that we tend to focus "on the trees of individual effects rather than on the forest of public health. In the final analysis, our mandate is not to use what-if toxicology to produce media headlines and stimulate funding for the investigation of phantom risks but to improve public health, and that should be the most basic principle of toxicology and all science."

The Environmental Protection Agency (EPA) established a range of 1 in 1 million (E-6 value) to 1 in 10,000 (E-4 value) for the incremental excess lifetime risk of cancer associated with possible exposures from contaminated sites, indicating the increased probability above background rate that someone could get cancer following repeated exposure. However, this ignores the risk from naturally present radiation, which is calculated to be 1 in 100. If natural risks are high, what future is there in attempting to manage lower risks from artificially introduced factors?

With increasing knowledge and understanding, methylene chloride has been increasingly demonized as harmful. In contrast TCDD, while universally acknowledged as extremely toxic, has become less threatening. During the last 30 years, one of the greatest advances in risk assessment has been the increased understanding of the toxicity of chemicals present naturally in the environment, whether in the atmosphere, water, or diet. Equally, there has been understanding of the balance of nature: that generally, the natural percentages of the individual chemicals to which we are exposed are not associated with any detectable epidemiological effect. Where that balance is disturbed and the percentages of certain chemicals increase beyond natural limits, as with increased solanine concentrations in insect-resistant potatoes, toxicity can result. MOSs in nature are frequently smaller than those set by regulators. For synthetic chemicals, the Delaney amendment may have looked sensible at one time but is clearly of questionable use now.

SUMMARY

Risk assessment for carcinogenicity should be based on the fact that cancer is multi-factorial, and the following points need to be taken into account:

- DNA, and the control of its expression, is central to carcinogenesis, either through direct attack or by changed regulation.
- DNA is vulnerable to many influences, including direct genotoxic action (endogenous as well as exogenous) and methylation of associated molecules, but is also able to repair itself.
- It is now realized that there are thresholds for genotoxicity; saturation of repair mechanisms appears to be an important factor in these.
- Low-dose extrapolation from high doses is complicated by the influence of DNA repair, carcinogenic mechanism, and the fact that there is no way of distinguishing between the linear and threshold models at low doses, due to background incidences and factors such as DNA repair.
- The tenuous basis of carcinogenicity risk assessment, using rodent bio-assay data, is the extrapolation of effect from the high dose levels used in the short-lived animal to the much lower doses expected in long-life-expectancy humans.
- For chemicals already in the market or in the environment, carcinogenic risk assessment is the subject of continuous evolution of understanding as new data emerge and understanding of specific and/or general mechanism deepens.
- For new chemicals, where there is no experience of human exposure, the main database is the safety evaluation conducted *in vivo* or *in vitro*.

For environmental risk assessment, much of the difficulty lies with the simplicity of the test designs and data, from which the risk assessment will be formed, compared with the complexity of the ecosystem and the difficulties encountered in assessing or predicting exposure. Single species tested in a laboratory environment do not necessarily give a sound basis for hazard characterization and risk assessment. As with carcinogenicity assessment, there is continuing evolution of understanding, which will affect existing conclusions, understanding, and future policy.

REFERENCES

Ahlberg R, Gennart J-R, Mitchell RE, Schulte-Koerne, Thomas ME, Va-hervuori H, Vrijhof H. ECETOC/EFOA Task Force Report on Environmental Risk Assessment of MTBE, 2001. Available at www.efoa.org/fr/mtbe_environment/EFOA-ECETOC%20report.pdf or through the EFOA web site (http://www.efoa.org).

Ames BN, Profet M, Swirsky Gold L. Nature's chemicals and synthetic chemicals: Comparative toxicology. Proceedings of the National Academy of Sciences of the USA. Revised 1990, and "Dietary Pesticides (99.99% All Natural)." Available at http://socrates.berkeley.edu/mutagen/ames.PNASII.html.

Dell LD, Mundt KA, McDonald M, Tritschler JP 2nd, Mundt DJ. Critical review of the epidemiology literature on the potential cancer risks of methylene chloride. *Int Arch Occup Environ Health* 1999; 72(7): 429–442.

Doll R, Morgan LG, Speizer FE. Cancers of the lung and nasal sinuses in nickel workers. *Br J Cancer* 1970; 24(4): 623–632.

Doull J. Chapter 1. In Salem H, Olajos EJ, Eds. *Toxicology in Risk Assessment*. London: Taylor & Francis, 2000.

Gibbs GW, Amsel J, Soden K. A cohort mortality study of cellulose triacetate-fiber workers exposed to methylene chloride. *J Occup Environ Med* 1996; 38(7): 693–697.

Gocke E, Wall M. In vivo genotoxicity of EMS: Statistical assessment of the dose response curves. *Toxicol Lett* 2009 Nov 12; 190(3): 298–302.

Green T. Methylene chloride induced mouse liver and lung tumors: An overview of the role of mechanistics studies in human safety assessment. *Hum Experimen Toxicol* 1997; 16(1): 3–13.

Hearne FT, Lifer JW. Mortality study of two overlapping cohorts of photographic film base manufacturing employees exposed to methylene chloride. *J Occup Environ Med* 1999; 41(12): 1154–1169.

Henderson L, Albertini S, Aardema M. Thresholds in genotoxicity responses. *Mutat Res/ Genet Toxicol Environ Mutagen* 2000; 464(1): 123–128.

Illing P. *Toxicity and Risk-Context, Principles and Practice*. London: Taylor & Francis, 2001.

Jenkins GJS, Doak SH, Johnson GE, Quick E, Waters EM, and Parry, JM. Do dose response thresholds exist for genotoxic alkylating agents? *Mutagenesis* 2005; 20(6): 389–398.

Kaldor J, Peto J, Easton D, Doll R, Hermon C, Morgan L. Models of respiratory cancer in nickel refinery workers. *J Natl Cancer Inst* 1986; 77(4): 841–848.

Kellow A. *International Toxic Risk Management*. Cambridge: Cambridge University Press, 1999.

Lutz WK. Dose–response relationships in chemical carcinogenesis: Superposition of different mechanisms of action, resulting in linear–nonlinear curves, practical thresholds, J-shapes. *Mutat Res* 1998; 405(2): 117–124.

Lutz WK, Kopp-Schneider A. Threshold dose response for tumor induction by genotoxic carcinogens modelled via cell-cycle delay. *Toxicol Sci* 1999; 49(1): 110–115.

Morgan LG, Usher V. Health problems associated with nickel refining and use. *Annals Occup Hygiene* 1994; 38(2): 189–198.

Ojajarvi A, Partanen T, Ahlbom A, Boffetta P, Hakulinen T, Jourenkov N, Kauppinen T et al. Risk of pancreatic cancer in workers exposed to chlorinated hydrocarbon solvents and related compounds: A meta-analysis. *Am J Epidemiol* 2001; 153(9): 841–850.

Riley DM, Fishbeck PS. History of methylene chloride in consumer products. In. Salem H, Olajos EJ, Eds. *Toxicology in Risk Assessment*. London: Taylor & Francis, 2000.

Tomenson JA. Epidemiology in relation to toxicology. In. Ballantyne B, Marrs T, and Syverssen T, Eds. *General and Applied Toxicology*, 2nd ed. London: Macmillan Reference, 2000.

Williams GM, Iatropoulos MJ, Wang CX, Jeffrey AM, Thompson S, Pittman B, Palasch M, Gebhardt R. Nonlinearities in 2-acetylaminofluorene exposure responses for genotoxic and epigenetic effects leading to initiation of carcinogenesis in rat liver. *Toxicol Sci* 1998; 45(2): 152–161.

19 Evaluation of Specific Classes of Chemical

INTRODUCTION

In regulatory terms, there are two divisions of chemical class: those that are centrally regulated and those that are subject to self-regulation by the relevant industry with central authority interest confined to monitoring, although monitoring may become management, such as banning or institution of restrictions. In Europe, the former are easily listed as pharmaceutical (human, veterinary, and some medical devices), agrochemicals, and what, for the sake of convenience, can be loosely categorized as industrial or general chemicals. For these classes, authorization to market a product is dependent on central authorization; someone has to say directly, "You may market this," often only for a specific use. A new use requires a new marketing authorization. The data sets for these chemicals have, historically, often been so large that they fitted with difficulty inside a 1 m cube.

Regulatory control of the second group is looser and usually requires a statement to a central authority that marketing of a product will start on a particular date. No positive feedback from the authority is expected, although a specific refusal is always possible. This group includes consumer products (such as cosmetics, toys, and household products), some medical devices, and the food industry. Although this apparent laissez-faire approach seems to be quite laid back, it comes with a number of heavy responsibilities on both the marketing company and the safety assessor. In Europe, it is usually the marketing company that bears the responsibility for the safety and correct documentation of their product; when things turn pear shaped, they are the ones who go to court and face penalties, which may include prison.

The circumstances of exposure to the different groups of chemicals also show marked differences. These can be summarized simply as follows:

- *Pharmaceuticals:* controlled high-level exposure of a defined population with intentional biological activity in humans
- *Medical devices:* controlled exposure, often to a relatively undefined population, usually without intentional biological activity in humans
- *Veterinary medicines:* controlled exposure of the animal to a biologically active chemical with potentially low, essentially uncontrolled, exposure of consumers
- *Agrochemicals and biocides:* uncontrolled, usually low-level exposure to agents with known biological activity that is not intended in man
- *Cosmetics:* uncontrolled, widespread, sometimes high-level exposure, usually with no intentional biological activity
- *General and industrial chemical:* the level of control depends on the user (worker or consumer) and the hazards of exposure

Control of exposure is exerted at several levels—regulators, physicians, users, and consumers. These different circumstances tend to drive the evaluation process in terms of testing objective and the subsequent risk assessment processes. For a drug, some risk of toxicological hazard is acceptable, depending on the indication for which it is to be used. Medical devices should be associated with low toxicological risk. It is clear that cosmetics should carry no significant risks of toxicity, although some individuals will prove sensitive even to the most "hypoallergenic" products. Agrochemicals, plant protection products, or biocides—the name tends to vary according to global location—are frequently extremely active materials that are, by definition, toxic to their target organism. There are two groups (apart from production workers) exposed to agrochemicals at different levels—the person using them and the consumer; the risks acceptable to these groups are very different and are tested for and regulated accordingly. An additional group of people exposed unintentionally is passersby exposed to spray from agrochemical machinery.

The following account is based mainly on the European situation. However, differences that are relevant in other major jurisdictions have been added where relevant information was readily available. Although each of the chemical classes considered here is regulated to a greater or lesser extent by individual governments, there is an overseeing role for the Organization for Economic Co-operation and Development (OECD), which, for instance, has published guidelines on Good Laboratory Practice (GLP), toxicity, and environmental testing. The OECD has 34 member countries at the time of writing, principally in Europe and North America together with the prominent Pacific Rim countries. While it is notable that Brazil, China, India, and Indonesia are not members, there are links with 70 other countries and nongovernmental organizations, giving the OECD a global reach and credibility. A study conducted to OECD guidelines should be acceptable globally.

The above listing is incomplete and, for instance, does not cover consumer products such as toys, herbal products, homeopathic medicines, and detergents. Toys, which have often been in the news due to their content of regulated or banned substances, are regulated through the EU Toy Directive 2009/48/EC on the safety of toys and in the USA by the Consumer Product Safety Commission via the American Section of the International Association for Testing Materials (ASTM) F963-11 Standard Consumer Safety Specification for Toy Safety.

REGULATORY INFLUENCES

Needless to say, there is a raft of regulations and standards to negotiate in testing any chemical. The principal standards are those of Good Manufacturing Practices (GMP), GLP for nonclinical tests, and Good Clinical Practices for clinical trials. These standards permeate through most of the testing paradigms; all are applicable to pharmaceuticals and cosmetics. While industrial chemicals and household products such as biocides (in the European sense) may not be subjected to clinical testing, any toxicity study should be conducted to GLP. While it might be argued that GMP is not strictly relevant to toxicity testing, the use of GMP-grade material is mandated in many test areas, and in any case, it makes good sense to be aware of the origins and purity of the material under test. An industrial chemical may not have a GMP certificate, but it may be available in different purities up to and including analytical grade.

These standards of study conduct are overseen and implemented by the major regulatory authorities, such as the Food and Drug Administration (FDA) and Environmental Protection Agency (EPA) of the United States, and the European Medicines Agency (EMA) and the European Chemicals Agency, the European Environmental Agency, and the Japanese Ministry of Health, Labour and Welfare, not to mention the individual authorities in member states of the European Union.

A further layer of oversight is provided by trade associations such as the US Cosmetics, Toiletries and Fragrance Association; the European Cosmetic Toiletry and Perfumery Association, known as Colipa; and the European Chemical Industry Council (Cefic). In the pharmaceutical industry, there are the Pharmaceutical Research and Manufacturers of America (PhRMA,), the European Federation of Pharmaceutical Industries and Associations (EFPIA), and national associations such as the Association of the British Pharmaceutical Industry.

Table 19.1 gives a small selection of websites that may be useful in assessing regulation and testing requirements.

TABLE 19.1
Useful Websites for Regulatory Standards and Requirements

Organization	Areas of Concern	Website
Cosmetics Europe (formerly Colipa)	Cosmetics	https://www.cosmeticseurope.eu/
Environmental Protection Agency, USA (EPA)	Pesticides, pollutants, any aspect of the environment	http://www.epa.gov
European Union Reference Laboratory for alternatives to animal testing (EURL-ECVAM)	Alternative methods	https://eurl-ecvam.jrc.ec.europa.eu/
European Chemicals Agency (ECHA)	REACH, GHS	http://echa.europa.eu/
European Commission	General site for European Union	http://ec.europa.eu/
European Medicines Evaluation Agency (EMA)	Human and veterinary medicines	http://www.ema.europa.eu/ema/
Food and Drug Administration, USA (FDA)	Medicines, some cosmetics, foods, medical devices	http://www.fda.gov
ICH	Human pharmaceuticals	http://www.ich.org
International Fragrance Association (IFRA)	Fragrance materials	http://www.ifraorg.org/
Japanese Ministry of Health, Labour and Welfare (JMHW)	Pharmaceuticals, cosmetics, and foods	http://www.mhlw.go.jp/english/index.html
Johns Hopkins Center for Alternatives to Animal Testing (CAAT)	Alternative methods	http://caat.jhsph.edu/
Personal Care Products Council (formerly CTFA)	Cosmetics	http://www.personalcarecouncil.org/
Organization for Economic Co-operation and Development (OECD)	Testing guidelines and GLP	http://www.oecd.org
VICH	Veterinary pharmaceuticals	http://www.vichsec.org/
UK Medicines and Healthcare Products Regulatory Agency (MHRA)	Medicines, medical devices	http://www.mhra.gov.uk

One aspect to consider in all this is that the regulatory landscape has changed over the past few years. It used to be the United States that was the most difficult market in which to launch a new chemical or application for an existing chemical. However, it may be that Europe is now the harder market to penetrate, and this may be due to a risk-averse approach to chemical regulation and an approach biased toward hazard rather than risk. There is always an equilibrium to be struck between risk and reality; both are difficult to define objectively, but it is possible that the European balance has been pushed toward avoidance of trivial risks. There is sometimes the suspicion, in any jurisdiction, that scientifically dubious decisions are taken in the name of consumer safety, while actually being politically motivated.

In Europe, there is frequent use of positive lists in annexes to the various directives, and it is often necessary to get a compound onto such an annex in order to be able to market it in the European Union. A further complication of EU regulatory practice is that, while the overall picture may appear harmonized, there may be local requirements or variations in definition, as with cosmetics and some pesticide products.

As an added layer of complexity for human and veterinary pharmaceutical in Europe, the potential residue of these products on manufacturing equipment must be taken into account and a permitted daily exposure (PDE) value generated (see Chapter 15, Case Study 15.1, for further details).

THE BASIC TOXICITY TEST PACKAGE

Although this chapter reviews the toxicity testing of a wide range of chemical types or classes, the toxicity package that is produced for most, with the notable exception of cosmetics, is broadly very similar. The basic elements consist of investigation of absorption, distribution, metabolism, and elimination (whether in an animal or in the environment), and toxicity testing (single and repeat dose) in animals, which can include evaluation of general and reproductive toxicity, carcinogenicity, genotoxicity, and, if relevant, local tolerance. To these basic elements may be added specialist pharmacology or environmental studies, so-called six-pack studies for worker safety assessment, or other more esoteric studies such as *in silico* predictions of toxicological effect.

In each case, the end use for these diverse data is a risk assessment, which may be at multiple levels taking differing circumstances and extent of exposure into account, for instance, workers and consumers exposed to a pesticide.

HUMAN PHARMACEUTICALS

Nonclinical testing of pharmaceuticals is directed toward elucidating hazards that are relevant to human clinical use. The objectives of the program of tests that is undertaken are similar to other areas of toxicology, in that dose response and mechanism of effect are important, and at the end of the program of evaluation, an overview is developed as to the significance of the various findings and whether the drug may be expected to be safe for its intended use. It is important to decide if the effects seen are due to mechanisms that are relevant to man and to dismiss those that are not.

One of the purposes of toxicity testing for pharmaceuticals is to support clinical trials in the target species, namely man. Generally, short toxicity studies support short clinical trials in man, the first human exposure generally being in healthy human volunteers who may receive single or repeated administration under controlled conditions. These early human studies are, essentially, toxicity studies, but conducted in a more inconvenient and less defined species than most laboratory animals and with more demanding husbandry requirements; also, necropsy and histopathology are not options. Up to the later phases of clinical development, toxicity studies are usually the same length or longer than the intended clinical trial. In addition, for intermittent treatment regimens, it is usual to mirror the clinical intention or to use a slightly more frequent administration schedule. Thus, for a clinical intention of once-weekly administration, it may be sensible to give the drug twice weekly in laboratory animals as this is a more rigorous examination of the drug. This also gives additional clinical flexibility in trials should more frequent administration in volunteers or patients become necessary. The route of administration is usually the same as that intended clinically, although one that gives greater systemic exposure than the clinical route may be used in some studies to evaluate a worst-case situation.

The test material should be the same as that used in clinical studies and should be produced according to GMP or be readily comparable. Having said that, there is some regulatory benefit in conducting early studies, especially the genotoxicity studies, with a batch that is less pure than intended for clinical use. This allows the *qualification* of impurities that may be present. Qualification is the process where an impurity is tested along with the main molecule at concentrations that are similar to or greater than those expected in the final active ingredient. Endless problems are created by new impurities emerging late in the life of a pharmaceutical, either as a result of new synthesis procedures or, more banally, as a result of more precise analytical procedures. Substituting a sophisticated technique, such as Liquid chromatography–mass spectrometry (LCMS), for a simple one like thin-layer chromatography (this has happened) causes all sorts of problems—none of them easy to resolve. Furthermore, as discussed previously (Chapter 5, Case Study 5.1), according to International Conference on Harmonisation (ICH) M7 guidelines, all such impurities in new drug substances should be examined, either *in cerebro* or *in silico*, for bacterial mutagenicity, or be shown to below the threshold of toxicological concern (TTC). This can cause yet more headaches and expense when an impurity must be synthesized for Ames test or an analytical method created for determination of exposure.

At the heart of pharmaceutical evaluation is a risk assessment process, which assesses hazard in relation to its acceptability in the patient population. The use of risk/benefit analysis gives an idea of the type of hazard and risk that is acceptable across a range of clinical indications. In other words, more toxicity is acceptable (and tacitly expected) in the case of a cytotoxic anticancer treatment than in a nonsteroidal anti-inflammatory drug (NSAID), which may eventually be sold over the counter at pharmacies. In the case of anticancer drugs, while cytotoxicity may be associated with acceptable toxicity, a receptor-targeted compound, such as imatinib mesylate, is likely to be more closely scrutinized.

For risk assessment, pharmaceuticals are (theoretically) an easy target, as they are used in a defined population at defined levels of exposure. However, as usual, this simplicity is skin deep, and the definitions swiftly become blurred by reality. The population is defined insofar as they have an indication or group of indications that the pharmaceutical has been identified for. Within that population, however, lurks a range of variations that affect the response of individuals to the drug. These include metabolic polymorphisms, comedication, diet, use of tobacco or alcohol, and a host of other factors such as intercurrent disease. The next indefinite is exposure. Although a drug may be prescribed, this does not equate to administration; patient compliance is always a problem, either in clinical trials or in day-to-day use. Additionally, clinicians tend to exercise their freedom of choice and judgment and may underprescribe or overprescribe, as well as use the drug "off-label" for indications for which it has not been clinically tested. These uncertainties have a considerable impact on epidemiological studies and postmarketing surveillance. For epidemiological studies, which may be undertaken as part of a postmarketing surveillance program, the problems may be compounded because meta-analysis of several studies may be prevented by poor comparability between protocols and other confounding factors.

For biotechnology products—generally proteins (e.g., monoclonal antibodies and cytokines)—the assessment is complicated by the need to ensure that the test species chosen for the evaluation is the most relevant available. The regulatory testing requirements for biotechnology product are outlined in Case Study 19.1. Although

CASE STUDY 19.1   REGULATION AND TESTING OF BIOLOGICS AND BIOSIMILARS

There is no one-size-fits-all approach to biotechnology-derived pharmaceuticals (colloquially known as biologics)—the EMA alone has produced over 20 separate guidance documents on various types of biologics. This does not even cover biological medical products that claim to be "similar" to a reference medicinal product (so-called biosimilars). To support clinical development, the nonclinical testing strategy must be flexible, case by case, and above all based on scientific rationale. The 1997 ICH S6(R1) guideline on preclinical safety evaluation of biotechnology-derived pharmaceuticals, and its 2011 addendum, provides a useful starting point, as it is the basis for the European, US, and Japanese guidelines.

The aims of the nonclinical evaluations are threefold: to identify a safe starting dose in humans, to identify potential target organs for toxicity (and its reversibility), and to identify safety parameters for clinical monitoring. One of the first steps in these goals is determining the biological activity of the test substance—this can generally be performed first *in vitro* using cell lines as well as primary cell cultures and thence *in vivo*. Such approaches allow examination of the direct effects on cellular phenotype, viability, and proliferation as well giving an indication of mechanism(s) of action and potential pharmacology. The selection of animal species is of vital importance in the testing of biologics; often, there may be only one relevant species (other than humans); in some cases, the only relevant species is human. The following points are relevant for consideration:

- Safety pharmacology—may be added to toxicity studies.
- Pharmacokinetics and toxicokinetics.

- Single-dose and repeat-dose toxicity studies; dosing regimen according to clinical intention.
  - Use of two species or tests in a single relevant species.
  - Use of homologous proteins (though this may not be the most regulatory-acceptable way to proceed).
  - Appropriate transgenic animals should be considered.
- Immunotoxicity studies—including neutralizing or nonneutralizing antibodies.
- Reproductive performance and developmental toxicity studies, according to product and intended patient population.
- Genotoxicity studies; ICH S6 states that these are not needed for biologicals. However, there may be circumstances when they are appropriate; linkers and nonnatural amino acids may require assessment (Thybaud et al. 2016).
- Carcinogenic potential should be assessed, but routine carcinogenicity studies may not be appropriate.
- Local tolerance is an important consideration for most biologics, which are mostly given parenterally.

Biosmilars: essentially a copy of an already marketed, and off-patent, biological drug

- A biosimilar is very similar to the comparator or originator product.
- This similarity extends to quality and chemistry, safety, activity, and efficacy.
- The amino acid sequence should match the originator compound.
- The biosimilar may not be completely identical in terms of tertiary structure, for instance, due to differences in folding of the protein.
- Despite any differences, the behavior of the biosimilar should match the originator.
- Chapman et al. (2016) indicated that, although nonclinical studies were conducted to support registration of biosimilars, these were not necessarily scientifically justified.
- If required, *in vivo* studies may be conducted in rats or mice, the important aspect being pharmacokinetics. The European Union is less likely to require *in vivo* studies than the United States at the time of writing, but this is changing; other jurisdictions may be less flexible.

Development of biosimilars is a complex area, and it is essential to discuss this with regulators before investing in expensive studies that may prove irrelevant or unacceptable.

testing of conventional small molecules usually demands two species, this may not be appropriate for a protein that is active in only one species. In these cases, it is usual to choose a species that is relevant to man or, in some cases to make a protein that is specific to the test species rather than man. This approach has been taken for the development of some interferons, where the mouse interferon was tested as an evaluation of the effects of the human protein.

One of the most important aspects of early toxicity studies with pharmaceuticals, whether conventional small molecules or biopharmaceuticals, is the choice of starting dose for the first clinical studies in humans [first in human (FIH)]. The FDA has produced a Guidance for Industry (FDA 2005) on selection of first-in-man doses for clinical trials. Partly because of its provenance, this is an accepted method of dose selection, but the use of a simpler safety factor approach, based on the no-observed-adverse-effect level (NOAEL) in the pivotal study or studies, is still an option. Another approach is the use of minimum anticipated biological effect level (MABEL), which has been used for biopharmaceuticals (Muller et al. 2009) to decide a safe starting dose in FIH clinical trials. The EMA has also published guidelines on this (EMA 2007).

The disastrous reaction of volunteers to administration of a novel monoclonal antibody, TGN1412, in 2006 is an example of how this can fall apart. However, one of the lessons learned may include the consideration of the number of target cells compared with the number of molecules of drug being administered; this was suggested as one reason for the spectacular effects seen, although it has been disputed. Another consideration is the appropriateness, or otherwise, of the species used to predict human effects and the differences in pharmacological potency between the two species. With biopharmaceuticals, choice of the most appropriate species for the toxicity testing is a risky operation if the effects of the compound are not fully understood and the physiological and pharmacological differences among species not appreciated.

Another aspect of pharmaceutical development is the requirement for many compounds for the conduct of toxicity studies in juvenile animals to allow their use in children. Until recently, drugs used in children had only been tested in adult animals and in adult patients. While some aspects of toxicity studies—in young rats or in perinatal and postnatal reproductive studies—may address some aspects of juvenile toxicity, they do not cover everything. Apart from the regulatory challenge of getting agreement to a program of such studies, the practical challenges are considerable, especially when administration to very young rats is required.

Pharmaceutical development is now effectively driven by the ICH guidelines—the International Conference on Harmonisation of Technical Requirements for Registration of Pharmaceuticals for Human Use, which are readily available on the Internet from the ICH, website, given in Table 19.1.

While all of the above have been driven by consideration of the process to get the drug into patients in clinical trials and thence to market, other studies are usually carried out on the drug substance and, in some cases, on intermediates that may be used outside closed production systems. These studies are aimed at evaluation of occupational health hazards and include studies such as acute inhalation exposure, dermal toxicity, and sensitization. Assessment of intermediates can be a tricky area, as there is usually not much money for a fuller toxicological assessment. In such cases, the proximity of the intermediate (in terms of structure and likely activity) to the final drug substance is important. This is a read-across exercise in which the structures of the intermediate and the final drug are compared, together with any information on structure–activity relationships. The use of *in silico* methods for the prediction of toxicity is another way of assessing possible effects and so indicating the types of precaution, for example, personal protection equipment levels, which should be taken to avoid adverse effects on workers' health.

Depending on the geographical area of interest and the type of substance, there may also be a requirement to evaluate the environmental impact of pharmaceuticals (e.g., EMA 2006). Certain classes of compounds are exempted from this, including vitamins, peptides or proteins, carbohydrates, vaccines, or herbal products, on the basis that they are unlikely to pose any significant environmental risk. In Europe, this is a two-phase procedure, of which the first estimates the environmental exposure to the drug and the second assesses fate and effects in the environment. The estimation of environmental exposure undertaken in phase I is based entirely on the drug itself, rather than on any metabolites or taking route of administration into account; it is also assumed that the major route of entry to surface water will be via the sewage system. Data relating to the dose per patient, the percent market penetration (to give an idea of how many people will use it), the amount of wastewater per person, and the dilution are used to produce a predicted environmental concentration (PEC) for surface water. If this falls below 0.01 µg/L for surface water and there are no other environmental concerns, it is assumed that there will be no risk to the environment if the drug is prescribed as expected. Substances that are potential endocrine disrupters, persistent, or highly lipophilic may need to be assessed in any case.

Phase II of the assessment is started if the PEC for surface water is more than 0.01 µg/L. This phase is itself in two tiers, A and B, in which a first base set of studies is conducted to assess aquatic toxicology and fate, and if indicated, a second tier in which more detailed study of emission, fate, and effects is conducted. The first part of tier A is to look at the fate and physicochemical properties of the drug; this includes an assessment of biodegradability and the sorption behavior of the drug, which is described by the adsorption coefficient (K_{OC}), defined as the ratio between the concentration of the substance in sewage sludge or sediment and the concentration in the aqueous phase at equilibrium. A substance with a high K_{OC}, retained in a sewage treatment plant, may reach the terrestrial compartment via spreading of sewage sludge.

The aquatic effect studies of tier A include long-term toxicity in *Daphnia* sp., fish, and algae to predict a concentration at which effects are not expected; this is the predicted no-effect concentration (PNEC), which is derived from no-observed-effect concentrations (NOECs) determined in the various studies. The ratio between the PEC and the PNEC is evaluated; and if this is less than 1, further testing in the aquatic compartment is not necessary. If this ratio is more than 1, further testing in tier B is needed. This phase includes investigation of sediment effects and effects on microorganisms. The concentration of the drug in the terrestrial compartment is calculated unless the K_{OC} is greater than 10,000 L/kg.

VETERINARY PHARMACEUTICALS

VICH, or the International Cooperation on Harmonization of Technical Requirements for Registration of Veterinary Medicinal Products, was officially launched in April 1996 and is in many ways similar in conception to ICH for human pharmaceuticals. There are, however, important differences in approach. For instance, there is emphasis on defining residues in food-producing animals, for the purposes of setting withdrawal intervals between treatment and harvest. This is interleaved with the need to set an acceptable daily intake (ADI), and clearly, if the ADI is lower than what can be achieved by the residues found in the various tissues of the animal, there is a problem.

In Europe, the process of registration of a veterinary pharmaceutical for use in food-producing animals begins with establishing maximum residue limits (MRLs), which is achieved by submission of a safety file (in essence, toxicity data) and a residues file. The residues file contains information on residue depletion, residue chemistry, and analytical methods for determination of residues in food. An MRL file has to consider aspects of the residues such as hormonal activity and, for antibiotics, whether the residues will have any impact on the human gut or on microorganisms used in industrial food processing. If no MRL is granted by the authorities, no use in food-producing animals will be possible. For companion animals, an MRL is not needed.

Tolerance studies in the target species are necessary for both companion and food-producing animals, and they are a fundamental part of the safety assessment in target species. However, they play a minor role in the setting of ADIs for veterinary medicines. They tend to differ slightly in design from standard toxicity studies. For instance, some of the animals may not be necropsied; this is an important consideration when the animal concerned is a cow or a horse. The study list for a companion animal treatment is less extensive than that for a food-producing animal.

There is also a requirement for a user safety report, which considers so-called six-pack studies. These are skin and eye irritation, dermal sensitization, acute oral and dermal toxicity, and, if relevant, inhalation toxicity.

Environmental impact studies for veterinary medicinal products have a more intuitive relevance than those for human pharmaceuticals, if only because they are often given to farm animals and, if to a large economic animal (such as a cow) probably in large doses. The effect of ivermectin given to cattle on the longevity of their fecal cowpats has already been mentioned in Chapter 10, and antiparasitic compounds are particularly examined in this process. The VICH guidelines on environmental impact assessment were published in two tranches (VICH 2000, 2006). In a similar manner to the procedure for human pharmaceuticals, there is a relatively straightforward phase I and a more complex phase II.

The guideline places emphasis on veterinary medicinal products that will be used in food-producing animals that may not be individual treatments but may, for example, be used for treating a whole herd or flock. A tacit assumption is made that a substance that is extensively metabolized will not enter the environment. Separate consideration is given to substances used in the aquatic environment, which may enter the wider aquatic environment, and those in terrestrial situations. Questions asked in the guideline include one about antiparasitic compounds, which may be a reaction in part to the environmental effects of ivermectin; antiparasitic agents—but not those acting against protozoans—advance automatically to phase II. If the concentration at which the product enters the aquatic environment is calculated to be less than 1 µg/L or the PECsoil is expected to be less than 100 µg/kg, environmental evaluation of the product may stop at phase I.

Phase II provides recommendations for standard data sets and conditions for determining whether more information should be generated for a given veterinary pharmaceutical. The tests are broadly similar to those indicated for human pharmaceuticals with appropriate adjustment for aquatic and terrestrial compartments. Animals that are reared in intensive conditions and those on pasture are given separate consideration, as are aquatic animals. The end process is calculation of the appropriate PECs followed by a risk assessment of the environmental impact.

MEDICAL DEVICES

This simple term covers a vast range of products that can be as simple as a walking stick, as complex as a cardiac pacemaker, or as mundane as a tongue depressor. While medical devices are classified, for regulatory purposes, according to the general level of risk associated with them, for toxicology purposes, they are classifiable by the extent to which they come into contact with the body. A device that will be implanted chronically requires more extensive evaluation than a temporary catheter or a needle and syringe for collecting a blood sample. The extent and duration of contact drive the testing and evaluation program that is required. The toxicity of medical devices is related to a number of aspects of their composition, as is the wider concept of biocompatibility, which relates to how they react with the tissues or fluids that come into contact with them. Biocompatibility can be defined as the ability of a biomaterial to promote a desirable tissue interaction. Since both the nature of the tissue and the response desired vary from case to case, it is a highly application-specific concept. Further layers of complexity are added when the medical device elutes a drug substance or contains an active power source (an active medical device).

Medical devices pose some interesting challenges in safety evaluation. They are supposed to be chemically inert with low biological activity, so no effect would be expected in a routine safety evaluation. Because a medical device is applied locally and has a discrete, usually solid form that is composed of a mixture of chemicals or ingredients, it is difficult to increase the dosage in a meaningful manner. In contrast, pharmaceuticals are usually single active substances, which are intended to exert defined biological activity and may be expected to have effects in other locations, according to dosage and route of administration. While medical devices may produce effects due to poor biocompatibility, this is not necessarily equivalent to a defined pharmacological action. While the process of safety evaluation for pharmaceuticals is relatively well defined, for medical devices, it is more diverse, and the choice of strategy is dependent on a range of factors, such as form, location and duration of application, and the degree of invasiveness of the device (cutaneous, subcutaneous, or deeply implanted).

In the course of contact with the body, there are a number of ways in which the device may elicit toxicity. These may be due to the chemical or physical characteristics of the material itself; reactions may be passive, as with toxicity due to chemicals, or active, as in attack by the immune system on the device. The presence of any leachate from the material (for instance, the leaching of a monomer from an incompletely reacted polymer or other chemical components of a plastic) should also be considered. Packaging may also have an effect on the device and thence on the body. Another layer of complexity is provided by methods of sterilization, which may react adversely with the material or the packaging. The use of ethylene oxide is an efficient, low-temperature method of sterilizing medical devices, such as nontextile drapes used in surgery. As ethylene oxide is toxic—it is a genotoxic carcinogen—the residues that are permitted are tightly controlled so that the daily dose to the patient of ethylene oxide derived from a device is restricted to a few milligrams, and levels should be controlled on the as-low-as-reasonably-practical (ALARP) principle.

In some cases, the location of an implanted device may mean that it is subject to gradual erosion with dispersion of particles into the surrounding tissues; this may be seen with metal hip prostheses. Depending on the design and precision of the interaction between the ball and acetabular cup of the device, small particles of metal alloy may be shed into the surrounding tissue and, ultimately, may produce increased concentrations of the metals in the patient's blood. There are two aspects to consider in this case: the local effects of the metal particles (it would be useful, if unlikely, to know the size of the particles produced and their rate of dissolution) and the systemic effects of increased concentrations of the metal. One aspect here is the specter of nanotoxicology; what are the effects of small amounts of metal produced at nanoparticle sizes, and where are they sequestered or distributed to? Many metals have toxicities at high concentration, and some may be associated with carcinogenicity; however, their presence at low concentrations does not mean that these hazards will automatically be expressed, and it is probable that the risks are very low. One of the concerns expressed about metal hip prostheses is that it has been shown that implanted patients have chromosomal aberrations in their peripheral lymphocytes. This does not mean, however, that similar aberrations are expressed in other tissues. The prostheses are set into the femur, and it is perhaps unsurprising that some effects should be expressed locally in the bone marrow. Furthermore, although such chromosomal changes might be taken as an indication of carcinogenic potential, there has been no epidemiological connection between the use of metal hip prostheses and cancer (or any other adverse effect), despite many years of use.

More modern ceramic joints tend to avoid this controversy, although that does not mean that they are necessarily better than metal ones. Once again, it must be remembered that lack of knowledge or understanding does not imply safety of use, and an incompletely understood substitute may ultimately prove to be less safe than a well-known standard material. The safety and suitability of most materials used in implants are based not on thorough mechanistic assessment but on years of clinical use without apparent ill effect. There is, therefore, a lack of knowledge and understanding of the biological effects of both novel and well-established materials. A clear determination of the ideal material for any given implant application remains an unattainable goal.

The extent of contact for a device may vary for different parts; at its simplest, a hypodermic needle attached to a syringe for a blood sample collection comes into transient contact with internal tissues, while the syringe itself only comes into contact with the liquid to be injected. For an infusion bag, the components of the device that need to be assessed include the material from which the bag is made, the catheter, and the needle used to effect the injection. If the bag is used for a blood transfusion, the interactions that are considered have to include those between the blood and the bag and the other components.

Evaluation of a device may be based on the 18 parts of ISO 10993—Biological Evaluation of Medical Devices, published by the British Standards Institute (2009). Broadly, the FDA accepts evaluations under ISO 10993, although there may be differences in test selection; as always, if there is any doubt about acceptability of a set of tests, it is best to ask. These cover all the expected aspects of toxicity, such as genotoxicity, carcinogenicity, and reproductive toxicity, and indicate that the following aspects should be considered:

- The materials used in the device
- Any intentional additives or unintentional contaminants resulting from the manufacturing process, or any residues such as monomers
- Substances that may leach from the device, including monomers or residual chemicals or degradation products
- Any other components and their interactions in the final product
- The properties and characteristics of the final product

The core of any evaluation of a medical device is the biological evaluation or safety report, as indicated in part 1 of these guidelines. Due to the diversity of patient contact scenarios that are possible, this guideline specifies consideration of a range of toxicological endpoints. Durations of contact—ranging from less than 24 hours to prolonged (up to 30 days) and permanent (more than 30 days)—are crucial in deciding which endpoints to include in the evaluation and whether the device is a surface device, an external communicating device, or an implanted device. Consideration is also given to the tissues contacted, whether the skin (breached or not), mucous membranes, or bone and other tissues.

For devices with transient contact with the patient, the minimum endpoints are given as cytotoxicity, sensitization, and irritation or intracutaneous reactivity. As the degree of contact increases, more endpoints are added to include, in approximate progression, systemic toxicity; subacute, subchronic, and chronic toxicity; genotoxicity; hemocompatibility; and implantation.

Implanted devices may be associated with several components that fall into different categories of patient contact. For example, an aortic stent would be a permanent implant for which the full range of endpoints should be considered; however, it may be supplied with a set of instruments used for the implantation via the femoral artery, which are in contact transiently (less than 24 hours) with the patient and so need only the basic endpoints considered together with hemocompatibility. Some implanted devices are associated with external electronic components for which different endpoints need to be considered.

However, it is important to note that ISO 10993 does not require that any particular tests must be carried out for any particular situation, simply that the toxicological evaluation as a whole, including any tests deemed necessary, must be designed, carried out, and evaluated by knowledgeable and informed individuals.

Toxicological testing of medical devices is usually based on extraction of a portion of the device into appropriate media, at defined weights and volumes. The eluate is then used in tests such as *in vitro* cytotoxicity using mammalian cells in culture or systemic toxicity in animals (for example by injection), or for genotoxicity assays. The problem is that these tests are crude in that it is an extract of the device being tested and that, consequently, the results may have more regulatory utility than scientific foundation. The most rigorous tests are those in which the device or representative portions are implanted into animals for assessment of local reactions and, if needed, systemic effects.

In evaluating a medical device, the prior use of the material in other devices is a powerful factor in reducing the amount of testing required and easing regulatory acceptance, always assuming that this use has not been associated with adverse reactions, of course. Needless to say, the simplest evaluation is that for a well-known material for use in a similar device to those already on the market; the next level of complication may be exemplified by a novel use for an existing material. A completely new material will need to be very carefully tested and evaluated before use in medical devices, especially if a new device is contemplated. One of the challenges in device evaluation lies in the normal process of evolution that takes place in device design and composition; in conducting an evaluation on a new version of a device, it may be difficult to reconcile existing test results with the materials used in the new version or in comparable devices. If this process of reconciliation fails—that it has to be concluded that the materials have not been tested previously, or that the tests and/ or results are now invalid—new testing should be considered.

The pitfalls present in medical devices are illustrated by the case of breast implants produced by Poly Implant Prothèse (PIP). In summary, these were implants in which the shell and silicone contents that were assessed for the initial evaluation were replaced with lower-quality and industrial-grade material respectively. The poor quality of the implant shell led to earlier failure of the implant compared with others and release of the silicone into the local tissues. A large number of patients reported reactions following rupture, resulting in extensive legal cases and actions. The situation was thoroughly reviewed under the auspices of the UK Department of Health and UK Medicines and Healthcare Products Regulatory Agency (MHRA) by Keogh (2012). The conclusions of this detailed review were that PIP implants were significantly more likely to rupture or leak silicone than other implants, and this difference was evident within 5 years of implantation. Rupture was associated with local reaction, but in the absence of rupture, there was no indication of increased clinical risk. There was no evidence of an increase in the incidence of breast cancer.

In summary, medical devices present a degree of challenge and fascination that is not seen with simple chemical toxicity. There are extra dimensions to consider apart from the straightforward issue of dose of chemical at the sight of contact. The components of the device external to the body and the methods of processing in manufacture and use have to be considered in arriving at a viable assessment of the risks involved in its use. Toxicity, or biocompatibility, has to be seen as just one component of a complex risk management program that weighs and balances a wide range of risks and benefits of often complex technology, over the entire lifecycle of a product.

AGROCHEMICALS/PLANT PROTECTION PRODUCTS

This is one of the areas of chemical legislation where a product may be regulated under two sets of guidelines or regulations. This is typically seen where a chemical has both professional, high-use, high-exposure applications, as with an agricultural herbicide, and lower-use domestic applications, for example, in vegetable gardens. They are also more complex, in a regulatory sense, than other classes, such as pharmaceuticals, in that their safety has to be considered at several levels. This may be seen in the following several levels of exposure in humans:

- In workers exposed to the unformulated compound at the production facility
- In users exposed to the formulated product before or during dilution for application
- In bystanders who may be inadvertently sprayed with a diluted pesticide during its application to a field
- In users exposed to the diluted product during use
- Unintentional exposure through contact with recently treated plants or soil
- Consumption of vegetables or crops that have been treated and have absorbed the compound

Routes of exposure, commonly, would be dermal or inhalation, but the use of personal protection equipment by production workers or professional users should reduce exposure in these groups of people. Production or professional use would normally be carefully controlled through workplace management, and professional users may be required by law to have certificates of competence in order to apply or use the pesticide. There is, however, no such confidence in the case of domestic users, and it is probably best to assume that instructions may not be followed and that the product is likely to be misused. In the United Kingdom, as a result, products destined for the home and garden market are subject to certain restrictions; e.g., unprotected use must not result in acceptable operator exposure levels being exceeded, and/or the product must not be damaging to eyes. The result of this is that the final concentrations of certain actives or coformulants in a home or garden product may be restricted.

In addition to the human exposure considered here, there is exposure to unintended parts of the environment such as beneficial insects (e.g., bees), nontarget plants, and the aquatic environment, and exposure to other species through the food chain. DDT is the classic example of this last possibility.

There are many emotive aspects of pesticides, and consumers often consider that their presence in foods is necessarily malign. However, the resulting drive toward organic produce does not take account of chemicals naturally present in food, many of which have not been tested to the rigorous standards of modern pesticides. In fact, those that have been tested have often been the subject of poor or incomplete study design and tendentious or at least questionable interpretation delivered as undeniable dogma. Organic production methods improve soil and animal husbandry but do not necessarily mean detectably safer food.

An open-access book published online by the UK Pesticides Safety Directorate (2010) gives a flavor of the data requirements for pesticides in the European Union and contains links to the EU directives mentioned. Broadly, the basic data set invoked at the start of this chapter is necessary, together with comprehensive environmental and ecotoxicological investigations. A notable difference from other chemical classes is seen in the reproductive part of the package, which usually includes a multigeneration study and avian reproductive evaluation, and there may be greater emphasis on irritation (dermal and ocular) and skin sensitization (usually only used for classification and labeling purposes). The overall objective is to establish NOAELs, which can be used in risk assessments in the various areas of concern. The different exposures of workers and consumers are typically addressed in 90-day studies for the former and studies of up to 2 years for the latter. One of the objectives is to calculate an ADI, against which modeled consumer intakes can be compared. One aspect to bear in mind is that there may be differences in definition across the European Union, for instance, between professional and domestic use. In the USA, pesticides and similar products are regulated by the EPA (https://www.epa.gov/).

Although environmental studies are now a part of pharmaceutical development, they were first conceived for agrochemicals and have reached a state of considerable refinement. While pharmaceuticals may be expected to reach the wider environment indirectly through the sewage system or, occasionally, by accidental spillage into a river or water course, pesticides are deliberately applied to large areas of the outdoors and so have much wider environmental access and potential ecotoxicological effects.

The studies (often termed fate and behavior studies) conducted are aimed at determining the fate of a chemical in the environment in terms of distribution, degradation (and mechanisms), and elimination from the ecosystem; this process is broadly analogous to the absorption, distribution, metabolism, and elimination (ADME) studies conducted for pharmaceuticals. Any indication that a chemical will persist unduly in the environment is a flag for more extensive (and expensive) studies and more difficult justification of its use. PECs are calculated, and persistence is assessed; degradation products are assessed to ensure that they do not have any adverse effects that add to those of the parent compound. The PECs for parent and degradation products are used to assess exposure of nontarget species in soil and water, potential contamination of drinking water or groundwater, and potential effects in crops, which follow on from the treated crop.

Potential toxicity to wildlife is assessed by standardized laboratory tests using nontarget organisms such as birds, bees and other insects, fish, and aquatic invertebrates; effects on environmental bacteria are also assessed. Values for acute median toxicity (Lethal Dose 50; LD_{50}) and acute median toxicity (Lethal Concentration 50; LC_{50}) are derived together with no-observed-effect level (NOELs) and NOECs and these are compared with the PECs. The overall goal is an indication of the overall toxicity of the material compared with the PECs to get an estimate of toxicity set against likely exposure levels. Internationally agreed-upon trigger values are used by the European Commission to decide whether the risk is acceptable.

While some of the studies are laboratory based and relatively easy to control, some are much larger and based outside in prepared containers or in the field. The container studies include microcosm and mesocosm studies; other studies may make use of artificial streams.

Ultimately, one of the species that could be exposed to pesticides or other agrochemicals is man, and it would seem sensible to obtain some information on the ADME of these substances in human volunteers. There has been much debate about the ethics of human studies with agrochemicals, and at the time of writing, there is considerable resistance to this, even to the extent of not using data when they have been generated. This does not seem to be entirely sensible. However, the recent advent of microdosing studies used for pharmaceuticals, where very small doses of radiolabeled compound are given to volunteers, may be relevant to agrochemical development. The use of small doses is consistent with normal expected exposure to pesticides, and it seems likely that these studies, with their complex and expensive analytical techniques, will prove to be more easily justifiable for low doses of pesticides than for pharmaceuticals, which are usually given at much higher doses than those studied in such experiments.

There is some overlap between agrochemicals (or plant protection products) and biocides, which are covered in the next section.

BIOCIDES

The name *biocide* has greater resonance in Europe than in the United States, where the definition tends to be a little narrower. Suffice it to say that the European legislation, enshrined in the EU Biocides Regulation 528/2012 (2012), is the toughest in the world.

A biocide can be a single chemical, a mixture of more or less known composition, microorganisms, extracts, and oils of plants. Each of these categories has its own challenges in terms of safety evaluation; single substances are relatively straightforward, but mixtures are notoriously difficult to assess, the complexity increasing with the number of ingredients or components. In some cases, the active agent is produced by mixing two or more components at the point of use and so may differ from the original components; in these cases, the toxicological assessment has to cover the original unmixed chemicals and the (probably) more biologically active final product.

From the above, it is evident that biocides have a wide variety of uses, and these come with differing levels of human exposure. In addition, some are relatively benign, while others are very toxic. The data requirements are dependent on the product type and expected exposure levels; in some cases, where very low human exposure is expected, data requirements may be minimal.

The objective of the directive is to show the following for each biocide:

- It is effective.
- Target organisms are not subject to unacceptable effects, such as unnecessary suffering in vertebrates.
- There are no unacceptable effects in nontarget organisms and in the wider environment generally.

- Fate and distribution have to be shown, especially with regard to groundwater and any consequent effects.
- There will be no harmful effects on human or animal health.

This information is contained in the dossier for each product, which is required under the European Directive. The following is a broad summary of the contents of Annex II of the directive, which outlines data requirements for active substances in the form of a core data set and an additional data set. In addition to information on the identity of the active substance and its physicochemical properties and analytical methods, and its efficacy, the following toxicity data (which are very similar to those required for pesticides) and ecotoxicological data are required:

Toxicity studies:
1. Acute toxicity; oral plus one other appropriate route.
2. Skin and eye irritation and skin sensitization.
3. Metabolism in mammals; toxicokinetics including dermal absorption.
4. Twenty-eight-day or 90-day toxicity studies in a rodent and a nonrodent.
5. Chronic toxicity studies in a rodent and a nonrodent.
6. Mutagenicity studies in bacteria and cytogenicity and mutation in mammalian cells in culture; if these are positive, an *in vivo* micronucleus test is required; further *in vivo* tests may be indicated.
7. Carcinogenicity study in one rodent and another mammalian species.
8. Reproductive toxicity to explore embryo development (in rabbits and a rodent) and fertility over at least two generations.
9. All available human data.

Ecotoxicological studies:
1. Acute toxicity in fish and *Daphnia magna.*
2. Inhibition of algal growth and microbiological activity.
3. Bioconcentration and extensive tests of fate in the environment, including degradation.

All the work should be conducted according to GLP. There are also separate annexes giving the requirements for products, fungi and microorganisms, viruses, and other categories.

Regulation of this type of product in the United States is carried out by the EPA using the Federal Insecticide, Fungicide and Rodenticide Act (FIFRA) as the legal basis. A full database, often required for biocides, is known as a chronic, oncogenicity, reproductive, and teratogenicity (CORT) database. The EPA uses a tiered approach to data requirements for biocides based on the levels of exposure from their use. While high-exposure agents have broadly similar, detailed, data requirements to those for other pesticides, low-exposure agents require only the minimum data set out in tier 1. This lowest set of data still requires acute toxicity, a 90-day-toxicity study by the most common route of exposure, teratogenicity in one species, and a battery of mutagenicity tests.

In the United States, high-exposure antimicrobial pesticides have the same data requirements as other pesticides, whereas low-exposure antimicrobial pesticides require only tier 1 minimum data. The tier 1 database includes an acute toxicity battery, one 90-day study (usually by most common route of exposure), teratogenicity (one species), and a mutagenicity battery. This might be an appropriate data set for a preservative for non-food-contact material. For low-exposure use including food contact, the data set swells to include an acute toxicity battery (three routes including inhalation), eye and dermal irritation and skin sensitization, and a two-generation reproductive study.

COSMETICS

This is an incredibly diverse group of substances and products that range from the application of small amounts to skin (perfumes and eau de toilette) that will not be rinsed off (leave-on cosmetics, as they are known) to products that will be rinsed off such as soaps and toothpastes (rinse-off products). They can be applied to skin in areas all over the body, carrying risks in increasing proportion to their intended area of use; for instance, a nail polish would be expected to carry less risk than a soap, which would be less risky to use than an underarm deodorant; products for use on the face and especially those around the eyes carry the highest risks and require special care in their evaluation. This simplistic list excludes products such as toothpaste (rinse off) or products intended for "intimate" contact, for instance, with genital mucosae (often leave on). In addition, the potential for an ingredient to be left on skin that is exposed to strong sunlight, especially sunscreens or tanning agents, leaves open the possibility of photoallergic reactions that can be extensive and disfiguring.

The safety evaluation of cosmetics used to follow the usual toxicological paradigm of tests in animals with some backup studies *in vitro*, genotoxicity studies, and the like. While this may continue to be the case in some areas of the world, in Europe, the testing of cosmetic ingredients and products in animals is being phased out. It is expected that this will eventually become a de facto worldwide ban given that the European Union is a significant market for cosmetics, and there will be little incentive to develop products solely for the European Union or to develop products that cannot be sold in the European Union. This has led to the problem that large multinational cosmetic manufacturers are required to have one product formulation for Europe and another formulation for countries where testing is required.

No cosmetic product may be tested in animals in the European Union, and in fact, it is not permissible to sell such new products in the European Union. For ingredients, the ban on the use of animals was enforced in 2013. However, the results of tests of ingredients conducted before the ban are accepted for safety evaluation.

Given this legislative landscape, there has been some diffidence about bringing new cosmetic ingredients to the marketplace. The legislation is somewhat unclear on the status of an ingredient, for example, a recently developed pharmaceutical excipient, which has been extensively tested in animals after the institution of the ban but which is then developed as a cosmetic ingredient.

The evaluation of cosmetic products is relatively straightforward, although there are legislative differences among the various areas of the world. In the European Union, the formulation of a cosmetic product is assessed by looking at the ingredients and their inclusion levels against the various annexes of the Cosmetics Directive

Council Directive 76/768/EEC. One of the challenges in cosmetic safety assessment (and a reliable source of incredible frustration) is the difficulty of getting sufficient trustworthy information about the ingredients to be used. Are they cosmetic grade or the scrapings from some chemical barrel in a disreputable country? Does the Material Safety Data Sheet provided actually refer to the ingredients to be used, or is it a version pirated from a more ethical company? With the introduction of the requirement that cosmetic ingredients be produced according to GMP, it is likely that this problem will recede with time. In this respect, the use of fragrances is fraught with difficulty, and it is important to ensure that the proposed fragrance comes with a manufacturer's certificate of purity, listing the known allergens (of 26 listed by the European Union) it contains and with safe-use limits; a statement of compliance with the Code of Practice of the International Fragrance Association (IFRA) is also useful.

However, assessment of a cosmetic formulation is not as simple as reading tables in lists of ingredients and checking inclusion levels. The components have to be assessed for their potential to interact when applied, for instance, for the occurrence of nitrosation. Greater harmonization across the various jurisdictions is taking place gradually; the International Nomenclature for Cosmetic Ingredients (INCI) system of cosmetic names is gaining ground, leading to greater conformity of labeling worldwide. Differences remain, however; in the United States, water is called water but is aqua in the European Union; plant names are usually given in English in the United States but in Latin in the European Union. In the United States, sunscreens are treated as pharmaceuticals but as cosmetics in the European Union.

Toxicity testing of cosmetic ingredients, at least for the EU market, therefore relies on nonanimal methods. Cosmetic ingredients are overseen in the European Union by the Scientific Committee on Consumer Safety (SCCS) and in the USA by the FDA, via sources such as the Cosmetic Ingredient Review, published by the *International Journal of Toxicology*.

Vinardell (2015) reported that the number of studies submitted to the SCCS that do not involve animals was still low and, in general, the safety of cosmetic ingredients was based on *in vivo* studies performed before the prohibition. Vinardell evaluated the *in vitro* methods reported in the dossiers submitted to the SCCS, from the published reports issued by the scientific committee of the Directorate General of Health and Consumers (DG SANCO). The information required includes acute toxicity, dermal and ocular irritation and corrosivity, skin sensitization, percutaneous absorption, repeat-dose toxicity, mutagenicity and genotoxicity, carcinogenicity, reproductive toxicity, toxicokinetics, photoinduced toxicity, and human data; studies should be conducted according to GLP (SCCS 2012).

Vinardell indicated that it would take more than 5 years to completely replace the animal-based tests used previously. However, the date by when replacements would be available for more complex endpoints, particularly repeat-dose toxicity, carcinogenicity, and reproductive toxicity, could not be estimated. Having said that, as indicated in Chapter 4, considerable progress has been made, and validated methods are now enshrined in OECD test guidelines and in continuing work published by the European Centre for the Validation of Alternative Methods (ECVAM). These include dermal and ocular irritation and genotoxicity; sensitization is under intensive investigation and methods are emerging that are likely to have regulatory acceptance in due course.

It is clear that the evaluation of new ingredients for cosmetics is one of the most swiftly changing areas of toxicity testing. It is also acting as a driver for the adoption of alternative methods that will, inevitably, have application in other areas, especially chemical testing demanded by Registration, Evaluation, Authorisation and restriction of Chemicals (REACH). There are some areas that are apparently intractable, or which have been seen to be intractable. These include *in vitro* investigation of subacute toxicity and carcinogenicity; however, as mentioned in earlier chapters, there is a good reason to believe that methods can be developed to assess these endpoints without the use of tests in living animals. The development of long-term hepatocyte culture and treatment systems is accelerating, and it is likely that carcinogenicity testing in 2-year studies in rodents will become increasingly discredited. It is difficult, at this time, to see how some endpoints, such as those in reproductive toxicity, can be developed credibly without the use of rodents or other animals. However, other animal systems such as invertebrates should offer some scope for investigation and testing.

All this activity will also have a knock-on effect on risk assessment, as the data available will be very different from those currently used. There is a possibility that risk assessment for cosmetics will err toward the more conservative side, and this is unlikely to be helpful in the long term. No cosmetic can be considered to be completely safe. A much-abused term, *hypoallergenic* means simply that the product is less likely to produce skin sensitization than more normal formulations. Wherever a formulation contains a fragrance or a plant extract, it is likely that at least a small number of people will react adversely.

One area that is attracting a lot of attention at the moment is the field of nanotechnology. We have been exposed to nanoparticles in the form of endogenous particles such as chylomicrons, macromolecules, and the like, ever since we crawled out of the prehistoric seas. However, the advent of designed synthetic nanoparticles, which hold a lot of attraction in cosmetics such as sunscreens, has brought about a new field of toxicology, which will pose new challenges.

GENERAL AND INDUSTRIAL CHEMICALS—REACH

In most countries, there has been a long-standing requirement that new chemicals should be notified to the authorities in whichever country the chemical will be marketed. In general, the extent of testing has been decided on the basis of the volume of production; a chemical produced at 10 tonnes/year requires a less extensive testing program than one produced, for instance, at more than 100,000 tonnes/year. This has long been the case for new chemicals; existing chemicals had effectively been grandfathered and, to some extent, ignored, unless some toxicological or other chemical crisis disturbed the even tenor of commercial existence. However, the European initiative, REACH, has attempted to redress this anomaly by setting out to register and, if necessary, authorize or restrict all chemicals produced at more than 1 tonne/year.

In Europe, existing chemicals were considered to be those put on the market before 1981; there were 100,106 of these. New chemicals (more than 4300) were brought to market after that date and were covered by the legislation relating to notification of new chemicals. Existing chemicals remained largely untested, and the resulting lack of knowledge of their properties was considered to slow the process of risk assessment and to make it more cumbersome. In contrast to the situation in the United States (where the National Toxicology Program has long had a testing program for chemicals coming to its notice), in Europe, there was no formal and centralized testing initiative. The promulgation of REACH in 2007 effectively (perhaps) removed the inertia that existed in Europe with respect to existing chemicals and was also intended to make registration of new chemicals easier. Particular attention was paid in the early stages of the process to high-volume chemicals and to those of particular concern—those that are carcinogenic, mutagenic, or reproductive toxicants (CMRs); those that are persistent, bioaccumulative, and toxic (PBT); and any that are very persistent and very bioaccumulative (vPvB).

The onus is on the manufacturers to provide or generate the data that will allow risk assessments of each chemical to be carried out; information will be sent to the European Chemicals Agency in Helsinki, Finland. Approximately 30,000 chemicals were affected by the legislation.

REACH stands for Registration, Evaluation, Authorisation and restriction of Chemicals. Producers and importers of chemicals in the European Union in quantities of more than 1 tonne/year are required to produce information that includes a technical dossier that assesses the risks due to use and managing them. For chemicals produced at 10 tonnes/year or more, and for chemicals of concern (CMRs, PBTs, and vPvBs), a chemical safety report is also needed. Evaluation of the submitted documentation is aimed at assessing its completeness as a dossier and on the need for further testing. Where a risk to health or the environment is suspected, action may be taken under authorization or restriction procedures. For instance, substances of very high concern (CMRs, PBTs, or vPvBs) may require authorization. Regulation of substances will extend to the pure chemicals and to their use in products. Restrictions on use may be put in place, and it is possible that a substance may be banned completely and/or that substitution will be required. The caveat on substitution should be considered here again; substitution of a known set of risks for a supposedly safe but novel and unknown chemical does not necessarily increase consumer safety.

While one of the stated goals of the REACH legislation was the reduction of animal tests, this has not necessarily been the result, due to requirements to conduct new tests for some substances. However, there has been some evolution in test design, with combination repeat-dose and reproductive toxicity studies (OECD test guidelines 421 and 422) and an extended one-generation reproductive toxicity study (OECD test guideline 443). The latter study will now be the information requirement for reproductive toxicity in REACH instead of the two-generation reproductive toxicity study (OECD test guideline 416).

Successful negotiation of REACH is based on preparation of a dossier, which should contain the information laid out in Focus Box 19.1.

While Focus Box 19.2 lays out the data requirements for each tonnage level, this may be better seen as a listing of the points that need to be covered in any dossier. Where information is not available, it is likely that additional testing will be

FOCUS BOX 19.1 REACH—DOSSIER CONTENT

The dossier that is sent to the European Chemicals Agency (ECHA) under REACH should contain the following information:

- Physical chemistry
- Toxicology
- Ecotoxicology
- Declaration of
  - PBT
  - CMR
  - vPvB
- Chemical safety assessment
- Chemical safety report—if tonnage is above 10 tonnes or safety is questionable

The chemical safety assessment includes the following:

- Human health hazard assessment
- Human health hazard assessment of physicochemical properties
- Environmental hazard assessment
- PBT and vPvB assessment
- If the manufacturer or importer concludes that the substance should be classified as dangerous or is PBT or vPvB, the chemical safety assessment shall include the following additional steps:
  - Exposure assessment
  - Risk characterization

The chemical safety report:

- Documents the chemical safety assessment (for all substances for registration if manufactured or imported at more than 10 tonnes/year).
- The main element of the "exposure" part describes exposure scenario(s) and the exposure scenario(s) recommended by the manufacturer or importer for the identified use(s).
- The exposure scenarios describe the risk management measures that the manufacturer or importer has implemented and recommends to be implemented by downstream users.
- These exposure scenarios including the risk management measures shall be summarized in an annex to the safety data sheet.

Source: Regulation (EC) No. 1907/2006 of the European Parliament and of the Council of 18 December 2006 concerning the Registration, Evaluation, Authorisation and restriction of Chemicals (REACH), establishing a European Chemicals Agency.

required. However, it is axiomatic of REACH that new tests in animals should be avoided as far as possible.

While each of the points would need to be addressed, part of the skill in dossier preparation will be to evaluate existing data and to say what is needed in addition in the form of new investigations. A typical data set may include documentation of years of production but probably little epidemiological or occupational information, unless it has been a problem chemical in the past. Many of the published documents are likely to be academic studies, which do not quite match the data requirements; for instance, there may be a limited genotoxicity assessment in the Ames test, which may also involve 70 other chemicals with little direct reference to the chemical of interest. Toxicity studies are likely to be old or very old and pre-GLP; furthermore, they are likely to have been published in obscure journals and to be available only in barely legible typefaces that have been photocopied 20 times.

However, if the chemical in question is a member of a group of other compounds that have been extensively investigated, it is possible to read across from one to the others. This also applies if the chemical is one of a series, for instance, alkylamides; this group includes acetamide, methylacetamide, and dimethylacetamide, and so on. The number of permutations possible is clearly quite large, and care should be exercised in choosing which chemicals to use as comparators. Clearly, if this initial choice is limited (or simply wrong), the whole assessment will be flawed. Two poor sets of data are unlikely to produce a good risk assessment. Such data weaknesses have to be taken into account in the final report. It follows that read-across needs to be supported fully by other information, including physicochemical data and, where possible, quantitative structure–activity relationship (QSAR) or structure–activity relationship (SAR) predictions of effect.

The quality of the available data has been a huge issue with REACH, when thousands of elderly chemicals have been assessed formally for the first time, despite decades of use. The principles of Klimisch et al. (1997) have been reviewed in some detail in Chapter 15 and will not be revisited in detail here. Suffice it to say, however, that reference to three of four papers conducted in the 1960s on issues of peripheral interest to the modern toxicologist is unlikely to be sufficient for an adequate assessment of risk. This applies just as much to the chemicals selected, as relevant for read-across, as to the subject of the assessment.

The process of assessment for REACH includes the compilation of a database of information that can be assessed for quality and gaps. The intention is to be able to propose a testing program to the European Chemicals Agency that will fill these gaps in the most expeditious manner, avoiding the use of animals wherever possible. The use of validated alternative techniques and of software to predict toxicity (QSAR, SAR) and other attributes such as metabolism will become vital parts of this process.

One of the cornerstones of REACH is the process of authorization, restriction, and substitution for hazardous chemicals such as CMRs. Persistent and bioaccumulative substances will only be authorized if no suitable substitute is available, and if it can be shown that the socioeconomic benefits from the particular use of the substance outweigh the risks to human health and the environment. However, as shown for methylene chloride in Chapter 15, substitution of a substance perceived as hazardous

FOCUS BOX 19.2   DATA OR EVALUATIONS
REQUIRED FOR REACH

These increase with increasing annual tonnage. The following is a sample of what is required and is not, necessarily, a full listing. The intention is to avoid new testing, especially in animals, and this should be seen as a list of points for evaluation. Each of the following stages is additive.

More than 1 tonne:
- Physicochemical data
- *In vitro* irritation and corrosion, skin sensitization [human data and local lymph node assay (LLNA)], *in vitro* mutation in bacteria, and cytogenicity
- Aquatic toxicity: *Daphnia* and algal growth inhibition
- Degradation: ready biodegradability is important

More than 10 tonnes:
- Physicochemical data, as above, plus light stability for polymers and leachates
- Toxicity, as above, plus *in vivo* irritation
- Gene mutation *in vitro* in mammalian cells and *in vivo* (if previous tests are positive)
- Acute toxicity by two routes (oral and dermal or inhalation)
- Toxicity to 28 days
- Developmental toxicity in two species, unless screening study is negative
- Toxicokinetic (TK) assessment derived from relevant available information
- Aquatic toxicity, as above, plus short-term toxicity in fish, activated sludge respiration inhibition, degradation, hydrolysis—as function of pH
- Environmental fate; absorption, desorption screening

More than 100 tonnes:
- Physicochemical data, as above, plus stability in organic solvents and relevant degradation products, dissociation constant, viscosity, and reactivity to container material
- Toxicity, as above, plus toxicity studies to 90 days, developmental toxicity
- Developmental toxicity in two species
- Aquatic toxicity, as above, plus long-term *Daphnia* toxicity and fish
- More extensive degradation studies including identification of products
- Environmental fate—accumulation, preferably in fish, plus more on adsorption/desorption

- Earthworm toxicity
- Soil microorganisms
- Short-term toxicity to plants
- Methods of detection and analysis

More than 1000 tonnes

- As above
- Confirmation of rates of biodegradation rates
- Additional environmental fate and behavior
- Long-term earthworm toxicity
- Long-term toxicity testing on other soil invertebrates
- Long-term toxicity to plants
- Long-term toxicity to sediment organisms
- Long-term or reproductive toxicity in birds

*Source: Regulation (EC) No 1907/2006 of the European
Parliament and of the Council of 18 December 2006 concerning
the Registration, Evaluation, Authorisation and restriction of
Chemicals (REACH), establishing a European Chemicals Agency.*

is itself hazardous if the substitute itself is not fully understood. For some hazardous chemicals, there may be circumstances where human or environmental exposure is very limited and risks can be controlled or managed. Authorization may be granted to CMRs, if it can be shown that there is a safe threshold below which there are no "negative" effects on humans or the environment; in other cases, the benefits have to outweigh the risks before authorization can be granted. It is clear that the process of authorization and restriction will have to be pragmatic; it remains to be seen how much pragmatism can be allowed in a society that is essentially risk averse.

Eventually, REACH will become integrated with the Globally Harmonized System (GHS) for the classification and labeling of hazardous chemicals. This seeks to classify chemicals by types of hazard and proposes harmonized communication of hazard including labels and safety data sheets.

TOBACCO PRODUCTS

There is an extensive and long history of toxicological evaluation of tobacco products, starting in the second half of the twentieth century. Early studies of the effects of tobacco smoke were poorly designed and reported and were limited by the disparity in toxicity of nicotine compared with the tars and associated chemicals that ultimately were considered to be responsible for the carcinogenicity in human lungs. The upshot was that there was no proof from animal studies that tobacco smoked caused lung cancer, especially the type of cancer found in human smokers' lungs.

There is no doubt that tobacco, especially in the form of cigarettes, is associated with a range of human illnesses, including lung cancer, and that these effects are largely attributable to the presence of tars and associated combustion products.

Nicotine itself is relatively "clean." The advent of the e-cigarette has opened up the debate on smoking (vaping) and its potential health effects. Typically, an e-cigarette contains simple solvents such as propylene glycol and glycerin, together with nicotine in carefully regulated concentrations and quantities. Intuitively, the absence of tars from the products should mean that the toxicity expressed should be limited to

CASE STUDY 19.2   TESTING AND ASSESSMENT OF TOBACCO PRODUCTS

Historically, toxicity tests of tobacco smoke, usually from cigarettes, were relatively crude and struggled with the disparity between the high acute toxicity of nicotine and the low chronic toxicity of the carcinogenic tars. Studies in rodents and dogs were poorly designed and executed and poorly reported, without any supportable evidence for carcinogenic effects. As study design improved and standards of good laboratory practice were introduced, there was increasing evidence of carcinogenic potential (Mauderly et al. 2004; Hutt et al. 2005); the results were not clear cut, and the most that could be said was that tobacco smoke was not a very potent carcinogen. Given the known effects of tobacco smoke, there are ethical pressures against the use of animals and, equally, incentives to produce *in vitro* methods that may replace them. However, some testing in animals may still be required, and it behooves toxicologists to ensure that they are used as efficiently as possible.

Dalrymple et al. (2016) explored the utility of reducing the exposure time to tobacco smoke from the usual 90 days to 3 or 6 weeks, and investigated genotoxicity and respiratory tract pathology induced by cigarette smoke; the effects of reducing the numbers of animals was also investigated. Separate lung lobes from exposed rats were used for histopathology and the comet assay, using alveolar type II cells. In addition, blood was collected for the Pig-a assay and quantitation of micronuclei. The results were as follows:

- Histopathology showed dose-dependent effects of cigarette smoke (1 or 2 hours for 5 days/week).
- The comet assay showed that DNA damage in the type II alveolar cells increased with exposure, with more damage evident at 6 weeks than at 3 weeks.
- However, there was no increase in Pig-a mutation levels or micronucleus counts.
- Differences between the two exposure groups—3 or 6 weeks— showed that the effects were cumulative.
- The conclusion was that the reduced exposure time of 3 weeks was enough to induce pathology in the respiratory tract and DNA damage in isolated type II alveolar cells. The additional conclusion was that this showed that reducing and refining animal use was possible using this approach.

the effects of nicotine. Part of the controversy has arisen from the use of flavors that may not have been properly assessed for this route of exposure and that may not have been produced to high standards associated with a regulated product. The lack of regulation has been key in the evolution of these products; however, the recent EU Tobacco Products Directive (European Commission 2014) has set out to regulate all tobacco products including e-cigarettes. The assessment of e-cigarettes is mostly by literature review as the main ingredients are well known, usually as pharmaceutical excipients. Actual testing in animals therefore is not likely to be required. Having said that, the testing of tobacco products is becoming more sophisticated, as indicated in Case Study 19.2.

SUMMARY

It is difficult to distil a summary out of such a diverse set of information, given the varying circumstances of exposure of each chemical class considered above. However, when the whole area is considered from a distance, as it were, some points emerge that are relevant to all:

- The basic driver is the identification of the toxicological hazard(s) associated with the chemical or product concerned.
- Are there any notable hazards associated with its breakdown products?
- The circumstances of use of the chemical and its purpose—human pharmaceutical, agrochemical, or cosmetic ingredient—should be considered.
- Likely human exposures should be considered; will this be direct or indirect; by which route; high or low level; short or long duration; intentional or accidental?
- Likely environmental exposure should be assessed along the same lines.
- The benefits of using the chemical/product must be assessed against the cost and risk of doing so, always remembering that nonuse may also have undesirable consequences, for instance, continuation of an insect infestation in a grain store.
- This risk assessment should indicate the acceptability or otherwise of the expected exposure to patients, users, consumers, the environment, and so forth.
- The risk assessment may be specific to a relatively small population or may give rise to several risk assessments relevant to different levels of use and exposure.

To this crude progression should be added the regulatory processes and requirements relevant to the chemical or product, and the ethical constraints that may exist in terms of conduct of animal testing. As indicated, the palette of tests available are broadly the same for every chemical class, but the ones conducted in the evaluation of any particular test material are driven by regulatory and legal guidance that not only indicates the test type, but also usually dictates minimum design and quality standards (meaning, effectively, that all new tests have to be GLP compliant).

REFERENCES

British Standards Institute. ISO 10993-1. Biological evaluation of medical devices—Part 1: Evaluation and testing. International Standards Organisation, Paris/European Committee for Standardisation, Brussels, 2009.

Chapman K, Adjei A, Baldrick P, da Silva A, De Smet K, DiCicco R, Hong SS et al. Waiving in vivo studies for monoclonal antibody biosimilar development: National and global challenges. *MAbs* 2016; 8(3): 427–435.

Dalrymple A, Ordoñez P, Thorne D, Walker D, Camacho OM, Büttner A, Dillon D, Meredith C. Cigarette smoke induced genotoxicity and respiratory tract pathology: Evidence to support reduced exposure time and animal numbers in tobacco product testing. *Inhal Toxicol* 2016; 28(7): 324–338.

EMA. Guideline on the Environmental Risk Assessment of Medicinal Products for Human Use. Doc. Ref. EMEA/CHMP/SWP/4447/00, 2006.

EMA European Medicines Agency Committee for Medicinal Products for Human Use (CHMP): Guideline on Strategies to Identify and Mitigate Risks for First-In-Human Clinical Trials with Investigational Medicinal Products. EMEA/CHMP/SWP/28367/07, 2007.

European Chemicals Agency. (ECHA). Read-Across Assessment Framework (RAAF) (ECHA-15-R-07-EN), 2015. Available at http://echa.europa.eu/.

European Commission. Regulation (EU) No 528/2012 of the European Parliament and of the Council of 22 May 2012, concerning the making available on the market and use of biocidal products. EU Biocides Regulation 528/2012.

European Commission. Directive 2014/40/EU of the European Parliament and of the Council of 3 April 2014 on the approximation of the laws, regulations and administrative provisions of the Member States concerning the manufacture, presentation and sale of tobacco and related products and repealing Directive 2001/37/EC. Official Journal of the European Union April 29, 2014.

FDA Center for Drug Evaluation and Research (CDER). Guidance for Industry: Estimating the Maximum Safe Starting Dose in Initial Clinical Trials for Therapeutics in Adult Healthy Volunteers, 2005.

Hutt JA, Vuillemenot BR, Barr EB, Grimes MJ, Hahn FF, Hobbs CH, March TH et al. Life-span inhalation exposure to mainstream cigarette smoke induces lung cancer in B6C3F1 mice through genetic and epigenetic pathways. *Carcinogenesis* 2005; 26(11): 1999–2009.

ICH S6 Guideline: Preclinical Safety Evaluation of Biotechnology-Derived Pharmaceuticals; July 1997 and Addendum: Preclinical Safety Evaluation of Biotechnology-Derived Pharmaceuticals, 2011.

Keogh B. NHS Poly implant Prothese (Pip) Breast Implants: Final report of the Expert Group. Department of Health, NHS Medical Directorate 2012.

Klimisch H, Andreae M, Tillmann U. A systematic approach for evaluating the quality of experimental toxicological and ecotoxicological data. *Regul Toxicol Pharmacol* 1997; 25: 1–5.

Mauderly JL, Gighotti AP, Barr EB, Bechtold WE, Belinsky SA, Hahn FF, Hobbs CA, March TH, Seilkop SK, Finch GL. Chronic inhalation exposure to mainstream cigarette smoke increases lung and nasal tumor incidence in rats. *Toxicol Sci* 2004; 81: 280–292.

Muller PY, Milton M, Lloyd P, Sims J, Brennan FR. The minimum anticipated biological effect level (MABEL) for selection of first human dose in clinical trials with monoclonal antibodies. *Curr Opin Biotechnol* 2009; 20(6): 722–729.

OECD. Guidelines. Available at http://www.oecd.org/.

Pesticides Safety Directorate. Data Requirements Handbook, 2010. Available at http://www.hse.gov.uk/pesticides/.

Scientific Committee on Consumer Safety (SCCS). The SCCS's Notes of Guidance for the Testing of Cosmetic Substances and their Safety Evaluation (8th Revision), 2012.

Thybaud V, Kasper P, Sobol Z, Elhajouji A, Fellows M, Guerard M, Lynch AM, Sutter A, Tanir JY. Genotoxicity assessment of peptide/protein-related biotherapeutics: Points to consider before testing. *Mutagenesis* 2016; 21. pii: gew013.

VICH GL6 (Ecotoxicity Phase I). Environmental Impact Assessments (EIA's) for Veterinary Medicinal Products (VMP's)—Phase I, 2000.

VICH GL38. Environmental Impact Assessments (EIAs) for Veterinary Medicinal Products—Phase II, 2006.

Vinardell, MP. The use of non-animal alternatives in the safety evaluations of cosmetics ingredients by the Scientific Committee on Consumer Safety (SCCS). *Regul Toxicol Pharmacol* 2015; 71(2): 198–204.

20 Errors in Toxicology

INTRODUCTION

Mistake, misunderstanding, misuse, and mismanagement have been pivotal in the evolution of toxicology as a discipline, with each of them contributing to regulation, testing, and understanding, sometimes with cataclysmic effect. Mistakes have included the authorization of thalidomide and registration of drugs, which subsequently have to be withdrawn due to unexpected effects in small percentages of the patient population. Misuse encompasses the flawed use of rodents in long-term carcinogenicity studies and the inappropriate use of statistical methods to further flawed agendas. Mismanagement includes the inappropriate use of the data from toxicity studies and risk assessment based on overconservative assumptions. Misunderstanding is the fact that understanding is transient, that as knowledge emerges, understanding evolves to bring new interpretations to existing knowledge and changes in regulation and risk management.

Much of toxicity testing and subsequent risk assessment is aimed at safety evaluation so that substances can be safely used in various target species or situations, for example, human consumers, the workplace, or the environment. Errors are likely to occur in this process because there is a mismatch between test systems and the target, and the understanding of the differences between them. Using the terms *mistakes*, *misuse*, *mismanagement*, and *misunderstanding* in this chapter will explore various errors and should encourage the evaluation of mismatches between test results and final outcome in the target species for which the safety evaluation was conducted. An error cannot always be defined as a single *M*, and often, a combination of multiple errors compound and combine to produce a much more serious problem. Consideration of the past may lead to better test systems, study design, conduct, and interpretation, and as a result, more rational risk assessment.

The history of errors in toxicology is, in many ways, the history of toxicology. Changes in legislation and practice are often led not by toxicologists themselves but by the effects of their mistakes; the results of misuse of substances by users; and mismanagement by manufacturers, academia, and government. Furthermore, public misunderstanding of consumer exposure has led the nontoxicologist population, via the popular (and sometimes scientific) media, to create perceived risks where risks sometimes do not exist. Another layer of misunderstanding is added by toxicologists themselves, who forget that understanding evolves and that if a situation is black and white, it means simply that the gray nuances between have not been discovered or understood. Equally, the opposite situations exist, where there are known hazards but they are ignored or their risks downplayed. A pleasurable risk, such as alcohol consumption or smoking, is more alluring than one perceived to be toxic, such as pesticides and the consequent irrational preference for organic wine or organic tobacco.

Perhaps the most important thing to note is that toxicological errors cannot always be prevented; even the most foolproof concept requires only a more ingenious fool to confound it. However, if we determine how the errors came (and come) about and learn from them, it is possible to reduce the likelihood of them occurring again. Of course, one problem is that an error of one type tends to blind people to the next that will arise.

ERRORS THROUGH THE AGES

There is a long and well-reported history of error in the appreciation of the toxicity of various substances and their misuse, which is also a product of lack of understanding of their effects. Qin Shihuang, the man who unified China and stood as its first Emperor, sought immortality. Logically, to do this, he required an elixir of life and hence settled (one assumes on the advice of his physicians) upon the health-giving properties of mercury. This rather unfortunate misunderstanding had exactly the reverse effect and led to his demise (Wright 2001).

While rulers and prominent people may have been motivated by the search for immortality, more mundane errors seem to be through the wide public. For example, Mithridatis attempted to avoid death by poison; he composed a universal antidote to poisons, which allegedly prevented his own death when he took poison in a suicide attempt to evade capture by the Romans. Prominent in this history of toxicological mistakes (confabulated with misunderstanding) is the misuse of various compounds of heavy metals in cosmetic products, starting with the Egyptian use of malachite (a green copper ore); lead sulfide; and a paste of soot, fat, and metals such as manganese, copper, or antimony. This area of misunderstanding may be offset by the evidence in the Ebers Papyrus that the Egyptians clearly had some medical knowledge, although the concept of toxicology was not yet conceived.

In a rather larger-scale example of misunderstanding, fashionable Elizabethan ladies (including Queen Elizabeth herself) applied a mixture known as "ceruse" to their faces in an effort to give the appearance of white facial skin and hide signs of ageing. Ceruse, a mixture consisting of vinegar and white lead, did have the desired effect if applied with regularity; however, the toxic effects of white lead meant that the "treatment" could not be stopped. The Victorians, and predecessors, continued the tradition of unwitting self-harm with the use of arsenic and white lead in skin products and mercury in the "cure" of syphilis. One of the more famous and ultimately erroneous examples of poisoning in the nineteenth century was the idea that Napoleon Bonaparte was poisoned by his green-colored arsenic wallpaper. However, this has since been discounted; he had 10.38 ppm arsenic in his hair, which was lower than in George III (17 ppm). In actuality, he died of stomach cancer, in common with his father and two sisters (Roberts 2014).

As the industrial revolution reached its zenith in the United Kingdom and United States of America, exploitation of workers and their exposure to noxious substances became a large public health issue. Examples of industrial mismanagement range from breaker boys, who separated impurities from coal by hand, and chimney sweeps where young children were employed, to radium girls and workers with phossy jaw. Such mismanagement could even be extended to workers in asbestos factories and its wider use in public buildings with the slow-burning consequences of respiratory disease and mesothelioma in subsequent decades.

The evolution of such errors in a particular example is illustrated by methylene chloride, which emerged as a replacement for lye (a caustic alkaline solution) as a paint stripper. The perception was that lye was unsafe (caustic) and that methylene chloride was characterized as very safe. Understanding of the toxicity of methylene chloride diluted this assumption of safety and eventually ran full circle to the proposition that lye is safe. This is covered in more detail in Chapter 18, Case Study 18.1; however, the take-home message from this history is that it is not sensible to substitute a supposedly toxic material with another for which the illusion of safety is based on ignorance of effect or lack of understanding. The substitution of benzene by toluene is another example of this type of substitution; benzene is certainly more toxic than toluene, but the latter has its own effects and problems.

A BRIEF HISTORY OF EARLY REGULATIONS

The earliest regulation that might be construed as relating to public health (and tenuously to toxicology) in the United Kingdom dates to 1266 and relates to punishment to bakers and brewers for overcharging or selling poor-quality products. The punishments were pillory for bakers and a journey in a tumbrel for brewers.

However, perhaps the first true attempt at regulation, in England at least, was in 1421, when physicians petitioned parliament for state control to prevent "the great harm and slaughter of many men" by "lewd" men who had not graduated from a school of "Physic" within a university. The physicians recommended that the punishment for such practice should be "pain of long imprisonment" and a fine of £40 to be paid to the king. Parliament responded by acknowledging that they should do something about it by limiting the practice to the appropriately qualified persons but did not indicate how this was to be regulated. It was not until the reign of Henry VIII, with the introduction of the Physicians and Surgeons Act (1511), that any actual regulations were put in place. Under this act, the practice of medicine was limited to those who had been examined. Such examination was regulated in London, with the assistance of qualified medical practitioners, by the Dean of St. Paul's Cathedral or the Bishop of London. Outside of London, regulation fell under the auspices of the bishops of each diocese. In 1518, regulation of the practice of medicine within the city of London was transferred to the newly founded Royal College of Physicians of London (Raach 1944; Warren 2000).

In the United States of America, the first federal regulation for medicines was the 1813 Vaccine Act (US Congress 1813). This act allowed for the appointment of "an agent to preserve the genuine vaccine matter and to furnish the same to any citizen of the United States…" The furnishing was to be via "the medium of the post office." The act was repealed in 1822 (with vaccination authority reverting to local officials) when the federal agent charged with distributing the vaccines accidentally mailed live smallpox to a physician in North Carolina, resulting in approximately 10 deaths (US Congress 1822; O'Malley 2001; Colgrove 2007).

In recent times, regulations have been driven by political and public perceptions of hazard and risk. The occurrence of public health incidents, whether it be the poor conditions of the American meat-packing industry or incidents such as elixir sulfanilamide, thalidomide, or the Santa Barbara oil spill, have spurred the creation of legalization to prevent the occurrence of similar events. Regulations, and by mild implication, regulators, are thus highly conservative and slow to change. The acceptance process for new methods (such as the local lymph node assay (LLNA) and replacements for 2-year carcinogenicity bioassays) is, rightly, drawn out as regulators must ensure that the new test is as least as good as the old—or better. This facet of testing is also complicated by varying requirements for studies and differing acceptance of certain assay types. That being said, regulators are, first and foremost, scientists and are thus open to argument and persuasion, provided that there is valid scientific reasoning behind decisions.

One piece of legislation that may be argued to have arisen proactively (as opposed to reactively) is the Registration, Evaluation, Authorisation and restriction of Chemicals (REACH) regulations enacted in Europe in 2007. This arose to increase the level of protection to humans and the environment from chemicals and was not the result of a significant health incident. That being said however, given the size and scope of the requirements—and associated expenses—some of the companies implementing the legislation might prefer to return to pillory and tumbrel.

MODERN MISTAKES AND MISMANAGEMENT

The basis of any scientist's work should be factual and unbiased. It is a sad feature of life that extremity of views often leads to distortion of perception and to biased misunderstanding and misinterpretation, which then leads to spurious extrapolation of exiguous data to erroneous conclusions about human safety. It is not unheard of for a controversial paper to spawn a flock of counterstudies disputing and/or refuting the data, methodologies, and conclusions of the original paper. In severe circumstances, such as those surrounding false linking of measles–mumps–rubella (MMR) vaccine to autism, a paper is not only redacted, but its lead author, struck off by professional bodies. Selective reporting of nonclinical data in an Edinburgh laboratory led to the first prosecution in the United Kingdom under the Good Laboratory Practice (GLP) regulations and was only brought to light when the laboratory itself reported its suspicions to the UK Medicines and Healthcare Products Regulatory Agency (MHRA).

It is notable that Séralini et al. (2012) (see Case Study 20.1) declared no conflict of interest in their paper on the purported effects of genetically modified (GM) corn without declaring that the main author was the founder of a virulently anti–genetically modified organism (GMO) organization [Committee of Research and Independent Information on Genetic Engineering (CRIIGEN)].

The methods employed by Séralini et al. are arguably examples of misuse of a range of disciplines: of valuable animal resources to pursue an agenda based on determined misunderstanding, of animal welfare standards, and of the use of inappropriate study design and the consequent hunt for statistical methods that would produce the desired statistically significant differences. The last point, the almost religious dependence on statistical significance in the absence of the more important biological or toxicological significance, is an example of misuse that is widely prevalent in toxicology and may result in overconservative risk assessments.

CASE STUDY 20.1 GENETIC MODIFICATION:
POTATOES AND GM CORN

Despite a long history of genetic manipulation by breeding for particular characteristics, the modern equivalent of genetic modification by insertion of foreign DNA has attracted extensive opprobrium arising from fear and misunderstanding of the wider context. It has to be said that some of the green lobby concerns have not been adequately addressed by the pro-GMO lobby, which has been tainted in public perception by the involvement of large companies such as Monsanto and Syngenta.

Potatoes have been a focus for toxicological interest for many years. They are from the same family of plants as deadly nightshade and contain related alkaloids such as solanine, which is found particularly in the green skin of potatoes that have been exposed to light. A review by Christie (1999) suggested that many cases of schizophrenia might be associated with consumption of potatoes. A selection of the alkaloids present in potatoes was assessed in the *Xenopus* assay *in vitro* for teratogenesis (Friedman et al. 1991), and their potential for teratogenicity was confirmed. The general toxicity of solanaceous alkaloids is well known. In other words, potatoes are a typical dietary constituent in that administration to animals of certain natural components at high-dose levels could be expected to be associated with undesirable effects. The addition, by genetic modification, of a new chemical entity to this existing cocktail could be expected to be of toxicological interest, depending on the expression levels and final content in a normal diet. This was investigated by Ewen and Pusztai (1999), but unfortunately, the study was poorly designed and rashly interpreted.

A more egregious example is provided by a paper on the purported effects of GM corn (maize), by Séralini et al. (2012). This paper added to the research portfolio of this group of authors and was consistent with other papers from this source, which had been the subject of skeptical criticism. In summary, they conducted a 2-year feeding study in groups of 10 male and 10 female Sprague Dawley rats, which received diets containing up to 33% of a Roundup-tolerant GM maize or drinking water containing Roundup (glyphosate) from 0.1 ppb to 2.25 g/L. GM maize was grown specifically for use in the study and was either treated or not treated with glyphosate.

The authors reported higher mortality in treated groups and development of large mammary tumors sooner than in controls, and identified the pituitary as a target organ due to hormonal imbalance. The liver and kidneys also showed differences from controls, which were attributed to treatment. Contrary to expectation, there was no dose relationship, and as a result of the small group size and mortality, differences from controls were by no means definitive. However, the differences were attributed to "non-linear endocrine-disrupting effects" of glyphosate, to overexpression of the transgene in the maize, and to the metabolic consequences. Although it was denied, in response to criticism, that this was a carcinogenicity study (for which groups of 50 males and 50 females are standard), the group size was still too small to provide any statistical confidence in evaluation of differences. In addition, it was evident that the statistics employed were complex and nonstandard, and involved considerable manipulation and transposition of the data. There was no suggestion that these methods had been planned before the study started, and the conclusion may be drawn that they were adopted as part of a statistical fishing initiative to create statistical significance where no biological or toxicological significance existed. There were serious reporting deficiencies, in that data for body weight and food and water consumption were not reported; administration in drinking water is a notoriously inaccurate method of treatment due to uncertainties in actual intake, and lack of reported record of consumption is a major deficiency in this study. There were also clear animal welfare issues in that animals with excessively large mammary tumors were retained on study for longer than normal husbandry and welfare practices would have tolerated. The pathology reporting and interpretation were also suspect.

The conclusions of an evaluation by the European Food Standards Agency (EFSA 2012) were that the study was "inadequately designed, analysed and reported" and "does not allow giving weight to their results and conclusions as published." The reported design, analysis, and results did not support the conclusions that were drawn. EFSA found that the study was not of sufficient quality for safety assessment.

The term *misuse* may equally be extended to the historic use of large groups of rodents to assess potential carcinogenicity and then to attempt to extrapolate the results from high doses in rats and mice to low doses in humans. The likely futility of this type of assay is neatly illustrated by the safety assessment of pharmaceuticals, and the fact that many marketed pharmaceuticals have been shown to be carcinogenic in rodents but by mechanisms that are irrelevant to humans.

Mismanagement of the data that have resulted from these tests is illustrated by the Delaney amendment in the USA, in which it was stated that a substance could not be used as a food additive if it was found to cause cancer in animals. Many substances may be shown to be carcinogenic in animals, often at extreme doses or through extreme treatment regimens; it is now generally understood that this does not necessarily translate to carcinogenic effect in humans. Quite apart from the more recent understanding that carcinogenicity is a function of toxicity, the effects found in such tests may well be due to mechanisms—adverse outcome pathways, to use the new terminology—that are found only in animals and are irrelevant to humans. Human relevance is defined by mechanism, and dose or systemic exposure; a mechanism responsible for cancer in animals may exist in humans but only at much less activity than in animals or at doses that are far higher than those actually encountered in real life. The dubious utility of these tests is only now being questioned and may result in the eventual discontinuation of their use.

CLINICAL TRIALS WITH NEW MEDICINAL PRODUCTS

Clinical trials are the cornerstone of pharmaceutical development, with the intention of demonstrating the safety and efficacy of new medicines. Each successive stage of development is supported by prior trial in nonclinical test systems—*in vitro, in silico*, and *in vivo*—with the intention of revealing any toxicity that may be relevant to human volunteers and/or patients. These are especially important in the phase I studies (first in man), where dose choice is critical; however, although single doses may be given, successfully repeated dosing may be associated with severe adverse effects that have not been predicted by the nonclinical work. This lack of predictive success may be due to factors such as misinterpretation of the data or to species differences. Adverse effects may manifest themselves even after extensive nonclinical and clinical study, as shown by the case of Vioxx (rofecoxib), a nonsteroidal anti-inflammatory drug that was used to treat conditions such as osteoarthritis and acute pain. Despite widespread acceptance, rofecoxib was withdrawn from the market due to increased risk of heart attack and stroke, associated with long-term use at high doses, effects that were not predicted by the nonclinical or clinical studies. One factor to bear in mind here is that the number of animals and patients studied before marketing a new medicine is authorized is far lower than the number of patients to whom the drug is actually given following open sale, meaning that the ability to detect rare events is limited.

Two cases where early studies failed to predict adverse effects are reviewed in Focus Box 20.1; that of TGN1412, in which single administration to human volunteers was associated with life-threatening effects, and of BIA-102474-101, in which single doses were shown to be safe but repeated administration was associated with severe effects and death.

These two cases underline the difficulties of risk assessment in the absence of full information. For TGN1412, the lack of understanding of the differences in homology of the receptor led to the cytokine storm in the trial volunteers. In the second case, the effects seen were unexpected because the scenario that emerged was outside previous experience; this suggests that it is only possible to predict effect from prior knowledge and that the unexpected may always be inaccessible to rational assessment.

CONTAMINATION

Most dictionaries define contamination along the lines of making something impure, unclean, soiled, corrupt, etc. Sometimes, contaminations are accidental, occurring as a result of a mistake or a misunderstanding, such as oil spills, contamination of rivers, or the unintended introduction of toxic substances into food or drugs. However, sometimes, mismanagement (and greed) plays a large role. One of the most recent and perhaps most shocking examples of contamination was that of the Chinese baby milk powder, which was deliberately contaminated with melamine to increase the nitrogen content, leading to widespread effects on the children's kidneys and even death. Diethylene glycol is another candidate for contamination that has been implicated in multiple health scares and tragedies. In a similar vein, as more and more people turn to alternative medicines, the likelihood of someone misusing or

FOCUS BOX 20.1   ERRORS IN HUMAN CLINICAL TRIALS

TGN1412

In 2006, six volunteers suffered multiple-organ failure due to rapid release of cytokines (a "cytokine storm") by activated T cells following single doses of TGN1412, a CD28 superagonist antibody, in a phase I clinical trial (Attarwala 2010; Eastwood et al. 2010).

- The first-in-man dose, 0.1 mg/kg, was calculated from the no-observed-adverse-effect level (NOAEL) from a 4-week study in non-human primates (NHPs), which was concluded to be 50 mg/kg per week, indicating a safety margin of 500-fold.
- Nonclinical tests were in NHPs on the basis that there was said to be 100% sequence homology of the extracellular domain of CD28 receptor and high conservation of Fc receptors between the species, indicating that similar antibody affinities and responses could be expected.
- However, subsequently, it was suggested that there were differences of up to 4% between the amino acid sequence of the C"D loop of CD28 receptor in rhesus and cynomolgus and that in humans.
- A review by the MHRA found that the results of the preclinical work were an accurate reflection of the raw data.
- Subsequent investigations *in vitro* showed that the key indicators of a TGN1412-type response were release of Interleukin-2 (IL-2) and Interferon gamma (IFN-γ) from CD4+ effector memory T cells.
- This mechanism of cytokine release differed from that of other therapeutic monoclonal antibodies (mAbs), which stimulate cytokine release primarily from natural killer cells.
- CD28 is not expressed on the CD4+ effector memory T cells of all species used for preclinical safety testing and so cannot be stimulated by TGN1412, unlike in humans.
- Eastwood et al. concluded that activation of CD4+ effector memory T cells by TGN1412 was probably responsible for the cytokine storm and that the absence of CD28 expression on the CD4+ effector memory T cells of NHPs used in nonclinical studies may explain the failure to predict a cytokine storm in humans.

This incident initiated new risk assessment procedures, including identifying drugs as high risk as appropriate and ensuring that first administration was to a single subject at a time to check for adverse events.

BIA-102474-101

BIA-102474-101 was a fatty acid amide hydroxylase (FAAH) inhibitor given to human volunteers in a phase I study in 2016; it had a similar mechanism of action to other compounds that had been studied in phase I and II clinical trials without adverse effects (Eddleston et al. 2016).

- Single doses between 0.25 mg and 100 mg were given to 48 patients (16 received placebo) without adverse events.
- Four cohorts received doses of between 2.5 and 20 mg for 10 days without effects. A fifth cohort was added at 50 mg because a maximum tolerated dose had not been achieved.
- One volunteer became ill after the first dose of 50 mg. Four other participants became ill subsequently and were hospitalized. The first volunteer died subsequently.
- Brain magnetic resonance imaging (MRI) of the five symptomatic participants showed deep-brain hemorrhage and necrosis. There was no such evidence in people who had received single doses or repeated exposure at lower doses.
- Nonclinically, higher doses had been safely given for longer durations to NHPs, with NOAELs of 100 and 75 mg/kg/day in 4-week and 3-month repeat-dose studies respectively, in NHPs. NOAELs of 50 and 20 mg/kg/day were found in similar dog studies. There was, however, no indication of effects that might have been seen at doses higher than these NOAELs.
- Pharmacokinetic data from the nonclinical studies did not suggest the need for waiting between doses before treating the next participant.
- The drug was not considered to be a high risk by the French authorities who authorized the trial.
- The findings of this trial were new in that the occurrence of adverse effects in subjects without any prior warning from previously treated cohorts, either single or repeat dose, or from nonclinical studies, had not been seen before.
- The total dose received by the affected participants (250 to 300 mg) was no more than 33% higher than the total dose (200 mg) received by the previous repeat-dose cohort, for which no toxicity was reported. It was considered that the available information was consistent with a dose effect and not idiosyncratic reaction.

Eddleston et al. concluded that assumptions about "high-risk" drugs should be changed, that risk assessment should be refined, and that sequential dosing should be used more frequently and should be guided by individual pharmacokinetic data. Final conclusions on the causes of this incident were not available at the time of writing and will no doubt emerge with future study and review.

misunderstanding how the herbal product is to be used increases—excessive green tea use, fake materials (such as fake ginkgo root), and incorrect use of particular herbs such as those containing aristocholic acid.

When a contamination event is noted in a medical product, an assessment must be made of the potential risk to patients. When Roche detected ethylmethane sulfonate in Viracept, a drug indicated for treatment of the HIV virus, they launched an investigation to determine how the patients might be affected by such potentially deleterious exposure (see Case Study 16.1, Chapter 16). As discussed, contamination, either deliberate or accidental, also occurs in the environment and can knock effects on both the human population and the ecosystem as a whole. Examples range from tetraethyl lead and chlorofluorocarbon (CFC; incidentally, both substances were created by the same man, Thomas Midgeley) to asbestos, heavy metals, and phthalates. Such toxicity is examined in Chapter 10.

MISUNDERSTANDING

Understanding is a function of existing knowledge and appropriate interpretation of the data; it is almost always transient. Today's understanding may be tomorrow's misunderstanding. The evolution of understanding is illustrated by the inception of a new test, for example, the LLNA in mice to assess sensitization potential. The sequence of events in the acceptance and evolution of a new test has often been repeated.

In the beginning is the conception of the test, which comes with a range of assumptions that underpin the conduct of the test and the interpretation of its output. Following the initial theory, hypothesis testing and validation of the test are implemented and the first data gathered. Risk assessments based on these early data are based on the assumptions made early in the history of the test, relating to its utility and relevance. This early phase is followed by realization that the method is fallible in certain circumstances. In the case of the LLNA, there was the discovery of false-positive results and the understanding of their origin; there was the suggestion that irritation of the ears to which the test chemical was applied might be the source of some positive results, although this is still a matter of debate.

One basic assumption that is made is that the time of day or month in which an animal is dosed does not affect the results of the study. This assumption may be challenged as greater understanding of the effects of biorhythms, such as circadian rhythms, on results is uncovered. Focus Box 20.2 discusses the potential effects of circadian rhythms on *in vivo* research. Knowledge of such effects is not new—the light–dark periodicity is a crucial element of all *in vivo* GLP protocols—but for the sake of cost, simplicity, and repeatability, it is standard practice in the majority of studies to dose only once in the day; furthermore, dosing is usually during the morning so that observations such as clinical signs that need to be related to time of dosing can be conducted to fit in with the normal working day.

CORRELATION, CAUSATION, AND STATISTICS

Although covered more than once in this book, the importance of statistical axiom "correlation does not equal causation" cannot be overstated. In short, the fact that a

FOCUS BOX 20.2   CIRCADIAN RHYTHMS

A biorhythm, within the context of scientific study, can be loosely defined as a recurring cycle in the physiology or functioning of an organism, stand-out examples being the menstrual cycle in female mammals and circadian rhythms. Circadian rhythms control a multitude of biological processes including body temperature, feeding, sleep–wake cycle, cell-cycle regulation, hormone secretion, and glucose homeostasis. The circadian clock itself is regulated by suprachiasmatic nuclei of the hypothalamus with inputs and outputs from and to multiple sources. Desynchronization of these physiologic and behavioral cycles (which tend to have a periodicity of around 24 hours) in humans can lead to conditions such as sleep disorders, depression, and the development of metabolic diseases. Jet lag is a common example of the effects of desynchonization of light–dark and feeding cycle. Furthermore, in 2007 International Agency for Research on Cancer (IARC) categorized "shift-work that involves circadian disruption" as "probably carcinogenic to humans (group 2A)" (IARC 2010; Arble et al. 2010; Lee et al. 2010; Zee et al. 2013).

Whether or not an assay or substance may show chronopharmacology or chronotoxicology is a matter for debate that is not easily solved and is highly dependent on the assay and substance under investigation. In a relatively simple experiment on this subject, Miura et al. (2013) injected groups of male C57BL/6J mice ($n = 5$) intraperitoneally with a single 6.4 mg/kg-bw dose of cadmium chloride at 6 different time points in the day (zeitgeber time [ZT]; ZT2, ZT6, ZT10, ZT14, ZT18, or ZT22). The latter three time points (ZT14, ZT18, and ZT22) were conducted during the dark phase and under red light of low lux (light time was 12 hours). Following dosing, the animals were then monitored for 14 days, after which they were killed. The first major difference noted between groups was the survival rate (up to 14 days). In the ZT2 group, the survival rate was 0% compared with 20% in the ZT6, ZT10, and ZT22 groups and 40% and 100% in the ZT14 and ZT18 groups respectively. Similar differences were noted in the mean survival time; at the extremes of the experiment, the ZT2 group's mean survival time was around 2 days, while the ZT18 group continued until the study's termination on day 14. Using a lower dose of 4.5 mg $CdCl_2$ kg/day and a similar protocol ($n = 5$), hepatotoxicity at the ZT6 and ZT18 time points was investigated. There appeared to be no difference in the accumulation and distribution of Cd in the liver; however, significant elevations of aniline transferase (ALT) and aspartame transferase (AST) were found in the ZT6, but not the ZT18, group. Basal hepatic metallothionein levels were similar in both groups, but the authors felt that there may be slight (significant at one time point) differences in the glutathione levels, with hepatic glutathione being lower in the ZT6 group. This simple example demonstrates that dosing time may have an effect under the correct circumstances and with the right chemical type.

The potentially carcinogenic effects on circadian rhythms notwithstanding, how a cell responds to DNA damage also appears to be linked to the circadian clock—which would have large implications on how *in vivo* geno-toxicity assays are conducted. DNA repair mechanisms such as nucleotide excision repair appear to be controlled by the genes related to the circadian clock. Furthermore, DNA damage control checkpoints and control of apoptosis have been shown to have links to the circadian clock (Antoch et al. 2005; Sancar et al. 2010).

Clearly, not every substance and not every test paradigm will be altered by circadian rhythm considerations—one or two examples does not a theory make. Equally, factoring such considerations into every single *in vivo* assay would be ludicrously expensive and unnecessary. As always each substance should be considered on its own merits, and if possible, the model exposure pattern should match that of the expected target.

difference is statistically significant, be it in an *in vivo* toxicology study, a clinical study, or a population analysis, does not necessarily mean that it shows a true biological effect or one of toxicological significance. In fact, it should be remembered that a difference may be of toxicological significance without achieving statistical significance. Equally, statistically significant differences should not be cavalierly dismissed without due care, and their relevance should always be assessed.

To illustrate the importance of properly applied statistics, a Canadian group evaluated for the occurrence of certain diseases in 12 randomly assigned groups taken from the Registered Persons Database of 10,674,945 residents from Ontario (aged between 18 and 100 years) for commonly diagnosed hospital admissions. Residents were then assigned to equal-sized derivation and validation cohorts and then subsequently classified according to their astrological sign. Within each astrological sign derivation group, two conditions were identified that were found to have a greater (and statistically significant) chance of occurring. These were then compared to the independent validation group. Surprisingly, or rather unsurprisingly depending on your point of view, the data suggested that if you are, for instance, a Pisces, you have a higher probability of being hospitalized with heart failure. If you happen to be a Libra, you are more likely to be admitted to hospital with a fractured pelvis. Disappointingly, when the statistics were adjusted to account for multiple comparisons, no significance was found in either cohort (Austin et al. 2006).

This study highlights a number of questions that should be applied to all statistical data:

- Firstly, if a statistically significant difference is observed, is it biologically plausible?
- If so and if relevant, has this sort of difference been observed in historical control data or previously with similar substances or classes of compound?
- Can a mechanism of action be proposed based on the action of the substance?
- If the result is not biologically relevant, it is important to justify why not with reasoned argument.

- Conversely, if the result is not statistically significant, it may still have biological relevance particularly if results are from a subchronic or subacute assay, appear reversible, or have a specific action in a diseased state.

This feeds into a second point; do not automatically compare all parameters and endpoints. The more parameters that are compared, the more likely one is to find a spurious, but statistically significant, result. Comparison should be rational and justified wherever possible; for example, changes in AST and ALT (markers of liver function change) may correlate with observable changes in weight and gross pathology or histopathology in the liver.

Finally, when assessing the applicability of your finding, look at the complete study and do not get bogged down in minutiae. If a significant result is found, assess how it relates to the whole—for example, is the effect dose dependent or only found in the mid-dose group, or do the controls have an abnormally low level of background occurrence?

THE PUBLIC AND TOXICOLOGY

In many cases, interactions between the public and toxicology are not based on actual risk but on the perception of hazard. In recent years, a number of health scares, perhaps encouraged by the popular press, have come to the fore. Some like the Poly Implant Prothése (PIP) breast implant scandal have legitimate grounds for worry, while others like bisphenol A at vanishingly small quantities are less cause for concern (see Case Study 16.2, Chapter 16). Antimicrobial resistance (see Focus Box 20.3) is a large area of concern in the public health arena, and its potential effects do not appear to be well appreciated by the public at large.

TOXICOLOGY AND THE INTERNET

There is no doubt that the Internet has changed society: how we interact, how we shop, and how we access information. In the same way, it has changed how we view (and assess) medical conditions; how medicines, tinctures, remedies, herbal preparations, and illicit substances are purchased; and how the public views toxicology. No longer does an interested party have go to a library and pick up a book or speak to a toxicologist; all information is freely accessible and in some cases with clickable blue nouns. While this allows greater understanding of toxicology and science as a whole, it also allows those who have no traditional toxicological training or knowledge to put across views that have a limited basis in science. Such views may not present any health issues and can be useful and informative, but they can exposure the unwary reader to potential risk and harm.

A case in point is that of the Miracle Mineral Solution (MMS, a solution of 28% sodium chlorite), which its makers claim can cure almost all known diseases including (but not limited to) autism, cancer, H1N1 flu virus, and malaria. It is recommended by its makers that it be acidified (generally with orange juice) to produce sodium dioxide, a potent bleach. As is pointed out on a number of the product's websites, sodium chlorite is used in hospitals as a floor cleaner and general disinfectant.

FOCUS BOX 20.3   ANTIMICROBIAL RESISTANCE

It is not an understatement to state that perhaps the greatest challenge facing humanity in the twenty-first century is the threat from antimicrobial resistance (AMR). According to the Review on Antimicrobial Resistance group (set up by the UK government in July 2014), AMR kills 700,000 people each year; by 2050, this has been estimated to increase to over 10 million. Such a death toll alone would reduce gross domestic product by 2% to 3.5%, costing up to $100 trillion worldwide.

There are many examples of AMR, including multidrug-resistant and extensively drug-resistant tuberculosis; resistance to best available malarial treatments in the Greater Mekong Subregion; drug-resistant bacteria (drug-resistant typhoid, drug-resistant gonorrhea, drug-resistant *Escherichia coli*); drug-resistant HIV; and drug-resistant influenza.

AMR is a natural phenomenon—at its heart, it is straightforward survival of the fittest—that occurs and has occurred for thousands of millennia in the absence of human intervention. What has changed, however, is the speed at which AMR is developing, in part due to our overuse and misuse of antimicrobial pharmaceuticals. In a recent WHO publication, it was found that in the Americas, 51% of the member states allowed access to antimicrobial medicine without a prescription—in the southeast of Asia, this increases to 64%. Colloquial reports of new superstores and pharmacies giving away antibiotics for a rainy day do not appear to be overstated. This does not even cover the use of antibiotics in animal husbandry or a lack of compliance from patients. Coupled with this is the fact that infectious diseases are not, for want of a better word, sexy. Between 2010 and 2014, the US National Institutes of Health spent $26.5 billion on cancer research, $14.5 billion on HIV/AIDS research, $5 billion on diabetes research, but only $1.7 billion on research for AMR. In addition, pharmaceutical companies are not keen on producing antimicrobial products such as antibiotics as there is a good chance that a few short months after they have released their vastly expensive product, it will become redundant.

To tackle this, the Review on Antimicrobial Resistance has given a summary of recommendations that need to occur to prevent the spread of AMR:

1. A massive global public awareness campaign
2. Improve hygiene and prevent the spread of infection
3. Reduce unnecessary use of antimicrobials in agriculture and their dissemination into the environment
4. Improve global surveillance of drug resistance in humans and animals
5. Promote new, rapid diagnostics to cut unnecessary use of antibiotics
6. Promote the development and use of vaccines and alternatives
7. Improve the numbers, pay, and recognition of people working in infectious disease

8. Establish a Global Innovation Fund for early-stage and noncommercial research
9. Better incentives to promote investment for new drugs and improving existing ones
10. Build a global coalition for real action—via the G20 and the UN

Whether these steps help alleviate this growing problem remains to be seen, however, without further investment in antibiotic development coupled with tightening on the controls of existing medications (AMR Review 2016; WHO 2016).

It can also be used for stripping textiles and industrial water treatment. High doses can lead to nausea, vomiting, diarrhea, and symptoms of severe dehydration. The extent of its use and its danger to the public are highlighted by the US Food and Drug Administration (FDA) issuing a consumer warning indicating that consumers "should stop using [MMS] immediately and throw it away." Some of the Internet's populace met this decision with derision and anger. Further anger was thrown upon the conviction of one MMS's sales people for "conspiracy, smuggling, selling misbranded drugs and defrauding the United States" and who could face a 34-year prison sentence.

In a similar vein, aspartame, a rigorously tested and widely used artificial sweetener, has been the subject of much public gnashing of teeth and brouhaha. In a comprehensive EFSA review of aspartame in 2013, it was found to be "not of safety concern" at EFSA's current acceptable daily intake of 40 mg/kg-bw per day. It is considered as "safe" for use as a sweetener by a number of national food standard and health agencies (EFSA 2013). If some of the websites devoted to its dangers are to be believed, this is not the case, and it is in fact directly responsible for vast number of medical conditions. These websites, however, pale in comparison to the extent that one American citizen went to prove its danger. Motivated by a need to show the adverse effects of aspartame, she conducted her own (non-GLP-compliant, one assumes) carcinogenicity bioassay in her garden. The validity of the positive outcome may have been somewhat tainted by her nonstandard techniques including (but not limited to) buying rats from a pet store (not a homogenous population with historical control values), selecting the control groups based on her favorite rats, dosing the animals with packets of NutraSweet (which contained substances other than aspartame), and a total lack of any pathology other than a visible external analysis of "tumor" (which were all assumed to be "cancerous").

SUMMARY

It is almost impossible to plan for every eventuality. It is unlikely that manufacturers of alcoholic hand gel considered that some members of the public, rather than sanitizing their hands with their product, would drink it. Decisions of where a toxicologist's personal and professional responsibility starts and stops depend heavily on the role played. However, all toxicologists should understand the following:

- The role of extrapolation, both intraspecies and interspecies, coupled with knowledge of the target population.
- The relationship of mechanisms in the test system relative to the target species to which the results will be extrapolated, thus avoiding inappropriate test systems.
- The effects of personal prejudices on how a study is conducted, interpreted, and reported.
- Integration of the postvalidation results with deeper understanding and better application of the method.
- Communication is key; informing and educating the lay public is a vital part of toxicology.

REFERENCES

Anti-Microbial Review, 2016. Available at http://amr-review.org/.

Antoch MP, Kondratov RV, Takahashi JS. Circadian clock genes as modulators of sensitivity to genotoxic stress. *Cell Cycle* 2005; 4(7): 901–907.

Arble DM, Ramsay KM, Bass J, Turek FW. Circadian disruption and metabolic disease: Findings from animal models. *Best Pract Res Clin Endocrinol Metab* 2010; 24(5): 785–800.

Attarwala H. TGN1412: From discovery to disaster. *J Young Pharm* 2006; 2(3): 332–336.

Austin PC, Mamdani MM, Juurlink DN, Hux JE. Testing multiple statistical hypotheses resulted in spurious associations: A study of astrological signs and health. *J Clin Epidemiol* 2006; 59(9): 964–969.

Christie AC. Schizophrenia—Is the potato the environmental culprit? *Med Hypotheses* 1999; 53(1): 80–86.

Colgrove, J. Immunity for the people: The challenge of achieving high vaccine coverage in American history. *Public Health Reports* 2007; 122(2): 248–257.

Eastwood D, Findlay L, Poole S, Bird C, Wadhwa M, Moore M, Burns C, Thorpe R, and Stebbings R. Monoclonal antibody TGN1412 trial failure explained by species differences in CD28 expression on CD4+ effector memory T-cells. *Br J Pharmacol* 2010; 161(3): 512–526.

Eddleston M, Cohen AF and Webb DJ. Implications of the BIA-102474-101 study for review of first-into-human clinical trials. *Br J Clin Pharmacol* 2016; 81: 582–586.

European Food Safety Authority: Final review of the Séralini et al. (2012a) publication on a 2-year rodent feeding study with glyphosate formulations and GM maize NK603 as published online on 19 September 2012 in *Food and Chemical Toxicology*. *EFSA J* 2012; 10(11): 2986.

EFSA ANS Panel (EFSA Panel on Food Additives and Nutrient Sources added to Food). Scientific Opinion on the re-evaluation of aspartame (E 951) as a food additive. *EFSA J* 2013; 11(12): 3496.

Ewen SWB, Pusztai A. Effect of diets containing genetically modified potatoes expressing *Galanthus nivalis* lectin on rat small intestine. *Lancet* 1999; 354: 1353–1354.

Friedman M, Rayburn JR, Bantle JA. Developmental toxicity of potato alkaloids in the frog embryo teratogenesis assay—Xenopus (FETAX). *Food Chem Toxicol* 1991; 29(8): 537–547.

International Agency for Research on Cancer. IARC monograph on the Evaluation of Carcinogenic Risks to Humans: Shiftwork. *World Health Organisation* 2010; 98.

Lee S, Jeong J, Kwak Y. Depression research: Where are we now? *Molecular Brain* 2010; 3: 8.

Miura N, Ashimori A, Takeuchi A, Ohtani K, Takada N, Yanagiba Y, Mita M et al. Mechanisms of cadmium-induced chronotoxicity in mice. *J Toxicol Sci* 2013; 38(6): 947–957.

O'Malley, M. Vaccine Law and Policy: A Question of "Compulsory Safety" Government Documents. Government Information Sources. 2001.

O'Neill J (chair). Tackling drug-resistant infections globally: Final report and recommendations. *The Review on Anti-Microbial Review*, 2016. Available at http://amr-review.org/.

Raach JH. English medical licensing in the early seventeenth century. *Yale J Biol Med* 1944; 16(4): 267–288.

Roberts A. *Napoleon the Great*. London: Allen Lane, 2014. ISBN 978-1846140273.

Sancar A, Lindsey-Boltz LA, Kang TH, Reardon JT, Lee JH, Ozturk N. Circadian clock control of the cellular response to DNA damage. *FEBS Letters* 2010; 584(12): 2618–2625.

Séralini E, Clair E, Mesnage R, Gress S, Defarge N, Malatesta M, Hennequin D, de Vendômois JS. Long term toxicity of a Roundup herbicide and a Roundup-tolerant genetically modified maize Gilles. *Food Chem Toxicol* 2012; 50: 4221–4231.

US Congress. "CHAP. XXXVII.—An Act to encourage vaccination" XXII Congress. 1813.

US Congress. "CHAP. L.—An Act to repeal the Act, entitled "An Act to encourage Vaccination" XVII Congress. 1822.

Warren MD. *A Chronology of State Medicine, Public Health, Welfare and Related Services in Britain 1066–1999*. Faculty of Public Health Medicine of The Royal Colleges of Physicians of The United Kingdom. 2000. ISBN 1 900273 06 3.

World Health Organisation. Antimicrobial resistance, 2015. Available at http://www.who.int/mediacentre/factsheets/fs194/en/.

Wright, DC. *The History of China*. Westport, US: Greenwood Publishing Group, 2001; p. 49. ISBN 0-313-30940-X.

Zee PC, Attarian H, Videnovic A. Circadian Rhythm Abnormalities. *Continuum (Minneap Minn)* 2013; 19(1): 132–147.

21 The Future of Toxicity Testing and Risk Assessment

INTRODUCTION

Toxicology is a dynamic discipline that is evolving rapidly, as new test methods and paradigms emerge and as understanding of data changes with the addition of new information. Good risk assessment is dependent on good toxicology and, crucially, on the correct understanding and interpretation of the toxicity data. However, as new information arrives, understanding grows and evolves to the point when today's understanding may be seen as tomorrow's misunderstanding or even mistake.

The principal pressures on toxicity testing and risk assessment have not changed since the first edition of this book and, in fact, have become more intense. Regardless of your point of view, the pressure to avoid the use of animals is growing year on year. This point is exemplified by the emphasis in Registration, Evaluation, Authorisation and restriction of Chemicals (REACH) legislation in Europe to avoid the use of animals wherever possible; however, a cynic or realist may dismiss this as pious hope rather than practicable expectation. The methods and philosophy of testing for toxicity are evolving constantly. The results of these tests form the foundations for toxicological risk assessment, and this too has undergone evolutionary change, although perhaps not as blatantly as in toxicology.

There is one unchanging aspect of toxicology and risk assessment, and that is the responsibility to the general public in terms of chemical safety—even if safety, as a negative concept, cannot be proved. The consequence of this is that toxicity studies and any subsequent risk assessment should be conducted ethically and to high standards, whether in industry in support of a new pesticide or in a university as part of a PhD thesis on a chemical naturally present in food. The conduct and results of toxicological study are under public scrutiny, unlike other sciences. As a result, there are pressures on how studies are conducted, how they are interpreted and the risks assessed, and how that is translated into risk assessment and management. Through all the pressures to change—use fewer animals, ignore that pressure group, keep those jobs, cure my baby, save my crops while not using toxic pesticides—the one thing that does not change is the unattainable public desire for a risk-free existence: the ability to use chemicals without any of the risks.

As discussed in Chapter 20, the regulation of safety evaluation and risk assessment has evolved in the light of periodic tragedies such as the thalidomide disaster and is focused by fear of insidious diseases such as cancer. This is a reactive approach rather than proactive, although to be proactive requires a degree of foresight and lateral thinking that may not be encouraged by the dynamic interactions between the three main stakeholders—the public, regulators, and industry.

The successes of safety evaluation and risk assessment are less trumpeted than the failures; these include the failure to predict an association in some patients between taking Vioxx and cardiac toxicity and the severe reactions in volunteers given TGN1412. Vioxx was, perhaps, an extreme example where there was no indication in the nonclinical studies in animals of any cardiotoxic potential. For TGN1412, the reactions in human volunteers were much greater than in the toxicity studies due to poor predictivity of the model chosen. In environmental terms, the problem is further illustrated by the difficulty of clearly assessing the effects of low, environmental exposure to endocrine disruptors and the consequences for human health; if there is a difference, what is its significance?

CHALLENGES FOR TOXICOLOGY

It is relatively easy to see a difference if it is clearly significant—statistically, biologically, and toxicologically; one of the main challenges for toxicology is to correctly assess small differences from controls or background. This may be exemplified by the sort of difference associated with a clear effect at the high dose and a dose relationship combined with the support of other findings. The assessment of difference becomes more complex and less certain as the values approach background or historical control data. This is a problem because some of the most significant differences may be small but be associated with long-term effect such as accelerated neurological decline or with cancer. These could include fractional changes in hormonal homeostasis or an insidious attack on renal function that may result in premature kidney failure in old age. Much chemically induced disease may be due to minor perturbation and imbalance in normal physiology. Such change is often only apparent late in life, and retrospective health and safety control is not possible. An additional complication is that epidemiological study of close-to-normal events requires vast numbers of subjects and may only identify a problem when there has been a significant effect on public health.

The toxicological challenge, then, is to detect small differences from normality and to assess them correctly in terms of their potential for long-term effect in the target species or in the environment. As indicated by Liebler (2006), toxicology has focused traditionally on exogenous agents; we should recognize the potential effects of long-term exposure to slightly varied concentrations of endogenous substances, although such change may be driven by exogenous substances.

Another challenge is the increasing pressure to ensure that test systems are relevant, with some groups insisting that animals are not useful for predicting human hazard. Implicitly, this means a reduction in the use of animals and increased use of alternative test systems or refined tests. Such refinement may mean the use of fewer animals, with a consequent reduction in the statistical power of experiments. A reduction to 10 rats per group from 15 may not wreck the statistical utility of an experiment. On the other hand, a reduction from four to three per group for large, genetically inhomogeneous animals such as dogs may fatally weaken the biological discrimination of an experiment. (At these group sizes you can nearly forget the statistics as only barn-door obvious differences will be flagged as statistically significant.) Clearly, it helps to have the correct test system in the first place.

The toxicology for any substance should identify hazard and dose response, for which the data are passed to risk assessment. This can be a delicate process, which has to be right; it follows from this that the risk assessment has to be correct too.

All toxicity testing, with the possible exception of mechanistic research, whether *in vitro*, *in silico*, or *in vivo*, is predictive; the intention is to predict safety, or liver toxicity, or genotoxicity, or any other endpoint. However, it is perceived by some that the current paradigm for toxicity testing is not sufficiently predictive, whichever test method is used. The predictions used may be correct for the majority of the target population but flawed for a few. The challenge therefore is to improve the predictivity of our toxicity test paradigm by better using and understanding the tools that are available to us.

EVOLUTION OF TOXICITY TESTING

Current practice is being changed by new techniques used in early development, which are not subject to regulatory guidance or Good Laboratory Practices (GLPs). These include the use of transgenic animals (knockout mice and rats, humanized mice) and microarray chips for the identification of patterns of gene expression and changes in protein synthesis, together with increasingly sophisticated analytical methods. The amount of data produced is phenomenal, and computational techniques are evolving to cope with the flood. The problem is not a paucity of new methods but selection of the technologies that will be useful in the medium to long term. The increase in knowledge and evolution of understanding will always tend to move the goalposts and make previous practice look dubious; there is no easy escape from this.

The use of such techniques may not be acceptable for mainstream regulatory toxicology, but it is likely that, even for pharmaceutical toxicology, they will become more widely accepted. For instance, the use of human hepatocytes in comparative *in vitro* metabolism studies is now routine. However, these suffer from inconsistency of product due to the diversity of people from whom the liver samples are taken. If an immortal, metabolically competent line of human hepatocytes could be developed and made widely available, its use would increase by default and might well become a regulatory requirement in due course.

The paradigms for toxicity testing and risk assessment are evolving continually as new techniques emerge. This is an evolution of both techniques and, in regulatory terms, study requirements. The advent of microdosing studies for pharmaceuticals, in which a dose of about 100 μg of a radiolabeled drug is given, has meant that a smaller set of studies may be acceptable before first administration in human volunteers. This type of study is dependent on highly sensitive analytical techniques such as accelerator mass spectrometry. The use of this type of study in human volunteers for new medicines should open the door for similar studies with agrochemicals in humans.

In practical terms, the development of new techniques and methods in toxicology will continue to refine the process of safety evaluation. It is likely that the *-omics* will become more widely used as they are better understood and become less expensive. Metabonomics and the closely related metabolomics, in particular, offer considerable scope for the noninvasive, *in vivo* investigation of animal responses.

We are a long way from replacing animals in toxicity testing, especially in studies of repeated administration. However, new techniques of culture are allowing increasingly long periods of exposure of cultivated cells or tissue slices, and these techniques will grow in acceptance as they become more widespread. The pressure to reduce the use of animals will always remain, just as animals are likely always to be used; the pressure to reduce may be augmented by the emphasis in REACH to avoid the use of animals.

Risk assessment is also developing with greater acceptance of concepts such as the benchmark dose and of TTCs. In addition, as understanding evolves, so does the precision with which risk assessment can be conducted. It is increasingly acknowledged that there are thresholds for genotoxicity, and this may be expected to have a large effect on risk assessments that have been driven by this central misunderstanding.

One aspect of risk assessment—a constant over many years—is the continual development of ever more sensitive analytical techniques. These can now detect levels of chemicals that are probably substantially lower than levels that pose any toxicological threat; however, the reaction is always, "It's there—save me!" Risk assessors sometimes have a rough relationship with the public, who may not understand the toxicological significance of low levels of chemicals—especially in relation to those occurring at higher concentrations in a normal diet.

The risk assessor is faced with pressures that require him/her to appreciate reality while adhering to the precautionary principle and to balance public perception and understanding. For this to be achieved, the exposure assessment has to be realistic, and the assumptions have to be assessed for relevance. Overhanging all this is the question, "What is an acceptable level of risk for the population concerned?"

All systems of toxicological evaluation and risk assessment when taken in isolation are fallible, and this situation is unlikely to change. The public are more likely to tolerate a false positive than a false negative where predicted safety dissolves into a toxicological disaster such as thalidomide, benoxaprofen, or Vioxx. There will always be public pressure for better test systems and data and for ethical conduct of safety evaluations. However, because safety cannot be proved but merely inferred, there will always be a possibility of error, whether in a general sense (thalidomide) or in sensitive individuals (Vioxx). Complete abolition of animal use in toxicology or complete removal of all restrictions will not produce better safety evaluation. A scientific compromise offers the best way forward but may be difficult to achieve without better communication with the public.

DEVELOPMENT OF NEW TEST METHODS AND MODELS

There are two broad approaches to development of test methods; one is to refine existing methodologies, and the second is to develop completely new test protocols or models. For the first approach, it is possible to subject the test system to a wider range of investigations; this approach is exemplified by the increasing use of rats in standard toxicity studies for bone marrow micronucleus assays or for behavioral tests for central nervous system (CNS) safety pharmacology. Other possibilities are use of the comet assay for DNA damage, use of the Pig-A assay, or the addition of new clinical pathology parameters such as troponin for cardiac damage; the examination of urine samples by metabonomics may also increase. The drawback of adding more and more investigations to the same study is that the complexity increases exponentially, making errors much more likely. These can be straightforward logistical errors of omission or misadministration or sampling, or more subtle ones where the conduct of one investigation impacts on the results of another.

Integration of new techniques into existing protocols is definitely going to be a growth area in the future. The work of Kramer et al. (2004) in integrating genomics and metabonomics with traditional toxicity endpoints was reviewed in Chapter 4, and it is clear that these techniques will develop and provide greater understanding of toxicity in standard laboratory models.

Historically, new methods have included the local lymph node assay and the comet assay; these are clear success stories. Methods that have fallen by the wayside include the use of hydra in reproductive developmental testing and the use of the chick chorioallantoic membrane test. These may still have potential in the new climate engendered by REACH. New methods under development include the slug mucosal assay for irritation, the long-term exposure of hepatocyte cultures and tissue slices, and further development of methods for testing for mechanisms of carcinogenicity, as discussed in Chapter 8.

The development of new models is also an important factor in the future of toxicity testing. These include the use of invertebrates, as with the slug for the assessment of eye irritation, and novel vertebrates such as zebra fish. Stem cells remain the great white hope of toxicity testing; their promise remains just that, at the moment, but may yet blossom.

In developing new test methods, the issue of reproducibility—within laboratories and between laboratories—must be considered. New tests must be robust enough to be transferred readily from one laboratory to another and also be capable of providing reproducible results. It is routine to repeat *in vitro* studies to confirm the results of the first test; at present, these tests are performed at the same laboratory. However, there may be an advantage in performing confirmatory studies in a second laboratory, especially where the data indicate a marginal effect, the reproducibility of which is subject to influences by statistical considerations and normal biological variation.

To be successful, a new toxicity screen should be

- Robust: The test should be relatively easy in technical terms; complication leads to error, and specialist equipment means expense. New animal models should not have overonerous husbandry requirements.
- Readily transferable between laboratories.
- Understood: There is little sense in producing data unless the mechanism of their generation and their significance is well understood.
- Reproducible: If not, its utility and relevance may be questioned. Baseline data for individual animals, plates, or replicates should not be so variable that change is indistinguishable from historical control data.

- Predictive: With good sensitivity and specificity.
- Quick: Lengthy experimental phases mean slowed development or lead candidate selection and additional expense.
- Cost effective: There is no future for any test if the costs outweigh the value of the results.

Ideally, any method should examine more than one endpoint. This is desirable even if it is addressed through several related models. It is expected that a transgenic mouse would be capable of expressing toxicity other than that shown through the gene of interest. All routine toxicity endpoints could be incorporated into transgenic assays. Any new method should be capable of showing a dose response.

Methods using transgenic animals have developed rapidly, and the utility and relevance of these models will become clearer during the next few years; however, they may not fulfill the criteria suggested above for ease and speed of technical performance and cost. They have significant potential in mechanistic studies, either for screening for an effect in a chemical class or series or for explanation of effect due to a single compound. There is also the possibility that a transgenic animal could be constructed specially to answer a particular question relating to toxicity.

THE FUTURE OF TOXICITY TESTING

THE TOOLS AVAILABLE AND THEIR PROBLEMS

The tools that we have for a safety assessment are, essentially, *in silico*, *in vitro*, and *in vivo*. Each of these has its issues and problems in terms of utility, credibility, and success. As the toxicologist armamentarium expands, the choices that are made in test system selection become ever more crucial.

The central problem is that no single system is entirely reliable for the prediction of toxicity and, thereby, for estimation of safe dose via a risk assessment process. Equally, the use of more than one system, which in fact is essential, can produce contradictory results. As new methods emerge, they have to undergo a rigorous process of validation that examines their predictivity, amongst other things. Validation of *in vitro* and *in silico* methods is routine, but successful validation should not be taken to imply that any test is infallible. It is often pointed out that tests using animals have never been validated; however, this is simplistic, as although specific validation studies have not been carried out, the test systems and test designs are now well understood and have effectively been validated by experience over years of use. With any validated test, the results have to be understood and interpreted appropriately; it can be argued that the results of countless carcinogenicity bioassays in rodents have been misunderstood and misinterpreted. While the tests may have been validated by duration of use, their utility has been overestimated and recently called into question.

As new techniques become available, such as genomics and metabonomics, the amount of data generated is likely to increase to the point where human review of a database is not possible within any sensible time frame. For this work there will have to be increasing reliance on bioinformatics and pattern recognition.

Toxicology In Silico

In silico toxicology is at the same point now that *in vitro* was when the Ames test was introduced during the 1970s. From the point of regulatory acceptance of that test as a predictor of genotoxicity, a range of other *in vitro* genotoxicity tests have been developed and accepted. In addition, other tests, such as the human Ether-à-go-go Related Gene (hERG) assay in safety pharmacology and the neutral red uptake assay amongst others, have been accepted, expanding the coverage of *in vitro* assays. The use and acceptance of *in silico* is at a similar point. Currently, data from defined *in silico* systems are accepted for the prediction of genotoxicity of impurities in active pharmaceutical ingredients under International Conference on Harmonisation (ICH) M7. *In silico* is also, theoretically, acceptable for use in REACH.

As with the simpler *in vitro* techniques, *in silico* tends to become less reliable as the complexity of the endpoint increases and the output needs to be supported by other streams of evidence: read-across and data from other tests and analyses. However, these systems are evolving rapidly and hold considerable promise. One potential use in the future could be to draw together all the strands of data from a test program and compare them with existing knowledge to suggest potential pitfalls that may result. This should include clinical data for pharmaceuticals.

In Vitro Systems

At the current stage of development, *in vitro* tests are good for mechanistic studies where single (or limited) endpoints are examined, although they are becoming more general in application. Investigation of a number of mechanistic endpoints, via a battery of tests, could be used to assess the presence of the individual mechanisms or events that lead to a complex conclusion. In terms of the three Rs, the most viable place for this could be assessment of the potential for human-relevant nongenotoxic mechanisms of carcinogenicity. They are also quick to perform and often inexpensive in comparison with traditional methods.

Individually, *in vitro* tests are poor for examining multiple endpoints or toxicities that are multifactorial, such as eye irritation or reproductive effects. In addition, due to the limited viability of the preparations, they are also poor for assessing the accumulation of effect that comes with repeated dosing over a long period, for example, the gradual but accelerated decline of functional reserve in nonrenewing or nonrepairing tissues like the CNS or the kidney. Although quick to perform, *in vitro* methods can be technically complex and, as a result, difficult to transfer between laboratories.

For an *in vitro* method to be accepted (especially for regulatory purposes), there must be understanding of the mechanisms and contributing or causative factors in the endpoint studied, plus understanding of how the *in vitro* data relate to the *in vivo* situation. From this position, it should be possible to make reliable predictions of human effect. In furthering this process of acceptance, the correlation of the data resulting from new systems *in vitro* with those derived from established methods must be considered, especially where classification is used to rank toxic hazard, but this should not necessarily be allowed to slow acceptance. This process of validation is highly contentious. While it is sensible to prove a concept with the use of chemicals known to target the test system under investigation (e.g., nephrotoxicity and mercuric chloride), use of the new test in parallel with the currently accepted methods is the surest way of achieving validation and acceptance. Using this approach, a percentage concordance with accepted methodology can be produced, and the utility of the test or test battery can be assessed objectively. Retrospective testing *in vitro* tends to produce a range of scientific "excuses" for lack of success in various circumstances, and in our view, this tends to muddy the waters to a point where the utility of the method becomes unclear. To say that a test is 90% successful in predicting neuropathy, providing that certain criteria are met, is the same as saying that the test is handicapped to a greater or lesser extent. A cynic might say that understanding when a test will give a negative result (when a positive result would be counterproductive in your development plans) might unduly influence choice of test and lead to a false indication of safety. There is no satisfactory way round this dilemma, other than careful scrutiny of test choice and results.

Although many people campaign aggressively for the use of animals in toxicity testing to be ended completely, it should be remembered that *in vitro* often means that animals are still used to provide organs, cells, or subcellular preparations. This is true for *ex vivo* assays or for the harvesting of tissues for *in vitro* tests in which primary cultures are essential, for instance, to retain metabolic capabilities.

In vitro toxicology has a great future for a host of reasons but has significant weaknesses, which means that complete replacement of animals in toxicity testing is unlikely at the current state of research. They offer potential for use in lead candidate screening assays and in mechanistic research, quite apart from their economic benefits in terms of space and speed of conduct. Organizations such as Fund for the Replacement of Animals in Medical Experiments (FRAME), the European Centre for Validation of Alternative Methods (ECVAM), and the Centre for Alternatives in Animal Testing at Johns Hopkins University in Baltimore play a significant role in furthering new methods of toxicological testing.

In Vivo: Animals in Toxicity Testing

In looking at the future of toxicity testing, the use of animals cannot be ignored. There is much debate about the utility of animal data in risk assessments intended for human use, much of it acrimonious and less than dispassionate.

There have been numerous studies that show that they either are an essential part of the process or are completely useless; the ultimate conclusion must be that if you select your studies or compounds according to your argument and ignore the others, you can prove what you like. Like all the toxicity studies on any chemical, whether *in vitro* or *in vivo*, animal studies are a tool to be used to achieve the objective of a realistic assessment of the compound. As with any tool, they need to be used correctly and appropriately in order to give the most accurate result. During this process, their limitations have to be realized and taken into account; this applies just as much to studies conducted in human volunteers or to reports of accidental exposure as to studies in transgenic mice or bacteria. No single study should be taken as the sole basis for a risk assessment; every study is one part of the database that is used and viewed as a whole package, and animal experimentation is simply one supporting aspect of this process. Although there may be human data that can carry more weight than other evidence, they may not be definitive for the target population of the risk assessment. Frequently, there are fewer human data than would be considered sufficient for a complete assessment, and support from other sources is essential. It is axiomatic that animal studies that have been badly designed and conducted should carry less weight in any risk assessment, and their data should be used (if at all) for support rather than definitive conclusion. However, this principle of data quality and integrity applies to any safety evaluation study in any test system, so animal studies are no different in that respect.

The addition of new and updated examinations to animal tests makes good sense, generally, as it saves animals, can give the results quicker, and may save time and cost. These additions now regularly include micronucleus assays to 28-day toxicity studies in rats, fertility elements for male rats in toxicity studies, safety pharmacology endpoints, and so forth. New endpoints are being introduced, such as tissue collection for RNA expression and more refined analyses of urine to detect kidney toxicity.

The sciences of genomics and proteomics continue to develop quickly and offer considerable utility in screening for lead candidates; they can be used in animals or cell cultures. Following a single dose, the number of genes expressed in mouse liver, for example, increases nearly exponentially during the few hours after administration, and the pattern of gene expression can be related to the toxicity manifested in the whole animal. The pattern of protein expression can be examined in an analogous way and, when used in conjunction with genomics, offers a powerful tool for assessment of toxicity in the short term. For prediction of long-term toxicity, these short-term methods may be limited by the difficulty of differentiating between normality and the slight changes that will result in long-term effects after prolonged exposure *in vivo*.

As the understanding of the significance of epigenetic change grows, it is likely that this will become a crucial arm in the investigation of carcinogenic potential, particularly nongenotoxic carcinogenesis. Developments of this type will assist the gradual decline in use of the 2-year rodent bioassay.

Another area that could use a "reimagining" is the area of dose choice. Too often, dose choice can appear arbitrary and may lack scientific relevance to the end application. One solution to this is the kinetically derived maximum dose (KMD), in which toxicokinetic data are used to aid in the selection. This method is particularly useful when a substance is expected to show nonlinear toxicokinetic behavior at high doses, and it is discussed in detail in Marty et al. (2013). Such an approach may not be applicable for all situations, however, particularly if one is obligated to demonstrate toxicity.

Given the complexity of the overall objective of toxicological testing, it is unlikely that any single system will be capable of giving a reliable and reproducible answer— now or in the future. At the current stage of technological development in toxicity testing and understanding of mechanism, it is clear that animal experiments that have been properly conducted and interpreted are an essential part of risk assessment.

This is not to say, however, that the use of animals will not decrease further or that their use cannot be further refined. There is, however, a clear need to move forward and to develop alternative strategies, through the use of new models such as invertebrates or other vertebrates such as fish.

A NEW PARADIGM FOR SAFETY EVALUATION

It is clear that new methods and models will play a far greater role in the future of safety evaluation and that there will be a continuation of the trend to reduce animal use. However, this should not be at the expense of credible experimental design. This area of toxicology has always been contentious, and there is no such thing as a definitive conclusion; each new set of data, assay, paradigm, study, or schema changes the debate and drives it forward.

Toxicity Testing in the 21st Century: A Vision and a Strategy, colloquially known as Tox21, is a 2007 National Research Council report that advocated "a far-reaching vision for the future of toxicity testing." Tox21 and its many progenies are discussed in Focus Box 21.1.

One area where animal use can clearly be lowered is in reducing the use of carcinogenicity bioassays, either in wild-type rodents or in transgenic mice. A cynic may say that the use of a transgenic mouse is a simple means of shortening development times because it allows you to identify irrelevant tumors more quickly. However, this comfortable assumption has been shaken by the increase in treatment period for some models from 6 to 9 months and by increasing the numbers of animals used. The main precursor and hurdle to abandoning such long experiments is, however, the need to develop credible tests for carcinogenic mechanisms that are relevant to humans. Advocates of the 2-year bioassay argue that these assays are known and can give predictive, valuable data. This may be true; however, it poses the question, Should one stagnate, fail to innovate, just because something is known and familiar? Provided that the alternatives can be proven to match (or exceed) the current standards, there is no reason to keep to the old paradigms just because they have been used ad nauseam.

The use of animals in toxicity testing is likely to continue for the foreseeable future because of the benefits they offer in examining a whole functioning organism over an extended treatment period, with all the interrelationships between tissues, blood supply, and absorption, distribution, metabolism, and elimination (ADME) that are currently not possible *in vitro*. For these reasons, as well as their multifactorial process controls, endpoints in reproductive toxicity, immunotoxicity, and general toxicology will continue to rely on animals. This does not mean, however, that the animal models currently in use cannot be refined and made more relevant to humans by the use of transgenic methods or by the investigation of new species. For instance, the metamorphosis from larval form to adult insect has some similarities to the processes of organogenesis in mammals; strains of *Drosophila* sp. have been developed that have some aspects of human metabolism.

The scrutiny of toxicology will continue and intensify due to the pressures on us all to make our assessments as safe and as accurate as possible, while maintaining the highest ethical standards in our work. Through all, we should be prepared

FOCUS BOX 21.1 TOXICITY TESTING
IN THE 21ST CENTURY (TOX21)

The initial aim of Tox21 was to provide a top-down vision of how the future of toxicity could and should be driven by *in vitro* assays in place of traditional animal testing models, in particular for environmental chemicals. Since Tox21's inception, the collaborative research team has developed and validated *in vitro* cell-based assays (tests) using quantitative high-throughput screening. The researchers have identified, developed, optimized, and screened more than 100 assays (tests). The Tox21 program includes two research phases, structured with guidance from two reports: Toxicology in the 21st Century: The Role of the National Toxicology Program and Toxicity Testing in the 21st Century: A Vision and a Strategy.

Tox21 is now a federal collaboration among the Environmental Protection Agency (EPA); National Institutes of Health (NIH), including the National Center for Advancing Translational Sciences and the National Toxicology Program at the National Institute of Environmental Health Sciences; and the Food and Drug Administration. Tox21 researchers aim to develop better toxicity assessment methods to quickly and efficiently test whether certain chemical compounds have the potential to disrupt processes in the human body that may lead to negative health effects. One of EPA's contributions to Tox21 is the chemical screening results from the Toxicity Forecaster (ToxCast).

In the guidance paper Toxicity Testing in the 21st Century: A Vision and a Strategy, a number of key and interconnecting elements are identified, which summarize the committee's vision for the future:

- Chemical characterization
- Toxicity testing
  - Toxicity pathways
  - Targeted testing (in animals)
- Dose response and extrapolation modeling
- Risk context
- Population and exposure data

Reaction from outside governmental organizations has been one of cautious optimism—a step in the right direction. A number of points have been raised and are summarized briefly below (Bus and Becker 2009):

- Appreciations of difference between chemical classes—one boot does not fit all.
- New tests and evaluations must account for biological complexity of target and be coupled with a greater understanding of target physiology.

- Such technologies should be used as opportunities to re-examine pre-conceived ideas and accepted norms.
- Finally, how dose affects toxicity in these systems and how this may be extrapolated to the target should be better understood.

Interest in Tox21 has not lessened in the intervening years, and the approach has been considered for both pharmaceutical (Rovida et al. 2015) and industrial chemicals (Settivari et al. 2015). Clearly, more research needs to be done into these assays types, but such an approach may represent a future paradigm for toxicology evaluation.

to change and not be ruled by the "we have always done it this way" philosophy; tradition is not necessarily science. Lack of change is not an option, but we should not forget that much toxicity testing is conducted to assess human safety; we should remember this as we attempt to reduce animal usage.

WHAT WILL THE FUTURE OF TESTING LOOK LIKE?

Burden et al. (2015) indicated, in a report on a workshop, which reviewed the current scientific, technical, and regulatory situation, that there are a number of factors producing movement away from relying on testing in animals for safety assessment of chemicals and that there has been progress in the development and validation of nonanimal methods.

In the future, safety testing and evaluation is likely to become more streamlined and to make better use of the tools available. It seems likely that this new process will place greater reliance on *in silico* and *in vitro* systems and, equally, reduce the use of and reliance on animals. There should be a careful use of data from all three sources to come to an overall conclusion. Safety evaluation is therefore likely to include the following:

- *In silico* assessment for a broad range of toxicity endpoints, but particularly for potential genotoxicity and DNA reactivity.
- *In vitro* assessment of toxicity, including extrapolation of concentrations *in vitro* to those expected *in vivo*. It is likely that a broader range of targeted endpoints will be evaluated to better characterize toxicity.
- Animals are likely to be used in a confirmatory role (although it has to be said that acceptance of this is some way off). For pharmaceuticals, after initial studies, it makes sense to test only in species that are relevant to the test material. With the exception of carcinogenicity assays, the longest studies are up to 6 months in rodents and 12 months in dogs or nonhuman primates with pressure, mounting to reduce this last to 6 months. For most purposes, it is possible that a 3-month study would be enough to characterize effects.

CHALLENGES FOR RISK ASSESSMENT AND MANAGEMENT

It could be argued that the ultimate aim of toxicology is to determine hazards and thence provide guidance for the assessment of risk. In turn, the challenge for risk assessors is to take the toxicological data and then to correctly assess the hazard's relevance for the target population. As before, the closer the differences are to normality, the more difficult it is to characterize any hazard and to evaluate the risks. The risk assessor has to take toxicological data from multiple sources and then assess the probability of the occurrence of the given hazard in the target population at a given exposure. This requires an appropriate understanding of the test paradigm, data quality, the potential mechanisms of effect, and finally, the relevance of any effects to humans. It must, of course, couple these considerations to exposure levels and any potential benefits of use. The continual problem for risk assessors is that the public (and hence regulators and politician) tend to be risk averse and demand "safety," when it cannot be guaranteed. Risk assessors themselves may also be burdened with particular assumptions—such as how the target will employ the product—that must be made to complete the assessment.

The advent of new assays and systems will not, or at least should not, change the process of risk assessment; Paracelsus's maxim that it is the dose that makes the poison still rings true. However, as the use of *in vitro* test batteries increases, the way in which hazard is assessed may alter how risks are identified, qualified, and quantified. Does a positive result in one *in vitro* test system outweigh the negative results of two other *in vitro* test systems for the same endpoint? Should human-based immortal cell lines be the ultimate arbiter of toxicity, or can non-human-derived primary cultures be considered alongside them? Can an *in silico* Ames negative result based on mechanistic interpretation dismiss a positive *in vitro* Ames positive in a single strain, in a single test?

It is self-evident that each risk assessment must be made on the available data and each case should be taken and understood on its own merits. What may be applicable for one chemical may not be valid for another. Following this premise, new test systems must be understood and their results for each individual chemical placed in context if any value is to be taken from them. A standard battery of *in vitro* and *in silico* test systems to replace *in vivo* methods completely may develop; however, the onus is upon the risk assessor to ensure that this battery is appropriate for the chemical of interest and the results produced relevant.

EVOLUTION OF RISK ASSESSMENT

Risk assessment is evolving as understanding of existing data sets and mechanisms grows; these processes may come together to reduce the inherent conservatism of risk assessors and to reduce the reliance on hazard as the starting point. Much of the new thinking revolves around increasing understanding of thresholds and expression of data to illustrate their presence. This understanding of thresholds is extending gradually to genotoxicity, for which thresholds have, traditionally, been discounted, on the one-hit model of carcinogenesis.

The area of greatest controversy has been in extrapolating the results of high-dose rodent carcinogenicity studies to real-life low-dose human exposure—quite apart from the understanding and/or appreciation of any mechanistic differences between species. Until recently, a nonthreshold approach was taken to extrapolate dose response data from animals to zero on the assumption that only a zero dose of the carcinogen would produce no tumors, or that the absence of tumors in the controls was due to the absence of test material.

Waddell (2002) reviewed carcinogenicity data from compounds approved as food additives by the Flavor and Extract Manufacturers Association (FEMA) as generally regarded as safe (GRAS), 15 of which were reported to be carcinogenic in rodents. The dose responses were explored by use of the Rozman scale: plotting the percent of animals with tumors against a log scale of dose expressed in molecules of compound per kilogram body weight per day. Three of the compounds evaluated in this manner had responses at three doses and fit a linear plot. The interception of these plots with the zero tumor percentage was at doses that were several orders of magnitude greater than human exposure. This was interpreted as showing clear thresholds for carcinogenicity. Critically, the point was made that these studies did not show danger but, rather, indicated the safety of these compounds at the current levels of human exposure.

In a further paper, Waddell (2004) made the point that use of log dose scales has been the norm for data for pharmacology and toxicology for noncancer toxicities and that there was no reason that this does not apply to chemical carcinogenicity. If dose–response curves from high-dose studies in animals are evaluated using a log scale for dose, it can be shown that there are clear thresholds for carcinogenicity. These observations call into doubt the relevance of the results of high dose studies in animals to risk assessment of much lower human exposures. Using this approach, it is possible to use the thresholds in animal studies to calculate safety margins for human exposures; Waddell also suggests that humans are more resistant to chemical carcinogenesis than animals.

This approach to thresholds in carcinogenicity has been extended to the results of *in vitro* genotoxicity tests, in which a positive result was traditionally taken to mean the potential for genotoxic effect at any dose in humans. MacGregor et al. (2015) reported on a working group on quantitative approaches to genotoxicity risk assessment, which examined the need for a quantitative approach to dose response, derivation of point of departure doses from dose response data, and other factors including the empirical relationship between mutation and cancer and extrapolation between test systems and species. This group considered the concept of the no-observed-genotoxic-effect level (NOGEL). The group recognized that thresholds probably exist below which genotoxicity does not occur for substances that do not react with DNA and for those that are DNA-reactive. They acknowledged, however, that the normal levels of such damage cannot be separated from what might be occurring due to the test substances and therefore that the thresholds are as much practical as empirical.

As discussed earlier, choosing a dose from which to start risk assessment calculations has been based largely on the no-observed-effect level (NOEL) and no-observed-adverse-effect level (NOAEL) that are defined by the pivotal toxicity studies. However, these are dependent on the doses tested, and numerical extrapolation from these is not an option. On the other hand, the benchmark dose uses expression of an effect to a defined level to calculate a dose for risk assessment. This was the approach used by the European Food Standards Agency (EFSA) for calculation of a new tolerable daily intake for bisphenol A (see Case Study 16.2 in Chapter 16). The drawback is that the benchmark dose is only calculable when there are a lot of data and if there is some confidence (or assumption) that the response is linear across the part of the dose curve being evaluated. As much of the discussion in this book has been on the presence of thresholds of effect and that response is unlikely to be completely linear across a complete dose range, assumptions of linearity need to be made with a lot of caution.

Another area in which progress is being made is in extrapolation from test concentrations *in vitro* to concentrations *in vivo*: quantitative *in vitro–in vivo* extrapolations (QIVIVE). Judson et al. (2011) described a framework for estimating the human dose at which a chemical significantly alters a biological pathway *in vivo*, using data from *in vitro* assays and pharmacokinetic model derived from *in vitro* data, together with estimates of population variability and uncertainty. This dose was designated the biological pathway altering dose or concentration (BPAD/C). This was then used with pharmacokinetic modeling to estimate the doses required *in vivo* to achieve the BPACs in the blood at steady state. The importance of the use of QIVIVE was underlined by Meek and Lipscomb (2015), who forecast that testing strategies will rely increasingly on *in vitro* data from which the early steps or key events in toxicity at relevant dose levels in humans may be characterized. This requires quantitative extrapolation from *in vitro* to *in vivo*, with a view to explore dose response as the base for comparison with exposure to estimate risk.

Adeleye et al. (2015), in a work in progress, explored the US National Research Council blueprint for change, Toxicity Testing in the 21st Century (TT21C): A Vision and Strategy, which called for the transformation of toxicity testing from reliance on high-dose studies in animals to the primary use of *in vitro* methods to evaluate changes in normal cellular signaling pathways using human-relevant cells or tissues. This can be done in the framework of adverse outcome pathways (AOPs), for which a molecular initiating event can be identified and then linked to outcomes *in vivo*. They explored *in vitro* data for quercetin, a flavonol found in plant food products, with the objective of seeing if the data could be used to prepare a risk assessment in the spirit of TT21C, without the use of data from rodent carcinogenicity studies. Broadly, they used high-throughput pathway biomarkers and markers of cell cycle, apoptosis, and micronuclei formation, plus gene transcription, to describe dose–response curves to calculate no-effect levels and benchmark doses, which were compared with biokinetic models, and then explored the potential for extrapolation from *in vitro* to *in vivo*.

SUMMARY

Toxicity testing and risk assessment are dynamic, constantly changing disciplines, although the pace of change is governed by regulatory conservatism. The future of toxicity testing is likely to include fuller use of the three main tool areas that are available currently:

- *In silico* assessment for a broad range of toxicity endpoints
- *In vitro* assessment of toxicity, including extrapolation of concentrations *in vitro* to those expected *in vivo*
- Animal tests, in appropriate, relevant species, in studies likely to be no longer than 13 or 26 weeks, used in a confirmatory role

Risk assessment is also evolving as new techniques arise and as understanding grows. There are a number of details that may help risk assessment, including the following:

- Better prediction of exposure both worst-case scenario and general use.
- New methods of dose expression—molar concentrations rather than mg/kg/day or numbers of molecules per kilogram—may assist high-end risk assessment, particularly in putting into perspective the significance of dose responses to genotoxic carcinogens relative to human exposure.
- Greater understanding of molecular initiating events, adverse outcome pathways, and mechanisms of action leading to more targeted testing.
- Extrapolation from *in vitro* to *in vivo* (QIVIVE), combined with pharmaco-kinetic modeling and use of adverse outcome pathways.
- Greater appreciation of relative potencies—use of NOGEL coupled with the use of the threshold of toxicological concern (TTC).
- Better context for low-level exposures and appreciation of the kinetics of the test substance.

For risk management, the challenge is to take the data from the risk assessment and manage the risks in a manner that does not impose pointlessly severe controls that require expensively unnecessary cleanup or impossibly low exposure limits. The obverse is the need to ensure that the process is not so lax that people or the environment is harmed as a result of faulty risk assessment and/or management.

REFERENCES

Adeleye Y, Andersen M, Clewell R, Davies M, Dent M, Edwards S, Fowler P et al. Quantitative in vitro to in vivo extrapolation (QIVIVE): An essential element for in vitro–based risk assessment. Implementing Toxicity Testing in the 21st Century (TT21C): Making safety decisions using toxicity pathways, and progress in a prototype risk assessment. *Toxicology* 2015; 332: 102–111.

Burden N, Mahony C, Müller BP, Terry C, Westmoreland C, Kimber I. Aligning the 3Rs with new paradigms in the safety assessment of chemicals. *Toxicology* 2015; 330: 62–66.

Bus JS, Becker RA. Toxicity testing in the 21st century: A view from the chemical industry. *Toxicol Sci* 2009; 112(2): 297–302.

EFSA. Scientific Opinion on the risks to public health related to the presence of bisphenol A (BPA) in foodstuffs. *EFSA J* 2015; 13(1): 3978.

Judson RS, Kavlock RJ, Setzer RW, Hubal EA, Martin MT, Knudsen TB, Houck KA, Thomas RS, Wetmore BA, Dix DJ. Estimating toxicity-related biological pathway altering doses for high-throughput chemical risk assessment. *Chem Res Toxicol* 2011; 24: 451–462.

Kramer K, Patwardhan S, Patel KA, Estrem ST, Lewin-Koh NJ, Gao H, Smyej IL, Colet JM, Jolly RA, Ganji GS et al. Integration of genomics and metabonomics data with established toxicological endpoints. A systems biology approach. *Toxicologist* 2004; 78(1-S): 260–261.

Liebler DC. The poisons within: Application of toxicity mechanisms to fundamental disease processes. *Chem Res Toxicol* 2006; 19: 610.

MacGregor JT, Frötschl R, White PA, Crump KS, Eastmond DA, Fukushima S, Guérard M et al. IWGT report on quantitative approaches to genotoxicity risk assessment I. Methods and metrics for defining exposure–response relationships and points of departure (PoDs). *Mutat Res Genet Toxicol Environ Mutagen* 2015; 1; 783: 55–65.

Marty MS, Neal BH, Zablotny CL, Yano BL, Andrus AK, Woolhiser MR, Boverhof DR et al. An F1-extended one-generation reproductive toxicity study in Crl:CD(SD) rats with 2,4-dichlorophenoxyacetic acid. *Toxicol Sci* 2013; 136(2): 527–47.

Meek ME, Lipscomb JC. Quantitative in vitro to in vivo extrapolation (QIVIVE): An essential element for in vitro–based risk assessment. Gaining acceptance for the use of in vitro toxicity assays and QIVIVE in regulatory risk assessment. *Toxicology* 2015; 332: 112–123.

Rovida C, Asakura S, Daneshian M, Hofman-Huether H, Leist M, Meunier L, Reif D et al. Toxicity testing in the 21st century beyond environmental chemicals. *ALTEX* 2015; 32(3): 171–181.

Settivari RS, Ball N, Murphy L, Rasoulpour R, Boverhof DR, Carney EW. Predicting the future: Opportunities and challenges for the chemical industry to apply 21st-century toxicity testing. *J Am Assoc Lab Anim Sci* 2015; 54(2): 214–223.

Waddell WJ. Thresholds of carcinogenicity of flavors. *Toxicol Sci* 2002; 68(2): 275–279.

Waddell WJ. Dose–response curves in chemical carcinogenesis. *Nonlinearity Biol Toxicol Med* 2004; 2(1): 11–20.

Glossary

α2u-Globulin: Protein produced in large amounts in the liver of male rats and excreted in the urine. Chemicals such as d-limonene and trimethylpentanol form slowly degraded complexes with it, which accumulate in the kidney and lead to a male rat specific nephropathy. The normal function of this protein may be to complex volatile pheromones and slow their release into the atmosphere. Depending on source, the "u" can stand for urinary, though some believe it to be a "μ."

α2u (or) microglobulin: See α2u-globulin.

Acute toxicity study: Single-dose study in which administration is normally followed by 14 days' observation and then macroscopic examination at necropsy.

ADI (acceptable daily intake): The daily intake of a chemical that is expected to be without adverse effect when ingested over a lifetime.

ADME (absorption, distribution, metabolism, and elimination): The basic processes that influence pharmacokinetic behavior and, hence, toxicity.

Angiogenesis: The formation of new blood vessels, seen in embryos and tumors.

Apoptosis: The process whereby cells are programmed to die. Reduced apoptosis can lead to tumor formation. It is an essential part of embryonic development, where effects may be associated with teratogenicity. Contrasts with necrosis (q.v.).

AUC (area under the concentration curve): A measure of systemic exposure via plasma concentrations. Short half-life generally leads to a low AUC.

Biocenosis: An integrated community of closely associated organisms.

Carcinogenicity bioassay: A study to assess potential for carcinogenic action when the test substance is administered for up to 2.5 years in rodents. Study duration with transgenic animals may be 6 or 9 months.

CAS no.: Chemical Abstracts Registry number; an identifying number for chemicals, used in literature searches and for defining the chemical assessed.

Chromosomal mutation: Any change in chromosome structure or number.

Chronic toxicity study: Usually a toxicity study of 26 weeks or longer.

Clastogen or clastogenic: Producing breakages in chromosomes.

Clearance: Measure of the removal of a substance from blood or plasma, expressed in units such as milliliter per minute. Clearance may differ among organs, and total clearance reflects all these values.

C_{max}: The maximum plasma concentration achieved after a given dose.

Corrosion: The production of irreversible damage at the site of contact as a result of chemical reaction with local molecules such as fats and proteins.

Dosage: Synonymous with dose level (q.v.)—a rate at which a test system is dosed, e.g., milligrams per kilogram bodyweight per day (mg/kg/day). See also dose.

Dose: An amount of compound administered on any one occasion, e.g., in milligrams per kilogram bodyweight (mg/kg) or, in clinical terms, milligrams (per person) per day (mg/kg). See also dosage.

Dose level: The rate at which a compound is dosed, e.g., milligrams per kilogram bodyweight per day (mg/kg/day).

Dose–response curve: The curve resulting when response is plotted against dose. A large increase in response for a small increase in dose indicates a steep dose–response curve. Some chemicals, such as paracetamol, show an early slow increase in this curve with a steep increase when a threshold of toxicity is exceeded. The dose–response curve may also be significantly affected by relatively small changes in factors, such as protein binding (see also therapeutic index).

Dose volume: Usually used to define the volume rate for oral studies, e.g., milliliters per kilogram.

EC3: The estimated concentration of the test item required to produce a 3-fold increase in draining lymph node cell proliferation.

EC_{50}: Effective concentration 50%—the concentration at which 50% efficacy is expressed.

ECHA (European Chemicals Agency): Purveyors of REACH.

EFSA (European Food Safety Authority).

EMA (European Medicine Agency).

EPA (Environmental Protection Agency).

FDA (Food and Drug Administration).

Gene mutation: Changes in the DNA at one or more bases; these may be insertion of a base (frameshift) or the substitution or misreading of one base for another.

Genotoxicity: Modification or damage to genetic material.

Half-life: The time taken for the concentration of a substance to reduce to half of the initial value. Usually measured in plasma but applicable to other matrices such as tissues, soils, water, and the atmosphere.

Healthy worker effect: The bias that can be introduced into epidemiological studies where a workforce is compared with the general population. The working population is expected to be healthier than the general population, which includes long-term sick and unemployed people as well as the healthy and other workers. This concept is now being questioned.

Hepatocytic hypertrophy: Increased size of the hepatocytes, typically around the central vein of the liver lobule (centrilobular). It is characterized by greater distance between nuclei (increased cytoplasm) and is usually due to enzyme induction.

hERG: Human Ether-à-go-go Related Gene—the subject of an *in vitro* safety pharmacology assay for cardiac function.

Hyperplasia: An increase in a normal cell population, which can be seen in response to hormonal disturbances, to changes in the control of apoptosis, or to increased cell turnover as a result of direct cellular toxicity.

IC_{50}: Inhibition concentration 50%—the concentration at which 50% inhibition is expressed.

Irritation: A reversible nonimmunologic inflammatory response at the site of contact with a test chemical. May be seen following various routes of administration including dermal, parenteral, gastric, or inhalational.

LC_{50}: Median lethal concentration—the concentration at which half of the test population is killed.

LD_{50}: Median lethal dose—dose required to kill half of the test population (LC_{50} median lethal concentration).

Limit dose: Usually, the highest practical dose used when no effect can be elicited in a particular study. For example, the limit dose for acute toxicity in pharmaceutical development is usually 2000 mg/kg; use of a single group or exposure is usually acceptable in these circumstances.

LOAEL (lowest-observed-adverse-effect level): The lowest dose level or concentration at which adverse effects were seen.

Log P: Octanol–water partition coefficient, a measure of lipophilicity that influences ADME. Skin permeability increases with log P over the midrange; low and high log P values are associated with lower skin permeability.

MHRA (Medicines and Healthcare Products Regulatory Agency).

MRL (maximum residue limit): The maximum acceptable concentration in foods for pesticides or veterinary drugs.

MTD (maximum tolerated dose): The MTD for a chemical depends to a large extent on the type of test contemplated. Although it has been used extensively in relation to carcinogenicity bioassays to indicate a 10% reduction in body weight gain, this is not appropriate in shorter studies where more severe toxicity is implied. Broadly, the MTD in any test type is one that elicits toxicity but does not compromise the survival of the test system during the course of the experiment. The MTD for a short exposure or single administration is likely to be significantly higher than that for a long exposure or chronic toxicity study.

Mutation: A change in the DNA that may be transmitted by division and give rise to heritable changes, if the initial change is not lethal. A reverse mutation causes a reversion to the wild type, as in the Ames test; a forward mutation test detects mutants in wild-type bacteria.

Necrosis: Death of tissues or individual cells within a tissue, for example, single-cell necrosis seen in liver. Unprogrammed cell death that contrasts with apoptosis (q.v.).

NOAEL (no-observed-adverse-effect level): The dose level or concentration that is associated with treatment-related change that is not considered to be adverse. This is a useful concept where there is no NOEL, but where effects are transient or due to intended properties of the compound. Often used as the starting dose for risk assessment, which is divided by uncertainty factors to determine "safe" exposure.

NOEL (no-observed-effect level): The dose level at which no treatment-related change was seen.

Nongenotoxic carcinogen: A chemical that causes cancer without directly damaging DNA.

Octanol–water partition coefficient: See log P.

OECD: Organization for economic co-operation and development.

OEL (occupational exposure limit): Average airborne concentrations of a chemical to which workers may be exposed over a defined period.

PDE: Permitted daily exposure.

Peroxisome proliferation: Increase in the numbers of peroxisomes—cellular organelles having high levels of oxidative enzymes and probably involved in lipid metabolism. This increase is induced by several chemical classes including some hypolipidemics and plasticizers such as diethylhexylphthalate, and chlorinated compounds such as trichloroethylene. This proliferation, particularly in the liver of rodents, is associated with nongenotoxic carcinogenesis of little relevance to humans.

Pharmacokinetics: The study of the time course of the absorption, distribution, and elimination of a compound from the body. This term usually refers to therapeutic doses; toxicokinetics being used for this in reference to toxicity studies.

Phase 1 metabolism: The process whereby molecules are made more polar to facilitate elimination, for instance, by hydroxylation or hydrolysis. This process of detoxification may backfire when reactive metabolites are produced that result in direct toxicity on cellular macromolecules such as proteins or DNA.

Phase 2 metabolism: Conjugation of metabolites from phase 1 with polar endogenous molecules such as glucuronide, glycine, sulfate, or glutathione to produce a more polar molecule that can be readily excreted in the urine or bile. Phase 2 metabolites are usually nontoxic, although there are several exceptions to this general rule of thumb.

pKa: The pH at which a molecule is 50% ionized. This affects absorption, particularly across the intestinal mucosa. For example, at low pH, benzoic acid is mostly nonionized; percent ionization increases as pH rises above 4, approaching 100 percent ionized at pH 7. Thus, benzoic acid is best absorbed from low-pH media such as those in the stomach.

QSAR (quantitative structural activity relationship).

REACH (Registration, Evaluation, Authorisation and restriction of Chemicals).

STEL (short-term exposure limit): The upper airborne concentration that is acceptable for short-term exposure (e.g., not longer than 15 minutes, experienced no more than four times in a day, at intervals of not less than 1 hour) without prolonged or unacceptable adverse effect.

Subacute toxicity study: Usually, a toxicity study of 28 days or less.

Subchronic toxicity study: Usually, a toxicity study of 13 weeks.

TDI (tolerable daily intake): Used in similar contexts to ADI, for residues and food contaminants.

Therapeutic index: A measure of the difference between therapeutic levels or doses of a drug and those that are associated with toxicity. This is often related to the plasma concentration of unbound drug or chemical, as with phenytoin or warfarin (see also dose–response curve).

TLV (threshold limit value): The upper permissible airborne concentration for occupational exposure.

TLV-C (threshold limit value—ceiling): An airborne concentration that should not be exceeded at any time.

Toxicodynamics: The study of the relationship between concentration at target tissues or plasma and toxic effect.

Toxicokinetics: The study of pharmacokinetics in toxicity studies. The knowledge of pharmacokinetics following single or repeated administration may be used to model or explain the effects expected in other species or in humans and may be related to toxicodynamics.

TWA (time-weighted average): The average concentration to which nearly all workers may be exposed repeatedly without adverse effect, during a working day of 8 hours or a 40-hour week.

WEL (workplace exposure limits): The concentration of a substance that, provided it is not exceeded, will not normally result in adverse effects to persons who are exposed; European equivalent to occupational exposure level (q.v.). See EH40/2005 Workplace Exposure Limits at http://www.hse.gov.uk/coshh /table1.pdf for recent listing of approved exposure limits in the United Kingdom.

Index

Page numbers followed by f, t, and b indicate figures, tables, and boxes, respectively.